Essential Neuromodulation

Essential Neuromodulation

Jeffrey E. Arle
Director, Functional Neurosurgery
and Research, Department
of Neurosurgery Lahey Clinic
Burlington, MA
Associate Professor of Neurosurgery
Tufts University School
of Medicine, Boston, MA

Jay L. Shils
Director of Intraoperative
Monitoring, Dept of Neurosurgery
Lahey Clinic Burlington, MA

AMSTERDAM • BOSTON • HEIDELBERG • LONDON
NEW YORK • OXFORD • PARIS • SAN DIEGO
SAN FRANCISCO • SINGAPORE • SYDNEY • TOKYO

Academic Press is an Imprint of Elsevier

Academic Press is an imprint of Elsevier
32 Jamestown Road, London NW1 7BY, UK
30 Corporate Drive, Suite 400, Burlington, MA 01803, USA
525 B Street, Suite 1800, San Diego, CA 92101-4495, USA

First edition 2011

Notice
No responsibility is assumed by the publisher for any injury and/or damage to persons
or property as a matter of products liability, negligence or otherwise, or from any use
or operation of any methods, products, instructions or ideas contained in the material
herein. Because of rapid advances in the medical sciences, in particular, independent
verification of diagnoses and drug dosages should be made

British Library Cataloguing-in-Publication Data
A catalogue record for this book is available from the British Library

Library of Congress Cataloging-in-Publication Data
A catalog record for this book is available from the Library of Congress

ISBN : 978-0-12-381409-8

For information on all Academic Press publications
visit our website at elsevierdirect.com

Typeset by Thomson Digital, B-10-11-12, Noida Special Economic Zone,
Noida-201 305, India, Website: www.thomsondigital.com.

Printed and bound in China
11 12 13 14 15 10 9 8 7 6 5 4 3 2 1

Working together to grow
libraries in developing countries

www.elsevier.com | www.bookaid.org | www.sabre.org

ELSEVIER   BOOK AID
            International   Sabre Foundation

Contents

Part III
The Neuromodulation Therapy Interface

     Daniel R. Merrill, PhD

     **Commentary on The Electrode – Materials
     and Configurations**   147
     Mark Stecker, PhD, MD

7.   The Electrode – Principles of the Neural Interface:
     Axons and Cell Bodies   153
     Cameron C. McIntyre, PhD

     **Commentary: Considerations for Quantitative
     Modeling of Excitation and Modulation
     of CNS neurons**   162
     Warren M. Grill, PhD

8.   The Electrode – Principles of the
     Neural Interface: Circuits   169
     Erwin B. Montgomery Jr, MD

     **Commentary on The Electrode – Principles
     of the Neural Interface: Circuits**   191
     Mark Stecker, PhD, MD

9.   Device Materials, Handling,
     and Upgradability   193
     John Kast, BSME, Gabi Molnar, MS and Mark Lent, BSME, MSMOT

     **Commentary on Device Materials, Handling,
     and Upgradability**   209
     Jay L. Shils, PhD

10. Electronics   213
     Emarit Ranu, MSEE, MSBS, EMT-B

     Commentary on Electronics   244
     Henricus Louis Journee, MD, PhD

11. Power   253
     Tracy Cameron, PhD, Ben Tranchina, MSEE and John Erickson, BSEE

     Commentary on Power   266
     Jay L. Shils, PhD

Part IV
Placing Neuromodulation in the Human Body

     Jeffrey E. Arle, MD, PhD

     Commentary on Surgical Techniques   281
     Phillip Starr, MD, PhD

Contributors

Ron Alterman, MD, Department of Neurosurgery, Mount Sinai School of Medicine, New York, USA

Jeffrey E. Arle, MD, PhD, Director, Functional Neurosurgery, and Research, Department of Neurosurgery, Lahey Clinic, Burlington MA Associate Professor of Neurosurgery, Tufts, University School of Medicine, Boston, MA, USA

Alim Louis Benabid, MD, PhD, CEA Clinatec, Clinatec, CEA Grenoble, Grenoble, France

Tracy Cameron, St Jude Medical Neuromodulation Division, Plano, Texas, USA

Sergio Canavero, MD (US FMGEMS), Turin Advanced Neuromodulation Group (TANG), Turin, Italy

Beatrice Cioni, MD, Department of Functional and Spinal Neurosurgery, Catholic University, Rome, Italy

Timothy R. Deer, MD, President and CEO, The Center for Pain Relief, Inc., Charleston, WV, Clinical Professor of Anesthesiology West Viriginia University, Charleston, WV, USA

John Erickson, BSEE, St Jude Medical Neuromodutation Division, Plano, Texas, USA

Chad W. Farley, MD, Department of Neurosurgery, University of Cincinnati (UC) Neuroscience Institute and UC College of Medicine, and Mayfield Clinic, Cincinnati, OH, USA

Steve Goetz, MS, Medtronic Neuromodulation, Minneapolis, MN, USA

Yakov Gologorsky, MD, Department of Neurosurgery, Mount Sinai School of Medicine, New York, USA

Warren M. Grill, PhD, Department of Biomedical Engineering, Duke University, Durham, NC, USA

Chris Hart, BA, Director of Urban and Transit Projects, Institute for Human centered Design, Boston, massachusetts, USA

John Heitman, Medtronic Neuromodulation, Cincinnati, OH, USA

Lisa Johanek, PhD, Medtronic Neuromodulation, Minneapolis, USA

Henricus Louis Journee, MD, PhD, Department of Neurosurgery, University Medical Center Groningen, Groningen, The Netherlands

John Kast, BSME, Medtronic Neuromodulation, Minneapolis, MN, USA

Joachim K. Krauss, MD, Medical University of Hannover/Neurosurgery, Hanover, Germany

Paul S. Larson, MD, Associate Professor, Department of Neurological Surgery, University of California, San Francisco, CA, USA Chief, Neurosurgery, San Francisco VA Medical Center, San Francisco, CA, USA

Mark Lent, BSME, MSMOT, Medtronic Neuromodulation, Minneapolis, MN, USA

Andres Lozano, Division of Neurosurgery, Toronto Western Hospital, Toronto, Ontario, Canada

George T. Mandybur, MD, Department of Neurosurgery, University of Cincinnati (UC) Neuroscience Institute and UC College of Medicine, and Mayfield Clinic, Cincinnati, OH, USA

Alastair J. Martin, PhD, Adjunct Professor, Department of Radiology and Biomedical Imaging, University of California, San Francisco, CA, USA

Cameron C. McIntyre, PhD, Cleveland Clinic Foundation, Department of Biomedical Engineering, Cleveland, OH, USA

Daniel R. Merrill, PhD, Vice President, Technical Affairs, Alfred Mann Foundation, Santa Clara, CA, USA

Y. Eugene Mironer, MD, 220 Roper Mountain Road Ext., Greenville, SC

Alon Y. Mogilner, MD, PhD, Assitant Professor of Neurosurgery, Chief, Section of Function and Restorative Neurosurgery, Hofstra–North shorf LIJ School of Madicine, President, AANS/CNS Joint section on Pain, Great Neck, NY, USA

Gabi Molnar, MS, Medtronic Neuromodulation Minneapolis, MN, USA

Guillermo A. Monsalve, MD, Department of Neurosurgery, University of Cincinnati (UC) Neuroscience Institute and UC College of Medicine, and Mayfield Clinic, Cincinnati, OH, USA

Erwin B. Montgomery Jr, MD, Dr Sigmund Rosen Scholar in Neurology, Professor of Neurology, University of Alabama at Birmingham, Birmingham, AL, USA

Richard B. North, MD, The Sandra, Malcolm Berman Brain & Spine Institute, Baltimore, MD Professor of Neurosurgery, Anesthesiology, Critical Care Medicine (ret.), Johns Hopkins University School of Medicine, Baltimore, MD

Francisco Ponce Division of Neurosurgery, Toronto Western Hospital, Toronto, Ontario, Canada

Emarit Ranu, MSEE, MSBS, EMT-B, Boston Scientific Neuromodulation, Fort Collins, CO, USA

Louis J. Raso, MD, CEO, Jupiter Intervention Pain Management Corp, Jupiter, FL, US

Ali Rezai, Ohio State University, Department of Neurosurgery, Columbus, OH, USA

S. Matthew Schocket, MD, Capital Pain Institute, Austin, TX, USA

Konstantin V. Slavin, MD, Department of Neurosurgery, University of Illinois at Chicago, Chicago, IL, USA

Philip A. Starr, MD, Phd, Department of Neurological Surgery, University of California, San Francisco, CA USA

Mark Stecker PhD, MD, Marshall University Medical Center, Huntington, WV, USA

Ben Tranchina, MSEE, St Jude Medical Neuromodulation Division, Plano, Texas, USA

Introduction

Andres Lozano and Francisco Ponce

Division of Neurosurgery, Toronto Western Hospital, Toronto, Ontario, Canada

The field of Functional Neurosurgery and Neuromodulation is experiencing a renaissance. The reasons for this are many. First, numerous patients with neurological and psychiatric disorders continue to be disabled despite the best available medical treatments. Second, there have been important advances in the understanding of the pathophysiology of these disorders. Third, there have been significant improvements in both structural and functional brain imaging, which make the identification of potential targets easier. Fourth, there have been significant improvements in the neurosurgical techniques, such as neuronavigation and microelectrode recording, as well as in the equipment, including the stimulating electrodes, the pulse generators, and the drug delivery pumps, that are being used in day-to-day treatment.

There are a large number of circuits in the brain, spinal cord, and peripheral nerves that are amenable to neuromodulation. Both constant electrical stimulation as well as responsive electrical stimulation are possible, in addition to modulation through the delivery of pharmacological agents. As this field evolves, we anticipate the further development and application of novel forms of modulation based upon techniques such as optogenetics and gene therapy, with the latter currently being evaluated in a number of trials in Parkinson's disease. In addition, there is some re-emerging activity in transplantation as an investigational therapy.

The types of pathologies that are being treated with neuromodulation include pain, movement disorders, psychiatric disease, and epilepsy, and the patients that could benefit from these therapies are many. The future is bright for this specialty, and we need to train young neurosurgeons to embark on this fascinating aspect of neurosurgery.

This book compiles a series of works by experts who discuss various aspects of this field. It provides an overview of the entire discipline, tells us where we have been, and also where we are heading.

Part I

The Neuromodulation Approach

Chapter 1

The Neuromodulation Approach

Jeffrey E. Arle, MD, PhD

Director Functional Neurosurgery and Research, Department of Neurosurgery, Lahey Clinic, Burlington, MA; Associate Professor of Neurosurgery, Tufts University. School of Medicine, Boston, MA

INTRODUCTION

Neuromodulation means many things to many people – but essential to any point of view is that the term implies some type of intervention that interfaces on some level with the nervous system of the patient and modifies function so as to effect benefit for the patient. What remains important to the definition, however, is a deeper belief that this therapeutic approach itself has greater merit, when chosen, than any of the alternatives. As a field of study, and as a burgeoning market in the vast expanse of health care overall, neuromodulation has taken several routes in achieving its current position – a position that has been estimated to be increasing from $3.0 billion worldwide to $4.5 billion worldwide in 2010 [1]. In contrast, the pharmaceutical industry has a market of approximately $20 billion/year in treating similar clinical conditions. This lopsided ratio is shifting in the direction of neuromodulation and, with continued innovation, favorable outcomes and a reasonable reimbursement context, neuromodulation stands to be one of the greatest sources of therapeutic intervention ever, in terms of numbers of people treated and overall contribution to quality of life.

It is not simply interesting, or honorable, to be involved in weaving the fabric of so widely applicable a cloth, but a responsibility as well. Our goals herein are to impart both basic and not-so-basic aspects of neuromodulation to the reader – in terms of design, application, revision and troubleshooting, the patient perspective, and the future. We focus primarily on electrical stimulation, with very limited discussions of other modulation therapies when they may support an important principle overall. Readers will be exposed not only to thorough descriptions of every facet of neuromodulation by some of the most expert names currently in the field, but also to commentary from additional experts on the same topics, lending perspective, raising questions. Whether design engineer, graduate student, post-doctoral fellow, resident, neurologist, pain specialist, neurosurgeon, or other interested party to neuromodulation, our goal is to provide the ability to carry that responsibility soundly into whatever endeavors they lead.

Advances and new applications continue apace, but it would not be out of order to consider what has happened in neuromodulation and call it a 'paradigm

Essential Neuromodulation. DOI: 10.1016/B978-0-12-381409-8.00001-2

shift' [2] in managing the clinical problems where it has been applied. This is a strong term, but emphasizes that, while previously the rampant belief has been that more and more precise pharmaceutical solutions could prevail for almost any clinical problem, this approach has had holes punched in it. Certainly, the success of the pharmaceutical paradigm over previous methods of treatment has been profound and has created its own paradigm. But it has also been shown to have weakness and outright failures, in the form of side effects, tolerances, and inability to account for the anatomical precision necessary in some cases to effect benefit. At the same time, surgical solutions for many of the same problems – specifically, using resections or lesions – have soared with some successes, and plummeted with failure as well in cases where morbidity, imprecision, or irreversibility have left patients without benefit and possibly harmed further.

Kuhn pointed out that: 'a student in the humanities has constantly before him a number of competing and incommensurable solutions to these problems, solutions that he must ultimately examine for himself' [2], but science is different in that, once a paradigm shift has occurred, one would find it completely incompatible to posit that flies spontaneously generate from rotting meat, the sun revolves around the earth, or that the principles of Darwinian natural selection have not replaced Lamarck's. Because of the wide successes now in neuromodulation, practitioners must recognize that this same transition, this paradigm shift, is occurring, or has occurred. It would be, at this point, reprehensible not to consider deep brain stimulation for a child with DYT-1 positive dystonia, a dorsal column stimulator for refractory CRPS-I in an extremity, or motor cortex stimulation for post-stroke facial or upper extremity pain. And these are but a few examples of how the neuromodulation approach has altered the algorithms of care. Neuromodulation has achieved this shift in every single field of application tried so far. One does not continue to ask: 'What do I try when other traditional approaches have failed for this patient?', one now asks instead: 'How can I use neuromodulation to help this patient?' — and this change in approach makes all the difference.

HISTORY

Several excellent reviews of our best knowledge of the history of therapeutic electrical stimulation [3–5] describe an early recognition of the potential benefits that electricity applied to human tissue could impart. As these authors have also appreciated, two earlier scholarly studies of this history [6,7], have brought out the ancient Egyptian references in hieroglyphics from the 3rd millennium BC on the use of the potent Nile catfish in causing fishermen to 'release the troupes' when they felt its strong current. These freshwater fish, and saltwater varieties of electric fish (e.g. torpedo fish) can generate up to about 200 volts at a time! The roots of several words in English have come down to the present day because of such phenomena (e.g. torpor, from the Roman name of the fish as 'torpedo' and narcosis from the Greeks naming the fish 'narke' [4]). A Roman text from 47 AD has suggested that multiple ailments (e.g. gout) were all treated by using the shocks from a torpedo fish. This electro-ichthyotherapy, as it is termed, has been noted by Kellaway [6] to have been used in various primitive African and American Indian tribes still into the 20th century.

To lend context to the development of therapeutic electrical devices, it is helpful to appreciate something of the development of more formal pharmaceutical therapies. The first drugstore as such is thought to have flourished from approximately 754 AD in Baghdad [8]. Most current larger pharmaceutical companies known today consolidated out of the drug store format throughout the 19th century, as refined ability to manufacture certain chemicals reliably on a large scale materialized – mostly in the Philadelphia area, it turns out [9]. This eventually completely displaced the owner/pharmacist with mortar and pestle individually filling his clients needs, and further allowed the widespread uniform access to standard formulations of pharmaceuticals and standards in the industry.

Further applications of electrical therapy however continued into the late 19th century, involving myriad devices that imparted shocks and other sensations to the ailing, including as mentioned above electro-ichthyotherapy, which was still used even in Europe into the mid-part of the century [10]. Perhaps the first device to reliably create man-made electricity though can be ascribed to von Guericke who, in 1662, created a generator of electrostatic discharges, among many other accomplishments. Over a hundred years later, following on from seminal work by Benjamin Franklin around 1774, who explored the phenomenon of muscle contraction following electrical shocks (even before Galvani more thoroughly examined it in the frog in 1780), many were quick to imbue the 'new' entity of electricity with magical healing powers, just as magnetite and amber had for many ages previously. It has been suggested that Christian A. Krantzenstein, however, was really the first to use electrical stimulation in a therapeutic manner [11], and this was before Franklin and others' observations. Somewhat of a polymath, Krantzenstein was appointed by the King of Denmark in 1754 (at the age of 31) to study electricity and the effects it might have on various ailments. (It seems the King of Denmark deserves some credit as well perhaps.) He had been already renowned for his studies of electricity and lectures in a wide range of subjects. The following is a description of the original Danish review of his work in 1924, from the British Medical Journal:

...he issued advertisements inviting all and sundry who hoped electricity might cure their ills to call at his lodgings between 4 and 6 in the evening, when 'everyone would be served according to the nature of the disease.' How he 'served' them is not quite clear. He used a rotatory apparatus with glass balls, and the sparks he drew out of his patients caused a penetrating pain which was worst in the toes; moreover, it was associated with a smell of sulphur, and he explained that the electrical vibrations put the minutest parts of the body in motion, driving out the unclean sulphur and salt particles; hence the smell.

Treatment with electricity, he said, made the blood more fluid, counteracted congestions, induced sleep, and was more effective than whipping with nettles in the treatment of paralysis.

Clearly, the bar was not high, as the therapy was competing with being whipped with nettles, for example. Kratzenstein, tangentially, has also been suggested as the basis for the character of Dr Frankenstein in the novel by Mary Shelley, first published anonymously in 1818 – a modern version of the classic Prometheus legend, stealing fire, the source of all creativity – in this case electricity, life, a cure of impossibly terrible ailments – from the gods, and the ruin it brings upon him by doing so.

There were several further key clinical observations through the end of the 19th century though insidiously at the same time, magnetic and electrical quackery became rampant on main street. Fritsch and Hitzig [12] showed that stimulating the cerebral cortex could elicit muscle contractions in dogs (1870) and then Bartholow [13] found it could be done in an awake human 4 years later. Sir Victor Horsely, one of the first few documented to perform what is considered a reasonable facsimile of a modern craniotomy in the 1880s, apparently tried to stimulate tissue within an occipital encephalocele, finding it produced conjugate eye movements [14]. This was one of the first real uses of an evoked response, remarkably prescient at the time, and a technique relied upon in so many ways today (see [15] for review).

Despite these noble attempts to make use of what was the most advanced information and insight into neural function to aid in patient care, little was otherwise advanced for decades with regard to neuromodulation or electrotherapeutics. In parallel course, several inventions worked off of rudimentary knowledge of batteries and insights of Faraday (Faraday's law which linked electricity and magnetism), and led to 'electrical therapies' such as the Inductorium, the Gaiffe electrical device, the Faradic Electrifier, and the Electreat, patented by Kent in 1919 [16]. The later device, similar to the present-day TENS unit, actually sold around 250 000 units over 25 years! Of note, these were promoted in ads such as the following:

All cases of Rheumatism, Diseases of the Liver, Stomach and Kidneys, Lung Complaints, Paralysis, Lost Vitality, Nervous Disability, Female Complaints...are cured with the Electrifier.

Subsequently, Kent was the first person prosecuted under the new Food, Drug and Cosmetic Act in 1938, because of unsubstantiated medical claims. The Electreat Company was forced to limit their claims to pain relief alone [16]. Early in the twentieth century, the maturing of a pharmaceutical industry and the disrepute of many practitioners of electrotherapy in general led to widespread abandonment in the use of electrical stimulation as a therapy.

That electrical stimulation has had detractors is an understatement, and early experience with dorsal column stimulators (first developed and implanted by

Shealy in 1967 [17]) in the neurosurgical community up until the 1990s high-lights this point of view. Shealy himself eventually abandoned the approach in 1973 [4] apparently because of frustrations with technique and technology. Many were discouraged either by the lack of efficacy, or by the short duration of efficacy. Unlike magnetic therapy, however, there is a strong grounding in the underlying biophysics of modulating neural activity using electrical fields. As a contrast on this point, it has been calculated that a typical magnetic therapy pad will generate a movement of ions flowing through a vessel 1 centimeter away by less than what thermal agitation of the ion generated by the organism itself causes, by a factor of 10 million [18]. Yet, claims of efficacy using magnetic therapy continue. An estimate of magnetic field strength required to produce potentially a 10% reduction in neural activity itself was calculated to be 24 Tesla [19]. Electrical stimulation on the other hand benefits from a deeper investigation and support of its principles, and technological advances continue to be made in refining appropriate applications.

The further details of the more recent history of neuromodulation devices has been well-documented elsewhere [4,20] but, importantly, the advances have come about by the continued collaborative efforts between industry and practitioners. This synthesis speaks to the current debates on conflict of interest that presently occupy much time and effort. In general, devices became more refined in terms of materials, handling characteristics, electrode design and implementation, power storage and management, and understanding of the mechanisms of action. They originally used RF transfer of power, and by the early 1980s had transitioned to multichannel and multiple-program devices. The first fully implantable generators (IPGs), however, came from advances in cardiac devices and, in 1976, Cordis came out with the model 199A that was epoxy-coated. It had limited capabilities and was marketed for treatment of spasticity primarily in MS for example. Eventually, a lithium ion-based battery was developed in their third generation device (the model 900X-MK1) and was hermetically sealed in titanium, ushering in what we now consider the standard platform of these devices. Rechargeability came about with competitive patents in the 1990s and all three major device companies (Medtronic, St Jude Medical, and Boston Scientific) make rechargeable IPGs for spinal cord stimulators that can last approximately 10 years with regular recharging. Closed-loop systems are being developed, wherein some type of real-time information about the system being stimulated can be incorporated into the function of the device. For example, a device in trials now for treating epilepsy (NeuroPace, Inc – [21,22]) analyzes cortical activity and can stimulate cortical regions or deeper regions to limit or stop a seizure. Further closed-loop applications are sure to become available in the near future, in deep brain stimulators (DBS), peripheral nerve stimulators (PNS), motor cortex stimulators (MCS), or spinal cord stimulators (SCS), or in other yet to be distinguished ways. All of these refinements, advances, and properties of these systems will be better characterized and elaborated in subsequent chapters in this text.

APPLICATIONS

Out of its early history, neuromodulation has now found a calling in numerous areas of care, and continues to be attempted in others. Although the main devices still include predominantly deep brain stimulators, dorsal column stimulators, vagus nerve stimulators, and peripheral nerve stimulators, modifications of these are establishing themselves and will likely see design refinements in the near future so as to optimize their application. Such modifications include motor cortex stimulators wherein standard dorsal column stimulator systems are used over the M1 region in the epidural space (cf. for review [23]), intradiskal stimulation for discogenic back pain [24] which has so far used a typical 4-contact DBS lead or an 8-contact percutaneous dorsal column lead, field stimulation for low back pain utilizing 4 or 8-contact percutaneous leads in the subcutaneous layers of paraspinal regions, and a variety of essentially peripheral nerve stimulation applications ranging from supraorbital nerve to occipital nerve to specific functional targets such as bladder or diaphragm modulation (see Chapter 5).

Beyond using one of the readily available products in a different application, there are also numerous applications of the devices in their intended locations but with different physiological or anatomical targets and clinical problems. So, for example, DBS is used to treat not only tremor, or Parkinson's disease, but also various forms of dystonia [25], Tourette's syndrome [26], obsessive–compulsive disorder [27], cluster headache [28], depression, obesity [29], epilepsy [30], anorexia nervosa, addiction [31], memory dysfunction [32], minimally-conscious states [33], and chronic pain [34]. Cortical stimulation is not only tried for post-stroke or other refractory forms of chronic pain, but also tinnitus [35], post-stroke rehabilitation [36], epilepsy [21] and depression. Dorsal column stimulation is not restricted to failed back surgery syndrome or CRPS, but can be used to treat anginal pain [37], post-herpetic pain [38], spasticity [39], critical-limb ischemia [40], gastrointestinal motility disorders [41], interstitial cystitis [42], or abdominal pain. Vagal nerve stimulation (VNS), typically used to treat epilepsy, has been successful in treating refractory reactive airway disorders [43]. Occipital nerve stimulation has found some success in treating some head pain, migraine, and other headache disorders [44].

What does this array of applications suggest about the overall approach of neuromodulation? Clearly, the methodologies already tried have met with a fair amount of success and innovative engineers and caregivers are seeking more. Additionally, it speaks to the often-espoused advantages of neuromodulation – reversibility, programmability, and specificity. Most of the disorders where it is routinely used are disorders that are notoriously difficult to treat otherwise. In the paradigm shift of our treatment algorithms, neuromodulation has become a tool of choice in addressing the trend to move from salvage operation to quality of life improvement. In neurosurgery, in particular, there is still an important need to retain the unique ability emergently to prevent herniation and impending death with certain decompressive procedures, secure vascular anomalies to prevent rebleeding and likely death or morbidity, or to resect enlarging masses of tumor to stave off impending herniation or impairment. Yet, as the population ages, and more people are faced with living with disabilities or discomfort for many years, the enhancement of quality of life has become a cause celèbre. Neuromodulation has risen to the fore in this regard. Patients with Parkinson's disease, tremor, dystonia, epilepsy and chronic pain of one sort or another, only rarely die from their disorders – but they live on with major difficulties and poor quality of life. Interventions that improve quality of life with comparatively little or no significant risk, such as neuromodulation, begin to make more and more sense – at least clinically.

ETHICS

Despite the hype and the promise, there might clearly be ethical issues raised when a therapeutic approach develops, such as neuromodulation, that can interface and modify the very function that determines our personalities, our thoughts, our perceptions, and our movements – surprisingly, there have already been several papers addressing this important issue [45–48]. The broad principles of *beneficence, non-maleficence, autonomy and justice* are the underpinnings of discussions on medical ethics. In writing on the ethical aspects of using transcranial magnetic stimulation (TMS), an intervention one might think is particularly safe and well-studied, Illes et al [46] point out that there are still outstanding questions that cannot be forgotten. They analyze the substantial support that single-pulse TMS appears to be safe and have no short or long-term effects on neural structure or function. But they still emphasize that concerns are debated as to whether patients are truly unaware of real versus sham stimulation when using TMS (in which case, whether or not informed consent is undermined), using TMS to treat psychiatric disorders when it is unclear what the precise target is, treating psychiatric disorders when there is an intended effect on the circuitry of the disorder (for benefit) without knowing fully the effects on other aspects of the circuit as well – permanent or temporary. They support the use of an ethical approach called casuistry, instead of the more typical approach describe above. Casuistry is essentially case and context-based practical decisions on the right or wrong of a particular procedure or other intervention. Most applications of neuromodulation involve conditions wherein the patient has little other option available – they have tried medication paradigms, less-invasive paradigms, non-invasive paradigms, and so forth, with no real benefit and still have a significantly compromised quality of life, loss of productivity or both, and the intervention at hand has little if any chance of making their situation worse, in addition to having often a moderate or high likelihood of helping them. Under such contexts, one might argue from a casuistry-based ethical framework that neuromodulation would always be acceptable.

Despite raising support for this perspective, however, Illes et al [46] question it as well, saying it would be imprudent to keep a scorecard of risk and benefit for each patient when (in the case of TMS) so much is unknown. Out of this deadlock, one might suggest, that because such unknowns can be cited for virtually any intervention, to varying degrees, and because typically no one has determined what degree of knowledge is acceptable before one can consider an intervention entirely safe, we should adopt a hybrid approach. Such an approach would use casuistry arguments under an umbrella of principle-guided ethics, but take as its reference points for safety and knowledge already agreed-upon interventions that have been considered safe enough. For example, electroconvulsive therapy (ECT) is considered safe enough to use routinely – it could be argued that there are at least as many unknowns with ECT in terms of long-term effects that are irreversible as there might be in TMS, and as such, this would bias individual studies or cases toward ethical grounding.

While TMS may be used beneficially to map functional brain regions before tumor surgery or to help victims obliterate memories for traumatic events like violent crime, it is also worth considering the potential commercial uses of this technology. TMS applications can impair memory in a confined experimental environment, but at high enough frequency, power and duration, TMS could more permanently disrupt or suppress memory formation, decrease sexual drive or possibly repress the desire to lie. TMS or other similar technologies have already been portrayed in film for these purposes, as in the movie Eternal Sunshine of the Spotless Mind (Focus Features, 2004) in which the protagonist seeks to have his memories of past romance erased from his mind. While advertising and sales of memory erasure technology are still absent from the open marketplace, we must consider means of ensuring that all frontier neurotechnology is reserved for responsible research and clinical use, and questionable uses kept at bay. The technology must never be used in coercive ways. We must also consider policy in the context of how our individual values come into play. For Illes et al [46] in an ethics perspective on transcranial magnetic stimulation (TMS) and human neuromodulation example, should society have unfettered access to this technology if it becomes available in the open market? What will protect consumers – especially the openly ill or covertly suffering – from marketing lures that, in the hands of non-expert TMS entrepreneurs, may be no more effective than snake oil?

Ethical issues in DBS surgery, particularly for disorders of mood, behavior, and thought (MBT) are potentially more problematic because DBS is overtly more invasive and riskier than TMS (see [49]). In this circumstance, usually (though not in every case), the exact target is reasonably well defined (more so than with TMS), and there are data on intervention of some sort in those areas from prior lesioning studies. But there are, of course, still unknowns as to what stimulation will bring about that lesioning did not, as to whether there are downstream effects with stimulation that do not occur with lesions, and whether or not long-term effects of stimulation are truly equivalent to lesioning. The oversight of a team including psychiatrists, bioethicists, and the neurosciences, in a center dedicated to embracing this intervention within the agreed upon ethical framework, is appropriately stressed. In cases where there are not prior lesion data to turn to, (area 25, for example, for refractory depression), then the ethical framework might be similar to the TMS case, with the enhanced aspect of risk with the procedure itself (hemorrhage, infection, stroke) taken into consideration within the consenting process, and with the oversight of the team and institution in place.

COST

While the preceding discussion suggests that neuromodulation can be spectacularly powerful, and relatively minimally invasive in its ability to achieve that benefit, it does come with cost, however, from a financial standpoint. With current health-care costs astoundingly eclipsing over 16% of the gross domestic product (GDP) in the USA, the following statement was made in a recent report on health care spending by the US Congressional Budget Office (CBO):

The results of CBO's projections suggest that in the absence of changes in federal law [50]:

1. Total spending on health care would rise from 16 percent of gross domestic product (GDP) in 2007 to 25 percent in 2025, 37 percent in 2050, and 49 percent in 2082.

2. Federal spending on Medicare (net of beneficiaries' premiums) and Medicaid would rise from 4 percent of GDP in 2007 to 7 percent in 2025, 12 percent in 2050, and 19 percent in 2082.

They emphasize, however, that the goal is not necessarily to limit or reduce costs, but to consider doing so if the ability to maintain or enhance health-care delivery, improved health care, can be achieved. As they note:

In itself, higher spending on health care is not necessarily a 'problem'. Indeed, there might be less concern about increasing costs if they yielded commensurate gains in health. But the degree to which the system promotes the population's health remains unclear. Indeed, substantial evidence exists that more expensive care does not always mean higher-quality care. Consequently, embedded in the country's fiscal challenge is the opportunity to reduce costs without impairing health outcomes overall.[50]

(CBO – The Long Term Outlook for Health Care Delivery, Nov, 2007)

So, in the current overhaul of health care reimbursement and health-care delivery, although no one can be sure what the future will bring, it does seem sensible to spend effort determining whether or not interventions using neuromodulation are in line with delivery of improved health care – because typically, these approaches are expensive. The cost of a DBS system for one side of the brain is approximately $25 000 for the electrode, securing burr hole cap, connecting extension wire, and the implanted pulse generator (IPG). This cost varies contextually with geography, third party payor contracts, whether or not the procedure is performed as an outpatient, 23-hour admission, or inpatient stay, one side or both sides are done in the same surgery, electrodes and IPG placements are split up in time, or whether or not a dual input IPG is used. This cost also does not factor in surgery, anesthesia, hospital and follow-up care fees, possible rehab stays, physical therapy, and neurology follow-up visits for medication adjustments. Nor does it consider IPG replacements needed in the future and the associated costs of removing the depleted or defective IPG and replacing it with a new one, usually within 3–5 years currently.

The economics of the current system in the USA at least, are unlikely to be able to sustain such device costs for long – even if efficacy is determined. Interestingly, several of the world's economies are intimately tied to medical device manufacture and derivative industries as well (e.g. packaging, plastics, metals, logistics, and marketing). Ireland, for example, has about one-third of all its exports related to medical products, many of which are tied to medical devices themselves (Medical Device Daily, Apr, 2005). Puerto Rico, a self-governing commonwealth associated with the USA, as of 2006, manufactured 50% of all pacemakers and defibrillators and 40% of all other devices purchased in the US market [51]. But one aspect of the debate often missing is the comparative cost of *not* using the neuromodulation device. There have been excellent studies in the previous 20 years, with several of the best in the last 5 years, which have evaluated exactly these aspects of the problem [52,53]. In related work, and as an important 'comparator', the publications from the NIHR HTA program in the UK, found in the international journal *Health Technology Assessment*, can be of value.

These studies predominantly hinge on QALY assessments and, if done well, can be used more or less in comparing one kind of treatment for a particular disorder with an entirely different treatment for a different disorder. QALY, of course, stands for Quality of Life Year, and has been refined over the years in the cost/benefit analyses since it first was put forth in an analysis of renal disease in 1968 [54] – it is the cost for a certain treatment or intervention at providing a single year of quality living for the patient. In general, most health-care systems agree that approximately $50 000 or less per QALY is acceptable from the standpoint of what that society would be willing to pay for [55]. This upper limit of acceptable cost per QALY may be in the midst of changing, but it has held up for many years across multiple economies and cultures to date [55]. It is also not a federal mandate – in other words, it is a value derived from the ebb and flow of the health-care structure itself, the reimbursement and utilization structure and the context of the culture itself. In the USA, for example, having air bags versus no air bags in the driving population and car passengers works out to be $30 000/QALY. It is unlikely now that anyone would dispute this intervention is worth such cost and, as a society, we have tacitly accepted this cost per QALY for air bags. Statin therapy versus usual care in patients between 75 and 84 years of age with a history of myocardial infarction adds up to $21 000/QALY. However, national regulation against using a cellular telephone while driving versus no regulation, in the US population in 1997 would have been \$350 000/QALY, annual screening for depression versus no screening in 40-year-old primary care patients is \$210 000/QALY, and even systematic screening for diabetes versus no screening in every individual over the age of 25 is \$67 000/QALY, according to [56].

An example from neuromodulation may help illustrate the value of this approach. Dudding et al, published an analysis of sacral nerve stimulation versus non-surgical management in patients who had undergone sacral nerve stimulation at a single institution over a 10-year period [57] (quality level 5 of 7). Fecal incontinence had been present for a median of 7 years before surgery, and all patients had failed to benefit from previous conservative treatments. Stimulation was effective in this most difficult group with a \$49 000/QALY – under the typical US acceptable level. But here is an additional key point – how does one factor in the *lost* QALY up to that point from *not* intervening with neuromodulation sooner? Certainly, some time might be spent evaluating less invasive treatments. And many patients will respond – but surely that could be done well within 7 years median time. This is a critical aspect of these analyses that is left out, or perhaps never even considered. What is a reasonable standard of care prior to considering neuromodulation? Quantification of such would likely swing the analysis much further in favor of neuromodulation.

DBS in the STN for Parkinson's disease has been studied twice in this way – 2001 and 2007 [58,59]. DBS provided 0.72 and 0.76 DALY respectively, though for slightly different costs/QALY (\$62 000 US in the earlier study and \$47 000/QALY in the more recent study, done in Spain), both very close to acceptable societal cost acceptance.

Spinal cord stimulation has been examined three times between 2002 and 2007 in this fashion, twice for treatment of failed back surgery syndrome and once examining physical therapy with and without SCS for CRPS in a single limb [60–62]. Again, it is important to consider that the patients in these studies are generally failures of conventional therapies already. All three of these studies showed not only QALY benefit, but at a *cost saving*.

Understanding both sides of the cost equation is paramount to the overall debate, even when considering the slant that QALY analyses have toward a rationing of health care. Such a view has, on the surface at least, not yet been emphasized. But the juggernaut of overall health-care costs over time will force some aspect of this perspective upon us. As a suggestion, cost of implants could be capped after research and development costs are recouped in a systematized manner. The advantage to this significant compromise from industry is that payment then is negotiated between government or third party payors and the device-makers directly – *all in exchange for less restriction on implant indications* – this will free up innovation and competition and reduce costs while broadening the beneficial impact for patients.

Without such changes, devices overall will become so restricted in use and their costs, and logistics, that to provide adequate Class I data to gain an indication will become so prohibitive, on top of already restricted schedules for clinicians and researchers, that the ability to sustain business may become impossible. Right now, the market is expected to grow at double digit rates for the next 5 years at a minimum, as it has for the preceding 10. But without the sustenance of a favorable reimbursement climate, that profitability would end quickly. The conclusion would not be that devices are implanted inappropriately because they are paid for; rather, in contradistinction, it would be that many patients who would benefit would be unable to get adequate treatment. As caregivers, and as the flag bearers of the neuromodulation approach, our responsibility is to bring these therapies safely to as many as is appropriate.

REFERENCES

1. Neurotech Report. The market for neurotechnology 2008-2012. Neurotech Report. 2007
2. Kuhn TS. *The structure of scientific revolutions.* Chicago: University of Chicago Press; 1962.
3. Gildenberg PL. Evolution of spinal cord surgery for pain. *Clin Neurosurg.* 2006;53:11–17.
4. Rossi U, The history of electrical stimulation and the relief of pain. In: Simpson BA, ed. *Electrical stimulation and the relief of pain*, vol. 15. New York: Elsevier Science; 2003:5–16.
5. Barolat G. History of neuromodulation. *Neuromod News.* 1999;2:3–9.
6. Kellaway D. The William Osler medal essay, The part played by electric fish in the early history of bioelectricity and electrotherapy. *Bull. Hist. Med.* 1946;20:112–137.
7. Kane K, Taub A. A history of local electrical analgesia. *Pain.* 1975;1:125–138.
8. Hadzovic S. Pharmacy and the great contribution of Arab-Islamic science to its development. *Med Arh.* 1997;51:47–50.
9. Liebenau J. *Medical science and medical industry.* Baltimore: Johns Hopkins University Press; 1987.
10. Stillings D. The first observation of electrical stimulation. *Med Instrum.* 1974;8:313.
11. Krantzenstein CA. A pioneer of electro-therapeutics. *Br Med J.* 1924;1:759–760.
12. Fritsch G, Hitzig E. The excitable cerebral cortex. Uber die elektrische Erregbarkeit des Grosshirns. *Arch Anat Physiol Wissen.* 1870;37:300–332.
13. Bartholow R. Experimental investigations into the functions of the human brain. *Am J Med Sci.* 1874;134:305–313.
14. Horsley, VA. Case of occipital encephalocele in which a correct diagnosis was obtained by means of the induced current. Pt xxvi. 1884.
15. Shils JL, Arle JE, Evoked potentials in functional neurosurgery. In: Lozano A, Gildenberg PL, Tasker., eds. *Textbook of stereotactic and functional neurosurgery*, vol. 1. Berlin: Springer-Verlag; 2009:1255–1282.
16. The Burton Report http://www.burtonreport.com/infspine/NSHistNeurostimPartI.htm.
17. Shealy N. Electrical inhibition of pain by stimulation of the dorsal columns: preliminary clinical report. *Anesth Analg.* 1967;46:489–491.
18. Ramey, personal communication. From Adair, Yale – cited on www.skeptically.org under Quackery.
19. Wikswo JP, Barach JP. An estimate of the steady magnetic field strength required to influence nerve conduction. *IEEE Trans Biomed Eng.* 1980;27:722–723.
20. Gildenberg, PL. Neuromodulation: A Historical Pespective. In Krames E, Peckham PH and Rezai AR, Neuromodulation, vol. 2. London: Elsevier; 2009:9–20.
21. Skarpaas TL, Morrell MJ. Intracranial stimulation therapy for epilepsy. *Neurotherapeutics.* 2009;6:238–243.

22. Fountas KN, Smith JR. A novel closed-loop stimulation system in the control of focal, medically refractory epilepsy. *Acta Neurochir Suppl*. 2007;97:357–362.
23. Arle JE, Shils JL. Motor cortex stimulation for pain and movement disorders. *Neurotherapeutics*. 2008;5:37–49.
24. Arle JE, Shils JL. *Intradiskal stimulation for refractory lower back pain*. Las Vegas: North American Neuromodulation Society meeting meeting; 2008 2008.
25. Krauss J. Surgical treatment of dystonia. *Eur J Neurol*. 2010;17(Suppl. 1):97–101.
26. Sassi M, Porta M, & Servello D. Deep brain stimulatoin for treatment refractory Tourette's syndrome: A review. Acta Neurochir (Wien), Sep 20.(epub ahead of print). 2010.
27. Mian MK, Campos M, Sheth SA, Eskandar EN. Deep brain stimulation for obsessive-compulsive disorder: past, present and future; review. *Neurosurg Focus*. 2010;2:E10.
28. Matharu MS, Zrinzo L. Deep brain stimulation in cluster headache: hypothalamus or midbrain tegmentum?; review. *Curr Pain Headache Rep*. 2010;14:151–159.
29. Pisapia JM, Halpern CH, Williams NN, Wadden TA, Baltuch GH, Stein SC. Deep brain stimulation compared with bariatric surgery for the treatment of morbid obesity: a decision analysis study. *Neurosurg Focus*. 2010;29:E15.
30. Boon P, Vonck K, De Herdt V, et al. Deep brain stimulation in patients with refractory temporal lobe epilepsy. *Epilepsy Curr*. 2007;48:1551–1560.
31. Lu L, Wang X, Kosten TR. Stereotactic neurosurgical treatment of drug addiction. *Am J Drug Alcohol Abuse*. 2009;35:391–393.
32. Laxton AW, Tang-Wai DF, McANdrews MP, et al. A phase I trial of deep brain stimulation of memory circuits in Alzheimer disease. *Ann Neurol*. 2010;68:521–534.
33. Schiff ND, Giacino JT, Kalmar K, et al. Behavioral improvements with thalamic stimulation after severe traumatic brain injury. *Nature*. 2007;448:600–603.
34. Cruccu G, Aziz TZ, Garcia-Larrea L, et al. EFNS guidelines on neurostimulation therapy for neuropathic pain. *Eur J Neurol*. 2007;14:952–970.
35. Litre CF, Theret E, Tran H, et al. Surgical treatment by electrical stimulation of the auditory cortex for intractable tinnitus. *Brain Stimul*. 2009;2:132–137.
36. Kim DY, Lim JY, Kang EK, et al. Effect of transcranial direct current stimulation on motor recovery in patients with subacute stroke. *Am J Phys Med Rehabil*. 2010;89:879–886.
37. Lanza GA, Grimaldi R, Greco S, et al. Spinal cord stimulation for the treatment of refractory angina pectoris: A multicenter randomized single-blind study (the SCS-ITA trial). *Pain*. 2010; epub ahead of print.
38. Meglio M, Cioni B, Rossi GF. Spinal cord stimulation in management of chronic pain. A 9-year experience. *J Neurosurg*. 1989;7:519–524.
39. Pinter MM, Gerstenbrand F, Dimitrijevic MR. Epidural electrical stimulation of posterior structures of the human lumbosacral cord: 3. Control of spasticity. *Spinal Cord*. 2000;38:524–531.
40. Klomp HM, Steyerberg EW, Habbema JD, et al. What is the evidence on efficacy of spinal cord stimulation in (subgroups of) patients with critical limb ischemia? *Ann Vasc Surg*. 2009;23:355–363.
41. Maher J, Johnson AC, Newman R, et al. Effect of spinal cord stimulation in a rodent model of post-operative ileus. *Neurogastroenterol Motil*. 2009;21:672–677.
42. Gajewski JB, Al-Zahrani. The long-term efficacy of sacral neuromodulation in the management of intractable bladder pain syndrome: 14 years of experience in one center. *Br J Urol Int*. 2010; epub ahead of pub, Sep 30.
43. Simon BJ, Emala CW, Lewis LM, et al. *Vagal nerve stimulation for relief of bronchoconstriction: Preliminary clinical data and mechanism of action*. Las Vegas: Oral presentation at North American Neuromodulation Society meeting; 2009.

44. Paemeleire K, Bartsch T. Occipital nerve stimulation for headache disorders. *Neurotherapeutics*. 2010;7:213–219.
45. Fins JJ. From psychosurgery to neuromodulation and palliation: history's lessons for the ethical conduct and regulation of neuropsychiatric research. *Neurosurg Clin N Am*. 2003;14:303–319.
46. Illes J, Gallo M, Kirschen MP. An ethics perspective on transcranial magnetic stimulation (TMS) and human neuromodulation. *Behav Neurol*. 2006;17:149–157.
47. Synofzik M, Schlaepfer TE. Stimulating personality: ethical criteria for deep brain stimulation in psychiatric patients and for enhancement purposes. *Biotechnol J*. 2008;3:1511–1520.
48. Lipsman N, Bernstein M, Lozano AM. Criteria for the ethical conduct of psychiatric neurosurgery clinical trials. *Neurosurg Focus*. 2010;29:E9.
49. Rabins P, Appleby BS, Brandt J, et al. Scientific and ethical issues related to deep brain stimulation for disorders of mood, behavior, and thought. *Arch Gen Psychiatry*. 2009;66:931–937.
50. CBO. The long term outlook for health care spending, Nov. 2007.
51. PRIDCO (2009). www.PRIDCO.com.
52. Taylor RS, Taylor RJ, Van Buyten J-P, et al. The cost effectiveness of spinal cord stimulation in the treatment of pain: a systematic review of the literature. *J Pain Symptom Manage*. 2004;27:370–378.
53. Simpson EL, Duenas A, Holmes MW, et al. Spinal cord stimulation for chronic pain of neuropathic or ischemic origin: systemic review and economic evaluation. *Health Technol Assess*. 2009;13, iii, ix-x:1–154.
54. Klarman HE, Francis JO, Rosenthal GD. Cost-effectiveness analysis applied to the treatment of chronic renal disease. *Med Care*. 1968;6:48–54.
55. https://research.tufts-nemc.org/cear4/SearchingtheCEARegistry/FAQs.aspx.
56. Harvard Center for Risk Analysis. Risk Perspec. 2003; 11.
57. Dudding TC, Meng Lee E, Faiz O, et al. Economic evaluation of sacral nerve stimulation for faecal incontinence. *Br J Surg*. 2008;95:1155–1163.
58. Tomaszewski KJ, Holloway RG. Deep brain stimulation in the treatment of Parkinson's disease: a cost effectiveness analysis. *Neurology*. 2001;57:663–671.
59. Valldeoriola F, Morsi O, Tolosa E, et al. Prospective comparative study on cost-effectiveness of subthalamic stimulation and best medical treatment in advanced Parkinson's disease. *Mov Disord*. 2007;22:2183–2191.
60. Kemler MA, Furnee CA. Economic evaluation of spinal cord stimulation for chronic reflex sympathetic dystrophy. *Neurology*. 2002;59:1203–1209.
61. Taylor RJ, Taylor RS. Spinal cord stimulation for failed back surgery syndrome: a decision-analytic model and cost-effectiveness analysis. *Int J Health Technol Assess Health Care*. 2005;21:351–358.
62. North RB, Kidd D, Shipley J, et al. Spinal cord stimulation versus reoperation for failed back surgery syndrome: a cost effectiveness and cost utility analysis based on a randomized, controlled trial. *Neurosurgery*. 2007;61:361–368.

Part II

Regions of Application

Chapter 2

Cerebral – Surface

Sergio Canavero, MD (US FMGEMS)
Turin Advanced Neuromodulation Group (TANG), Turin, Italy

The goal of cortical stimulation (CS) is to change the excitability or activity of cortical and related subcortical networks involved in pathophysiological processes. Any neurological or psychiatric disorder can be affected by CS, either by reactivating hypoactive neuronal structures, as first proposed by us ('whenever SPECT [single photon emission computed tomograhy] shows cortical disactivation, the therapeutic rationale would be trying to stimulate it')[1] or inhibiting overactive structures (epilepsy, auditory hallucinations, tinnitus), or both, such as in depression and stroke, i.e. by activating one side and simultaneously inhibiting the contralateral one [2,3].

Considering the risk, albeit small, of serious intracerebral hemorrhages and mortality attendant to electrode insertion in deep brain stimulation (DBS), it seems surprising that the much more benign procedures involved in CS have not given the latter the edge in the field of brain stimulation. While DBS for movement disorders may confer a superior benefit (although this awaits head-to-head trials for confirmation), CS outdoes DBS for neuropathic pain, stroke rehabilitation, tinnitus, and probably coma rehabilitation and epilepsy. Importantly, Extradural CS and transcranial direct current stimulation (tDCS) have been proven better than placebo stimulation – given the lack of physiologic effects elicited, whereas DBS cannot be evaluated with the same degree of confidence for several applications. Finally, CS has the potential for neuroprotection (by hyperpolarization of neurotoxic currents) and has clear neuroplasticity-promoting effects.

Several reasons can be adduced:

1. DBS is approved for the treatment of central nervous system disorders; the huge marketing efforts from the manufacturers may have 'swamped' other experimental procedures. However, approval by regulatory bodies of repetitive transcranial magnetic stimulation (rTMS) for depression in the past few years might help reverse the trend
2. Not many neurosurgeons have experience with invasive cortical stimulation. Even worse, in view of the supposed 'simplicity' of such procedures, some surgeons simply rushed in without an adequate competence and came away with negative results
3. A philosophical reason: neurosurgeons are both enamored of their ability to be precise (as required by the small size of DBS targets) and the

Essential Neuromodulation. DOI: 10.1016/B978-0-12-381409-8.00002-4

19

empowering high-tech glittering technology involved. Contrast this with the relative low-tech simplicity of CS, which does not necessitate stereo-tactic equipment and allied paraphernalia

4. Results with cortectomy were attempted for pain and motor disorders in years gone by but results were less than compelling

5. The daunting vastness of the cortical mantle and the astonishing structural intricacy thereof: suffice to say that only in 2009 we finally learned the number of neurons in the human brain (86 billion neurons – 16 billion in the cerebral cortex and a mere 85 billion non-neuronal cells, one tenth of previous estimates) [4]. Also, much of our knowledge of cortical micro-anatomy and corticocortical connections is based on non-human primates.

HISTORY

Systematic application of electromedical equipment for therapeutic use started in the 1700s. Although clearly any form of electricity applied to the head also stimulates the cortex (including the discharge from electric fish used to therapeutic effects since 4000 BCE), CS was applied for the first time by Giovanni Aldini (1762–1834), Luigi Galvani's nephew, at the end of the 1700s and it was his demonstrations (and the sensationalist newspaper reports) in London that spurred Mary Shelley's highly successful novel 'Frankenstein, or the modern Prometheus'. Aldini stimulated the cerebral cortex of one hemisphere in criminals sacrificed about an hour earlier and obtained contralateral facial muscular contractions [5]. This finding was not exploited and had to be rediscovered by Fritz and Hitzig in the second half of the 19th century. Despite attempts by others (including John Wesley and Benjamin Franklin), Aldini was the first to develop transcranial direct current brain stimulation by exploiting Alessandro Volta's bimetallic pile (Fig. 2.1) and apply it to psychiatric patients, in particular depressed ones, by stimulating the shaved and humidified parietal area. Sir Victor Horsley (1888–1903) triggered movements in the extremities of human patients by electrically stimulating the cerebral cortex. Keen (1887–1903) did the same with a rubberized handpiece with two partially isolated end poles fed by a battery. Others followed, in particular Penfield and Boldrey in the 1930s. In the 1890s, Jacques d'Arsonval induced phosphenes in humans when their heads were placed within a strong time-varying magnetic field which stimulated the retina. This was the first magnetic stimulation of the nervous system. In 1985, Barker and colleagues introduced the first TMS apparatus and transcranial direct current stimulation (tDCS) was 'rediscovered' at the end of the 1990s (for historical reviews see [3,6,7]).

In the 1970s, Alberts reported that stimulation at 60 Hz with a 7-contact Delgado cortical plate electrode of an area near the rolandic fissure between motor and sensory sites (SI) could initiate or augment parkinsonian tremor in patients, while Woolsey temporarily alleviated parkinsonian rigidity and tremor in two patients by direct acute intraoperative stimulation in the primary motor cortex (MI). He wrote:

FIGURE 2.1   First patient ever to be submitted to non-invasive therapeutic cortical stimulation (Aldini 1803).

...marked tremor and strong rigidity...The results suggest the possibility that subthreshold electrical stimulation through implanted electrodes might be used to control these symptoms in parkinsonian patients.

However, it was only 10 years later that Tsubokawa's group in Japan applied extradural motor cortex stimulation for the treatment of central pain and another 10 years passed before the same technique was brought to bear on Parkinson's disease and then other neural disorders (see historical review [3]). On the whole, the progress of therapeutic cortical stimulation has been slow and only gained momentum in the first decade of the 21st century.

ANATOMICAL CONSTRAINTS ON TARGETING

The neocortex is a dishomogeneous, ultracomplex, six-layered structure (Fig. 2.2), and is strongly folded: in humans almost two thirds of the neocortex is hidden away in the depth of the sulci. The individual sulci vary in position and course

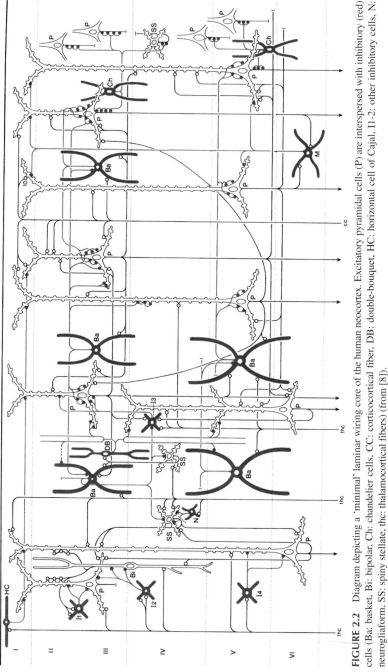

FIGURE 2.2 Diagram depicting a 'minimal' laminar wiring core of the human neocortex. Excitatory pyramidal cells (P) are interspersed with inhibitory (red) cells (Ba: basket, Bi: bipolar, Ch: chandelier cells, CC: corticocortical fiber, DB: double-bouquet, HC: horizontal cell of Cajal, I1-2: other inhibitory cells, N: neurogliaform, SS: spiny stellate, thc: thalamocortical fibers) (from [8]).

among subjects, but also between the two hemispheres in the same subject, may show one or several interruptions and some may be doubled over a certain part of their trajectory [8]. There are also several cortical hemispheric structural asymmetries [9]. This severely limits the possibility to make overarching generalizations as of targeting.

Cytoarchitectonically, the cortex has been divided into 44 sharply delineated areas by Brodmann a century ago, whose boundaries generally do not coincide with the sulci on the cerebral surface. This areal distribution has been revised by several authors, but the result has added more confusion: anatomical exploration with basic histological stains gives little insight on functional subdivisions. Numerous attempts at defining functionally segregated areas (including electrical stimulation) are on record, with a harsh conflict between localizationists (neo-phrenologists) and anti-localizationists. Based on neuroimaging data, it can be estimated that about 150 juxtaposed structural and potentially functional entities are present in the human neocortex (e.g. areas 9/46 and 44/45 are distinct architectonic entities). Each cortical area has a unique pattern of corticocortical/corticosubcortical connections (connectional and functional fingerprint). Yet, since the neocortical wiring is characterized by a distributed hierarchical network that contains numerous intertwined, cross-talking processing streams, the identification of functionally segregated domains remains a difficult problem. Moreover, most of the human neocortex is occupied by association areas of various kinds and the boundaries between these areas do not closely correspond to those of cytoarchitectonic fields as delineated by Brodmann and others. Additionally, all cortical areas (primary and association) show considerable intersubject variability: this appears to be a general feature of neocortical architectonic areas, a microstructural variation superimposed upon the also considerable macrostructural variation pertaining to the overall size and shape of the hemispheres, as well as the sulcal and gyral pattern. This variability seriously hampers structural–functional correlations. This means that simply transferring 'hot spots' in brain imaging studies to a 3D version of Brodmann's chart incorporated in the stereotaxic atlas of Talairach and Tournoux is apt to lead to erroneous conclusions, since the atlas neglects variability [10], imposing a serious limit on CS procedures. Spatial normalization procedures are thus necessary.

In most cognitive tasks, two or more cortical areas are activated and these may be considered as nodal points in the networks underlying the process. At the same time, cortical regions (e.g. prefrontal cortex, posterior parietal cortex) are engaged in a wide variety of cognitive demands. The most parsimonious explanation is that they reflect cognitive processes that are tapped by tasks in different domains. This, unfortunately, makes the selection of cortical targets for psychiatric neuromodulation, for example, problematic. The dorsolateral prefrontal cortex (DLPFC), the approved primary target for the treatment of depression, is mainly a cognitive, not a limbic area (and might provide benefit by restoring cognitive control over affect): it is quite large and actually there may be subareas whose stimulation would result in stronger effects [11,12]. In the end, functional localization and specialization are important principles, but do not offer a complete or sufficient explanation of cortical organization. Rather, a process should be explained in terms of distributed patterns of changing neural activity in networks of interconnected functionally specialized areas. In other words, *cognitive and mental abilities result from the functional integration of the elementary processing operations occurring in a smaller or larger number of functional areas*. In practice, given the inter-areal connectedness, it is logical to conclude that whatever nodal point is stimulated will entrain the whole network. This has been cogently shown for Parkinson's disease [3]. Recently, a rostrocaudal gradient model of frontal lobe function has been elaborated upon, undermining the discrete model of frontal functions compartmentalized to highly demarcated zones [13]. The rostrocaudal axis (BA10 to BA9/46 to BA8 to BA6) forms a coherent functional network with longer connections being unidirectional: this implies that adjacent regions along the rostrocaudal axis are connected to one another, but do not project to more rostral regions beyond those immediately adjacent. This has a great importance when one considers possible targets in psychiatric CS (i.e. BA10 would stand out as a primary focus for CS attempts).

Hemispheric specialization must also be accounted for: the right hemisphere is tasked with processing negative affect (and vice versa for the left one), an important consideration for psychiatric ECS: interestingly, parameter modulation (e.g. changing frequency) may 'recode' the target function and obtain the sought-after clinical benefit.

CAN CORTICAL STIMULATION BE OPTIMIZED THROUGH MODELING?

Recently, attempts to model cortical structure and function to fine-tune cortical stimulation efforts have been attempted, in the tracks of what has been done for spinal cord and deep brain stimulation (see [3]). In practice, they are of little help to practitioners. Why?

For starters, there is very little evidence in favor of the concepts that:
1. the entire neocortex is composed of radially oriented columnar units or modules
2. all of these entities represent variations on one and the same theme
3. all of these entities essentially have the same structure and
4. they all essentially subserve the same function [8].

This represents a major hurdle by factoring out cortical homogeneity as a foundation for understanding electric field effects. Add to this the dazzling intricacy of cortical cyto- and myelo-architecture [8]. Also, electrical resistance is four to six times higher in the gray than in the white matter.

1. Cells

The cortex accommodates pyramidal (typical and atypical) – 60–85% of all neocortical neurons – and non-pyramidal cells (15–40%) (PC and NPC), with a total number of neocortical synapses numbering at about 300 000 billion.

A. The somata of PC are not under the direct influence of any extrinsic afferent system, but only of local circuit neurons (basket cells) and other NPC. PC somata projecting to particular cortical or subcortical targets are preferentially located in particular cortical layers and sublayers. Corticocortical and callosally projecting fibers arise from both LII–III and infragranular PC. The smaller, more superficially situated PC tend to project to ipsilateral cortical areas situated nearby, whereas the larger, more deeply placed cells to contralateral and to more remote ipsilateral cortical areas. Lamina V PC project subcortically to multiple targets: the smallest and more superficial project to the striatum, the largest and most deeply situated to the spinal cord, the intermediate ones to the remaining sites including the thalamus. The projections to the specific thalamic relay nuclei project exclusively from layer V PC.

Although most cortical neuronal populations projecting to a particular cortical or subcortical target show a distinct laminar specificity, it is not uncommon to find some degree of overlap in the boundaries demarcating different populations of projection neurons. Importantly, the degree of subcortical collateralization of corticofugal fibers is limited. *The axons of all typical PC release a number of intracortical collaterals: together they constitute the largest single category of axons in the neocortex.* Apart from local collaterals, PC axons may also give rise to one to five long, horizontally disposed branches (6–8 mm). These long-range collaterals do not remain within the cytoarchitectonic area in which their parent soma lies but project to adjacent cortical areas. They give off secondary branches in regularly spaced, perpendicularly oriented clusters (column-like) which contact dendrites of other PC but also non-PC. The collaterals of one PC contact numerous other PC and, conversely, one PC receives the converging input of numerous other PCs. Thus, neocortical γ-aminobutyric acid (GABA) interneurons receive input directly from PC axon collaterals and, in turn, synapse with PC, accounting for PC feed-forward/back inhibition. The branching process of axons allows for easier activation by stimulation in comparison to axons without branching.

PC show ample structural diversity: size, laminar position, branching pattern of dendrites, density of spines along apical dendrites, affinity to particular afferent systems, cortical or subcortical target regions, distribution of axon collaterals and patterns of intracortical synaptic output. The somata of PC projecting to a particular target are located in one and the same layer or sublayer and show striking similarities in dendritic morphology, thalamocortical connectivity and distribution of axon collaterals and are in receipt of similar extra and intracortical inputs. Likely, all PC projecting to a particular target are in receipt of similar inputs and have similar functions.

B. Non-PC, especially spiny stellate cells, are equally vital. Their axons may descend superficially or to deeper layers and contact PC, whereas their short collateral branches likely contact similar cells. Spiny stellate cells play a crucial role in the radial propagation of the activity fed by thalamocortical afferents into layer IV of primary sensory areas. Local circuit neurons are, with a single exception, GABAergic (inhibitory); 25–30% of these cells also express one or several neuropeptides. There are different subpopulations based on morphology and neurochemistry:

1. stellate neurons (in all layers), including neurogliaform cells in sensory areas
2. chandelier cells (especially layer II) which especially influence corticocortical activity
3. basket cells (large, small and nest), making up about 50% of all inhibitory neocortical interneurons, with the axon giving rise to 4+ horizontal branches and contacting hundreds of PC and tens of other basket cells
4. vertically oriented neurons (bipolar, bitufted, including double bouquet cells, Martinotti cells (all layers except LI)
5. horizontal cells (layers I, or of Cajal, layer VI and NOS).

Interneurons are contacted and contact other interneurons, forming an intricate network which includes electrical coupling, autaptic innervation and specific extrathalamic input. Chandelier cells terminate on the PC axon hillock, basket cells target the somata and proximal dendrites of PC; both classes control output and oscillatory synchronization of groups of PC. Unfortunately, it is not known whether these interneuronal networks extend indefinitely across the neocortex or have distinct boundaries and this makes modeling a desperate enterprise.

To sum up, it can safely be said that each particular neocortical area contains a number of networks of interconnected, type specific PC. The number and extent of pyramidal networks present within a given cortical area is unknown. Likely, the various PC belonging to a particular network are in receipt of afferents from cohorts of inhibitory interneurons, each cohort contacting a specific domain of the receptive surface of the PC involved. The inhibitory cells forming these cohorts are all of the same type and are generally reciprocally connected by chemical and electrical synapses. Thalamic inputs selectively contact and strongly excite the interneurons belonging to particular cohorts, while others receive weaker or no thalamic inputs. Each of the various cohorts of inhibitory interneurons impinging on a particular pyramidal network is specifically addressed by one or more of the extrathalamic modulatory systems. Not only the inhibitory input but also the excitatory input to PC belonging to the same network may be specific. Although the degree of separation among pyramidal and interneuronal networks is largely unknown, likely the abundant double bouquet cells with their vertically oriented axonal systems contact PC belonging to different networks and the neurogliaform cells form gap junctions with several other types of inhibitory interneurons.

2. Fibers

Myeloarchitectonically, the myelinated fibers in the cortex show two principal orientations, tangential and radial. Tangential fibers tend to form laminae which, in general, can be readily identified in conjunction with the corresponding layers observed in Nissl preparations. The radially oriented fibers are arranged in bundles (radii) which ascend from and descend to the subcortical white matter. However, the number and distinctness of the tangential fiber layers show considerable local differences in the cortex and the same holds true for the extent to which the radii penetrate into the cortex. Moreover, our knowledge of the fiber connections is almost entirely based on studies in non-human primates (particularly the rhesus macaque) and fiber tracking with diffusion tensor imaging in the human has yet to bear substantially on this problem. This is a major point in CS models.

Specifically, there is a horizontal axonal system contacting the basal dendrites of PC situated at specific levels, but the cortex also contains vast numbers of vertically oriented axonal elements (columnar radial coupling), including thalamocortical and corticocortical association fibers, axons and recurrent collaterals of PC and the vertically elongated axonal systems of some types of cortical local circuit (bipolar) neurons. The latter two classes assemble in highly characteristic radially oriented bundles.

In view of variations in length and in position of their apical dendrites, different PC may receive different samples of lamina-specific extracortical and intracortical afferents and apical dendrites of different PC may exhibit different specific affinities to particular afferent systems. Plus, there are distinct lamina-specific differences in the density of spines along the apical dendrites, lamina-specific side branches on the apical dendrites are present and apical dendritic segments of different PC passing through a particular layer may receive highly different numbers of synapses from the afferents concentrated in that layer. There is also the apical dendritic tuft extending into lamina I to be considered which is contacted by thalamic, monoaminergic, recurrent LII–III PC, ascending deep multipolar/bitufted neuron and horizontal lamina I neuron axons. The afferents from different thalamic nuclei which, after having traversed the cortex, spread in lamina I terminate in different subzones of that layer and the apical dendritic tufts of the pyramids thus receive stratified input from different sources. Extrinsic afferent fibers follow a radial course and most distribute themselves in layered arrays. Different (groups of) thalamic nuclei project in a particular laminar fashion to smaller or larger parts of the neocortex. Importantly, more than 10 different extrathalamic subcortical structures projecting to the neocortex have been identified. The effects of the cholinergic,

GABAergic and monoaminergic systems are not generalized excitation or inhibition, but rather region-specific enhancement or diminution of activity in limited neuronal ensembles during certain stages of information processing. Additionally, each particular neocortical area also receives a strong input from other neocortical ipsi- and contralateral areas ending in layers III and IV.

3. Association Fibers

Cascades of short association fibers interconnect modality-specific primary with secondary sensory association areas and these latter with multimodal sensory areas located at the borders. They may remain within the gray matter of the cortex or pass through the superficial white matter between neighboring cortical areas as U fibers and are believed to play a starring role in the mechanism of action of CS [14] (see also in [3]). Long association systems connect the modality-specific parasensory association cortex and the multimodal areas in the occipital, temporal and parietal lobes with the premotor and prefrontal cortex [15]. Short association fibers interconnect the prefrontal cortex, the premotor area and the motor cortex with the primary somatosensory cortex. Connections from parasensory and multimodal association cortices and prefrontal cortex (PFC) to limbic structures pass via the cingulum to the medial temporal lobe; other fibers originating from parasensory association cortices reach limbic structures via the insula. Most association connections are reciprocal. Connections from the primary sensory areas to their neighboring association areas usually originate from the supragranular layers and terminate in/around layer IV (forward connection). Feedback connections originate in the infragranular layers and terminate in layers I and VI. The laminar analysis of association connections may therefore reveal the direction of information transfer.

In sum, the apical dendritic branches of neocortical PC receive input from various sources, but corticocortical projections constitute by far the largest neocortical input system, making these one of the obvious candidates in the mechanism of action of CS. Thus, it can be safely stated that the neocortex communicates first and foremost with itself [8]. An important consideration: the literature on CS often quotes distant effects (for instance in the case of chronic pain) on limbic areas and brainstem as paramount in the mechanism of action, but these must actually be understood as 'knock-on' effects (see a critique of these studies in [3]).

There are also differences in laminar electrophysiology. For instance, there exists a major difference between sensory-evoked and spontaneous activity in primary sensory cortical regions, namely the site of initiation (layer IV but also upper layer VI versus layer V). Layer V neurons are intrinsically more depolarized than layers II–III, on average being about 10 mV closer to action potential threshold (i.e. more excitable). In addition, layer V neurons are strongly synaptically coupled to other nearby layer V neurons in a highly recurrent excitatory microcircuit (making spontaneous waves of excitation more easily spread).

Lamina V neurons have relatively weak connections to layers II–III. Both evoked and spontaneous activities have a relatively limited horizontal spread in superficial layers (i.e. more localized coding) and a more extended propagation in deep layers. But this applies only to action potentials: subthreshold activity propagates widely in superficial layers. How is information encoded? Layer V pyramidal cells fire at a higher rate during both spontaneous and evoked activity (dense firing or population code), whereas lamina II–III pyramidal neurons overall fire at low rates during both types of activity (sparse firing or cell-specific temporal code). There appear to be some neurons in each layer that are orders of magnitude more active than other nearby neurons; perhaps the less active neurons provide a reserve pool to become active at the appropriate moment [16]. How these can all be accommodated inside a model seems a daunting task with current tools.

In the end, this discussion highlights the extreme aspecificity of current cortical stimulation paradigms, since stimulation tends to affect the cortex across the board. A first step would be complexity analysis with closed-loop stimulation devices (e.g. the NeuroPace device for epilepsy control), but it is moot that this alone may circumvent the amazing intricacy of cellular architecture [3]. Does cortical stimulation affect differentially positioned cells in the same way? Does a homogeneous wave of excitation create intracortical conflicts (e.g. two self-effacing inhibitions)? Should dendrites, soma, axon hillocks, nodes, internodes and unmyelinated terminals, all having different electrical properties, be stimulated differentially? This is way beyond current technology. When it comes to details, the only currently feasible approach is to consider the cortex a sort of black box, from which a net effect is sought through trial and error.

MI AS A PARADIGM OF CORTICAL STIMULATION

The primary motor cortex (MI) has been the first target of CS endeavors, especially for chronic pain and control of movement disorders [3,17]. Understanding it may help bring out general principles which can then be applied to other areas and disorders. The upshot can be anticipated: MI is less straightforward than previously thought.

MI is far from the passive servant of higher motor structures. It performs a complex integration of multiple influences, originating in both cerebral hemispheres, in a role as the ultimate gate-keeper that is carefully and differentially tuned to generate well-defined motor behaviors [18]. The discharge pattern of individual MI neurons conveys a bewildering diversity of information. Thus, some neurons receive strong sensory input, whereas others do not. Some neurons respond to contralateral, ipsilateral or bilateral movements; some neurons even reflect sensory signals used to guide action [19]. Many pyramidal tract neurons respond with a wide range of peripheral inputs (visuo-audio-vestibular) [20].

MI has two subdivisions. A rostral region lacks monosynaptic corticomotoneuronal cells (evolutionarily old MI) – descending commands are mediated through spinal circuitry, and a caudal region (evolutionarily new MI) with mono-synaptic cortico-motoneuronal cells which have direct access to motoneurons in the ventral horn essential for highly skilled movements [21]. Neurons in the rostral portion of MI may be more related to kinematic variables, such as velocity and movement direction, than more caudally placed cells [22].

MI is partially sensory due to the coexistence within the same neurons of motor and sensory properties. In particular, *MI and SI hand cortices overlap and are not divided in a simple manner by the central sulcus and sensory responses are elicitable well outside the classically accepted anatomical borders* (see references in [3]). In functional magnetic resonance imaging (fMRI) studies, the motor hand area may extend to (50% of cases), or be located exclusively, in SI (20% of cases), even during the simplest motor tasks [23]. Apart from intrinsic responses, MI and SI are so tightly interconnected by short corticocortical U-fibers that arborize over a considerable rostrocaudal distance in MI to make them almost a unique struc-ture [20]. SI is a major source of somatosensory input to MI and MI is strongly modulated by sensory flow (and vice versa) [18,24]. Clearly, uniformly targeting MI in ECS efforts for chronic pain and Parkinson's disease may be misplaced: SI could be another potential target. Also, BA44 (found 2 cm anterior to MI tongue area) has direct fast conducting corticospinal projections with a role in voluntary hand movements [25], confirming the haziness of MI borders.

Evidence shows a rough body-centered map of MI that matches the tra-ditional motor homunculus. This map extends to nearby premotor areas. Yet, rather than discrete regions of MI controlling different parts of the arm, control of each part is mediated by an extensive territory that overlaps with the ter-ritories controlling other parts [26,27]. Whereas the prior view suggested that stimulation of different regions of MI should elicit movement of different body parts, it is now clear that stimulation can elicit movement of a given body part from a broad region, i.e. MI has a broadly overlapping mosaic of points where stimulation elicits movements of different body parts. Any given MI neuron may influence the motoneuron pools of several muscles (not just one). Selective stimulation of different regions in MI can produce the same movement, due to intra-MI dense bi-directional projections of up to 1 cm. Limb joints are repre-sented in the cortex more than once, but with different contiguity (shoulder to wrist, shoulder to elbow) [26]. Rather than simply controlling different body parts, MI directs a host of body parts to assume complex postures. The map appears to be organized not just according to muscle groups, but to the posi-tions in space where the movements conclude [28]. Two dissociable systems for motor control (one for the execution of small precise movements – especially distal muscles – and another for postural stabilization – especially proximal muscles) coexist in MI, with the representation of distal and proximal mus-cles substantially intermingled within the MI arm representation. Depending on duration of stimuli applied on MI, simple or complex movements can be elicited. This clearly proves the difficulty of modeling even such an apparently known cortical area.

The picture gets even more complex. In one out of five patients, there are variations in the organization of MI, i.e. mosaicism (overlapping of functional areas), variability (inverted disposition of MI functional areas) or both [29,30]: for instance, the sensory hand area may be found between 1 and 7 cm from the sylvian sulcus and leg sensation can be found within 3 cm of the sylvian fissure. These findings suggest that individual neurons over the postcentral gyrus responding to a specific stimulus may appear to be arranged randomly rather than grouped together. There is significant intermixing of sensory neurons that respond to different sensory modalities and similar results apply to MI [29,31]. Moreover, the local mosaic-like topography (somatotopy) of individual distal arm representations is highly idiosyncratic, with wide variability among subjects [32]. Finally, somatotopic differences not only exist between subjects, but also between hemispheres in the single case. In Parkinson's disease (PD) specifically, map shifts are found in the majority of the patients, both in untreated early cases and treated cases of long duration, with a correlation between inter-side differences in the severity of PD symptoms and inter-hemispheric map displacement [33].

The left and right hemispheres are specialized for controlling different features of movement. In reaching movements, the non-dominant arm appears better adapted for achieving accurate final positions and the dominant arm for specifying initial trajectory features (e.g. movement direction and peak acceleration) [34]. Also, the area of hand representation is greater in the dominant (left) than in the non-dominant hemisphere, with greater dispersion of elementary movement representations and more profuse horizontal connections between them, thus leading to more dexterous behavior of the dominant hand [35]. Stronger beta rebound after right median nerve stimulation is observed in the left compared with the right hemisphere [36]. This suggests that left MI ECS may be expected to have different effects.

In sum, MI is not just classical Brodmann's area 4: more anterior and posterior areas must be investigated. Premotor cortex BA6 lies on the crown of the precentral gyrus, thus needing less energy for activation, while MI is mostly within the central sulcus. SI is another option for both pain and Parkinson's disease.

MECHANISM OF ACTION AND PARAMETERS CONSIDERATIONS

1. Neural changes during stimulation include excitation, inhibition (Fig. 2.3), oscillatory changes in corticosubcortical loops and intracortical layers and neuroplastic changes. Despite several authors suggesting an exclusive subcortical action of CS, neuroimaging and electrophysiological data confirm that the primary locus of action is the cortex itself (see discussion in [3]). This applies to both extradural and non-invasive CS. For instance, the analgesic effects of rTMS of both MI and DLPFC do not depend on the activation

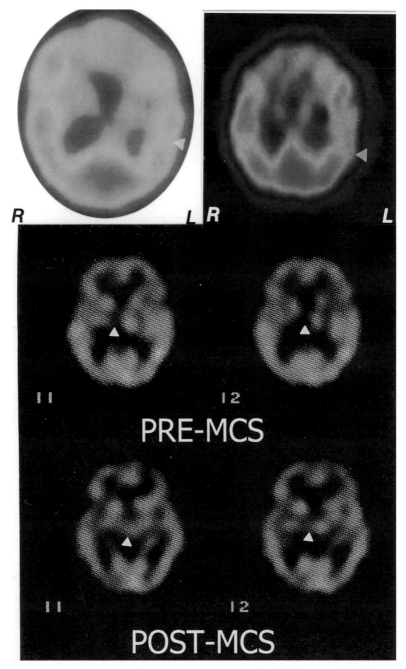

FIGURE 2.3   SPECT imaging showing normalization of cortical (top) and thalamic (bottom) hypoperfusion in a central pain patient.

of descending inhibitory systems [37]. CS renormalizes a disrupted intracortical function (disinhibition, as demonstrated in the setting of central pain with GABAergic–propofol challenge [23,38]: by acting on small inhibitory axons (probably Golgi-II cells with long axons) and, via U-fibers, modulates nearby areas, specifically SI in pain patients. At the same time, disrupted oscillatory patterns between cortex and thalamus (e.g. central pain) or basal ganglia (e.g. Parkinson's disease) are shifted towards more normal patterns [23,38], also by way of antidromic effects [39]. On the other hand, the Neuropace apparatus appears to be purely cortical when delivered through cortical paddles.

The predominant idea in the field is that stimulation leads to a sphere of activated neurons around the electrode tip that increases in size with increasing current [40–42], but this has little experimental support. For instance, while chronaxie measurements suggest that axons have the lowest threshold as compared to somas and dendrites [41–43], it is unclear whether initial segments have lower thresholds (especially for corticocortical axons which are often unmyelinated) which would cause preferential activation of cells near the electrode tip. Previous work relied on the idea that increasing current activates neurons whose cell bodies are located at an increasing distance from the tip. A recent study [44] found that, during intracortical microstimulation, instead of activating a group of cell bodies with different thresholds that increases in size and distance as current is increased, the activated neurons are simply those whose axons or dendrites (neuropil) pass very locally through a small volume (15 μm) around the electrode tip, but whose cell bodies are sparse and widely distributed, in a pattern that is highly sensitive to the exact location of the electrode in the neuropil. This makes it impossible to activate a set of cells restricted to a small spatial volume, and only areas where neurons of similar function lie near one another can be homogeneously stimulated. The mechanism of activation is local and direct (direct depolarization); moving the electrode by 30 μm completely changes the patterns of activated cells. The pattern of activated cells, moreover, is likely to reflect the pattern in which axons project through the cortex.

Near-threshold activation is mediated primarily by axons, due to their wider extension and lower threshold than somas and dendrites. While some cells are likely to be activated through their dendrites at higher currents, axons are likely to be recruited first. Axons are the main neural elements activated by stimulation, with smaller diameter axons having higher thresholds than large axons [42,43]. Postsynaptic effects are far weaker than direct effects: larger currents can recruit inhibitory neurons, cortical synapses are weak and a postsynaptic spike requires many presynaptic inputs and synaptic depression is often seen in cortex. Low currents result in activation of a set of directly driven neurons which induce only a small number of spikes in their connected partners. At higher currents, direct activation still predominates, but postsynaptic effects may play a relatively more important role. Postsynaptic summation might also occur in subcortical areas to which stimulated axons project. It is nearly impossible to stimulate single cells using microstimulation. These data have important implications for cortical visual neuroprosthetics [3]. Because stimulation of a single site in the cortex activates neurons that are spread widely from that site, achieving high resolution rasterized visual percepts by electrical stimulation through high density arrays may not be possible, unless the brain can learn to interpret these distributed patterns. Of course, microstimulation is very different from DBS and ECS. ECS may induce spikes in layer I axons. The pattern in which cells are activated will depend on projection patterns in the cortex. Different cortical areas with different axonal anatomy and projection patterns may respond differentially to stimulation.

2. General principles of parameter selection are difficult to come by and the literature contains some contradictory and potentially confusing findings. Identical stimulation parameters can excite, inhibit or both, depending on the brain region, even close ones [45] and elicit opposite effects in different subjects [46,47]: in one study, rTMS increased raclopide binding by 58% in the caudate of one patient, but decreased it by 43% in another [48]. Even within the same subject, the effects of CS appear to depend on the initial cortical activation state and specific neuronal populations [49]. Even relatively small variations in parameters may result in unintended effects locally and remotely [50]. In tDCS, excitation or inhibition depends on the placement of the reference electrode or the intensity of the stimulation.

Factors bearing on responsiveness to CS include genetic factors, hormonal factors, attention, inter-individual differences in anatomy and shift of cortical areas, medications (type and serum levels), and prior state of activation of the recruited circuits, i.e. baseline inhibitory tone (less inhibition, more effect): higher pre-TMS spontaneous activity predicts greater post-TMS activity [51]. The processes leading to depression of synaptic transmission are more effective when postsynaptic activity is high, whereas potentiation of synaptic transmission is more likely when postsynaptic activity is low [52]. Previous neuronal activity also modulates the capacity for subsequent plastic changes [53,54]. Thus, priming cortical stimulation aimed at modulating the initial state of cortical excitability could influence subsequent ECS-induced changes in cortical excitability (see references in [55]). Intrinsic excitability (sensory and motor thresholds) also changes from day to day in relation to time of day, mood, last meal and hours of last sleep. *Variations in existing activity levels contribute to the variability of CS responses, explaining, in part, the discrepancies between subjects and trials.* The direct monitoring of neural activity (power EEG) could arguably guide the empirical use of CS in the clinic.

The geometry of the electrical field induced into the brain and then the nature of the activated structures depend on the waveform of the magnetic pulse (mono/biphasic, sinusoidal) and on the type and orientation of the coil/paddle. The direction of the excitability changes may vary according to the characteristics of the cortical target (e.g. MI versus DLPFC) and current flow from cathode (−), which is depolarized by the outward flow of current, to anode (+), which is hyperpolarized by its inward flow, and vice versa, can differentially affect the neuron response [56]. Also, the selective activation of neuronal cell bodies should require asymmetrical charge-balanced biphasic stimuli which is not provided by current techniques.

As for TMS, the effects of ECS highly depend on various parameters: frequency, amplitude, pulse width, duty cycle, montage – mono versus bipolar CS, polarity (anodic versus cathodic CS) and the distance between electrodes and the neural elements (basically depending on the thickness of CSF layer). In ECS, a further confounder is due to the wide spacing between contacts, resulting in bifocal monopolar stimulation (both anode and cathode are active).

Modeling suggests that, at least in the case of MI ECS, a cathode excites preferentially the fibers that run horizontally (tangentially) under it, whereas an anode excites perpendicular (radially) to cortex fibers. A bipolar stimulus is more effective with the stimulation electrodes aligned transversally, rather than longitudinally, to the axon [57,58]. Yet, in the cortex, as discussed, fibers are not straight and uniformly oriented, but curved and bend in various directions. Thus, even though stimulation may have the lowest activation threshold at fiber ending, the bend acts as a focal point for excitation. Unfortunately, it is presently impossible to factor in the thousands of bends in a stimulation algorithm.

The search for effective parameters must also allow for the different pathophysiologies underlying different symptoms of the same disorder. Case in point: Parkinson's disease with its three defining axes (rigidity, akinesia, tremor). Here, the final choice must take into account the most disabling symptom. Also, MI ECS effects, unlike DBS, are almost never immediate. Intervals of assessment after a change of parameters must take into account that, after about 2–4 weeks, a long after effect sets in as a result of neuroplastic changes. Moreover, effects, particularly on akinesia, grow over time.

A further example comes from rTMS employed for depression. The rationale for its application comes from a belief that depression is accompanied by right prefrontal hyperactivity and left hypoactivity. Yet, low-frequency right DLPFC stimulation appears to be equally effective as the approved high-frequency left DLPFC protocol and better tolerated, and bilateral approaches may prove more effective, given the individual variation in laterality [11].

This calls for extensive parameters search and customization in the single patient.

3. ECS can be continuous or intermittent, but this depends on the treated disorder and, importantly, *after effects*. While a post effect (i.e. effect outlasting the end of stimulation) is seen for all neural stimulation techniques, it seems particularly strong in CS, building up to days and even weeks, depending on

the subject [3]. Whereas for chronic pain, after effects tend to diminish in time, in the setting of PD and the vegetative state [59], this grows in time, preventing the sort of blinded studies possible for DBS. After effects may be seen after even a few minutes of acute stimulation, but are most marked after weeks (e.g. in PD). Post effects are evidence that CS alters brain plasticity: ECS can boost drug effects and stroke rehabilitation [2], such as constraint induced therapy; CS may also accelerate the onset of benefit of antidepressant drugs [60]. TMS can elicit reverberating excitatory potentials in postsynaptic cells producing a persistent bursting response that outlasts the TMS pulse train. Higher baseline excitability leads to recurrent excitation (i.e. bursting) upon stimulation, whereas lower baseline excitability signifies a greater inhibitory tone that dampens recurrent excitation [51].

4. In the course of CS, bilateral effects can be observed clearly, as, for instance, shown in the setting of CS for Parkinson's disease, in which unilateral extradural CS relieves both hemibodies [3]. Transcallosal pathways are responsible for the effect (e.g. [48,61]). Transcallosal fibers connect homotopic as well as heterotopic areas [62]. Effective inter-hemispheric conduction pathways exist between the hand representations of MI [63], but weaker transcallosal connections for body parts outside hand areas [64], which explains why the hand area should be targeted for MI ECS in PD. Most of the association areas are strongly interconnected by callosal fibers. Heterotopic commissural connections connect a cortical area with non-corresponding areas in the contralateral hemisphere, but along a similar pattern as per its connections to ipsilateral association connections. Association and commissural connections often originate from and terminate in strips which, in turn, are separated from each other by strips lacking these particular connections, with a periodicity of 0.2–1 mm, and applies to primary sensory areas, but also to multimodal, frontal and paralimbic association cortices. Cells of origin and their homotopic terminations are located in the same strips.

MI has also ipsilateral projections which are important for axial muscles and muscles supplied by cranial nerves and more generally in the generation of bilateral synergistic movements [65]. 0.3 Hz rTMS of the right MI also inhibits contralateral SI [50]. Thus, MI stimulation has effects that extend to both contralateral MI and SI via the corpus callosum.

Anyway, bilateral stimulation is warranted in failures or failing cases: continuous stimulation may lead to 'cortical habituation' and alternate stimulation may be a solution.

COMPARING TECHNIQUES OF CORTICAL STIMULATION

The cortex can be stimulated both invasively (with surgically positioned stimulating paddles, i.e. extradural cortical stimulation) and non-invasively (TMS, tDCS). Presently, ECS is superior to both TMS and tDCS in terms of relief and number of relieved patients suffering central (50% of all patients relieved over years) and peripheral (particularly trigeminal: 60–80% of patients relieved) neuropathic pain; the same applies for Parkinson's disease [3] and likely all other current applications (depression, chronic tinnitus and perhaps stroke rehabilitation). Some authors even believe that regulatory-body approval of rTMS for depression was too quick [66–68], given the moderate degree of response to high frequency left DLPFC stimulation (d = 0.39) [69,70]. tDCS has also been found effective for depression, but not if severe, with current paradigms [71].

A major advantage of ECS is that it can be applied continuously without interfering with everyday activities and, in the future, may be boosted by closed-loop capabilities. Also, it remains in place for future relapses (e.g. depression) and multiple paddles can simultaneously excite or inhibit different areas, a feat not possible with non-invasive CS.

Invasive CS is generally carried out extradurally since, compared to sub-dural stimulation, ECS increases the activation threshold and reduces the risk of induced seizure [3]. Stimulating paddles can be inserted either via one or two burr holes or a craniotomic flap. This author strongly argues for a one to two burr holes approach (Fig. 2.4), for two reasons: it carries no risk of causing a clinically apparent extradural (or subdural, if stitches are used to anchor the plate to the dura) hematoma and results are not different from more invasive

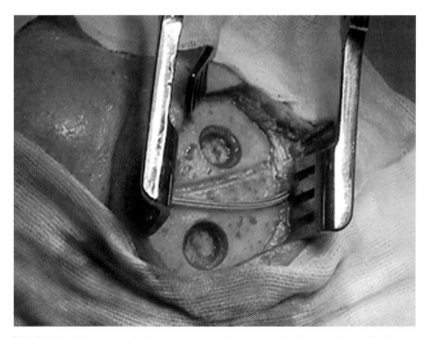

FIGURE 2.4 The two-burr hole approach to positioning extradural stimulating paddles favored by the author.

positionings. It is often said that targeting must be accurate to the millimeter for benefit to be seen. As shown in previous sections, the area of cortex to be covered to see a response is often wide (e.g. in case of movement disorders, depression, stroke rehabilitation and the vegetative state), which is why such a coarse technique as tDCS can provide benefit. Even in the case of pain, effects well beyond the expected somatotopy are on record [3] and targets may differ for each patient (e.g. MI versus SI, or right DLPFC versus left DLPFC). Most importantly, since we do not know beforehand which exact subareas (including the hazy MI) and the final extent of cortex to stimulate to achieve a benefit, bringing to bear such techniques as evoked potentials seems unfounded. Even fMRI guidance has several limits, as, for instance, demonstrated by the failure of a stroke rehabilitation trial which based targeting on 'hot spots' whose significance is questionable (see discussion in [3]). Again, tDCS, with all its coarseness, can achieve similar, or even better results than neuronavigated TMS (and neuronavigation is not feasible in the ordinary clinical context, except for ECS). Some authors strongly suggest using large coils that cover wide swaths of PFC rather than trying to target small areas with sophisticated techniques in treating depression [68].

A few differences must be mentioned. tDCS is considered neuromodulatory, TMS and ECS stimulatory.

Chronic ECS consists of continuous trains of stimuli delivered all day long at 1–130 Hz. TMS uses a large, rapidly changing magnetic field to induce electrical stimulating currents in the brain that are similar to those that are produced by a conventional electric nerve stimulator. These short pulses initiate action potentials. Stimulators can deliver either single or repeated pulses at frequencies of 0.2–50 Hz. rTMS consists of daily sessions lasting less than 1 hour and repeated for only several weeks at best. There may be a difference between descending volleys elicited by the two [55]. TDCS delivers weak direct currents – 1–2 mA – through a sponge electrode placed on the scalp for 4–5 seconds to 20–30 minutes. A portion of the applied current enters the skull where it is thought to polarize cortical neurons. Depending on the orientation of the cells with respect to the current, the membrane potentials may be hyperpolarized or depolarized by a few millivolts [72]. TDCS is applied daily for 20–30 minutes and repeated for days to weeks. TMS is heavy, large and expensive, whereas tDCS is small, light and much cheaper, portable and battery driven. Although TMS is more focal than tDCS, for most therapeutic purposes – as stated – focality is not a major issue (MI and premotor areas or SI in stroke rehabilitation, DLPFC in depression). Priming stimulation and theta burst stimulation have as yet an unknown role in boosting effects and rTMS and tDCS may not achieve the same benefit in the same subject in some individuals.

Side effects include seizures but are rare; hearing loss with TMS is a possibility (use earplugs for temporal stimulations). Intracranial ferromagnetic material contraindicates TMS. TDCS can be associated with scalp burns and ECS with infection, but these are generally treatable.

REFERENCES

1. Canavero S, Pagni CA, Castellano G, et al. The role of cortex in central pain syndromes: prelim-inary results of a long-term Tc-99 HMPAO SPECT study. *Neurosurgery*. 1993;32:185–191.

2. Canavero S, Bonicalzi V, Intonti S, Crasto S, Castellano G. Effects of bilateral extradural corti-cal stimulation for plegic stroke rehabilitation. *Neuromodulation*. 2006;9:28–33.

3. Canavero S, ed. *Textbook of therapeutic cortical stimulation*. New York: Nova Science Biomedical; 2009.

4. Azevedo FA, Carvalho LR, Grinberg LT, et al. Equal numbers of neuronal and nonneu-ronal cells make the human brain an isometrically scaled-up primate brain. *J Comp Neurol*. 2009;513:532–541.

5. Aldini J. *An account of the late improvements in galvanism, with a series of curious and inter-esting experiments performed before the commissioners of the French national institute, and repeated lately in the anatomical theatres of London, to which is added an appendix containing experiments on the body of a malefactor executed at Newgate, and dissertations on animal electricity 1793 and 1794*. London: Cuthell & Martin and J. Murray; 1803.

6. Mottelay PF. *Bibliographical history of electricity & magnetism*. London: Charles Griffin & Co. Ltd; 1922.

7. Reilly JP. *Electrical stimulation and electropathology*. Cambridge: Cambridge University Press; 1992.

8. Nieuwenhuys R, Voogd J, Van Huijzen C. *The human central nervous system*. Berlin: Springer Verlag; 2008.

9. Lyttelton OC, Karama S, Ad-Dab'bagh Y, et al. Positional and surface area asymmetry of the human cerebral cortex. *Neuroimage*. 2009;46:895–903.

10. Uylings HBM, Sanz-Arigita E, De Vos, et al. The importance of a human 3D database and atlas for studies of prefrontal and thalamic functions. Progr. *Brain Res*. 2000;126:357–368.

11. Nahas Z, Anderson BS, Borckardt J, et al. Bilateral epidural prefrontal cortical stimulation for treatment-resistant depression. *Biol Psychiatry*. 2010;67:101–109.

12. Eskandar, E. et al. *Congress communication*. AANS 2009 annual meeting, A803; 2009.

13. Badre D, D'Esposito M. Is the rostro-caudal axis of the frontal lobe hierarchical?. *Nature Rev Neurosci*. 2009;10:659–669.

14. Muenchau A, Bloem BR, Irlbacher K, Trimble MR, Rothwell JC. Functional connectivity of human premotor and motor cortex explored with repetitive transcranial magnetic stimulation. *J Neuroscience*. 2002;22:554–561.

15. Schmahmann JD, Pandya DN. *Fiber pathways of the brain*. New York: Oxford University Press; 2006.

16. Crochet S, Petersen CH. Cortical dynamics by layers. *Neuron*. 2009;64:298–299.

17. Arle JE, Shils JL, Canavero S. Extradural cortical stimulation for central pain. In: Canavero S, ed. *Textbook of therapeutic cortical stimulation*. New York: Nova Science Biomedical; 2009.

18. Reis J, Swayne OB, Vandermeeren Y, et al. Contribution of transcranial magnetic stimula-tion to the understanding of cortical mechanisms involved in motor control. *J Physiol*. 2008;586:325–351.

19. Scott SH. The role of primary motor cortex in goal-directed movements: insights from neuro-physiological studies on non-human primates. *Curr Opin Neurobiol*. 2003;13:671–677.

20. Canedo A. Primary motor cortex influences on the descending and ascending systems. *Progr Neurobiol*. 1997;51:287–335.

21. Rathelot J-A, Strick PL. Subdivisions of primary motor cortex based on cortico-motoneuronal cells. *PNAS USA*. 2009;106:918–923.

22. Hatsopoulos N, Xu Q, Amit Y. Encoding of movement fragments in the motor cortex. *J Neuroscience*. 2007;27:5105–5114.

23. Canavero S, Bonicalzi V. Extradural cortical stimulation for movement disorders. In: Sakas D, Simpson B, Krames E, eds. *Operative neuromodulation. II Neural networks surgery*. Vienna: Springer-Verlag; 2007:223–232.

24. Enomoto H, Ugawa Y, Hanajima R, et al. Decreased sensory cortical excitability after 1Hz rTMS over the ipsilateral primary motor cortex. *Clin Neurophysiol*. 2001;112:2154–2158.

25. Uozumi T, Tamagawa A, Hashimoto T, Tsuji S. Motor hand representation in cortical area 44. *Neurology*. 2004;62:757–761.

26. Schieber MH. Constraints on somatotopic organization in the primary motor cortex. *J Neurophysiol*. 2001;86:2125–2143.

27. Beinsteiner R, Windischberger C, Lanzenberger R, Edward W, Cunnington R, Erdler M. Finger somatotopy in human motor cortex. *NeuroImage*. 2001;13:1016–1026.

28. Graziano M. The organization of behavioral repertoire in motor cortex. *Annu Rev Neurosci*. 2006;29:105–134.

29. Branco DM, Coelho TM, Branco BM, et al. Functional variability of the human cortical motor map: electrical stimulation findings in perirolandic epilepsy surgery. *J Clin Neurophysiol*. 2003;20:17–25.

30. Tanriverdi T, Al-Jehani H, Poulin N, Olivier A. Functional results of electrical cortical stimulation of the lower sensory strip. *J Clin Neurosci*. 2009;16:1188–1194.

31. Farrell DF, Burbank N, Lettich E, et al. Individual variation in human motor-sensory (Rolandic) cortex. *J Clin Neurophysiol*. 2007;24:286–293.

32. Penfield W, Boldrey E. Somatic motor and sensory representation in the cerebral cortex of man as studied by electrical stimulation. *Brain*. 1937;60:389–443.

33. Thickbroom GW, Byrnes ML, Walters S, Stell R, Mastaglia FL. Motor cortex reorganisation in Parkinson's disease. *J Clin Neurosci*. 2006;13:639–642.

34. Schaefer SY, Haaland KY, Sainbur RL. Ipsilesional motor deficits following stroke reflect hemispheric specializations for movement control. *Brain*. 2007;130:2146–2158.

35. Hammond G. Correlates of human handedness in primary motor cortex: a review and hypothesis. *Neurosci Biobehav Rev*. 2002;26:285–292.

36. Salenius S, Hari R. Synchronous cortical oscillatory activity during motor action. *Curr Opin Neurobiol*. 2003;13:678–684.

37. Nahmias F, Debes C, Ciampi De Andrade D, Mhalla A, Bouhassira D. Diffuse analgesic effects of unilateral rTMS in healthy volunteers. *Pain*. 2009;147:224–232.

38. Canavero S, Bonicalzi V. Central pain sydnrome. Cambridge: Cambridge University Press; 2011, 2nd ed.

39. Bishop PO, Burke W, Davis R. Single-unit recording from antidromically activated optic radiation neurons. *J Physiol*. 1962;162:432–450.

40. Stoney SD, Thompson WD, Asanuma H. Excitation of pyramidal tract cells by itracortical microstimulation: effective extent of stimulating current. *J Neurophysiol*. 1968;31:659–669.

41. Ranck JB. Which elements are excited in electrical stimulation of mammalian central nervous system. A review. *Brain Res*. 1975;98:417–449.

42. Tehovnik EJ, Tolias AS, Sultan F, Slocum WM, Logothetis NK. Direct and indirect activation of cortical neurons by electrical microstimulation. *J Neurophysiol*. 2006;96:512–521.

43. Nowak LG, Bullier J. Axons but not cell bodies are activated by electrical stimulation in cortical grey matter. *Exp Brain Res*. 1998;118:477–488.

44. Histed MH, Bonin V, & Reid RC. Direct activation of sparse, distributed populations of cortical neurons by electrical microstimulation. *Neuron*. 2009; 63, 508-522.

45. Speer AM, Willis MW, Herscovitch P, et al. Intensity-dependent regional cerebral blood flow during 1-Hz repetitive transcranial magnetic stimulation (rTMS) in healthy volunteers studied with H215O positron emission tomography: I. Effects of primary motor cortex rTMS. *Biol Psychiatry*. 2003;54:818–825.

46. Gangitano M, Valero-Cabré A, Tormos JM, Mottaghy FM, Romero JR, Pascual-Leone A. Modulation of input-output curves by low and high frequency repetitive transcranial magnetic stimulation. *Clin Neurophysiol*. 2002;113:1249–1257.

47. Sommer M, Sommer M, Wu T, Tergau F, Paulus W. Intra- and interindividual variability of motor responses to repetitive transcranial magnetic stimulation. *Clin Neurophysiol*. 2002;113:265–269.

48. Kim JY, Chung EJ, Lee WY, et al. Therapeutic effect of repetitive transcranial magnetic stimulation in Parkinson's disease: analysis of [11C] raclopride PET study. *Mov Disorders*. 2007;23:207–211.

49. Silvanto J, Muggleton NG. A novel approach for enhancing the functional specificity of TMS: revealing the properties of distinct neural populations within the stimulated region. *Clin Neurophysiol*. 2008;119:724–726.

50. Seyal M, Shatzel AJ, Richardson SP. Crossed inhibition of sensory cortex by 0.3 Hz transcranial magnetic stimulation of motor cortex. *J Clin Neurophysiol*. 2005;22:418–421.

51. Pasley BN, Allen EA, Freeman RD. State-dependent variability of neuronal responses to transcranial magnetic stimulation of the visual cortex. *Neuron*. 2009;62:291–303.

52. Bienenstock EL, Cooper LN, Munro PW. Theory for the development of neuron selectivity: orientation specificity and binocular interaction in the visual cortex. *J Neurosci*. 1982;2:32–48.

53. Abraham WC, Tate WP. Metaplasticity: a new vista across the field of synaptic plasticity. *Prog Neurobiol*. 1997;52:303–323.

54. Turrigiano GG, Nelson SB. Homeostatic plasticity in the developing nervous system. *Nat Rev Neurosci*. 2004;5:97–107.

55. Lefaucheur JP. Principles of therapeutic use of transcranial and epidural cortical stimulation. *Clin Neurophysiol*. 2008;119:2179–2184.

56. Rattay F. Analysis of electrical excitation of CNS neurons. *IEE Trans Biomed Eng*. 1998;45:766–772.

57. Rise MT. Instrumentation for neuromodulation. *Arch Med Res*. 2000;31:237–247.

58. Testerman RL, Rise MT, Stypulkowski PH. Electrical stimulation as therapy for neurological disorder. *IEEE Eng Med Biol Mag*. 2006;25:74–78.

59. Canavero S, Massa-Micon B, Cauda F, Montanaro E. Bifocal extradural cortical stimulation-induced recovery of consciousness in the permanent post-traumatic vegetative state. *J Neurol*. 2009;256:834–836.

60. Rumi DO, Gattaz WF, Rigonatti SP, et al. TMS accelerates the antidepressant effect of amitriptyline in severe depression: a double-blind, placebo-controlled study. *Biol Psych*. 2005;57:162–166.

61. Plewnia C, Lotze M, Gerloff C. Disinhibition of the contralateral motor cortex by low-frequency rTMS. *Neuroreport*. 2003;14:609–612.

62. Cavada C, Goldman-Rakic PS. Posterior parietal cortex in rhesus monkey: II. Evidence for segregated corticocortical networks linking sensory and limbic areas with the frontal lobe. *J Comp Neurol*. 1989;287:422–445.

63. Meyer BU, Röricht S, Gräfin von Einsiedel H, Kruggel F, Weindl A. Inhibitory and excitatory interhemispheric transfers between motor cortical areas in normal humans and patients with abnormalities of the corpus callosum. *Brain*. 1995;118:429–440.

64. Komssi S, Aronen HJ, Huttunen J, et al. Ipsi- and contralateral EEG reactions to transcranial magnetic stimulation. *Clin Neurophysiol*. 2002;113:175–184.

65. Muellbacher W, Boroojerdi B, Ziemann U, Hallett M. Analogous corticocortical inhibition and facilitation in ipsilateral and contralateral human motor cortex representations of the tongue. *J Clin Neurophysiol.* 2001;18:550–558.
66. Murphy DN, Boggio P, Fregni F. Transcranial direct current stimulation as a therapeutic tool for the treatment of major depression: insights from past and recent clinical studies. *Curr Opin Psych.* 2009;22:306–311.
67. Hines JZ, Lurie P, Wolfe SM, et al. Post hoc analysis does not establish effectiveness of rTMS for depression. *Neuropsychopharmacology.* 2009;34:2053–2054.
68. Fitzgerald PB. R-TMS treatment for depression: lots of promise but still lots of questions. *Brain Stimulation.* 2009;2:185–187.
69. Schutter DJ. Antidepressant efficacy of high-frequency transcranial magnetic stimulation over the left dorsolateral prefrontal cortex in double-blind sham-controlled designs: a meta-analysis. *Psychol Med.* 2009;39:65–75.
70. George MS, Aston-Jones G. Noninvasive techniques for probing neurocircuitry and treating illness: VNS, TMS and tDCS. *Neuropsychopharmacology.* 2010;35:301–316.
71. Palm U, Keeser D, Schiller C, et al. tDCS in therapy-resistant depression: preliminary results from a double-blind, placebo-controlled study. *Brain Stimulation.* 2008;3:241–242.
72. Priori A, Hallett M, Rothwell JC. Repetitive transcranial magnetic stimulation or transcranial direct current stimulation? *Brain stimulation.* 2009;2:241–245.

Commentary on Cerebral – Surface

Beatrice, Cioni MD

Department of Functional and Spinal Neurosurgery, Catholic University, Rome, Italy

The chapter by Canavero deals with the stimulation of the cortex, either transcranially or extradurally or direct. The author gives an extensive description of the structural and physiological complexity of the cerebral cortex, analyzing in detail the role of cells, fibers, and association fibers, focusing mainly on neuronal activity. However, a major portion of the brain is made of glial cells, including astrocytes that may release neurotransmitters such as glutamate through a vesicular non-synaptic mechanism [1]. The released glutamate may act on adjacent neurons through their pre- and postsynaptic glutamate receptors, leading to a synchronization of the neural firing pattern; blocking synaptic transmission does not abolish this synchronization [1]. The interaction between glial cells and neurons probably plays a major role in the abnormal synchronization of neuronal firing pattern underlying many brain disorders (epilepsy, Parkinson's disease) and disrupting this abnormal synchronization may be a mechanism of action of neuromodulation at cortical level. Fregni et al. in a paper published online report a phase II sham-controlled clinical trial assessing the clinical effect and brain metabolic correlate of low frequency TMS targeting SII in patients with visceral pain due to chronic pancreatitis [2]. Modulation of right SII with 1 Hz TMS was associated with a significant analgesic effect and this effect was correlated with a change of glutamate and N-acetyl aspartate levels as measured in vivo by single voxel proton magnetic resonance spectroscopy.

The cortex shows a morphological high variability both at micro- and macroscopic level between different subjects and in the same subject, between different areas and same areas of different hemispheres. It shows also a great functional variability. Most of the studies on the activity of the cerebral cortex come from non-human experiments. In humans, the functions of the cortex are studied utilizing non-invasive stimulation, neuroradiology and electrophysiology during surgery. Non-invasive surface transcranial magnetic stimulation is widely spread; its limitation is the poor spatial resolution unless neuronavigated; the motor area is defined as the spot with the lowest threshold for the activation of that specific muscle, regardless of the actual position of M1. Functional MRI (fMRI) is rarely performed with pure tasks: finger tapping is not a pure motor task, sensory proprioceptive inputs play an important role and this may explain why the fMRI 'motor area' is often placed more posteriorly than the electrophysiologically identified motor area. And we have to keep in mind that both fMRI and DTI are mathematical probability functions and not 'true' images. Electrophysiological studies during surgery or utilizing surgically placed electrodes give more precise results. However, the technique used to stimulate the cortex is important. Penfield first systematically stimulated the sensory–motor cortex and described the sensory and motor 'homunculus' [3]. He utilized a bipolar direct stimulation of the cortex, applying 50–60 Hz stimuli up to 20 mA for 1–4 seconds, in the awake patient and looked for movements or sensations. This technique often induced complex motor or sensory responses and provoked epileptic seizures in a high percentage of cases (20–25%). Motor responses (as well as sensory responses) involving more than one joint may be obtained both from precentral and postcentral gyrus. Negative motor points have been identified as well as spots interfering with the production of language. Penfield and Jasper wrote [3]:

Electrical stimulation (like local epileptic discharge) may produce movement... Stimulation produces toe movement within the longitudinal fissure. But ankle movement is elicited half of the time within the fissure and half of the time on the lateral aspect...the location of any given movement response may vary several centimetres...although the order of representation remains constant...Movements may be elicited also by stimulation of the post-central gyrus. In general, when the total motor responses are considered, it is found that 80% of them resulted from precentral stimulation and 20% from postcentral stimulation...of the total sensory responses, 25% resulted from precentral stimulation as compared with 75% from postcentral.

Nowadays, we have the possibility to stimulate the cortex with a short train of high frequency pulses. A train of 3–5 pulses at 250–500 Hz is delivered directly over the cortex in anesthetized patients and motor responses are recorded from muscles of the controlateral hemibody The motor threshold for each muscle may be established [4]. We believe that in such a complex scenario as the cortex, it is of paramount importance to collect all the possible electrophysiological and biochemical information on the electrode position and interference with the underlying cortex and with the whole brain. In cases of chronic pain, motor cortex stimulation is performed placing the extradural electrode paddle perpendicular to the central sulcus and the collection of neurophysiological data to be compared to the clinical results led to the same conclusion in different groups of surgeons. In 2007, Yamamoto et al. [5] found a significant direct correlation between D-wave amplitude recorded by a cervical spinal epidural electrode following motor cortex stimulation and VAS reduction. Holsheimer et al. [6] found that the anode providing the largest muscle response in the area of pain gave the best pain relief. Our group, after a small craniotomy, stimulated the motor cortex by a monopolar handheld probe using the high frequency short train technique with increasing current. We recorded the muscle response in order to select the area somatotopically corresponding to the body region of pain with the lowest motor threshold. Here we placed the electrode paddle perpendicular to the central sulcus with at least one contact over the sensory cortex. We stimulated the spot with the lowest motor threshold as a cathode, while the contact over the sensory cortex was used as an anode. The patients implanted with such a technique, got excellent long-lasting pain relief.

At the moment, we use cortex stimulation in an empirical way: we do not know which neurons or cells or fibers should be activated, inhibited or tonically polarized; we do not know which electrode geometry or combination is the best or which parameters of stimulation (pulsewidth, frequency, voltage) are optimal. We select parameters of stimulation through a process of trial and error. An analysis of the distribution of electrical field and current density generated in the brain during stimulation and of the effect of such an electrical field on different neurons is important to optimize the delivery of cortical stimulation. (For a detailed description of the fundamental principles governing cortex stimulation see [7].) The authors in this paper address many common misconceptions. It is a common belief that bipolar stimulation will target the area between the two contacts because current flows from the anode to cathode through this region, but the area where the activating function is greatest is directly beneath the contacts, whereas exactly in the middle between anode and cathode the activating function is minimal. Computer modeling may be of help in enhancing neuromodulation of cerebral cortex.

Manola et al. developed a computer model of how extradural cortical stimulation for pain treatment may work [8,9]. The model consists of a 3D volume conductor model and a nerve cell model and their modeling results may be summarized as follows: the bipolar stimulation between two contacts of the electrode paddle corresponds to a bifocal monopolar stimulation, due to the wide distance between contacts. The anode cannot be considered an indifferent contact, it excites the fibers that run perpendicular to the electrode surface, while the cathode excites the fibers running horizontally under the paddle. They studied how different contact combinations influenced different fibers (assuming the same diameter for all). A bipolar combination with the cathode over the precentral gyrus will excite fibers parallel to the cortical laminae, being intrinsic cortical fibers or bifurcations/collaterals of ascending cortical afferents. Only axons can be excited, not cell bodies/dendrites; antidromic propagation of stimulus-induced axon potentials by thalamocortical fibres may be possible. The distance between the electrodes and the neural elements is important; for every 1–mm of CSF we need 6.6 V to obtain the same effect on the neural tissue. Probably this problem was overestimated; in clinical practice, the straight paddle implanted over the convex surface of the dura will squeeze the CSF underlying the paddle. These modeling predictions match with experimental and clinical data.

In the near future, improvements in electrode technology, namely nanotechnologies, will allow better stimulation/recording with fine tuning of complex neural networks, closed loop stimulation and influence on neurotransmitters concentration and distribution [10].

REFERENCES

1. Ni Y, Malarkey EB, Parpura V. Vesicular release of glutamate mediates bidirectional signaling between astrocytes and neurons. *J Neurochem.* 2007;103:1273–1284.
2. Fregni F, Potvin K, DaSilva A, et al. Clinical effects and brain metabolic correlates in non-invasive cortical neuromodulation for visceral pain. *Eur J Pain.* 2010; doi: 10.1016/j.e.pain.2010.08.002.
3. Penfield W, Jasper H. *Functional anatomy of the human brain.* Boston: Little and Brown Company; 1937.
4. Cioni B, Meglio M, Perotti V, De Bonis P, Montano N. Neurophysiological aspects of chronic motor cortex stimulation. *Neurophysiol Clin.* 2007;37:441–447.
5. Yamamoto T, Katayama Y, Obuki T, et al. Recording of cortico-spinal evoked potential for optimum placement of motor cortex stimulation electrodes in the treatment of post-stroke pain. *Neurol Med Chir.* 2007;47:409–414.
6. Holsheimer J, Lefaucheur JP, Buitenweg JR, Guijon C, Nineb A, Ngujen JP. The role of intra-operative motor evoked potentials in the optimization of chronic cortical stimulation for the treatment of neuropathic pain. *Clin Neurophysiol.* 2007;118:2287–2296.
7. Wongsarnpigoon A, Grill WM. Optimization of parameter selection and electrical targeting. In: Canavero S, ed. *Textbook of therapeutic cortical stimulation.* New York: Nova Science; 2009:69–89.
8. Manola L, Roelofsen BH, Holsheimer J, Marani E, Geelen J. Modelling motor cortex stimulation for chronic pain control: electrical potential field, activating functions and responses of simple nerve fibre models. *Med Biol Eng Comput.* 2005;43:335–343.
9. Manola L, Holsheimer J, Veltink P, Buitenweg JR. Anodal vs cathodal stimulation of motor cortex: a modelling study. *Clin Neurophysiol.* 2007;118:464–474.
10. Andrew RJ. Neuromodulation. Advances in the next decade. *Ann NY Acad Sci.* 2010;1199:212–220.

STUDY QUESTIONS

1. Considering the functional localization and network circuitry within the cerebral cortex, what approach or details in modeling its activity might be most helpful?
2. What might account for the delay that often occurs in clinical benefit from motor cortex stimulation?

Chapter 3

Cerebral – Deep

Yakov Gologorsky, MD and Ron Alterman, MD
Department of Neurosurgery, Mount Sinai School of Medicine, New York, USA

INTRODUCTION

The notion that functional brain disorders can be treated by modulating the activity of subcortical brain regions is as old as modern neurosurgery. Meyers is credited with performing the first transventricular lesions of the basal ganglia [1], but it was not until the invention of ventriculography and the human stereotactic frame that surgical approaches to deep brain targets became routine. Ablation represents the simplest means by which to modulate neural activity but, given the risks associated with creating irreversible brain lesions, the horrific experience of trans-orbital frontal lobotomy in America, and the introduction of chlorpromazine in the 1950s and levodopa in the 1960s, neuroablative procedures fell into disfavor until their resurrection in the late 1980s.

Over the last two decades, the field of deep cerebral neuromodulation has developed rapidly (Table 3.1). Chronic electrical deep brain stimulation (DBS) has supplanted neuroablation as the primary neuromodulatory technique and has become a standard treatment for medically refractory essential tremor, Parkinson's disease, and primary dystonia. The treatment of obsessive–compulsive disorder (OCD) with DBS has been approved in the USA and pivotal trials of DBS for epilepsy and major depressive disorder (MDD) are either completed or in progress. In addition, alternatives to electrical neuromodulation are being developed including gene therapy directed at both neuroprotection/restoration and neuromodulation. In this chapter, we provide an overview of the various deep cerebral targets currently being employed for neuromodulatory therapy. The scientific/physiologic rationale for modulating these targets will be discussed and key clinical research findings will be highlighted. Due to space constraints, we will focus on electrical neuromodulation as this is currently the most widely employed modality, but any of these sites may be targeted with novel neuromodulatory techniques in the future.

THE THALAMUS

Following the pioneering work of Hassler in Germany [2], Cooper in the USA [3], and Narabayashi in Japan [4], the thalamus was the favored target of functional neurosurgeons in the pre-computed tomography (CT), pre-microelectrode, pre-levodopa era. The reasons are obvious:

TABLE 3.1 Summary of deep cerebral targets and indications for neuro-modulation

Disease/disorder	Target
Pain	
Nocioceptive	Periventricular/periaqueductal gray (PVG/PAG)
Neuropathic	Ventrocaudal thalamus
Tremor	
Essential tremor*	Ventrolateral thalamus#
Parkinsonian tremor*	Zona incerta/pre-lemniscal radiation
Intention tremor	
Parkinson's disease*	
Rigidity	
Bradykinesia	Posteroventral globus pallidus pars internus #
Levodopa-induced dyskinesia	Subthalamic nucleus#
Motor fluctuations	
Tremor	
Gait akinesia and postural instability	Pedunculoponinte nucleus (PPN)
Dystonia	
Primary generalized dystonia*	Posteroventral globus pallidus pars iInternus#
Secondary dystonia	Subthalamic nucleus#
	Ventolateral tThalamus
Epilepsy	
Remote from the epileptogenic focus	Cerebellum
	Centromedian nucleus of the thalamus
	Anterior nucleus of the thalamus
	Subthalamic nucleus
	Head of the caudate nucleus
At the epileptogenic focus	Cortical
	Mesial temporal lobe (MTL)
Tourette's syndrome	Centromedian nucleus of the thalamus
	Posteroventral globus pallidus pars interna
	Anteromedial globus pallidus pars interna
	Nucleus accumbens (NAc) and anterior limb of internal capsule (IC)
Obsessive–compulsive disorder*	Ventral capsule/ventral striatum (VC/VS)#
	Nucleus accumbens
Depression	Subgenual cingulate cortex (Brodmann's area 25)
	Rostral cingulate cortex (Brodmann's area 24a)
	Ventral striatum/nucleus accumbens
	Inferior thalamic peduncle
	Lateral Habenula
Addiction	Nucleus accumbens
Obesity	Ventromedial hypothalamus

The current list of proposed indications and potential deep cerebral targets for neuromodulation are presented.
*Indicates an approved indication.
#Indicates an approved target for the given indication. (NB: Dystonia and obsessive–compulsive disorder are approved in the USA under a 'Humanitarian Device Exemption')

1. in structures such as the ventrocaudal (Vc) and ventrolateral (VL) nuclei, neurons are arranged in a clear topographic manner, simplifying electrophysiological mapping with the cruder macroelectrode techniques of the day
2. the effects of stimulation at these targets typically are immediate, allowing the surgeon to feel comfortable about electrode position prior to performing an irreversible ablation
3. functions are relatively compartmentalized in the thalamus so that one may treat movement, for example, without affecting sensation.

Today, the thalamus is targeted less frequently than other deep cerebral structures, but a working knowledge of thalamic anatomy and physiology remains essential. The interested reader is directed to Dr Ronald Tasker's classic work on thalamic physiology [5] and Dr Patrick Kelly's detailed description of his ventrolateral thalamotomy technique, employing semi-microelectrode recording [6].

Pain

Deep brain stimulation-derived analgesia was first observed and reported by Pool [7] and Heath [8] who found that stimulating the septal nuclei, including the diagonal band of Broca anterolateral to the forniceal columns, resulted in significant pain relief in psychiatric patients. Mazars reported that thalamic stimulation produces paresthesias with simultaneous long-lasting relief of deafferentation pain [9]. As a direct extension of Melzack and Wall's 'Gate theory'[10], Reynolds reported on the analgesic effect of aqueductal stimulation in rats [11]. Analogous work by Hosobuchi [12] and Richardson [13,14] first demonstrated the efficacy of thalamic and periventricular/periaqueductal gray (PVG/PAG) stimulation for the relief of pain. Since then, the Vc and PVG/PAG have been the most studied sites of DBS for pain in humans.

The mechanisms underlying pain relief via stimulation at these sites appear to be different yet are still not completely understood. Hosobuchi proposed that pain relief derived from PVG/PAG stimulation is mediated by opioid release following the observation that stimulation-induced analgesia at this site is blocked with naloxone [12]. Current thought maintains that the analgesic effect of PVG/PAG stimulation is mediated by multiple opioid- and biogenic amine-dependent supraspinal descending pain modulatory systems. In addition, ascending pathways from the PVG to the medial dorsal nucleus of the thalamus, an area associated with the limbic system and with extensive connections to the amygdala and cingulate cortex, have been identified, raising the possibility that stimulation of the PVG may also modify the patient's emotional response to pain [15]. Consequently, the majority of PVG stimulation studies have concentrated on its utility in treating intractable nociceptive rather than neuropathic pain. The results of many individual studies and pooled meta-analyses of PVG/PAG DBS for nociceptive pain have demonstrated success rates as high as 63%, depending on the etiology [16].

In contrast, pain relief from Vc thalamic stimulation is thought to be mediated by activation of the nucleus raphe magnus of the rostro-ventral medulla as well as descending inhibitory pain pathways [17]. Ventrocaudal thalamic stimulation has been applied most frequently in the setting of neuropathic/deafferentation pain syndromes, including anesthesia dolorosa, post-stroke pain, brachial plexus avulsion, post-herpetic neuralgia, and post-cordotomy dysesthesia. In general, deafferentation pain syndromes respond less well to stimulation than do nociceptive syndromes, with relief in a mean of 47% of patients [16]. Of these, 31% of patients with a central pain etiology (e.g. thalamic post-stroke pain) respond to thalamic DBS, while 51% of those with a peripheral etiology (e.g. post-herpetic neuralgia) experience a meaningful response. Interestingly, the rate of long-term pain alleviation is highest in those patients undergoing DBS of the PVG/PAG alone (79%), or the PVG/PAG plus the thalamus (87%). Stimulation of the thalamus alone is less effective (58%) than stimulation of the PVG/PAG±thalamus ($P < 0.05$) [16]. Many studies have thus concluded that DBS is more effective in treating nociceptive pain syndromes and that stimulation at both the PVG/PAG and thalamus may be most effective [16]. Presently, DBS is not approved in the USA at either target for the treatment of refractory pain.

Tremor

Tremor is a rhythmic, involuntary oscillation of the musculature that can affect the head, extremities, and/or trunk. Tremor is characterized by its clinical manifestations (i.e. resting, postural, action, and/or intention) and may be caused by multiple neurological disorders including Parkinson's disease (PD), essential tremor (ET), traumatic brain injury, stroke, and multiple sclerosis. In the 1950s, Cooper serendipitously discovered that ligation of the anterior choroidal artery ameliorated tremor, though paresis could also result [18]. Further research by Narabayashi [4], Hassler [2], Cooper [3], and others identified the ventrolateral nucleus of the thalamus as the primary target for eliminating tremor and employed ventriculography-based stereotaxis to ablate this site directly. Thereafter, thalamotomy remained the most commonly performed procedure for involuntary movement disorders until the late 1980s when Benabid developed DBS [19].

The junction of the ventral intermediate (Vim) and ventral oralis posterior (Vop) subnuclei of the VL thalamus is the most commonly targeted site for treating disabling parkinsonian and essential tremor with DBS [19,20]. The Vim and Vop are histologically distinct subnuclei located posteriorly in the VL nucleus. The Vim receives excitatory cerebellar input and projects to the motor cortex. The zona incerta, which is often included in the stimulaton field, contains the thalamic fasciculus and is partly made up of dentatothalamic and pallidothalamic projections.

Multicenter trials in North America [21,22] and Europe [23,24] as well as smaller case series report excellent results with unilateral and bilateral thalamic

DBS for tremor. Taken together, these studies report significant improvement of hand tremor in up to 75% of unilaterally and 95% of bilaterally stimulated patients, respectively [19]. Axial tremor (head, voice) is improved in up to 50% and 100% of unilaterally and bilaterally stimulated patients, respectively [19]. These effects appear to be long lasting, though tremor recurrence due to stimulation tolerance has been reported. Studies comparing DBS to radiofrequency thalamotomy demonstrate equivalent tremor suppression but a lower risk of neurological complications in patients treated with DBS [25–27]. The most common deficits related to thalamic interventions are hemiparesis, dysarthria, ataxia, and sensory deficits, most of which abate over time. Suppression of tremor results in significant reductions in functional disability in patients with ET [28]. In contrast, patients with advanced PD do not realize significant functional improvements following thalamic DBS because their other more disabling symptoms (e.g. rigidity, bradykinesia, motor fluctuations, levodopa-induced dyskinesia, and gait disturbance) are not improved. Consequently, DBS at other targets is more commonly employed for patients with advanced PD (see below).

Epilepsy

By its very nature, epilepsy would appear to be the ideal disorder to treat with electrical neurostimulation and, in particular, intermittent responsive stimulation. Toward that end, neurosurgeons have targeted a number of deep cerebral, cerebellar, and brainstem sites with the hope of controlling seizure disorders. These include the corpus callosum, caudate nucleus, centromedian thalamus, posterior hypothalamus, subthalamic nucleus and the hippocampus. Stimulation at these targets has often appeared efficacious in small open-label studies, but failed to achieve significant seizure control when tested in a controlled fashion [29]. Consequently, most of these deep brain stimulation strategies have been abandoned and the substantial population of medically refractory epilepsy patients who are not candidates for resective/ablative surgery are currently treated with vagus nerve stimulation. Nevertheless, two neurostimulation strategies, one 'open-loop' and one responsive, are in advanced stages of clinical testing and might be commercially available by the time this textbook is published.

Anterior Nucleus of the Thalamus (ANT)

The ANT is a component of Papez' circuit and is thought to play a central role in the propagation of seizure activity. Its small size, surgical accessibility, and direct connection to limbic structures, make it an attractive target for neuromodulation. High frequency stimulation of the ANT has been found to raise seizure thresholds in animal models of epilepsy and preliminary open-label clinical trials have demonstrated significant reductions in seizure frequency in small numbers of patients [30–34]. Based on these successes, a 110-patient, double-blind, multicenter trial of ANT DBS for medically refractory epilepsy was completed in 2008 [35]. Cycled stimulation at the ANT resulted in a statistically significant

40% reduction in the median seizure frequency of the treatment group versus a 14% seizure reduction in the control group. After 2 years of open-label stimulation, the median seizure frequency was reduced 56%, with 54% of patients achieving seizure frequency reductions of 50% or more [35]. Based on these results a United States Food and Drug Administration (FDA) advisory board has recommended approval of ANT DBS for the treatment of medically refractory complex partial epilepsy.

Responsive Neurostimulation (RNS)

A second approach to therapeutic neurostimulation for epilepsy involves the use of a 'closed loop' or responsive system, which detects seizures before they manifest clinically, and disrupts them with a short burst of electrical stimulation. Neuropace, Inc. recently presented the results of their multicenter, double-blind trial in which 191 patients with medically refractory partial epilepsy were randomized to therapeutic or sham stimulation for a 3-month period following implantation of the device [36]. At the conclusion of the 3-month blinded phase of the study, patients who received therapeutic stimulation experienced a mean 29% reduction in disabling seizures versus a 14% reduction in the sham-stimulation control group [36] (interestingly, the identical placebo response observed in the ANT/DBS trial [35]). A ruling from the FDA is pending.

Tourette's Syndrome

Tourette's syndrome (TS) is a chronic complex neuropsychiatric disorder characterized by sudden, repetitive, stereotyped motor or vocal tics. Tourette's syndrome is often co-morbid with OCD, attention deficit hyperactivity disorder (ADHD), and/or self-injurious behavior [37,38]. Similar to Parkinson's disease (see below), disordered cortico-striato-pallido-thalamo-cortical circuitry may be responsible for the motor and non-motor manifestations of TS. Hyperactivity within the dopaminergic system may lead to excessive thalamocortical drive, resulting in hyperexcitability of cortical motor areas and the release of tics. Hyperactivity in Broca's area, the frontal operculum, and the caudate nucleus may underlie vocal tics, while abnormal activation of the orbitofrontal region (as is observed in OCD), may underlie the compulsions that patients with TS experience [37,38].

Based on Hassler and Dieckmann's success with thalamic lesioning for TS [39], Visser-Vandewalle et al (1999) performed the first thalamic DBS for TS in a 42-year-old male, achieving complete resolution of his tics one year postoperatively [40]. Since then, several small series have reported success with DBS for TS at four different targets:

1. the centromedian nucleus including either the substantia periventricularis and nucleus ventro-oralis internus (CM–SPv–Voi) [40–43] or the parafascicular nucleus (CM-Pf) [44]
2. the posteroventral globus pallidus pars interna (GPi) [45–47]

3. the anteromedial GPi [44]
4. the nucleus accumbens (NAc) and anterior limb of internal capsule (IC) [37,48,49].

Preliminary data suggest that the efficacy of thalamic versus pallidal DBS in TS is similar, though pallidal stimulation may attenuate tics more abruptly and thalamic stimulation may yield better effects on mood and impulsivity [38,44]. According to a recent review by Porta et al, tic reduction ranging from 25 to 100% has been reported in a total of 39 TS patients with follow-up periods of 3–60 months [50]. Additional research will be necessary to determine whether an optimal target for TS exists and whether efficacy can be demonstrated in larger case series.

GLOBUS PALLIDUS PARS INTERNUS (GPi)

Spiegel and Wycis may be credited with inventing electrical pallidotomy (or rather ansotomy) for the treatment of parkinsonian tremor and rigidity in 1947 [51]. Their groundbreaking work was independently confirmed by Narabayashi, who performed procaine oil-induced transient pallidotomies in a series of PD patients [52,53]. Over the next decade, Fenelon, Leksell, Guiot, and Cooper all advocated mesial pallidotomy and ansotomy for the treatment of tremor and rigidity [54–57]. Interestingly, Cooper noted superior results from pallidotomy in patients with dystonia musculorum deformans (now known as DYT1-associated torsion dystonia) [58–60]. In 1960, Svennilson reported improved results for pallidotomy in Parkinson's disease when the lesion was placed more ventrally, posteriorly, and laterally in the GPi [61]. A quarter century later, DeLong and colleagues employed microelectrode recording techniques to demonstrate that the neurons in the posteroventral GPi subserve sensorimotor functions, and that this region becomes hyperactive in primates with MPTP-induced parkinsonism [62]. These findings provided the scientific underpinnings for the resurgence of posteroventral pallidotomy for medically refractory PD, which was championed by Laitinen and colleagues in Sweden [63,64]. Moreover, DeLong and colleagues' work resulted in the more widespread use of microelectrode recording as a localization technique during stereotactic targeting in patients. Though controversial, microelectrode techniques provide useful information that macroelectrode techniques do not, and may impact targeting in a significant proportion of pallidal interventions [65].

Posteroventral pallidotomy effectively alleviates tremor, rigidity, bradykinesia, and levodopa-induced dyskinesias in the contralateral hemibody of PD patients, however, the performance of bilateral procedures may be associated with serious speech, swallowing, and cognitive complications, so that the overall utility of pallidotomy in patients with advanced symmetric disease, severe motor fluctuations, and gait disturbance is limited [66]. However, the successes of both unilateral pallidotomy and thalamic DBS as a replacement for thalamotomy set the stage for the use of bilateral DBS at both the GPi and the subthalamus (see below) for the treatment of Parkinson's disease and primary generalized dystonia.

Parkinson´s disease

The serendipitous discovery that MPTP (1-methyl-4-phenyl-1,2,3,6-tetrahydro-pyridine) poisoning can induce a parkinsonian state in humans and non-human primates has contributed greatly to our current understanding of PD and basal ganglia physiology as it pertains to motor function. In the classic model proposed by DeLong and Crutcher [67], the striatum (caudate nucleus and putamen) receives broad input from the cortex and the intralaminar nuclei of the thalamus as well as dopaminergic input from the substantia nigra pars compacta (SNc). The globus pallidus pars interna (GPi) and substantia nigra pars reticulata (SNr) generate the dominant motor output of the basal ganglia, extending projections to the ventroanterior and ventrolateral (VA/VL) nuclei of the thalamus via the ansa lenticularis and fasciulus lenticularis (Fig. 3.1). The VA/VL nuclei, in turn,

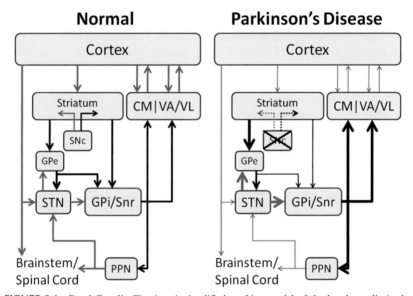

FIGURE 3.1   Basal Ganglia Circuitry A simplified working model of the basal ganglia in the normal and Parkinsonian states is presented. Blue arrows denote excitatory projections; black arrows denote inhibitory projections. The thickness of the arrow represents the strength of the neural activity. The striatum influences the GPi and SNr via two pathways, a direct pathway, which inhibits GPi/SNr activity, and an indirect pathway through GPe and STN that stimulates GPi/SNr output to the motor thalamus. The projections from GPi/SNr to motor thalamus are inhibitory. In the Parkinsonian state, loss of dopaminergic input to the striatum causes an imbalance in these two pathways the net result of which is increased activity of the STN, which in turn drives GPi/SNr. The increased inhibitory activity to motor thalamus reduces motor cortical activity, presumably resulting in the rigidity and bradykinesia that partially characterize Parkinson;s disease. Ablation or high frequency stimulation of the STN or GPi is thought to reduce this hyper-inhibition of the motor thalamus thereby improving motor function. Abbreviations: CM-Centromedian Nucleus; VA-Ventral Anterior Nucleus; VL-Ventro-Lateral Nucleus; SNc-Substantia Nigra pars Compacta; SNr-Substantia Nigra pars Reticulata; GPi-Globus Pallidus pars Interna; GPe-Globus Pallidus pars Externa; PPN-Pedunculopontine. Adapted from DeLong M. Primate models of movement disorders of basal ganglia origin. Trends Neurosci. 1990; 13:281–285.

project to supplemental motor regions anterior to the primary motor cortex. In addition, there exist projections from the GPi/SNr to the pedunculopontine nucleus (PPN), which is important in locomotion, and to the superior colliculus, which is involved with eye movements. The striatum modulates output from the GPi/SNr via direct inhibitory axonal projections and an indirect pathway via the globus pallidus par externus (GPe) and the subthalamic nucleus (STN).

In PD, neuronal degeneration within the SNc reduces dopaminergic innervation of the putamen, throwing off this finely balanced system. This imbalance is characterized by hyperactivity of the indirect pathway and hypoactivity of the direct pathway resulting in excessive inhibition of the thalamus by the GPi and the paucity of movement characteristic of PD [67]. Since both the GPi and the STN are overactive in PD, inactivation of either may improve motor function. While the subthalamus has been the most widely employed target for DBS in PD over the last 15 years (see below), there are some who believe that pallidal DBS can achieve similar results with fewer cognitive side effects [68,69]. The results of pallidal DBS for PD will be discussed in the section on the subthalamus (see below).

Dystonia

Perhaps the most impressive results thus far achieved with neuromodulation involve the treatment of primary generalized dystonia. The rationale for treating dystonia at the GPi is purely empiric and results directly from Cooper's work in the 1950s and 1960s [3,57–59], as well as the more contemporary observation that off-state dystonia in PD improves with pallidotomy [70]. A number of retrospective open-label studies report dramatic improvement in motor function in dystonia patients following pallidotomy, however, surgeons are loathe to make irreversible destructive lesions in the brains of children. Instead, pallidal DBS has been employed with remarkable results. In 1999, Coubes et al described the case of an 8-year-old girl with generalized torsion dystonia whose symptoms were so severe she required sedation and mechanical ventilation [60]. Thirty-six months after DBS, she had returned to school with near normal neurologic function [60]. In a subsequent study of 31 patients with primary generalized dystonia (PGD), Coubes reported a mean 79% improvement in the BFMDRS (Burke–Fahn–Marden dystonia rating scale) motor subscore, 2 years after surgery [71]. Vidailhet prospectively examined 22 patients with PGD noting a mean 51% improvement in the BFMDRS scores one year after surgery [72]. Kupsch et al performed the only double-blind, sham stimulation-controlled study of pallidal DBS for PGD and found that patients who received therapeutic stimulation for 3 months exhibited a statistically significant improvement in their motor function that those who received sham stimulation did not [73]. Kiss et al performed the only prospective multicenter trial of DBS for cervical dystonia, demonstrating both improved motor function and decreased pain following pallidal DBS [74]. The response to pallidal DBS in dystonia patients may be influenced by patient age and disease duration, the presence or absence of fixed skeletal deformities, phasic versus fixed motor symptoms, and the anatomical position of the implanted DBS electrode [75,76].

The physiological mechanism(s) through which DBS acts remain obscure but appear to be different for PD and dystonia. Of primary importance is the observation that while PD symptoms often improve within minutes after the onset of stimulation, the effects of pallidal DBS for dystonia may not be realized for days or weeks and improvement may not be complete for a year or more, raising questions about the neuroplastic changes that may be induced by chronic stimulation. Moreover, Alterman et al [77] demonstrated that some forms of dystonia may respond as well to stimulation at 60 Hz as at the higher frequencies (i.e. > 100 Hz) that are required to treat PD. This corresponds to intraoperative microelectrode recording (MER) data, which demonstrate a mean internal pallidal neuronal firing rate of 90–100 Hz in PD but only 50–60 Hz in dystonia. Though the firing rates may differ, both disorders are characterized by aberrancies in the pattern of neuronal firing including increased synchronous oscillatory activity in the basal ganglia circuitry. Disruption of these aberrantly patterned signals within the basal ganglia may underlie the effects of both pallidotomy [78] and pallidal DBS [77].

SUBTHALAMIC NUCLEUS (STN)

Key elements of the current working model of the parkinsonian basal ganglia are neuronal hyperactivity, increased firing pattern aberrancy, and greater synchronous beta band firing within the STN and GPi. Moreover, DeLong and colleagues demonstrated that ablating the hyperactive STN in primates with MPTP-induced Parkinsonism results in a reversal of the parkinsonian phenotype [79–81]. Clinicians were initially wary of lesioning the STN in humans for fear of inducing hemiballism, a well-known complication of stroke in this region. Instead, Benabid extended his success with thalamic stimulation, employing DBS at the STN bilaterally in the early 1990s [82]. Since that time, STN DBS has become the preferred surgical treatment for advanced Parkinson's disease, with tens of thousands of implants worldwide.

Scholarly publications consistently demonstrate the safety and efficacy of STN DBS in properly selected patients. At a 2009 consensus conference, leading American and European DBS practitioners from various related subspecialties agreed that STN DBS was effective in PD patients who remain responsive to levodopa but have developed marked fluctuations in their motor response with or without medication induced dyskinesiae [83]. Patients should undergo detailed neurocognitive testing prior to surgery in order to rule out significant dementia, which is a contraindication to the procedure. These experts agree that, on average, patients can expect an improvement in functional 'on' time of 4–6 hours, a significant decrease in on-state dyskinesia, and reduced rigidity, tremor, and bradykinesia [83]. Medication reductions are often significant but the goal of therapy is to maximize performance, not minimize medications. The best reported long-term results suggest that a positive response to STN DBS one year after surgery will be maintained for at least 5 years, on average [84]. Gait and balance difficulties that are not responsive to levodopa are unlikely to be improved by STN DBS [83].

Potential complications of DBS procedures include hemorrhage, device breakage, and infection [85,86]. Complications related specifically to STN DBS for PD include transient or permanent postoperative confusion (which is uncommon in patients who are pre-screened for neurocognitive decline) and decreases in verbal fluency [87]. A few small retrospective studies have reported that these neurocognitive difficulties occur less frequently in PD patients who undergo bilateral GPi DBS [68]. Confirmation of these observations in controlled studies is pending [88].

In a recent meta-analysis of 37 cohorts comprising 921 patients treated over a span of nearly two decades, the estimated improvement in unified Parkinson's disease rating scale (UPDRS) II (activities of daily living) and UPDRS-III (motor) scores after surgery in the stimulation on/medication off state compared to the preoperative medication off state were 50% and 52%, respectively [85]. The average reduction in L-dopa equivalents, dyskinesia, and daily off periods following DBS surgery was 55.9%, 69.1%, and 68.2%, respectively. The average improvement in quality of life as measured by the 39-item Parkinson's disease questionnaire (PDQ-39) was 34.5%. Independent predictors of a greater improvement in motor score postoperatively were preoperative L-dopa responsiveness, higher baseline motor scores, and disease duration [85].

Recently, Follett et al [69] compared pallidal versus subthalamic deep brain stimulation for advanced Parkinson's disease. After 24 months, there was no difference between mean changes in motor function, as blindly assessed on the UPDRS-III, between sites ($P = 0.5$), although patients undergoing subthalamic stimulation required a lower dose of dopaminergic agents than did those undergoing pallidal stimulation ($P = 0.02$). Visuomotor scores declined more after subthalamic stimulation than after pallidal stimulation ($P = 0.03$), and depression levels worsened after subthalamic stimulation and improved after pallidal stimulation ($P = 0.02$). Overall, similar improvements in motor function were observed after either pallidal or subthalamic stimulation [69].

PEDUNCULOPONTINE NUCLEUS (PPN)

Akinesia of gait and postural instability are crippling and sometimes life-threatening features of advanced PD. In contrast to appendicular symptoms, such as rigidity or tremor, these axial motor disturbances are often resistant to both dopaminergic therapy and DBS at either the STN or GPi [89]. Consequently, investigators have turned to a new target, the pedunculopontine nucleus (PPN) in an attempt to address these disabling symptoms.

Converging evidence implicates the PPN in the control of gait and posture in humans [90–94]. The PPN is believed to be part of the mesencephalic locomotion center, a functionally defined area within which it is possible to elicit stimulation-induced, dose-escalating locomotion in the decerebrate cat and possibly the monkey [90,91,93]. In humans, the PPN is a cluster of cells located in the caudal mesencephalic tegmentum that extends rostrally to the dorsomedial aspect of the posterolateral substantia nigra, and to the retrorubral field dorsally. It is bound medially by the fibers of the superior cerebellar peduncle and its decussation, and laterally by the medial lemniscus. The PPN is subdivided into the pars compacta (PPNc) and pars dissipatus (PPNd) on the basis of cell density. In humans, the PPNc is comprised primarily of cholinergic neurons (>90%), while the PPNd has a significant glutamatergic population [90,91,93].

Like the motor thalamus, the PPN receives GABAergic projections from the GPi and SNr. It is hypothesized that, in advanced PD, hyperinhibition of the PPN may underlie gait-related akinesia [90,91,93]. The pallidal/nigral projections terminate preferentially on the non-cholinergic, glutamatergic neurons of the PPNd and largely avoid neurons of the PPNc. These glutamatergic PPNd neurons provide descending projections to the spinal cord.

Preliminary studies demonstrate the potential of PPN DBS to improve both axial stability and freezing of gait in advanced PD patients [95–97]. While the majority of these results were recorded in small numbers of patients via open-label evaluations, Moro et al (2010) recently reported results of a prospective, double-blind pilot study of six PD patients with debilitating gait and postural abnormalities, finding significant reductions in falls 3 and 12 months following unilateral PPN DBS [94]. Stimulation at the PPN improved posture and gait only; appendicular symptoms were unaffected, suggesting that unilateral PPN DBS may serve as an adjunct to bilateral STN or GPi DBS. Multicenter trials in larger patient cohorts will be required to confirm these results.

DEEP CEREBRAL TARGETS FOR PSYCHIATRIC ILLNESS

It is often forgotten that the very first stereotactic procedures in humans were performed for psychiatric disorders. These early 'psychosurgeries' were undertaken in an attempt to identify subcortical targets for ablation that might prove safer and/or more efficacious than the frontal lobotomies of the day. The history of cortical and subcortical ablative surgery for psychiatric illness is both dramatic and extensive but is beyond the scope of this chapter. Instead, we will focus our discussion on targets for neuromodulation under active investigation. The interested reader can learn more about the history of subcortical ablation for psychiatric illness from the following references [98,99].

Ventral Capsule/Ventral Striatum (VC/VS)

Similar to the targets employed to treat movement disorders, VC/VS was chosen as a target for neuromodulation therapy in psychiatric disease as a direct extension of its use as a target for ablation in the past. Anterior capsulotomy was developed by Leksell in the 1940s with the intention of interrupting thalamocortical projections thought to be involved in affective disorders [100]. Numerous studies have subsequently demonstrated that bilateral capsulotomy, whether by radiofrequency or radiosurgical ablation, alleviates symptoms of OCD with symptomatic improvements of up to 62% as measured with the Yale-Brown obsessive–compulsive scale (YBOCS) [101,102].

Nuttin et al published the first report of anterior capsular stimulation for OCD in 1999 [103]. Subsequent studies have confirmed the safety and efficacy of DBS at the VC/VS in patients with refractory OCD [104–106]. In a large multicenter trial, Greenberg et al reported that 50% of patients experience a >35% improvement in their YBOCS scores 36 months postoperatively [106]. Based on these results in particular, the United States Food and Drug Administration granted a Humanitarian Device Exemption for the use of the Medtronic Reclaim™ DBS system for the treatment of refractory OCD in 2009. In an update to this work, Greenberg et al reported that 58% of patients experience a >35% improvement in their YBOCS scores 36 months postoperatively, and 61.5% experience a >35% improvement in their YBOCS scores at their last follow-up visit [107].

The potential application of DBS at VC/VS for major depressive disorder derives directly from the study of VC/VS DBS in OCD. While conducting these studies, the investigators noted that OCD patients with co-morbid depression experienced improvement in both disorders after the onset of stimulation [106,108–110]. The respective improvements in OCD and depression exhibited different time courses, with mood improving weeks to months before the improvement in OCD symptoms, response times that are consistent with the clinical responses of mood and OCD to medications, and suggest that the improvement in mood is a direct effect of stimulation and not a secondary effect of alleviating the symptoms of OCD.

Malone et al conducted the first prospective, open-label feasibility trial specifically examining the safety and efficacy of VC/VS DBS in MDD [111]. Fifteen patients were studied, 14 of whom met DSM-IV criteria for MDD. Study participants had been treated with an average of six different antidepressant medications, six augmentation/combination trials, and a mean of 30.5 electroconvulsive therapy (ECT) treatments. The patients were severely depressed as evidenced by a mean baseline 24 item Hamilton depression rating scale ($HDRS_{24}$) score of 33.1 ± 5.5. Patients were followed for a mean of 23.5 months (range: 6–51 months). Both the mean $HDRS_{24}$ and Montgomery-Asberg depression rating scale (MADRS) scores were reduced by ≈45% 6 months after the onset of stimulation. In patients who responded, scores improved steadily over the initial 6-month period and were maintained for up to 3 years. Approximately half of the patients were categorized as responders (defined as a 50% or greater reduction in the $HSDR_{24}$ and/or MADRS) and approximately one third were categorized as achieving remission (defined as an absolute MADRS or $HSDR_{24} < 10$). Overall, the DBS procedure was well tolerated, though one patient experienced two stimulation-related episodes of hypomania [111].

Subgenual Cingulate Cortex (Cg25)

The subgenual cingulate target is unique in that it is the first to be identified via functional neuroimaging, which demonstrates metabolic overactivity of Cg25 during both acute sadness and clinical depression and a decrease in Cg25 activity that coincides with improved mood, whether spontaneous or in response to anti-depressant treatment [112–114]. This region is a component of the corticolimbic circuits that are putatively disrupted in patients with mood disorders and several anatomic studies confirm this target's intricate connections to the nucleus accumbens, amygdala, hypothalamus, and orbitofrontal cortex [115–120].

In 2005, Mayberg et al published the first clinical report of DBS of the subgenual cingulate white matter for treatment-resistant depression (TRD), reporting that Cg25 DBS induced remission in four of six patients that persisted at 6 months [114]. They noted striking acute effects of stimulation such as 'sudden calmness or lightness', 'disappearance of the void', a sense of heightened awareness, increased interest, and sudden room brightening. These phenomena were reproducible and time-locked with stimulation. Improvements in psychomotor performance such as motor speed, volume and rate of spontaneous speech, and enhanced prosody were observed as well [114]. In a larger open-label trial, Lozano et al extended these findings in 20 TRD patients. Six months after surgery, 60% of patients were classified as responders and 35% met criteria for remission. These results remained stable for up to one year [121].

Other Deep Cerebral Targets for Psychiatric Illness

Four other potential DBS targets for TRD have recently been described based on anatomic and neuroimaging research: the ventral striatum/nucleus accumbens [109,122], the inferior thalamic peduncle [123,124], rostral cingulate cortex (area 24a) [125], and the lateral habenula [126]. Treatment at three of these four targets has been tested clinically and has proven safe and effective, albeit in small numbers of patients [127].

Finally, multiple avenues of research implicate the nucleus accumbens (NAc) and the frontal dopaminergic projections in the biology of addiction. Consequently, the NAc has been proposed as a potential neuromodulatory target for the treatment of various forms of addiction. One case has been reported that highlights the possibilities. The patient underwent ventral capsule/NAc DBS for the treatment of TRD. One year later, his depression remained severe, but he had spontaneously abstained from drinking after many years of alcoholism [128].

HYPOTHALAMUS

Recently, attention has been turned to the lateral and ventromedial hypothalamus as potential DBS targets to treat obesity [129]. Classically, food intake is thought to be controlled by a feeding center in the lateral hypothalamus, while a satiety center exists in the ventromedial hypothalamus. The lateral hypothalamus may regulate appetite by the production of peptides such as neuropeptide Y, agouti-related protein, melanin concentrating hormone (MCH) and orexins (hypocretins) [130–134]. Direct lesioning of the lateral hypothalamus (LH) in obese humans leads to transient weight loss and appetite suppression [135], while bilateral DBS of the LH in rats resulted in significant weight loss 24 days postoperatively [136]. Lesions of the ventromedial hypothalamus (VMH) are associated with weight gain while low-frequency VMH stimulation suppresses feeding in rats [137,138]. DBS of the hypothalamus is complicated by the potential spread of stimulation to nearby hypothalamic nuclei and other structures including the fornix, mamillary bodies, and the optic nerve [129]. Stimulation of the LH may affect the nuclei responsible for critical physiologic functions such as body temperature regulation, sexual activity, reproductive endocrinology, and the sympathetic response. Given the proximity of the VMH to the mamillary bodies and its connections to the cicrcuit of Papez, its stimulation may induce seizures [129]. Finally, concurrent stimulation of the LH and the VMH may be directly antagonistic [129]. In one interesting case report, Hamani et al (2008) relate the case of patient who underwent implantation of hypothalamic DBS leads to treat morbid obesity. Hypothalamic DBS failed to induce weight loss, however, activation of the more dorsal contacts, which were located near the fornices, significantly enhanced the patient's memory [139].

CONCLUSION

The success of treating movement disorders with DBS has generated great enthusiasm for the possibility of applying this reversible technology to the treatment of other disorders of brain function. This enthusiasm is reflected in the rapidly expanding indications and targets for DBS currently in use or under active investigation (see Table 3.1). Moreover, the success of DBS makes real the conception that functional brain disorders can in fact be treated by modifying neural function at discrete sites. One must merely identify the key node in the aberrant circuit to be modified and understand how it is misfiring in order to 'normalize' abnormal function. Given the complexities of the human brain, it is astonishing how successful the very simplistic DBS device currently in use has been; a fact that raises the hope of even greater success as our understanding of normal and abnormal neural signaling progresses.

A critical shortcoming of DBS relates to the chronically implanted hardware itself, which can become infected, causes scalp/skin erosions, limits access to magnetic resonance imaging for other ailments, is costly, and, at the very least, requires additional surgery at varying intervals to replace exhausted devices. Consequently, other means of modulating neural function will be sought in an attempt to overcome these obstacles. Gene therapy is the most exciting of the neuromodulatory techniques currently under development; but it is likely that other approaches will be created. What is certain is that we are at the beginning of a very exciting era in neurotherapeutics and human neuroscience research.

REFERENCES

1. Meyers R. The modification of alternating tremors, rigidity and festination by surgery of the basal ganglia. *Res Publ Assoc Res Nerv Ment Dis.* 1942;21:602–665.

2. Hassler R, Riechert T, Mundinger F, Umbach W, Ganglberger JA. Physiological observations in stereotaxic operations in extrapyramidal motor disturbances. *Brain.* 1960;83:337–350.

3. Cooper I, Bravo G. Chemopallidectomy and chemothalamectomy. *J Neurosurg.* 1958;15:244–250.

4. Ohye C, Kubota K, Hongo T, Nagao T, Narabayashi H. Ventrolateral and subventrolateral thalamic stimulation. motor effects. *Arch Neurol.* 1964;11:427–434.

5. Emmers R, Tasker RR. *The human somesthetic thalamus, with maps for physiological target localization during stereotactic neurosurgery.* New York: Raven Press; 1975.

6. Kelly P. Contemporary stereotactic ventralis lateral thalamotomy in the treatment of parkinsonian tremor and other movement disorders. In: Heilbrun M, ed. *Stereotactic neurosurgery, volume 2: Concepts in neurosurgery.* Baltimore: Williams & Wilkins; 1988:133–148.

7. Pool J, Clark WD, Hudson P, Lombardo M. *Hypothalamus-hypophyseal interrelationships.* Springfield: Charles C. Thomas; 1956.

8. Heath R. *Studies in schizophrenia: a multidisciplinary approach to mind-brain relationships.* Cambridge: Harvard University Press; 1954.

9. Mazars G, Roge R, Mazars Y. [Results of the stimulation of the spinothalamic fasciculus and their bearing on the physiopathology of pain]. *Rev Prat.* 1960;103:136–138.

10. Melzack R, Wall PD. Pain mechanisms: a new theory. *Science.* 1965;150:971–979.

11. Reynolds D. Surgery in the rat during electrical analgesia induced by focal brain stimulation. *Science.* 1969;164:444–445.

12. Hosobuchi Y, Adams JE, Linchitz R. Pain relief by electrical stimulation of the central gray matter in humans and its reversal by naloxone. *Science.* 1977;197:183–186.

13. Richardson DE, Akil H. Pain reduction by electrical brain stimulation in man. Part 1: Acute administration in periaqueductal and periventricular sites. *J Neurosurg.* 1977;47:178–183.

14. Richardson DE, Akil H. Pain reduction by electrical brain stimulation in man. Part 2: Chronic self-administration in the periventricular gray matter. *J Neurosurg.* 1977;47:184–194.

15. Rezai A, Lozano AM, Crawley AP, et al. Thalamic stimulation and functional magnetic resonance imaging: localization of cortical and subcortical activation with implanted electrodes. Technical note. *J Neurosurg.* 1999;90:583–590.

16. Bittar R, Kar-Purkayastha I, Owen SL, et al. Deep brain stimulation for pain relief: a meta-analysis. *J Clin Neurosci.* 2005;12:515–519.

17. Rasche D, Rinaldi PC, Young RF, Tronnier VM. Deep brain stimulation for the treatment of various chronic pain syndromes. *Neurosurg Focus.* 2006;21:E8.

18. Cooper I. Ligation of the anterior choroidal artery for involuntary movements; parkinsonism. *Psychiatr Q.* 1953;27:317–319.

19. Lyons K, Pahwa R. Deep brain stimulation and tremor. *Neurotherapeutics.* 2008;5:331–338.

20. Israel Z. Surgery for tremor. *Isr Med Assoc J.* 2003;5:727–730.

21. Koller W, Pahwa R, Busenbark K, et al. High-frequency unilateral thalamic stimulation in the treatment of essential and parkinsonian tremor. *Ann Neurol.* 1997;42:292–299.

22. Pahwa R, Lyons KE, Wilkinson SB, et al. Long-term evaluation of deep brain stimulation of the thalamus. *J Neurosurg.* 2006;104:506–512.

23. Limousin P, Speelman JD, Gielen F, Janssens M. Multicentre European study of thalamic stimulation in parkinsonian and essential tremor. *J Neurol Neurosurg Psychiatry.* 1999;66:289–296.

24. Sydow O, Thobois S, Alesch F, Speelman JD. Multicentre European study of thalamic stimulation in essential tremor: a six year follow up. *J Neurol Neurosurg Psychiatry*. 2003;74:1387–1391.

25. Tasker R. Deep brain stimulation is preferable to thalamotomy for tremor suppression. *Surg Neurol*. 1998;49:145–153; discussion 153–144.

26. Schuurman P, Bosch DA, Bossuyt PM, et al. A comparison of continuous thalamic stimulation and thalamotomy for suppression of severe tremor. *N Engl J Med*. 2000;342:461–468.

27. Pahwa R, Lyons KE, Wilkinson SB, et al. Comparison of thalamotomy to deep brain stimulation of the thalamus in essential tremor. *Mov Disord*. 2001;16:140–143.

28. Lyons K, Pahwa R, Busenbark KL, Tröster AI, Wilkinson S, Koller WC. Improvements in daily functioning after deep brain stimulation of the thalamus for intractable tremor. *Mov Disord*. 1998;13:690–692.

29. Ellis T, Stevens A. Deep brain stimulation for medically refractory epilepsy. *Neurosurg Focus*. 2008;25:E11.

30. Hodaie M, Wennberg RA, Dostrovsky JO, Lozano AM. Chronic anterior thalamus stimulation for intractable epilepsy. *Epilepsia*. 2002;43:603–608.

31. Kerrigan J, Litt B, Fisher RS, et al. Electrical stimulation of the anterior nucleus of the thalamus for the treatment of intractable epilepsy. *Epilepsia*. 2004;45:346–354.

32. Lee K, Jang KS, Shon YM. Chronic deep brain stimulation of subthalamic and anterior thalamic nuclei for controlling refractory partial epilepsy. *Acta Neurochir Suppl*. 2006;99:87–91.

33. Lim S, Lee ST, Tsai YT, et al. Electrical stimulation of the anterior nucleus of the thalamus for intractable epilepsy: a long-term follow-up study. *Epilepsia*. 2007;48:342–347.

34. Osorio I, Overman J, Giftakis J, Wilkinson SB. High frequency thalamic stimulation for inoperable mesial temporal epilepsy. *Epilepsia*. 2007;48:1561–1571.

35. Fisher R, Salanova V, Witt T, et al. Electrical stimulation of the anterior nucleus of thalamus for treatment of refractory epilepsy. *Epilepsia*. 2010;.

36. Morrell M, and the RNS System Pivotal Investigators. Results of a multicenter double blinded randomized controlled pivotal investigation of the RNS™ system for treatment of intractable partial epilepsy in adults. 63rd Annual Meeting of the American Epilepsy Society Boston, MA, USA 2009.

37. Ackermans L, Temel Y, Visser-Vandewalle V. Deep brain stimulation in Tourette's syndrome. *Neurotherapeutics*. 2008;5:339–344.

38. Mukhida K, Bishop M, Hong M, Mendez I. Neurosurgical strategies for Gilles de la Tourette's syndrome. *Neuropsychiatr Dis Treat*. 2008;4:1111–1128.

39. Hassler R, Dieckmann G. [Stereotaxic treatment of tics and inarticulate cries or coprolalia considered as motor obsessional phenomena in Gilles de la Tourette's disease]. *Rev Neurol (Paris)*. 1970;123:89–100.

40. Visser-Vandewalle V, Temel Y, Boon P, et al. Chronic bilateral thalamic stimulation: a new therapeutic approach in intractable Tourette syndrome. Report of three cases. *J Neurosurg*. 2003;99:1094–1100.

41. Vandewalle V, van der Linden C, Groenewegen HJ, Caemaert J. Stereotactic treatment of Gilles de la Tourette syndrome by high frequency stimulation of thalamus. *Lancet*. 1999;353:724.

42. Bajwa R, de Lotbiniere AJ, King RA, et al. Deep brain stimulation in Tourette's syndrome. *Mov Disord*. 2007;22:1346–1350.

43. Servello D, Porta M, Sassi M, Brambilla A, Robertson MM. Deep brain stimulation in 18 patients with severe Gilles de la Tourette syndrome refractory to treatment: the surgery and stimulation. *J Neurol Neurosurg Psychiatry*. 2008;79:136–142.

44. Houeto J, Karachi C, Mallet L, et al. Tourette's syndrome and deep brain stimulation. *J Neurol Neurosurg Psychiatry*. 2005;76:992–995.

45. Van der Linden C, Colle H, Vandewalle V, Alessi G, Rijckaert D, De Waele L. Successful treatment of tics with bilateral internal pallidum (GPi) stimulation in a 27-year-old male patient with Gilles de la Tourette's syndrome. *Mov Disord*. 2002;17(Suppl. 5):S241.

46. Diederich N, Bumb A, Mertens E, Kalteis K, Stamenkovic M, Alesch F. Efficient internal segment pallidal stimulation in Gilles de la Tourette syndrome: a case report. *Mov Disord*. 2004;19(Suppl. 9):S440.

47. Shahed J, Poysky J, Kenney C, Simpson R, Jankovic J. GPi deep brain stimulation for Tourette syndrome improves tics and psychiatric comorbidities. *Neurology*. 2007;68:159–160.

48. Flaherty A, Williams ZM, Amirnovin R, et al. Deep brain stimulation of the anterior internal capsule for the treatment of Tourette syndrome: technical case report. *Neurosurgery*. 2005;57(4 Suppl):E403 discussion E403;

49. Kuhn J, Lenartz D, Mai JK, et al. Deep brain stimulation of the nucleus accumbens and the internal capsule in therapeutically refractory Tourette-syndrome. *J Neurol*. 2007;254:963–965.

50. Porta M, Sassi M, Ali F, Cavanna AE, Servello D. Neurosurgical treatment for Gilles de la Tourette syndrome: the Italian perspective. *J Psychosom Res*. 2009;67:585–590.

51. Spiegel E, Wycis HT, Marks M, Lee AJ. Stereotaxic apparatus for operations on the human brain. *Science*. 1947;106:349–350.

52. Narabayashi H, Okuma T. Procaine oil blocking of the globus pallidus for treatment of rigidity and tremor of parkinsonism: Preliminary report. *Proc Jpn Acad*. 1953;29:134.

53. Narabayashi H, Okuma T, Shikiba S. Procaine oil blocking of the globus pallidus. *AMA Arch Neurol Psychiatry*. 1956;75:36–48.

54. Fenelon F. [Neurosurgery of parkinsonian syndrome by direct intervention on the extrapyramidal tracts immediately below the lenticular nucleus. Communication followed by film showing patient before and after intervention.]. *Rev Neurol (Paris)*. 1950;83:437–440.

55. Fenelon F. [Account of four years of practice of a personal intervention for Parkinson's disease]. *Rev Neurol (Paris)*. 1953;89:580–585.

56. Guiot G, Brion S. [Treatment of abnormal movement by pallidal coagulation]. *Rev Neurol (Paris)*. 1953;89:578–580.

57. Cooper I. Chemopallidectomy: an investigative technique in geriatric parkinsonians. *Science*. 1955;121:217–218.

58. Cooper I. Relief of juvenile involuntary movement disorders by chemopallidectomy. *J Am Med Assoc*. 1957;164:1297–1301.

59. Cooper I. Dystonia musculorum deformans alleviated by chemopallidectomy and chemopallidothalamectomy. *AMA Arch Neurol Psychiatry*. 1959;81:5–19.

60. Coubes P, Echenne B, Roubertie A, et al. [Treatment of early-onset generalized dystonia by chronic bilateral stimulation of the internal globus pallidus. Apropos of a case]. *Neurochirurgie*. 1999;45:139–144.

61. Svennilson E, Torvik A, Lowe R, Leksell L. Treatment of parkinsonism by stereotatic thermolesions in the pallidal region. A clinical evaluation of 81 cases. *Acta Psychiatr Scand*. 1960;35:358–377.

62. DeLong M. Primate models of movement disorders of basal ganglia origin. *Trends Neurosci*. 1990;13:281–285.

63. Laitinen L, Bergenheim AT, Hariz MI. Leksell's posteroventral pallidotomy in the treatment of Parkinson's disease. *J Neurosurg*. 1992;76:53–61.

64. Laitinen L, Bergenheim AT, Hariz MI. Ventroposterolateral pallidotomy can abolish all parkinsonian symptoms. *Stereotact Funct Neurosurg*. 1992;58:14–21.

65. Alterman RL, Sterio D, Beric A, Kelly PJ. Microelectrode recording during posteroventral pallidotomy: impact on target selection and complications. *Neurosurgery.* 1999;44:315–321.

66. Guridi J, Lozano AM. A brief history of pallidotomy. *Neurosurgery.* 1997;41:1169–1180 discussion 1180-1163..

67. DeLong M, Crutcher MD, Georgopoulos AP. Primate globus pallidus and subthalamic nucleus: functional organization. *J Neurophysiol.* 1985;53:530–543.

68. Hariz M, Rehncrona S, Quinn NP, Speelman JD, Wensing C. Multicentre Advanced Parkinson's Disease Deep Brain Stimulation GroupMulticenter study on deep brain stimulation in Parkinson's disease: an independent assessment of reported adverse events at 4 years. *Mov Disord.* 2008;23:416–421.

69. Follett K, Weaver FM, Stern M, CSP 468 Study Groupet al. Pallidal versus subthalamic deep-brain stimulation for Parkinson's disease. *N Engl J Med.* 2010;362:2077–2091.

70. Lozano AM, Lang AE, Galvez-Jimenez N, et al. Effect of GPi pallidotomy on motor function in Parkinson's disease. *Lancet.* 1995;346:1383–1387.

71. Coubes P, Cif L, El Fertit H, et al. Electrical stimulation of the globus pallidus internus in patients with primary generalized dystonia: long-term results. *J Neurosurg.* 2004;101: 189–194.

72. Vidailhet M, Vercueil L, Houeto JL, et al. Bilateral deep-brain stimulation of the globus pallidus in primary generalized dystonia. *N Engl J Med.* 2005;352:459–467.

73. Kupsch A, Benecke R, Muller J, et al. Pallidal deep-brain stimulation in primary generalized or segmental dystonia. *N Engl J Med.* 2006;355:1978–1990.

74. Kiss Z, Doig-Beyaert K, Eliasziw M, Tsui J, Haffenden A, Suchowersky O. The Canadian multicentre study of deep brain stimulation for cervical dystonia. *Brain.* 2007;130:2879–2886.

75. Isaias IU, Alterman RL, Tagliati M. Outcome predictors of pallidal stimulation in patients with primary dystonia: the role of disease duration. *Brain.* 2008;131:1895–1902.

76. Tisch S, Zrinzo L, Limousin P, et al. Effect of electrode contact location on clinical efficacy of pallidal deep brain stimulation in primary generalised dystonia. *J Neurol Neurosurg Psychiatry.* 2007;78:1314–1319.

77. Alterman RL, Miravite J, Weisz D, Shils JL, Bressman SB, Tagliat M. Sixty hertz pallidal deep brain stimulation for primary torsion dystonia. *Neurology.* 2007;69:681–688.

78. Vitek J, Chockkan V, Zhang JY, et al. Neuronal activity in the basal ganglia in patients with generalized dystonia and hemiballismus. *Ann Neurol.* 1999;46:22–35.

79. Bergman H, Wichmann T, Karmon B, DeLong MR. The primate subthalamic nucleus. II. Neuronal activity in the MPTP model of parkinsonism. *J Neurophysiol.* 1994;72:507–520.

80. Wichmann T, Bergman H, DeLong MR. The primate subthalamic nucleus. I. Functional properties in intact animals. *J Neurophysiol.* 1994;72:494–506.

81. Wichmann T, Bergman H, DeLong MR. The primate subthalamic nucleus. III. Changes in motor behavior and neuronal activity in the internal pallidum induced by subthalamic inactivation in the MPTP model of parkinsonism. *J Neurophysiol.* 1994;72:521–530.

82. Benabid A, Pollak P, Gross C, et al. Acute and long-term effects of subthalamic nucleus stimulation in Parkinson's disease. *Stereotact Funct Neurosurg.* 1994;62:76–84.

83. Bronstein, J., Tagliati, M., Alterman, R.L., et al. *Deep brain stimulation for Parkinson's disease: an expert consensus and review on key issues.* New York, USA; 2009.

84. Krack P, Batir A, Van Blercom N, et al. Five-year follow-up of bilateral stimulation of the subthalamic nucleus in advanced Parkinson's disease. *N Engl J Med.* 2003;349: 1925–1934.

85. Kleiner-Fisman G, Herzog J, Fisman DN, et al. Subthalamic nucleus deep brain stimulation: summary and meta-analysis of outcomes. *Mov Disord.* 2006;21(Suppl. 14):S290–304.

86. Benabid A, Chabardes S, Mitrofanis J, Pollak P. Deep brain stimulation of the subthalamic nucleus for the treatment of Parkinson's disease. *Lancet Neurol.* 2009;8:67–81.

87. Parsons T, Rogers SA, Braaten AJ, Woods SP, Tröster AI. Cognitive sequelae of subthalamic nucleus deep brain stimulation in Parkinson's disease: a meta-analysis. *Lancet Neurol.* 2006;5:578–588.

88. Okun M, Fernandez HH, Wu SS, et al. and mood in Parkinson's disease in subthalamic nucleus versus globus pallidus interna deep brain stimulation: the COMPARE trial. *Ann Neurol. Cognition.* 2009;65:586–595.

89. Rodriguez-Oroz M, Obeso JA, Lang AE, et al. Bilateral deep brain stimulation in Parkinson's disease: a multicentre study with 4 years follow-up. *Brain.* 2005;128:2240–2249.

90. Aziz T, Davies L, Stein J, France S. The role of descending basal ganglia connections to the brain stem in parkinsonian akinesia. *Br J Neurosurg.* 1998;12:245–249.

91. Pahapill P, Lozano AM. The pedunculopontine nucleus and Parkinson's disease. *Brain.* 2000;123:1767–1783.

92. Jenkinson N, Nandi D, Miall RC, Stein JF, Aziz TZ. Pedunculopontine nucleus stimulation improves akinesia in a Parkinsonian monkey. *Neuroreport.* 2004;15:2621–2624.

93. Matsumura M. The pedunculopontine tegmental nucleus and experimental parkinsonism. A review. *J Neurol.* 2005;252(Suppl. 4):IV5–IV12.

94. Moro E, Hamani C, Poon YY, et al. Unilateral pedunculopontine stimulation improves falls in Parkinson's disease. *Brain.* 2010;133:215–224.

95. Mazzone P, Lozano A, Stanzione P, et al. Implantation of human pedunculopontine nucleus: a safe and clinically relevant target in Parkinson's disease. *Neuroreport.* 2005;16:1877–1881.

96. Plaha P, Gill SS. Bilateral deep brain stimulation of the pedunculopontine nucleus for Parkinson's disease. *Neuroreport.* 2005;16:1883–1887.

97. Stefani A, Lozano AM, Peppe A, et al. Bilateral deep brain stimulation of the pedunculopontine and subthalamic nuclei in severe Parkinson's disease. *Brain.* 2007;130:1596–1607.

98. Sakas D, Panourias IG, Singounas E, Simpson BA. Neurosurgery for psychiatric disorders: from the excision of brain tissue to the chronic electrical stimulation of neural networks. *Acta Neurochir Suppl.* 2007;97:365–374.

99. Tye S, Frye MA, Lee KH. Disrupting disordered neurocircuitry: treating refractory psychiatric illness with neuromodulation. *Mayo Clin Proc.* 2009;84:522–532.

100. Lipsman N, Neimat JS, Lozano AM. Deep brain stimulation for treatment-refractory obsessive-compulsive disorder: the search for a valid target. *Neurosurgery.* 2007;61:1–11; discussion 11-13.

101. Oliver B, Gascón J, Aparicio A, et al. Bilateral anterior capsulotomy for refractory obsessive-compulsive disorders. *Stereotact Funct Neurosurg.* 2003;81:90–95.

102. Liu K, Zhang H, Liu C, et al. Stereotactic treatment of refractory obsessive compulsive disorder by bilateral capsulotomy with 3 years follow-up. *J Clin Neurosci.* 2008;15:622–629.

103. Nuttin B, Cosyns P, Demeulemeester H, Gybels J, Meyerson B. Electrical stimulation in anterior limbs of internal capsules in patients with obsessive-compulsive disorder. *Lancet.* 1999;354:1526.

104. Anderson D, Ahmed A. Treatment of patients with intractable obsessive-compulsive disorder with anterior capsular stimulation. Case report. *J Neurosurg.* 2003;98:1104–1108.

105. Nuttin B, Gabriëls LA, Cosyns PR, et al. Long-term electrical capsular stimulation in patients with obsessive-compulsive disorder. *Neurosurgery.* 2003;52:1263–1272; discussion 1272-1264.

106. Greenberg B, Malone DA, Friehs GM, et al. Three-year outcomes in deep brain stimulation for highly resistant obsessive-compulsive disorder. *Neuropsychopharmacology.* 2006;31:2384–2393.

107. Greenberg B, Gabriels LA, Malone Jr DA, et al. Deep brain stimulation of the ventral internal capsule/ventral striatum for obsessive-compulsive disorder: worldwide experience. *Mol Psychiatry*. 2010;15:64–79.

108. Gabriëls L, Cosyns P, Nuttin B, Demeulemeester H, Gybels J. Deep brain stimulation for treatment-refractory obsessive-compulsive disorder: psychopathological and neuropsychological outcome in three cases. *Acta Psychiatr Scand*. 2003;107:275–282.

109. Aouizerate B, Cuny E, Martin-Guehl C, et al. Deep brain stimulation of the ventral caudate nucleus in the treatment of obsessive-compulsive disorder and major depression. Case report. *J Neurosurg*. 2004;101:682–686.

110. Abelson J, Curtis GC, Sagher O, et al. Deep brain stimulation for refractory obsessive-compulsive disorder. *Biol Psychiatry*. 2005;57:510–516.

111. Malone DJ, Dougherty DD, Rezai AR, et al. Deep brain stimulation of the ventral capsule/ventral striatum for treatment-resistant depression. *Biol Psychiatry*. 2009;65:267–275.

112. Mayberg H, Liotti M, Brannan SK, et al. Reciprocal limbic-cortical function and negative mood: converging PET findings in depression and normal sadness. *Am J Psychiatry*. 1999;156:675–682.

113. Seminowicz D, Mayberg HS, McIntosh AR, et al. Limbic-frontal circuitry in major depression: a path modeling metanalysis. *Neuroimage*. 2004;22:409–418.

114. Mayberg H, Lozano AM, Voon V, et al. Deep brain stimulation for treatment-resistant depression. *Neuron*. 2005;45:651–660.

115. Vogt B, Pandya DN. Cingulate cortex of the rhesus monkey: II. Cortical afferents. *J Comp Neurol*. 1987;262:271–289.

116. Carmichael S, Price JL. Connectional networks within the orbital and medial prefrontal cortex of macaque monkeys. *J Comp Neurol*. 1996;371:179–207.

117. Ongür D, An X, Price JL. Prefrontal cortical projections to the hypothalamus in macaque monkeys. *J Comp Neurol*. 1998;401:480–505.

118. Freedman L, Insel TR, Smith Y. Subcortical projections of area 25 (subgenual cortex) of the macaque monkey. *J Comp Neurol*. 2000;421:172–188.

119. Barbas H, Saha S, Rempel-Clower N, Ghashghaei T. Serial pathways from primate prefrontal cortex to autonomic areas may influence emotional expression. *BMC Neurosci*. 2003;4:25.

120. Haber S. The primate basal ganglia: parallel and integrative networks. *J Chem Neuroanat*. 2003;26:317–330.

121. Lozano A, Mayberg HS, Giacobbe P, Hamani C, Craddock RC, Kennedy SH. Subcallosal cingulate gyrus deep brain stimulation for treatment-resistant depression. *Biol Psychiatry*. 2008;64:461–467.

122. Schlaepfer T, Cohen MX, Frick C, et al. Deep brain stimulation to reward circuitry alleviates anhedonia in refractory major depression. *Neuropsychopharmacology*. 2008;33:368–377.

123. Jiménez F, Velasco F, Salin-Pascual R, et al. A patient with a resistant major depression disorder treated with deep brain stimulation in the inferior thalamic peduncle. *Neurosurgery*. 2005;57:585–593; discussion 585–593.

124. Velasco F, Velasco M, Jiménez F, Velasco AL, Salin-Pascual R. Neurobiological background for performing surgical intervention in the inferior thalamic peduncle for treatment of major depression disorders. *Neurosurgery*. 2005;57:439–448 discussion 439-448.

125. Sakas D, Panourias IG. Rostral cingulate gyrus: A putative target for deep brain stimulation in treatment-refractory depression. *Med Hypotheses*. 2006;66:491–494.

126. Sartorius A, Henn FA. Deep brain stimulation of the lateral habenula in treatment resistant major depression. *Med Hypotheses*. 2007;69:1305–1308.

127. Hauptman J, DeSalles AA, Espinoza R, Sedrak M, Ishida W. Potential surgical targets for deep brain stimulation in treatment-resistant depression. *Neurosurg Focus.* 2008;25:E3.

128. Kuhn J, Lenartz D, Huff W, et al. Remission of alcohol dependency following deep brain stimulation of the nucleus accumbens: valuable therapeutic implications? *J Neurol Neurosurg Psychiatry.* 2007;78:1152–1153.

129. Halpern C, Wolf JA, Bale TL, et al. Deep brain stimulation in the treatment of obesity. *J Neurosurg.* 2008;109:625–634.

130. de Lecea L, Kilduff TS, Peyron C, et al. The hypocretins: hypothalamus-specific peptides with neuroexcitatory activity. *Proc Natl Acad Sci USA.* 1998;95:322–327.

131. Marsh D, Hollopeter G, Kafer KE, Palmiter RD. Role of the Y5 neuropeptide Y receptor in feeding and obesity. *Nat Med.* 1998;4:718–721.

132. Peyron C, Tighe DK, van den Pol AN, et al. Neurons containing hypocretin (orexin) project to multiple neuronal systems. *J Neurosci.* 1998;18:9996–10015.

133. Sakurai T, Amemiya A, Ishii M, et al. Orexins and orexin receptors: a family of hypothalamic neuropeptides and G protein-coupled receptors that regulate feeding behavior. *Cell.* 1998;92:573–585.

134. Bewick G, Gardiner JV, Dhillo WS, et al. Post-embryonic ablation of AgRP neurons in mice leads to a lean, hypophagic phenotype. *FASEB J.* 2005;19:1680–1682.

135. Quaade F, Vaernet K, Larsson S. Stereotaxic stimulation and electrocoagulation of the lateral hypothalamus in obese humans. *Acta Neurochir (Wien).* 1974;30:111–117.

136. Sani S, Jobe K, Smith A, Kordower JH, Bakay RA. Deep brain stimulation for treatment of obesity in rats. *J Neurosurg.* 2007;107:809–813.

137. Hoebel B, Teitelbaum P. Hypothalamic control of feeding and self-stimulation. *Science.* 1962;135:375–377.

138. Krasne F. General disruption resulting from electrical stimulus of ventromedial hypothalamus. *Science.* 1962;138:822–823.

139. Hamani C, McAndrews MP, Cohn M, et al. Memory enhancement induced by hypothalamic/fornix deep brain stimulation. *Ann Neurol.* 2008;63:119–123.

Commentary on Cerebral – Deep

Joachim K. Krauss, MD
Professor of Neurosurgery, Chairman and Director, Medical School Hanover, Germany

The authors of the chapter *Cerebral – Deep* are to be congratulated for their succinct and yet comprehensive overview on contemporary deep brain stimulation (DBS) for a variety of indications in the fields of neurology and psychiatry. In contrast to the usual approach considering different disorders, discussing treatment options and choosing appropriate targets, the authors choose to develop their views according to the thalamic, basal ganglia and upper brainstem target structures that are being used nowadays.

Movement disorders have been the most frequent indications for DBS since its inception a couple of decades ago. We all owe credit to the Grenoble group for their scientific strength and their endurance that DBS has become popular worldwide, in particular with regard to thalamic DBS for tremor and subthalamic nucleus (STN) DBS for advanced Parkinson's disease [1,2]. It was only a few years later that the globus pallidus internus (GPi) was introduced as a target for DBS in dystonia [3,4], and it took several years before GPi DBS was accepted widely as a valuable treatment for dystonic disorders [5]. The practice of DBS, and the variety of disorders that are treated by DBS vary considerably from country to country. This is no longer an issue of distribution and education in Western countries, it reflects merely national and societal health insurance practices, socioeconomic issues and reimbursement strategies. Patients with severe movement disorders nowadays may have access to DBS operations in developed countries worldwide backed up by local regulations. The same is not true, however, for other indications such as chronic pain and psychiatric disorders. It is very difficult to obtain appropriate reimbursement for new indications, and that fact actually delays the more widespread application of DBS techniques for common disorders such as obsessive–compulsive disorder or depression.

The chapter on deep cerebral targets for neuromodulation provides more ample discussion on the thalamus than on other targets, which stresses that the 'oldest target' in functional neurosurgery is still relevant. Until today, we have had no nomenclature of the thalamus that is consistently used by each researcher and by each clinician [6]. Daily use most frequently combines terminology from the nomenclatures of Hassler and of Walker. With that regard it should be noted that according to the 'unified concept' of the new nomenclature of Jones, the ventral oralis posterior (Vop) is not identified as a nucleus per se any longer, but rather as a transition zone between the ventral oralis anterior (Voa) and the more posteriorly located ventral intermediate (Vim) [6]. The zona incerta, however, should be identified as a proper nucleus which is an extension of the reticular thalamus, a shell-like structure covering the lateral thalamus, while the fiber tracts run immediately adjacent to the zona incerta in Forel's fields H, H1 and H2.

Cooper frequently has been credited erroneously both for the introduction of pallidotomy and of thalamotomy for the treatment of movement disorders [7]. Considering pallidotomy, the story goes that Cooper accidently severed the anterior choroidal artery while performing a pedunculotomy on a patient with parkinsonian tremor in 1952. Postoperatively, the patient s tremor was much improved, and it was concluded that this resulted from an ischemic infarction of the pallidum which is partly supplied by the anterior choroidal artery. This event led Cooper to clip the artery in a series of patients to treat parkinsonian tremor. It has been stated that he then, in 1954, started to consider to target the pallidum directly, and this often has been cited as the birth of pallidotomy. It is unclear why Cooper ignored the work of Spiegel and Wycis in Philadelphia and if he was aware of the work of the French and German pioneers who had performed pallidotomies at that time for more than 5 years. Considering thalamotomy, it has been said that it was introduced by Cooper after he inadvertently misplaced a pallidal lesion in the thalamus in 1955. He claimed that this 'remarkable act of serendipity' was due to an error 'in trigonometric calculations' of one of his associates. At that time, however, details of the first thalamotomy, which had been performed by the young Mundinger in 1952 in Freiburg, Germany, had already been published by Hassler and Riechert in *Der Nervenarzt*, in 1954 [8].

Thalamic stimulation for pain has been abandoned in the USA after preliminary reports of studies yielding not clear-cut benefit were published which were used by health insurance carriers not to provide further reimbursement. Nevertheless, thalamic deep brain stimulation is still being performed in specialized European Centers, and besides the periventricular gray and the VPL (or the Vc as used in this present chapter on neuromodulation), the intralaminar nuclei of the thalamus, the CM-Pf, are being re-explored [9].

In the late 1990s, both GPi and STN stimulation were used to treat advanced Parkinson s disease. The STN, however, subsequently became the preferred target and only rarely was the GPi considered a valuable target in single patients by most neurosurgeons over the next few years. Only recently, randomized studies have become available which show that the motor benefits of GPi and STN stimulation show little differences between the two targets [10], with the main difference being marked reduction of medication with STN stimulation.

Lately, the PPN has received much attention as a possible target for the treatment of gait disorders that are refractory to dopaminergic medication in patients with Parkinson's disease or with progressive supranuclear palsy [11,12]. This is a problematic field regarding several issues including the morphology and the anatomical nomenclature of the PPN area, the optimal target site for placing the electrode and the interpretation of DBS effects [13]. It may be, that PPN DBS will not primarily serve as a target to treat gait disorders per se but rather to treat freezing and subsequent falls.

The percentage of those patients with dystonia who are candidates for DBS surgery and who actually undergo DBS probably is higher nowadays than that of patients with Parkinson s disease. While the beneficial effects of DBS have been consistently shown in patients with primary generalized, segmental or cervical dystonia, its effects on secondary dystonia, however, are less clear. The concept that pallidal DBS provides only little benefit in secondary dystonia must be reviewed with regard to recent findings showing that it is a most useful treatment for tardive dystonia and that the effects on quality of life in patients with infantile cerebral palsy are much larger than the mere improvement on the motor scales [14,15].

Certainly, one of the most exciting fields in contemporary medicine is the application of DBS to psychiatric disorders. With that regard, however, we have to admit that limbic and cognitive–associative circuitries connecting the basal ganglia and other deep structures with the cortex are much less well understood than those for movement disorders.

We have seen tremendous progress in techniques and in technology over the past few years. Microelectrode recording is a wonderful technique which may help further to refine the target and which provides additional information on the pathophysiology of the disorders being treated. It is by itself not a controversial technique. The controversy is rather whether it is definitely needed or not in routine surgery, and with that regard of course opinions differ. I am sceptical that we will be able to identify 'key nodes' in the complex networks of brain circuitries for each specific disorder. Instead of having one key node, there might be several key structures in different disorders. With that regard, multitarget strategies might become more relevant in the near future.

Although the chapter covers most of the present and forthcoming indications for DBS, it can of course not cover all indications. I just would like to add two disorders, that is cluster headache which can be successfully treated by stimulation of the posterior hypothalamic area [16,17] and Alzheimer s disease which may be treated by stimulation of the fornices when applied in the early stages [18]. Neuromodulation by DBS and by other techniques certainly has a bright future, in particular regarding the ongoing technological development and the openness of our societies to accept technological progress. Let me conclude by agreeing wholeheartedly with the authors' last statement that we are at the beginning of an exciting era in neurotherapeutics – it does not matter how far we come, we will always be beginners.

REFERENCES

1. Benabid AL, Pollak P, Gervason C, et al. Long-term suppression of tremor by chronic stimulation of the ventral intermediate thalamic nucleus. *Lancet.* 1991;337:403–406.

2. Limousin P, Pollak P, Benazzouz A, et al. Bilateral subthalamic nucleus stimulation for severe Parkinson's disease. *Mov Disord.* 1995;10:672–674.

3. Krauss JK, Pohle T, Weber S, Ozdoba C, Burgunder JM. Bilateral deep brain stimulation of the globus pallidus internus for treatment of cervical dystonia. *Lancet.* 1999;354:837–838.

4. Coubes P, Roubertie A, Vayssiere N, Hemm S, Echenne B. Treatment of DYT1-generalised dystonia by stimulation of the internal globus pallidus. *Lancet.* 2000;355:2220–2221.

5. Kupsch A, Benecke R, Muller J, Deep-Brain Stimulation for Dystonia Study Groupet al. Pallidal deep-brain stimulation in primary generalized or segmental dystonia. *N Engl J Med.* 2006;355:1978–1990.

6. Jones EG. Morphology, nomenclature, and connections of the thalamus and basal ganglia. In: Kraus JK, Jankovic J, Grossman RG, eds. *Surgery for Parkinson's disease and movement disorders.* Philadelphia: Lippincott, Williams & Wilkins; 2001:24–47.

7. Krauss JK, Grossman RG. Historical review of pallidal surgery for treatment of parkinsonism and other movement disorders. In: Kraus JK, Jankovic J, Grossman RG, eds. *Surgery for Parkinson's disease and movement disorders.* Philadelphia: Lippincott-Raven; 1998:1–23.

8. Hassler R, Riechert T. Indikationen und Lokalisationsmethode der gezielten Hirnoperationen. *Nervenarzt.* 1954;25:441–447.

9. Weigel R, Krauss JK. Center median-parafascicular complex and pain control: review from a neurosurgical perspective. *Stereotact Funct Neurosurg.* 2004;82:115–212.

10. Follett KA, Weaver FM, Stern M, CSP 468 Study Groupet al. Pallidal versus subthalamic deep-brain stimulation for Parkinson's disease. *N Engl J Med.* 2010;362:2077–2091.

11. Moro E, Hamani C, Poon YY, et al. Unilateral pedunculopontine stimulation improves falls in Parkinson's disease. *Brain.* 2010;133:215–224.

12. Ferraye MU, Debû B, Fraix V, et al. Effects of pedunculopontine nucleus area stimulation on gait disorders in Parkinson's disease. *Brain.* 2010;133:205–214.

13. Alam M, Schwabe K, Krauss JK. The pedunculopontine nucleus area: critical evaluation of interspecies differences relevant for its use as a target for deep brain stimulation. *Brain.* 2011; 134:11–23.

14. Capelle HH, Blahak C, Schrader C, et al. Chronic deep brain stimulation in patients with tardive dystonia without a history of major psychosis. *Mov Disord.* 2010;25:1477–1481.

15. Vidailhet M, Yelnik J, Lagrange C, French SPIDY-2 Study Groupet al. Bilateral pallidal deep brain stimulation for the treatment of patients with dystonia-choreoathetosis cerebral palsy: a prospective pilot study. *Lancet Neurol.* 2009;8:709–717.

16. Leone M, Franzini A, Broggi G, Bussone G. Hypothalamic stimulation for intractable cluster headache: long-term experience. *Neurology.* 2006;67:150–152.

17. Bartsch T, Pinsker MO, Rasche D, et al. Hypothalamic deep brain stimulation for cluster headache: experience from a new multicase series. *Cephalalgia.* 2008;28:285–295.

18. Laxton AW, Tang-Wai DF, McAndrews MP, et al. A phase I trial of deep brain stimulation of memory circuits in Alzheimer's disease. *Ann Neurol.* 2010; 68:521–534.

STUDY QUESTIONS

1. Briefly review the differences in 'open loop' and 'closed loop' stimulation systems and consider advantages and disadvantages of both.
2. In what ways could more refined or 'steerable' current delivery in DBS be useful and what might be the trade-offs in developing such technology?

Chapter 4

Spinal – Extradural

Timothy R. Deer[1] MD and S. Matthew Schocket MD[2]

1. *Presidnet and CEO, The Center for Pain Relief, Inc., Charleston, WV*
   Clinical Professor of Anesthesiology West Virginia University, Charleston, WV
2. *Capital Pain institute, Austin, Tx*

INTRODUCTION

History

Spinal injections were first used in the mid-19th century to produce anesthesia for surgical procedures. Over the next century, new needles were developed and our anatomical understanding of the spine was enhanced. This mixture of progressive thoughts led to the development of epidural stimulation in the mid-1960s. This new therapy involved placing a lead with electrodes over the spine and creating an electrical current with a power source. Dr Norman Shealy reported the first successful case in 1967, in a patient with neuropathic cancer pain. In that case, a lead was placed in the intrathecal space. The lead was crudely designed, the energy source was archaic and the overall system was rudimentary, but the outcome was positive. These are important considerations since we are now dealing with much more advanced tools. Critical historic steps since that time have included the development of multicontact leads, a better understanding of anode and cathode field shaping, better computer modeling, totally implantable generators, rechargeable systems, miniaturization, and the development of prospective controlled studies. It is the purpose of this chapter to give an overview of the use of spinal cord stimulation (SCS) in modern medicine, and to examine the place of this therapy in the treatment algorithm.

General indications

An implantable neuromodulation system is indicated for spinal cord stimulation (SCS) as an aid in the management of chronic, intractable pain of the trunk and/or limbs. The best outcomes occur in patients with neuropathic, as opposed to nociceptive, pain syndromes, although many patients have mixed pain patterns that require both SCS, and additional treatments of muscle, joint and visceral pain. In the past, the best outcomes were confined to patients with unilateral

Essential Neuromodulation. DOI: 10.1016/B978-0-12-381409-8.00004-8

pain in a single extremity. Fortunately, recent improvements in technology, specifically the development of multiple lead systems with greater ability to steer current, drive current deeper to the cord layers, and impact new neural pathways, have enhanced outcomes for more complex pain presentations, including axial low back pain, bilateral extremity pain, visceral pain from the chest, abdomen and pelvis, and vascular pain syndromes.

Patient selection for spinal cord stimulation

Patient selection is the most important aspect for impacting a good outcome in those who undergo SCS. The most relevant issues are patient characteristics, and disease state causing the pain syndrome.

Patient characteristics

Several factors are predictive of a potential poor outcome in SCS. While these factors do not indicate an absolute contraindication, they should be considered carefully in the decision process to implant a patient.

1. Abuse or abnormal behavior with opioids suggesting abuse or diversion.
   a. Dose escalation without doctor's consent.
   b. Lost or stolen prescriptions.
   c. Early refill requests.
   d. Doctor shopping.
2. Presence of psychiatric and psychological disease.
   a. Untreated severe depression or anxiety.
   b. Untreated psychosis.
   c. Personality disorders such as borderline disorder.
3. Inability to understand risks and benefits of SCS.
4. Presence of bleeding abnormalities.
   a. Presence of drugs that impact bleeding.
   b. Disease states that lead to increased risk.
5. Presence of infection at the site of implant or systemically.
6. Physician impression that the patient is a poor candidate.

Disease-specific characteristics

The other major issue involved in the selection process is choosing the patient with the proper disease state and indication. Published literature has evolved on several patient groups who have undergone stimulation with both successful and less than optimal results. We can learn from this information to select patients in a more informed manner. This is also helpful to identify patients who are less likely to have an optimal outcome. The best outcomes may be expected in patients who have pain characterized by burning, crawling, stabbing, or shooting pain in the extremities after spinal surgery, those with spinal nerve entrapment from mechanical spinal diseases, complex regional pain syndrome type I and II, peripheral nerve injury, and painful neuropathies of various causes. Another group who have shown great promise are those that suffer from refractory angina, ischemic pain of the extremity, and pain related to peripheral vascular disease or vasospasm. Axial back pain once mystified the interventional pain physician and played a major role in SCS failure. With the advent of new percutaneous arrays, paddle multicolumn arrays, and combined epidural and peripheral nerve stimulation, more modern studies are reporting improved results, decreased opioid consumption, and improved return to work and to active duty in the armed services. Many other patient groups have been reported to be successfully treated with spinal cord stimulation, including intercostal neuralgia, spinal cord injury, focal peripheral nerve injury, phantom pain or neuropathic pain after trauma, and chest wall pain.

The achievement of parasthesia in the area of pain is thought to be an essential component in achieving a good outcome with relief of pain. In some patients, a good area of stimulation coverage is achieved, but pain relief is not achieved to a degree that would lead to an acceptable outcome by the patient. In some cases, the physician may be able to predict a probable failure of the implantable theory. Patient groups who have a less than optimal chance of a good outcome include those with spinal cord injury, central pain after stroke or traumatic brain injury, perirectal pain, pelvic pain, and nerve root transaction or brachial plexus or lumbar plexus injury.

Once the patient is selected for an epidural stimulator placement, the physician should be adept at placing the lead, anchoring the lead, and placing the generator. The next section will review these concepts.

EPIDURAL SPINAL CORD STIMULATION: THE PROCEDURE

Patient education is an important part of the procedure. The patient and their caretaker should be made aware of the risk of infection, bleeding, epidural hematoma, epidural abscess, lead failure, generator failure, and failure of the therapy. This education process can be part of the informed consent process. It is helpful for the patient to consult with anesthesiologists prior to moving forward with the procedure. Prior to the trial, preoperative antibiotics are given thirty to sixty minutes prior to incision, or skin puncture. This antibiotic regimen is recommended prior to the trial and standard for the permanent [14]. The implanter should consult with the local infectious disease physician for recommendations regarding local antibiotic resistant organisms. Some clinicians have chosen to use chlorohexidine baths or intranasal bacitracin prior to implant, particularly in high risk patients.

The anesthesiologist is helpful in achieving the implant. Ideally, the patient should be comfortable, but remain conversant during the procedure. In the cervical spine, the need to have the patient alert and conversant should be the standard of care.

The optimal positioning for the patient for lumbar thoracic implants is to have the abdominal area padded to alleviate lumbar lordosis. This position allows for ideal opening of the intralaminar spaces. Cervical placement is assisted by proper positioning which involves having the arms to the side, padding of the chest, and slight neck flexion. The well-positioned patient makes the procedure easier to achieve, reduces risks, and improves fluoroscopic guidance.

After proper positioning, achieving sterile technique is critical. Standard prep solutions include alcohol, povidone-iodine or chlorhexidine. Some clinicians prefer to finalize their preparation with a binding type of prep stick that clings to the skin or with clinging impregnated drapes.

Fluoroscopic guidance is utilized throughout the procedure, and an attention to safety to radiation exposure is very important. This is very important for the long-term health of the doctor. The use of fluoroscopy can be used in a pulsed fashion as opposed to continuous exposure, which will limit the exposure. Lateral and anterior-posterior views are critical to assure proper placement of the needle and leads. The spinal level of entry depends on the patient's anatomy and doctor preference. Previous back surgery usually precludes epidural entry at lower lumbar levels. Entry at L1–L2, T12–L1 may facilitate better lead control when placing the leads at the desired level (Fig. 4.1). Entry into the epidural space above the level of the conus medullaris may facilitate easier lead placement, however, there is potentially greater risk of spinal cord injury at any level above L2.

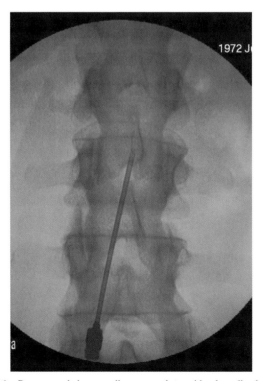

FIGURE 4.1   Recommended paramedian approach to epidural needle placement.

Once the entry level is determined, the skin is anesthetized with local anesthetic. Common choices include bupivicaine 0.25% or more commonly lidocaine 1% with epinephrine and sodium bicarbonate. The sodium bicarbonate
is often added with a 1:9 ratio (i.e. 1 ml of sodium bicarbonate in with 9 ml of
lidocaine). The sodium bicarbonate hastens the onset of topical analgesia and
decreases the burning sensation of the local anesthetic. The use of epinephrine
optimizes vasoconstriction and reduces bleeding. Careful attention should be
used to avoid deep local infiltration which can lead to an unintentional spinal
injection.

A 14-gauge modified-Tuohy (provided with the lead kit) or bent tip needle
should be used to enter the epidural space with a loss-of-resistance (LOR) technique, or hanging drop technique. Some instructors have recommended using
a lead wire to identify the epidural space, but this method has not been studied
and may increase the risk of wet tap or accidental spinal cord injury. There is no
literature to support the use of air, saline or a combination in the syringe used
to find the epidural space by loss of resistance. Some have theorized that saline
may lead to current disbursement and change programming, but that has never
been shown to be the case in a prospective fashion. [15,16]

The needle-entry point at the skin should be just medial to the pedicle, one
and one-half to two vertebral bodies below the intended interlaminar entry site
(Fig. 4.2). The ideal needle angle should be less than 30 degrees to the skin, but

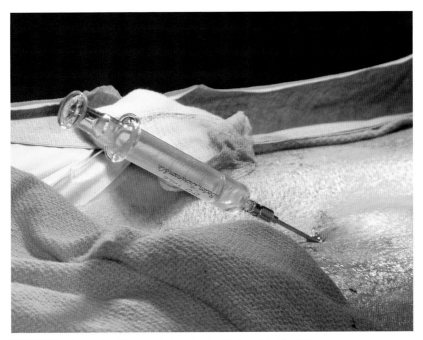

FIGURE 4.2   Loss of resistance technique to identify the epidural space.

this may vary based on body habitus and spinal anatomy. The orientation of the bevel on needle entry into the epidural space has been debated, but no clear instruction has been documented in the literature.

Placement of the lead(s) must be performed with fluoroscopic guidance. It is recommended that the practitioner use a low-dose, pulsed technique or intermittent fluoroscopy to reduce radiation exposure. The implanter should always keep their hands out of the direct path of the fluoroscopic beam when passing the leads. The initial target for the leads should be based on the pain pattern (Table 4.1). The use of a single lead, dual leads, or tripolar arrays are at the discretion of the implanter. Studies have shown the ability to treat axial and bilateral radicular pain with a single lead, but these studies have been short term, and have been criticized by some because of the inability to maintain stimulation on a long-term basis with the minimal capability to change programming [17]. Recent studies have shown improvements in axial back and extremity coverage using dual eight leads, staggering lead arrays, and targeting based on anatomical targets. [10,18,19]

New smaller rechargeable generators have made the percutaneous epidural lead more attractive, since the energy requirements can be increased without exhausting the battery as quickly. Some older studies showed that paddle leads were superior to percutaneous leads because of limited ability to increase energy and programs. Newer generators allow for multiple high energy programs that can be cycled in continuous patterns. These changes have negated some, but not all advantages of paddle leads. Persistent advantages include the capability to place the lead in areas where a percutaneous lead could not be easily placed, the ability to overcome scar tissue, and the ability to target deeper spinal cord structures with programming.

Placement of the internal programmable generator and tunneling of the leads is an important part of the percutaneous permanent implant. The physician should closely examine the patient to determine the ideal location for the generator. Factors to consider include size of the generator, location of bony landmarks,

TABLE 4.1 Sensory mapping in spinal cord stimulation

Location of pain	Approximate lumbar spinal cord stimulator lead placement
Low back	T7–T10
Buttock	T10–T12
Anterior thigh	T10–T12 slightly lateral
Posterior thigh	T10–L1
Foot	T11–L1

and patient skin condition such as lesions or infected areas. With all factors being equal the generator should be as close to the lead implant site as possible.

The pocket is made by making an incision to the subcutaneous tissue and then by blunt dissection to the appropriate size. The pocket should be 110 to 120% of the volume of the device to allow room to close without excessive dead space for fluid or seroma accumulation. The tunneling rod should be used with care to avoid inadequate or excessive tissue depth. Once tunneled a strain relief loop should be at both the lead placement site and the pocket.

EPIDURAL STIMULATION TO TARGET SPECIFIC DISEASE STATES

Specific disorders

The patient with failure of surgery of the lumbar or cervical spine; the patient with inoperable lumbar or cervical radiculopathy

Failed back surgery syndrome (FBSS) is defined as persistent or worsening pain of the trunk, back, neck, arms, legs, or multiple areas after attempted surgical correction of spinal disease. This diagnosis, commonly used to describe multiple patients with varying pain patterns, represents a diverse group of patients. These patients may have varying pain generators and mechanisms of pain production including nerve injury or mechanical pain. The patient may also suffer from chemical radiculitis, mechanical pain from joints or muscle, spinal or foraminal narrowing, scar around a nerve or spinal structure or inflamed arachnoid tissues.

Comparative prospective randomized evidence-based studies support the effectiveness of spinal cord stimulation as a comparative treatment of failed back surgery syndrome. In a very well done study, North identified failed back surgery patients with recurrent disk disease that were felt to be surgically correctible or an acceptable candidate for spinal cord stimulation. Randomization treatment was grouped into repeat surgery or spinal cord stimulation. The results of this study were favorable for SCS. SCS proved to be superior to repeat surgery when measured by global satisfaction and analgesia ($P > 0.01$). The crossover analysis also favored the stimulation group versus the repeat surgery group ($P = 0.02$) [38]. This study suggests that spinal cord stimulation is an effective alternative to repeat surgery in patients who have failed previous lumbar surgery, and should be considered to be a first line treatment in this complex group of patients.

Reviewing past studies in SCS one finds that the success rate has been very positive in those suffering from radicular neuropathic limb pain. Radiculopathy is a prime indication for the procedure. Axial back pain stimulation has proven to be more challenging. The difficulty in relieving axial low back pain centers on the fact that the nerve fibers that must be stimulated are located in the deep lateral areas of the dorsal columns near the dorsal root entry zone and the nerve root. In order to achieve stimulation of these deep lateral fibers, the nerve fibers of the dorsal nerve roots at the level of the stimulator are often activated causing painful nerve stimulation and involuntary motor function. This problem has been lessened by new multicolumn stimulation patterns that allow depolarization of the nerve roots, and focused current into the deeper lateral fibers. This gives a greater ability to focus current on the midline fibers and lateral fibers without subsequent stimulation of the nerve roots. In a prospective study, the combination of an increased number of leads and advanced programming led to improved success in patients suffering from axial pain [37]. Continued work on lead constructs and engineering models for both computer analysis and clinical studies are critical to future advancement.

Studies have shown that the earlier the patient is implanted after the failure of back or neck surgery, the better the chance of a good outcome [5].

Lead placement

In patients with single limb radicular pain, the lead may be placed to the midline or slightly off midline to the effected side between T8 and L1 to achieve good coverage in the lower body. The targets in the neck are often C2 to C7. The risk of one lead is migration or scarring under the lead, with limited programming to correct for these changes in the coverage. Dual lead systems may improve the overall long-term outcome [35]. Crossing over the midline with one or two leads may allow for coverage of the axial region and the limbs (Fig. 4.3).

Complex regional pain syndrome (CRPS)

Complex regional pain syndrome was formally known as reflex sympathetic dystrophy, or causalgia. This problem has led to loss of function, severe pain, and tremendous expense to society [40–42]. The goal of SCS in this population is multifaceted and includes pain relief, improved blood flow in those who have vasoconstriction, global satisfaction, and increased ability to tolerate rehabilitation. Achieving these goals will lead to a reduction in muscle atrophy, preservation of movement, and maintenance of strength via physical therapy and home exercise.

The success of spinal cord stimulation for CRPS is well supported by high powered statistical studies. In a prospective, randomized trial of 36 CRPS patients, the patients that were implanted with a permanent stimulator showed long-term pain reduction and improvements in quality of life, and global satisfaction. At long-term follow up of 5 years, the group undergoing SCS and physical therapy had persistent good outcomes with pain reduction [69–71].

One common thread in all studies regarding SCS for CRPS includes the importance of moving forward with the therapy early in the course of the disease. Once the process spreads to other body parts, or the patient develops contractures, the chance of a good outcome diminishes [43–45].

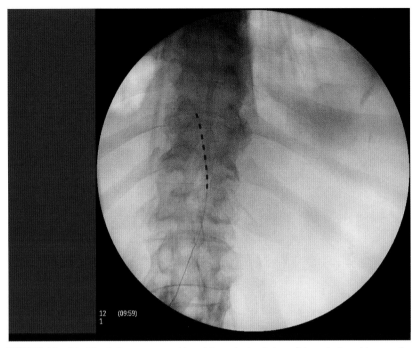

FIGURE 4.3 Ideal lead placement to achieve bilateral stimulation of the spinal cord by crossing the radiological and anatomical midline.

Lead placement

The lead target depends on the site of the CRPS symptoms. For the upper extremity, placement is recommended just ipsilateral to midline at the C2–C7 levels. For the lower extremity, placement is recommended just ipsilateral to midline at the T8–L1 levels. In patients where the foot is an involved area, the implanter should consider placing the lead at the T12–L1 level. This may require crossing the midline, and entering the spine two to three levels lower with the needle. Sensory mapping studies can be used to target initial lead placement, but the lead should be adjusted based on patient response (see Table 4.1).

Examples of this concept would be to place the lead at T12/L1 for the foot or T10/T11 for the knee. In complex cases involving multiple body parts, the implanter may choose multiple leads and, in some cases, more than one generator. The use of high-frequency stimulation (greater than 500 Hz) may also add some benefit [46].

Peripheral neuropathy

The burning, stinging pain of peripheral neuropathy is very amenable to SCS. Many of these patients respond to anticonvulsants or selective serotonin reuptake inhibitor (SSRI) medications. These drugs are very costly and often lose effect over time or cause unacceptable side effects. SCS can be much more cost effective over time, and can avoid the issue of end-organ effects of systemic medications. Many studies showing improvement of pain in neuropathic limb pain include these patients with neuropathies [29].

Lead placement

The targets for these leads for peripheral neuropathy are similar to that of the other syndromes noted above and are based on pain pattern. In some cases, in order to achieve stimulation, the lead must be placed in the area of the nerve root at L5–S1.

Post-herpetic neuralgia (PHN)

Severe nerve pain can develop in the area of previous herpes zoster activation. This chronic and severe problem is caused by an eruption of dormant Herpes varicella virus living in the dorsal root ganglia. This can lead to a chronic disruption of the nerve with abnormal activation of the A-delta and C fibers. Studies on the efficacy of SCS have been mixed in this condition. It does appear that the outcomes have improved over time with evolution of new technology and better programming. The efficacy may be due to direct stimulation at the cord level, but it also has been theorized to be due to restoration of blood flow due to vasodilatation, changes in the sympathetic nervous system, or improved blood flow to the nerve [48,49,50]. In patients who fail SCS for this condition, the implanter may consider peripheral nerve stimulation, or intrathecal drug delivery.

Lead placement

The most common array for this condition in the epidural space involves placement of one lead off midline in the ipsilateral side two levels above the lesion with a second lead in the lateral ipsilateral space one level above the lesion. New targets are needed to improve the outcome in this patient group.

Peripheral vascular disease (PVD) and ischemic pain

Neuropathic pain secondary to ischemia is a Food and Drug Administration (FDA) approved indication in the USA. The use of SCS for treatment of peripheral vascular disease has been a common use in Europe. The mechanism of action for the ability of spinal cord stimulation to relieve ischemic pain is not proven, but many have theorized it causes a change in sympathetic tone and thus increases blood flow [49,50]. Spinal cord stimulation improves microcirculation and increases capillary density and increases red blood cell velocity through capillary beds [51,52]. The current literature does show improved function in walking and function with SCS and may also improve wound healing in lesions of equal to or less than $3\,cm^2$ [53]. The presence of wet gangrene is a relative contraindication to placement of a SCS device in these patients.

Studies in the USA have used SCS with mixed results. The European evidence indicates improved function, improved wound healing, and improved pain scores when SCS is used early in the treatment protocol. Considering this information, the use of SCS for ischemic limb pain and peripheral vascular disease should be considered earlier in the course of treatment.

Lead placement

Leads should be placed to target the entire extremity. The most common locations would be T10–T12 for lower extremity ischemia, or C3–C6 in the upper extremity.

Angina

SCS has anti-anginal and anti-ischemic effects on the myocardium and may impact survival and function in the patient with intractable angina. Some have theorized the mechanism to be segmental inhibition of the activity on the sympathetic nervous system to the heart, causing an increase in microcirculation, improved metabolism, and a reduction in myocardial demand of oxygen [54]. SCS has been shown to improve achievable cardiac work load, increase time to ischemia and angina and improve function while not blunting the patient's ability to identify significant ischemic symptoms [55,56,57].

Lead placement

The lead for angina is placed at C7–T2 in most cases with a goal of producing a parasthesia in the chest wall, and left arm.

Visceral pain

Recent work by Kapural, Deer, and colleagues has shown improved pain relief with SCS in those suffering from visceral abdominal pain syndromes. Pain generation may develop secondary to ischemia to the bowel, adhesions, chronic pancreatitis, or post operative pain syndromes [58,72]. Other reports have shown an improvement in symptoms using SCS with or without opioids than to opioids alone in patients with chronic pancreatitis [59].

Lead placement

The lead for abdominal visceral pain has most commonly been reported at T5 or T6.

Pelvic pain

Common causes of pelvic pain include interstitial cystitis, endometriosis, and post-surgical scarring. Many of these patients are treated with high dose opioids with poor results. SCS has been used successfully in these cases after failure of more urological or gynecological treatments. The success of treatment for pain in the interstitial cystitis patient has led to an expansion of these therapies to patients with other causes of pelvic pain.

Lead placement

Most implanters now use dual octapolar leads placed over the S2, S3, and S4 nerves bilaterally. In some patients this will lead to improved pain and improved bladder volumes. Leads can be placed antegrade through the sacral hiatus, retrograde from the lumbar spine downward, or directly through the sacral hiatus (Figs 4.4 and 4.5). Leads can be placed by a paddle approach via sacrotomy at the upper sacrum [61,62].

Based on the information noted above we can make conclusions about patients who may have the best chance of a good outcome [10]:

High probability of a good outcome:

- Chronic cervical or lumbar radicular pain syndromes
- Complex regional pain syndrome, types 1 and 2
- Painful peripheral mononeuropathies
- Angina pectoris refractory to conventional surgical bypass and medical management
- Painful ischemic vascular disease refractory to medical management or surgical intervention.

Moderate chance of success with SCS:

- Axial low back pain
- Pelvic pain
- Visceral pain syndromes of the abdomen
- Post-herpetic neuralgia.

Difficult to achieve good outcomes with SCS:

- Neuropathic pain following spinal cord or brain injuries, nerve root avulsions
- Iatrogenic nerve root destruction
- Phantom limb pain.

CONTRAINDICATIONS TO SCS

Prior to moving forward with SCS the implanter should consider the contraindications to implanting a SCS device.

Contraindications

- Uncorrected coagulopathies
- Current sepsis/infection with fever
- Implantable cardiac defibrillator
- Inability to control device or lack of patient cooperation
- Thoracic syrinx.

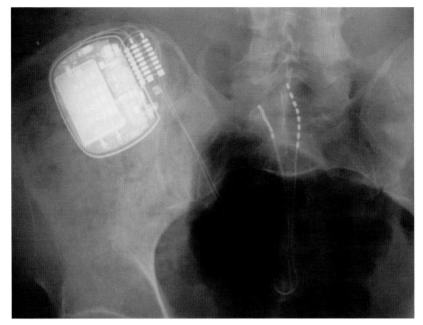

FIGURE 4.4   Stimmulation of the sacral nerve roots.

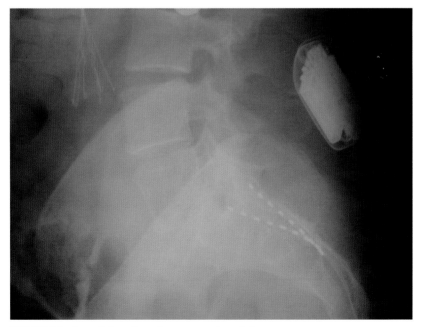

FIGURE 4.5   Lateral view of sacral nerve root stimmulation.

Relative contraindications

- Thoracic stenosis (if <10mm for a percutaneous lead [10])
- Patients who may require serial MRI evaluations (e.g. multiple sclerosis)
- Demand cardiac pacemaker: most pacemakers are now compatible with SCS but the cardiologists should be consulted prior to permanent implant.

PROGRAMMING THE EPIDURAL LEADS

With each implant several factors are critical to success. These include patient selection, proper needle and lead placement, and proper programming. The ability to program a lead depends on the position of the lead and the number of contacts. The detailed description of programming is very complex, but each implanter should know basic concepts. The cathode is the negatively charged electrode that impacts the shape and depth of the field. The anode is the positively charged electrode that disperses current and makes the field broader, but less focused. In order to drive current towards a specific portion of the cord, the cathode should be activated over the target. The target can be more specifically impacted by minimizing the number of cathodes and surrounding this electrode with two or more anodes.

Complications

Like other surgical procedures, the implantation of SCS systems can be associated with complications. The risks of the procedure must be weighed against the potential benefits. The majority of devices are placed and maintained without complications but, when they do occur, the most common complications are infection, post dural puncture headache, increased impedance from epidural scarring, bleeding, spinal cord injury, nerve injury, lead fracture, and lead migration.

Of these complications, the most common problem is lead migration. The movement of the lead laterally or vertically may result in loss of stimulation, and the need for surgical revision. The incidence of these complications has varied in the reported literature, but appears to range from 1% to 23% with the most likely number incidence being 13.5%. Lead fracture appears to be the second most common complication. [20].

Inadvertent dural puncture appears to occur in less than 1% of patients, and leads to post dural puncture in less than 1% of patients. There appears to be no reason to abandon the procedure in cases where a dural puncture has occurred as long as the patient is stable and has no parasthesias.

Infection is a potential complication of SCS. Patients with high risk of infection should be optimized medically prior to implant. Preoperative blood sugar control, preoperative chlorhexidine baths, intravenous antibiotics 30 minutes prior to incision or needle placement, intraoperative antibiotic irrigation, and careful wound closure may be helpful in reducing these risks. Early identification and aggressive treatment of superficial wound infections may prevent more extensive infection and help avoid expensive loss of the device.

Bleeding is a rare but serious complication of SCS. The majority of bleeds are superficial in the wound or pocket of insignificant consequence. In the event of an epidural bleed, a hematoma can develop. This can lead to serious injury to the neural structures and paraplegia if not addressed rapidly. Diagnosis is made by clinical suspicion and confirmed by CT. Treatment is surgical drainage.

Prevention of significant bleeding is based on reducing trauma and modification of oral medications that can impact bleeding in the perioperative period. Many patients are on these drugs because of cardiac, neurological or hematological problems. The treating cardiologists, neurologist, or family doctor should make decisions on the appropriateness of discontinuing these medications. The American Society of Regional Anesthesia and Pain Medicine establish and update guidelines on the issue of anticoagulants at regular intervals [21].

Spinal cord injury, nerve injury, and cord contusion are risks of spinal cord stimulation. These risks can be reduced by using a shallow needle entry, gentle lead placement, and avoidance of forcing leads past areas of resistance. Keeping the patient alert and conversant may also reduce the risk of injury.

Complications around the generator are another potential source of difficulty for the implanted patient. Fluid collection around the generator is a common complication. This problem, called seroma, can lead to swelling, redness and pain. The problem can be differentiated from infection by lack of fever, minimal white blood cell elevation, and lack of malaise. The incidence of seroma formation may be reduced by minimizing tissue trauma by blunt dissection techniques, by making the pocket prior to lead placement so the wound can be packed to reduce small venous bleeders, by creating an appropriately sized device pocket (generator blanks or spacers may help in this process), and by maximizing health in those with protein deficiency prior to surgery. Seroma may be treated by observation, pressure dressings, aspiration or by incision and drainage.

Another rare but important complication is placing the generator at a depth that is not optimal. The generator that is placed to deep will be unable to communicate with the transdermal telemetry equipment and a generator that is too superficial may lead to erosion. Hand held telemetry should be performed on the generator prior to leaving the sterile environment of the operating room. Since manufacturers vary on recommended depth, the corporate technician should be consulted if any questions exist regarding appropriate depth.

The other area of concern in the immediate postoperative period is wound dehiscence. The implanting physician should be vigilant in closing the tissue levels carefully to avoid lack of tissue congruency. The occurrence of wound dehiscence often leads to loss of the generator and, in most cases, the entire system.

CONCLUSION

The placement of computerized leads into the epidural space above the nerve and spinal cord can lead to changes in the patient's neurophysiology that eventually leads to changes in the pain perception and other neural reactions that can lead to major changes in the disease process and functional loss resulting from the rampages of pain and ischemia. These devices should be considered to be a major part of the treatment algorithm for chronic pain.

REFERENCES

1. Melzack R, Wall P. Pain Mechanisms: A new theory. *Science.* 1965;150:971–979.
2. Shealy CN, Mortimer JT, Reswick JB. Electrical inhibition of pain by stimulation of the dorsal columns: Preliminary clinical report. *Anesth Analg.* 1967;46:489–491.
3. North R. Spinal cord stimulation for intractable pain: Indications and technique. In: North RB, Kidd DH, Zohavak M, eds. *Current Therapy in Neuro-logical Surgery.* Philadelphia, PA: BC Decker; 1989:297–301.
4. Tasker R, de Carvalho G, Dolan E. Intractable pain of spinal cord origin: Clinical features and implications for surgery. *J Neurosurg.* 1992;77:373.
5. Kumar K, Hunter G, Demeria D. Spinal cord stimulation in treatment of chronic benign pain: Challenges in treatment planning and present status, a 22-year experience. *Neurosurgery.* 2006;58(3):481–496.
6. Alo K, Yland M, Charnou JH, Redko V. Multiple program spinal cord stimulation in the treatment of chronic pain: Follow up of multiple programs SCS. *Neuromodulation.* 1999;2:266–275.
7. Oakley JC, Francisco E, Bothe H, et al. Spinal cord stimulation: Results of an international multicenter study. *Neuromodulation.* 2006;9:192.
8. North RB, Kidd DH, Farrokhi F, Piantadosi SA. Spinal cord stimulation versus repeated lumbosacral spine surgery for chronic pain: A randomized controlled trial. *Neurosurgery.* 2005;56:98–107.
9. Kemler MA, Barendse GA, van Kleef M, et al. Spinal cord stimulation of patients with chronic reflex sympathetic dystrophy. *N Engl J Med.* 2000;343:618–624.
10. Barolat G, Oakley JC, Law JD, et al. Epidural spinal cord stimulation with multiple electrode paddle lead is effective in treating low back pain. *Neuromodu-lation.* 2001;4:59–66.
11. Oakley JC, Espinoso F, Bothe H, et al. Transverse tripolar spinal cord stimulation: Results of an international multi-center study. *Neuromodulation.* 2006;9:192–203.
12. Comiter CV. Sacral neuromodulation for the symp-tomatic treatment of refractory interstitial cystitis: A perspective studies. *J Urol.* 2003;169:1369–1373.
13. Harke H, Gretenkort P, Ladleif HU, Koester P, Rahman S. Spinal cord stimulation and postherpetic neuralgia and an acute herpes zoster pain. *Anesth Analg.* 2002;94:694–700.
14. Harke H, Gretenkort P, Ladleif HU, Rahman S. Spinal cord stimulation in sympathetically maintained complex regional pain syndrome type I with severe disability. A prospective clinical study. *Eur J Pain.* 2005;9:363–373.
15. Burchiel KJ, Anderson VC, Brown FD, et al. Pro-spective, multi-center study of spinal cord stimulation for relief of chronic back and extremity pain. *Spine.* 1996;21:2786–2794.
16. Calvillo O, Racz G, Didie J, Smith K. Neuroaug-mentation in the treatment of complex regional pain syndrome of the upper extremity. *Acta Orthop Belg.* 1998;64:57–63.

17. Kumar K, Malik S, Demeria D. Treatment of chronic pain with spinal cord stimulation versus alternative therapies: Cost effectiveness analysis. *Neurosurgery*. 2002;51:106–116.

18. Ohnmeiss DD, Rashbaum RF, Bogdanffy GM. Pro-spective outcome evaluation of spinal cord stimulation in patients with intractable leg pain. *Spinal*. 1996;21:1344–1351.

19. Villavicencio AT, Leveque JC, Rubin L, Bulsara K, Gorecki JP. Laminectomy versus percutaneous electrode placement for spinal cord stimulation. *Neurosurgery*. 2000;46:399–406.

20. Mekhail NA, Aeschbach A, Stanton-Hicks M. The cost benefit analysis of neurostimulation for chronic pain. *Clin J Pain*. 2004;20:462–468.

21. De La Porte C, Van de K. Spinal cord stimulation in failed back surgery syndrome. *Pain*. 1993;52:55–61.

22. North RB, Kidd DH, Zahurak M, James CS, Long DM. Spinal cord stimulation for chronic intractable pain: Experience over two decades. *Neurosurgery*. 1993;32:384–394.

23. Kumar K, Hunter G, Demeria D. Spinal cord stimulation and treatment of chronic benign pain: Challenges and treatment planning and present status, 22-year experience. *Neurosurgery*. 2006;58:481–496.

24. Kahn YN, Shariq SR, Kahn EA. Application of spinal cord stimulation for the treatment of abdominal visceral pain syndromes: Case reports. *Neuro-modulation*. 2005;8:14–27.

25. Chandler III GS, Nixon B, Stewart LT, Love J. Dorsal column stimulation for lumbar spinal stenosis. *Pain Physician*. 2003;6:113–118.

26. Feler CA, Whitworth LA, Brookoff D, Powell R. Recent advances: Sacral nerve root stimulation using a retrograde method of lead insertion for the treatment of pelvic pain due to interstitial cystitis. *Neuromodulation*. 1999;2:211–216.

27. North RB, Ewend MG, Lawton MT, Kidd DH, Piantadosi S. Failed back surgery syndrome: 5-year follow-up after spinal cord stimulator implantation. *Neurosurgery*. 1991;28:692–699.

28. Taylor RS. Spinal cord stimulation and complex regional pain syndrome and refractory neuropathic back and leg pain/failed back surgery syndrome: Results of a systematic review and metaanalysis. *J Pain Symptom Manage*. 2006;31:S13–S19.

29. Cameron T. Safety and applicacy of spinal cord stimulation for the treatment of chronic pain: A 20 year literature review. *J Neurosurg*. 2004;100:254–267.

30. Taylor RS, VanBuyten JP, Buchser E. Spinal cord stimulation for complex regional pain syndrome: A systematic review of the clinical and cost effectiveness literature and assessment of prognostic factors. *Eur J Pain*. 2006;10:91–101.

31. Taylor RS, VanBuyten JP, Buchser E. Spinal cord stimulation for chronic back and leg pain and failed back surgery syndrome: A systematic review and analysis of prognostic factors. *Spine*. 2005;30:152–160.

32. Taylor RS, Taylor RJ, Van Buyten JP, et al. - Deer and Masone S90 The cost effectiveness of spinal cord stimulation in the treatment of pain: A systematic review of the literature. *J Pain Symptom Manage*. 2004;27:370–378.

33. Alo K, Yland M, Kramer D, et al. Computer assisted and patient interactive programming of dual octrode spinal cord stimulation in the treatment of chronic pain. *Neuromodulation*. 1998;1:30–46.

34. North R, Kidd D, Olin J, et al. Spinal cord stimulation for axial low back pain: A prospective, controlled trial comparing dual with single percutaneous electrodes. *Spine*. 2005;30(12):1412–1418.

35. Van Buyten J, Zundert J, Milbouw G. Treatment of failed back surgery syndrome patients with low back and leg pain: A pilot study of a new dual lead spinal cord stimulation system. *Neuromodulation*. 1999;2:258–266.

36. Law J. Clinical and technical results from spinal stimulation for chronic pain of diverse pathophysiologies. *Stereotact Funct Neurosurg*. 1992;59(1–4):21–24.

37. Oakley JC, Krames ES, Prager JP, et al. A new spinal cord stimulation system effectively relieves chronic, intractable pain: A multi-center prospective clinical study. *Neuromodulation.* 2007;10:262–278.
38. North RB, Kidd DH, Farrokhi F, Piantadosi SA. Spinal cord stimulation versus repeated lumbosacral spine surgery for chronic pain: A randomized, controlled trial. *Neurosurgery.* 2005;56(1):98–107.
39. Deer T. Catheter tip associated granuloma: Inflammatory mass with intrathecal drug delivery. *Pain Med.* 2004;27(6):540–563.
40. Barolat G, Schwartzman R. Epidural spinal cord stimulation in the management of reflex sympathetic dystrophy. *Stereotact Funct Neurosurg.* 1989;3:29–39.
41. Oakley J, Weiner R. Spinal cord stimulation for complex regional pain syndrom: A prospective study of 19 patients at two centers. *Neuromodulation.* 1999;2:47–51.
42. Bennett D, Alo K, Oakley J, Feler C. Spinal cord stimulation for complex regional pain syndrome (RSD): A retrospective multicenter experience from 1995-1998 of 101 patients. *Neuromodulation.* 1999;3:202–210.
43. Stanton-Hicks M. Consensus report on complex regional pain syndrome guidelines. *Clin J Pain.* 1998;14:155–166.
44. Stanton-Hicks M. Spinal cord stimulation in the management of complex regional pain syndrome. *Neuromodulation.* 1999;3:193–201.
45. Kemler MA, Barendse GA, van Kleef M, et al. Spinal cord stimulation in patients with chronic reflex sympathetic dystrophy. *N Engl J Med.* 2000;343(9):618–624.
46. Alo KM, Redko V, Charnov J. Four year follow-up of dual electrode spinal cord stimulation for chronic pain. *Neuromodulation.* 2002;5:79.
47. Jones M. Effects of dorsal column stimulation on three aspects of pain: Burning, shooting and steady aching. *Acta Neurochir.* 1992;117:95.
48. Bonica J. The Management of Pain, Vol II, 2nd edition.. Philadelphia, PA: Lea & Febiger; 1990.
49. Fedoresak I, Linderoth B, Bognar L, et al. Peripheral vasodilation due to sympathetic inhibition induced by spinal cord stimulation. *Pro IBRO World Congress Neurosci.* 1991;:126.
50. Lilley J, Su D, Wang J. Sensory and sympathetic nerve blocks for post herpetic neuralgia. *Reg Anesth.* 1986;11:165.
51. Jacobs J, Slaaf D, et al. Dorsal column stimulation in critical limb ischemia. *Vasc Med Rev.* 1990;1:215–220.
52. Jacobs MJ, Jorning P, Joshi S, et al. Spinal cord electrical stimulation improves microvascular blood flow in severe limb ischemia. *Ann Surg.* 1988;207:179–183.
53. Augustinsson L, Holm J, Carl A, et al. Epidural electrical stimulation in severe limb ischemia: Evidences of pain relief, increased blood flow, and a possible limb saving effect. *Ann Surg.* 1985;202:104–111.
54. Augustinsson L. Spinal cord stimulation in severe angina pectoris. *Pain Clin.* 1995;8(2):161–165.
55. Bagger J, Jensen B, Johannsen G. Long term outcome of spinal cord stimulation in patients with refractory chest pain. *Clin Cardiol.* 1998;21:286–288.
56. Deer TR. Current and future trends in spinal cord stimulation for chronic pain. *Curr Pain Headache Rep.* 2001;5(6):503–509.
57. Yu W, Maru F, Edner M, et al. Spinal cord stimulation for refractory angina pectoris: A retrospective analysis of efficacy and cost-benefit. *Coron Artery Dis.* 2004;15(1):31–37.
58. Kahn Y. Presentation at the International Neuromodulation Society, Rome, Italy, June 2005.
59. Tiede J, Ghazi S, Lamer T, Obray J. The use of spinal cord stimulation in refractory abdominal visceral pain: Case reports and literature review. *Pain Pract.* 2006;6:197.

60. Paicius RM, Bernstein CA, Lempert-Cohen C. Peripheral nerve field stimulation in chronic abdominal pain. *Pain Physician.* 2006;9(3):261–266.

61. Feler CA, Whitworth LA, Fernandez J. Sacral neuromodulation for chronic pain conditions. *Anesthesiol Clin North America.* 2003;21(4):785–795.

62. Borawski K, Foster R, Webster G, Amundsen C. Predicting implantation with a neuromodulator using two different test stimulation techniques: A prospective randomized study in urge incontinent women. *Neurourol Urodyn.* 2007;26(1):14–18.

63. Horlocker T. ASRA Consensus. June, 2003.

64. Deer T, Raso L. Spinal cord stimulation for refractory angina pectoris and peripheral vascular disease. *Pain Physician.* 2006;9(4):347–352.

65. North R, Kidd D, Wimberly R, Edwin D. Prognostic value of psychological testing in patients undergoing spinal cord stimulation: a prospective study. *Neurosurgery.* 1996;39(2):301–310.

66. Kupers R, Van den Oever R, Van Houdenhove B, et al. Spinal cord stimulation in Belgium: A nationwide survey of the incidence, indications and therapeutic efficacy by the health insurer. *Pain.* 1994;56(2):211–216.

67. Williams D, Gehrman C, Ashmore J, Keefe F. Psychological considerations in surgical treatment of patients with chronic pain. *Tech Neurosurg.* 2003;8(3):168–175.

68. Villavicencio A, Burneikiene S. Elements of the preoperative workup, case examples. *Pain Med.* 2006;7(suppl 1):535–546.

69. Kemler MA, De Vet HC, Barendse GA, Van Den Wildenberg FA, Van Kleef M. The effect of spinal cord stimulation in patients with chronic reflex sympathetic dystrophy: two years' follow-up of the randomized controlled trial. *Ann Neurol.* 2004 Jan;55(1):13–18.

70. Kemler MA, de Vet HC, Barendse GA, van den Wildenberg FA, van Kleef M. Effect of spinal cord stimulation for chronic complex regional pain syndrome Type I: five-year final follow-up of patients in a randomized controlled trial. *J Neurosurg.* 2008 Feb;108(2):292–298.

71. Harke H, Gretenkort P, Ladleif HU, Rahman S. Spinal cord stimulation in sympathetically maintained complex regional pain syndrome type I with severe disability. A prospective clinical study. *Eur J Pain.* 2005 Aug;9(4):363–373.

72. Kapural L, Deer T, Yakolev A, et al. Technical aspects of spinal cord stimmulation for managing chronic visceral abdominal pain: the results from the National Survey. *Pain Medicine.* 2010;11(5):685–691.

Commentary on Spinal – Extradural

Jeffrey E. Arle, MD, PhD,

Director Functional Neurosurgery and Research, Department of Neurosurgery, Lahey Clinic,
Burlington, MA; Associate Professor of Neurosurgery, Tufts University School of medicine Boston, MA

Dr Deer and colleague have written a classic version of background, indications, techniques, and complication avoidance for epidural spinal cord stimulation. It is difficult to bring out any deficiencies in their review, or to highlight particular aspects of SCS that they *have* covered as their work is reasonably thorough, though by design limited in depth on any one topic. Importantly, some of the breadth of indications for SCS have been represented as well (e.g. angina and visceral pain) and future work in the field will likely clarify these applications.

There are two aspects I wish to comment on in the SCS field of neuromodulation, however, that will lend a certain degree of roundedness for the reader. These are with regard to the underlying physiological mechanisms of SCS and how they might help refine the therapy in future iterations, and with regard to practice structure and how this may contribute to choice of device or patient that receives therapy.

MECHANISMS

The underlying mechanism for creating analgesia with stimulation of the dorsal column fibers is not yet well understood, although progress is being made. It turns out that our relative ease in accessing the dorsal columns of the spinal cord affords us a lucky conduit to influence relevant cord circuitry and the processing of pain itself. This is because the IA and IB fibers that predominate in the dorsal columns form branches to many types of cells within the cord as they enter, including to alpha and gamma motor neurons themselves, while the same fibers as well then ascend in the dorsal column itself. Multiple loops of circuitry in a dizzying array of connections thus allow us to send retrograde action potentials from above these circuits (with the SCS electrodes) back into the cord through these fibers, ultimately influencing the output of the wide dynamic range (WDR) cells, the source cells for the fibers of the spinothalamic tract that carries most pain signals to the brain. There are also some descending influences from the brain that are activated by the stimulation of the dorsal columns and these may have influence as well. But other than stimulating the spinothalamic tract directly (which may in fact *create* pain), there are few, if any, other means of accessing the cord circuitry from a single location to cover multiple dermatomes and/or myotomes than the dorsal columns, and in this sense, we are 'lucky'. This anatomical arrangement is what allows for the entire therapy to exist at all.

Some wonder whether or not stimulating the dorsal root itself would work better, as all the fibers are already contained in one location. This concept misses the point of the anatomical serendipity, however. A single dorsal root being stimulated will work well – for only an isolated region. So, for example, if one knows the patient needs a specific area of the L5 distribution covered, on only one side, and nothing else, then placing an electrode over this single root can be very helpful. But most stimulation involves multiple distributions and this one aspect of patient symptoms more than anything leads us to require somehow getting into the system from a fairly local source, yet activating many areas – the dorsal columns satisfy this. Otherwise, we would need to place an electrode around several dorsal roots at several levels or sides or both. It would become impractical and increase risk, not to mention be logisitically challenging to run the wires into connectors and IPGs.

Linderoth and colleagues have explored and reviewed much of the data supporting the mechanisms involved in stimulating dorsal column fibers and I suggest the reader take a look at some of their papers for further study (e.g. Meyerson and Linderoth, 2006).

PRACTICE STRUCTURE CONTRIBUTION

The second aspect of comment on the chapter involves the consideration of various aspects of practice resources and their relationships. These might include:

- OR and/or surgeon availability
- Prior experience with trials/techniques/complications
- Patient volume in general
- Referral volume of difficult cases
- Cost constraints within practice or institution (related to reimbursement).

I would suggest that cases be divided into only a few categories and difficulties, in order best to determine method and outcome:

- Standard single lower leg dominant pain
- Back and/or lower extremity from FBSS or other (might require consideration of field stimulation or eventual use of an intrathecal medication pump)
- Upper extremity single limb
- Neck and/or single or bilateral upper extremity
- Prior surgeries in area of access (whether percutaneous or laminotomy)
- Prior lead implantation with or without IPG complication, with or without success.

When such categories are combined with the different types of practice structure, it becomes easier to see how cases are planned and aligned. This interplay of clinical indication and resource availability and mission, along with reimbursement structure, largely determines how SCS is utilized, throughout the USA at least.

REFERENCES

1. Meyerson BA, Linderoth B. Mode of action of spinal cord stimulation in neuropathic pain. *J Pain Symptom Manag.* 2006;31(4 suppl):S6–12.

STUDY QUESTIONS

1. SCS therapy currently stimulates dorsal column fibers. Consider other components of the spinal cord and how they might be feasibly targeted with the same or new technology.
2. What aspects of the surgical placement of SCS therapy might be streamlined or improved upon?

Chapter 5

Peripheral Nerve

Konstantin V. Slavin MD

Department of Neurosurgery, University of Illinois at Chicago, Chicago, IL, USA

INTRODUCTION

One of the less common areas of neuromodulation, peripheral nerve stimulation (PNS) is probably its fastest growing direction, at least in the field of pain treatment. The approach, which is still considered novel and experimental by many, is in fact neither – PNS was introduced before the much more accepted modality of spinal cord stimulation (SCS) and there are devices on the market today that are fully approved for PNS applications.

The history of PNS goes back to early 1960s. Few cases of electrical device implantation next to a nerve for control of neuropathic pain were performed even before the groundbreaking 'gate-control' theory of pain was proposed by Melzack and Wall [1]. Shelden and colleagues implanted PNS devices around the trigeminal and other nerves for control of pain as early as 1962 [2,3]. Since then, however, PNS has been through a period of relative growth followed by almost complete abandonment and then was reborn with the introduction of the percutaneous implantation approach in the late 1990s [4].

Today, PNS is used for variety of painful conditions in almost every part of the human body. Interestingly enough, the rapid growth in the number of reports and studies on PNS occurred despite the lack of dedicated and approved devices and, in almost every case, PNS is performed using hardware designed for SCS applications.

GENERAL PRINCIPLES OF PNS

The basis of PNS is considered a reversible suppression of pain due to production of concordant paresthesias. Similar to SCS, this effect may be supported by the above-mentioned 'gate-control' theory of pain. PNS does not alter sensation in the zone supplied by the stimulated nerve – but it usually does not work when such sensation is already altered or absent, meaning that it is pretty much impossible to obtain a pain-relieving effect in the area of complete numbness, such as in extreme cases of diabetic neuropathy or true anesthesia dolorosa.

Essential Neuromodulation. DOI: 10.1016/B978-0-08-089064-7.00005-X

There are two distinct technical principles of PNS use and they dictate the choice of equipment and surgical approach. Both of them follow the same goal of delivering repetitive electrical stimulation to the nerve that is involved in pain production or transmission. For this to happen, one has to know which nerve or nerves are involved in that particular patient, and the best way to confirm PNS usefulness is to perform a trial of stimulation. Although nerve blocks and transcutaneous electrical nerve stimulation (TENS) were considered as possible predictors of PNS success, their effects did not correlate with PNS results.

For the first PNS approach, the nerve in question is surgically exposed and the electrode is placed directly over or under it (or around it in the case of wrap-around electrodes). This is an older technique – the first clinical series of the successful application of this approach was published in early 1970s [5–7]. The use of flat (paddle-type) electrodes [8] allowed the elimination of some concerns of nerve injury from the scar around the electrode, and the subsequent suggestion of putting a thin layer of fascia between the electrode contacts and the nerve itself was aimed at further reduction in nerve irritation from the presence of the large electrode nearby. Right around that time, a special electrode was developed and approved for use in PNS – a paddle electrode with a mesh attached to it (OnPoint, Medtronic, Minneapolis, MN). Also, even though not a single implantable pulse generator (IPG) is officially approved for PNS, the radiofrequency-coupled systems manufactured by Medtronic and St Jude Medical (Plano, TX) have PNS among their approved uses. The use of paddle-type electrodes for PNS continues to be an accepted means of PNS delivery and multiple recent reports document its success and reliability in a variety of clinical settings [9–12].

The second way of PNS application involves percutaneous insertion of stimulating electrodes. The initial technique of percutaneous PNS was used to prove the concept when Wall and Sweet stimulated their own infraorbital nerves to confirm development of analgesia during the stimulation [13]. This approach, however, was not used clinically until the mid-1990s when Weiner and Reed described their technique of percutaneous insertion of PNS electrodes for treatment of occipital neuralgia [14]. Since then, this approach has been successfully used in many anatomical locations and for different clinical conditions. Although associated with a high rate of complications and frequent need of surgical revisions [15], the percutaneous PNS approach is very appealing due to its technical ease and low invasiveness. A recent suggestion to use ultrasound guidance for location of the nerve to be stimulated [16] now allows implanters to target many peripheral nerves along their subfascial or epifascial course. This includes occipital nerves and nerves in the trunk and extremities [17–19].

Somewhat similar to the percutaneous PNS approach is the so-called peripheral nerve field stimulation (PNFS). The difference in PNS and PNFS is the substrate of stimulation [20]. Although both modalities definitely stimulate the nerve fibers that carry nociceptive information from the periphery to the central processing areas, PNS works with visible and identifiable nerves, usually the named ones, whereas PNFS works on unnamed, frequently multiple, smaller nerves that are hard to identify within subcutaneous tissues.

Overall, however, all these approaches are quite similar to each other. The goal of stimulation remains the same as it is necessary to produce non-painful sensation – paresthesias – in the area of pain, and maintain these paresthesias, usually on a continuous basis, for the pain to subside.

From a technical point of view, PNS implantation somewhat resembles the SCS procedure as it is usually performed in stages. During the first stage, the electrodes are implanted for trialing purposes. If the plan is to keep the trialing electrode in place for subsequent permanent use, the electrode has to be anchored and connected to a temporary extension cable that is then tunneled away from the insertion point. Care is taken to avoid damaging the nerve during electrode insertion. If the electrode is inserted with an open technique, the nerve is identified and dissected so that the electrode can be placed in its immediate vicinity. Anatomical location of the stimulation site should take into consideration the size of the electrode and the room for anchors, connectors and extensions. There is usually no need for any adjunctive imaging technique as the large nerves are easily identifiable in their expected anatomical locations.

With the use of the percutaneous technique, the trial electrodes are frequently discarded upon completion of the trial, so different electrodes can be implanted during the second stage of the PNS procedure. Here, there is no need to visualize the stimulated nerve directly. Instead, one may use standard anatomical landmarks, trusting limited variability in the nerve course and an ability to capture the nerve with multiple contacts of the stimulating electrode, particularly if the electrode is placed perpendicular to the course of the nerve. In addition to this, both fluoroscopy and ultrasound have been used intraoperatively to check the position of the nerves (ultrasound) and direction of the electrode path (fluoroscopy).

Following trial electrode insertion, the patient goes through a testing period that varies from 2–3 days to a week or even longer. During this time, the patient is encouraged to evaluate the effectiveness of PNS in terms of pain suppression, and to note any of the side effects that PNS may produce (pain, discomfort, spasms, etc.) so the decision can be made on whether the overall benefits of PNS justify the trauma and expense associated with permanent implantation.

At the time of permanent implantation, an IPG is implanted away from the area of stimulation, and the electrode(s) may be either directly connected to the IPG or connected to it via extension cables of appropriate length. As opposed to SCS, where electrodes are almost uniformly inserted into posterior epidural space, PNS electrodes may be implanted anywhere in the body (face, head, neck, trunk and extremities) and therefore one has to be creative in choosing the IPG site and the path for the electrodes and/or extension cables in order to minimize the chance of hardware migration (if the anchors are loose) or fracture (if the anchors are too tight) whenever excessive mobility is encountered. The thickness of tissues overlying the electrodes, anchors, connectors and generators should also be taken into consideration so that erosion and infections are avoided.

Since the decision on PNS effectiveness and side effects may only be made by the patient, it is important to go through all the logistics of appropriate patient selection. In addition to confirmation of severity and chronicity of the pain, it is important to check whether less invasive modalities have already been tried, and whether the patient's psychological condition makes him/her an appropriate surgical candidate. The importance of psychological evaluation has been shown from the very beginning of PNS use as those with untreated depression, psychosis, major secondary gains and somatization disorder had overall unsatisfactory results in the long term [5,21]. And similar to all other neuromodulation applications, it is also important to set realistic expectations as PNS does not cure the underlying pain syndrome and rarely eliminates pain completely but, with appropriate use in selected patients, it decreases pain levels and improves or normalizes their functionality. The patients also have to be prepared for a high chance of needing some kind of reoperation during the follow up as statistics show that, although most PNS complications are minor, they appear to be much more common compared to other neuromodulation procedures.

SPECIFIC PNS APPLICATIONS

The indications for PNS evolved over time. If the initial indications concentrated around isolated peripheral neuropathies, post-traumatic and post-surgical, and complex regional pain syndromes type 2, the more current ones include craniofacial pain syndromes, such as occipital neuralgia, transformed migraines, cervicogenic headaches, trigeminal neuropathic pain, post-herpetic neuralgia, and various localized pain syndromes in the trunk and extremities.

The immediate and long-term results of PNS in the extremities were encouraging in the earlier series. The success rate was between 50 and 60% in most series and the complications were rare. But the procedure did not become universally accepted for two main reasons. First was the lack of appropriate, specially designed electrodes for this application, and the ones that were used were usually custom made as small cuffs or buttons. Second was the need to expose the nerve for electrode implantation and only a few centers had enough interest and expertise to do it on regular basis. In addition to this, multiple reports indicating development of perineural fibrosis after a long-term PNS use cooled down the enthusiasm for PNS application. Most importantly, however, was the wide acceptance of SCS as a pain-relieving surgical procedure, particularly for those very indications where PNS was used in the past. Therefore, a change in practice came with the change in indications and PNS was tried for those conditions where SCS had not been successful.

Following the pioneering work of Weiner and Reed [14], multiple other centers started using PNS for stimulation of the greater and lesser occipital nerves. The electrodes are usually implanted perpendicular to the course of these nerves on one or both sides of the patient's head at the level of craniocervical junction. The direction of insertion and the anchoring points vary from center to center. We routinely prefer anchoring the occipital PNS electrodes in the retromastoid region and tunneling them toward IPG located in the infraclavicular region [22]. The results of our early experience with occipital PNS have been repeated in many other centers [23]. Moreover, after the initial suggestion of Popeney and Aló [24], occipital PNS has been explored as a treatment for drug-resistant migraines [25,26]. With very high prevalence of migraine headaches and significant proportion of drug- and treatment-resistant cases, occipital PNS for migraines may be one of the most common applications of neuromodulation.

The other, much less prevalent but perhaps even more debilitating than severe migraine condition, is the cluster headache. Occipital PNS has been used with a great degree of success in these patients [27–30] and even those centers that use hypothalamic deep brain stimulation for treatment of this condition would frequently prefer occipital PNS as a first surgical option [31].

In addition to using percutaneous cylindrical electrodes for occipital PNS, paddle-type electrodes have been used for similar indications. The benefit of using paddle-type electrodes is their better stability and lower risk of migration. On the other hand, the insertion of this type of electrode is more invasive as it requires tissue dissection, whereas percutaneous electrodes are inserted through a spinal needle. Another difference that may be of particular importance in the case of occipital PNS is the direction of stimulation: paddle-type electrodes provide a unidirectional stimulation compared to circumferential in cylindrical wire-like percutaneous electrodes. This difference positively differentiates paddles in the case of SCS where the stimulation field is aimed at the posterior columns of the spinal cord. But, in the case of semi-blind positioning of PNS electrodes, this circumferential stimulation may be of benefit and the cylindrical electrodes may be tried in some cases of paddle electrode ineffectiveness.

Trigeminal branch stimulation is somewhat technically similar to occipital PNS. Electrodes are placed based on anatomical landmarks crossing the course of the supraorbital, infraorbital or auriculotemporal nerve. The epifascial location of the electrode makes it important to keep it at a sufficient depth – primarily to avoid electrode erosion. The anchoring point for these electrodes is usually placed in the retroauricular region and, from there, the electrode or extension cable is tunneled toward the infraclavicular IPG. Trigeminal PNS seems to work best for post-traumatic neuropathic pain but is not so good for post-herpetic neuralgia [32,33]. It is also now being explored for treatment of migraines and cluster headaches [34,35]. PNS for pain outside the craniofacial region may be divided into several subgroups. Back pain, by far the most prevalent chronic pain location, has been successfully treated with PNFS and the location of electrodes included all kinds of combination of vertical and horizontal paraspinal positions [36]. In addition to this, there are now reports of a so-called 'cross-talk approach' with stimulating electrodes placed far from each other and stimulation delivered from anodes to cathodes located on the opposite sides of the lumbar region [37], and of a 'hybrid stimulation' where PNFS is used in conjunction with PNS [38].

Similar to lower back pain, PNS has been used for the treatment of neck pain [39], chest wall pain [40], inguinal pain [41] and abdominal pain [42,43]. In each of these scenarios, PNFS electrodes are placed into the middle of the painful area or around it, and PNS electrodes are inserted in the vicinity of the nerve responsible for the pain. Finding inguinal nerves with ultrasound may be helpful for correct electrode placement [44] but, in other parts of trunk (back, neck, chest wall, etc.), PNS electrode positioning is performed based on the patient's description of the painful area: the area of pain is usually drawn on the patient's skin and the electrode position is chosen in such a way that the entire painful area is covered with a stimulation field from its center or from the edges.

Finally, pain in the extremities may be successfully treated with PNS. In the original series of the 1970s and 1980s, PNS of the upper extremities resulted in better pain relief compared to that of the lower extremities – partly due to the mixed nature and larger size of the commonly stimulated sciatic nerve. Recently, however, percutaneously inserted PNS electrodes have been used for the control of pain in both upper and lower extremities [18,45].

VAGAL NERVE STIMULATION

The principle of vagal nerve stimulation (VNS) is in many ways different from all other PNS applications. This refers to indications, devices, single-stage implantation approach, and even the pattern of stimulation.

First of all, the indications for VNS are not related to pain – it is mainly used for treatment of epilepsy and depression. The antiepileptic effects of vagal stimulation have been known for a long time and, after convincing anticonvulsant effects of VNS in experimental animal models in the late 1980s [46], this approach was used in human patients with epilepsy [47]. Within a few years, a dedicated VNS device (Cyberonics, Houston, TX) became officially approved for the treatment of refractory epilepsy. The antiepileptic effects of VNS are well documented, particularly for patients older than 12 with refractory partial seizures [48,49].

Several years later, based on pharmaceutical data showing antidepressant effects of multiple anticonvulsants, and clinical experience with VNS improving mood in epilepsy patients independently of seizure reduction, VNS was tried for treatment of refractory major depression [50] and, based on limited encouraging results, this modality became approved for this indication as well [51]. Unfortunately, the antidepressant effects of VNS are less impressive than the anticonvulsant and, therefore, at the time of this writing, VNS procedures are not covered by many insurance carriers when it comes to treatment of depression, although depression remains an approved indication for this neuromodulation approach.

VNS implantation is done in one stage as no trial period is required. During the surgery, the left vagal nerve is exposed at mid-cervical level and a special wrap-around electrode is placed over the nerve as it travels within the carotid sheath. Both contacts of this bipolar lead and the plastic anchor are implanted by wrapping the plastic spiral-shaped strips around the exposed segment of the nerve. The VNS IPG is implanted in the infraclavicular region through a separate incision. The leads come in two sizes that refer to the diameter of these spiral-shaped contacts, but overall design and technical characteristics remain the same for both lead models.

The stimulation pattern in VNS is quite different from other PNS applications. PNS is usually delivered on a continuous basis with relatively high frequency and the amplitude of stimulation is chosen that so the patient feels the stimulation-induced paresthesias. VNS parameters are chosen and titrated based on clinical effect but, instead of continuous stimulation, VNS is usually given over several seconds followed by a few minute long intervals. Although some patients experience paresthesias and speech alterations during these few seconds of stimulation, there are some others who do not perceive VNS at all.

Due to differences in function of the vagal nerve and those sensory nerves that are targeted for PNS, the side effects of stimulation are also quite different and may include speech and swallowing difficulties, usually directly related to the stimulation parameters and patterns. Overall, however, the VNS procedure is very safe and well-tolerated by the patients.

Research is now concentrating on more esoteric indications for VNS, including refractory migraines and cluster headaches [52] and Alzheimer's disease [53] but, so far, the results indicate great safety but not a convincing efficacy for these newer, potentially groundbreaking, indications.

CONCLUSION

Overall, it appears that PNS is indeed the fastest growing segment of neuromodulation for pain. However, the challenges in the development of this approach remain the same – there are no dedicated devices, there is no regulatory approval and, most importantly, there is no sound science in terms of its mechanism, patient selection, or the long-term outcome. Multiple anecdotal reports and small series do not provide sufficient evidence for PNS efficacy – and until such evidence is created, regulatory approval will be unlikely.

REFERENCES

1. Melzack RA, Wall PD. Pain mechanisms: a new theory. *Science*. 1965;150:971–979.
2. Shelden CH. Depolarization in the treatment of trigeminal neuralgia. Evaluation of compression and electrical methods; clinical concept of neurophysiological mechanism. In: Knighton RS, Dumke PR, eds. *Pain*. Boston: Little, Brown; 1966:373–386.
3. Shelden CH, Paul F, Jacques DB, Pudenz RH. Electrical stimulation of the nervous system. *Surg Neurol*. 1975;4:127–132.

4. Slavin KV. Peripheral nerve stimulation for neuropathic pain. *Neurotherapeutics*. 2008;5:100–106.

5. Nashold Jr BS, Goldner JL. Electrical stimulation of peripheral nerves for relief of intractable chronic pain. *Med Instrum*. 1975;9:224–225.

6. Picaza JA, Cannon BW, Hunter SE, Boyd AS, Guma J, Maurer D. Pain suppression by peripheral nerve stimulation. II. Observations with implanted devices. *Surg Neurol*. 1975;4:115–126.

7. Sweet WH. Control of pain by direct electrical stimulation of peripheral nerves. *Clin Neurosurg*. 1975;23:103–111.

8. Racz GB, Browne T, Lewis Jr R. Peripheral stimulator implant for treatment of causalgia caused by electrical burns. *Tex Med*. 1988;84:45–50.

9. Hassenbusch SJ, Stanton-Hicks M, Schoppa D, Walsh JG, Covington EC. Long-term results of peripheral nerve stimulation for reflex sympathetic dystrophy. *J Neurosurg*. 1996;84:415–423.

10. Oh MY, Ortega J, Bellotte JB, Whiting DM, Aló K. Peripheral nerve stimulation for the treatment of occipital neuralgia and transformed migraine using a C1-2-3 subcutaneous paddle style electrode: a technical report. *Neuromodulation*. 2004;7:103–112.

11. Kapural L, Mekhail N, Hayek SM, Stanton-Hicks M, Malak O. Occipital nerve electrical stimulation via the midline approach and subcutaneous surgical leads for treatment of severe occipital neuralgia: a pilot study. *Anesth Analg*. 2005;101:171–174.

12. Mobbs RJ, Nair S, Blum P. Peripheral nerve stimulation for the treatment of chronic pain. *J Clin Neurosci*. 2007;14:216–223.

13. Wall PD, Sweet WH. Temporary abolition of pain in man. *Science*. 1967;155:108–109.

14. Weiner RL, Reed KL. Peripheral neurostimulation for control of intractable occipital neuralgia. *Neuromodulation*. 1999;2:217–221.

15. Falowski S, Wang D, Sabesan A, Sharan A. Occipital nerve stimulator systems: review of complications and surgical techniques. *Neuromodulation*. 2010;13:121–125.

16. Huntoon MA, Hoelzer BC, Burgher AH, Hurdle MF, Huntoon EA. Feasibility of ultrasound-guided percutaneous placement of peripheral nerve stimulation electrodes and anchoring during simulated movement: Part two, upper extremity. *Reg Anesth Pain Med*. 2008;33:558–565.

17. Narouze SN, Zakari A, Vydyanathan A. Ultrasound-guided placement of a permanent percutaneous femoral nerve stimulator leads for the treatment of intractable femoral neuropathy. *Pain Physician*. 2009;12:E305–308.

18. Huntoon MA, Burgher AH. Ultrasound-guided permanent implantation of peripheral nerve stimulation (PNS) system for neuropathic pain of the extremities: original cases and outcomes. *Pain Med*. 2009;10:1369–1377.

19. Skaribas I, Aló K. Ultrasound imaging and occipital nerve stimulation. *Neuromodulation*. 2010;13:126–130.

20. Abejon D, Krames ES. Peripheral nerve stimulation or is it peripheral subcutaneous field stimulation; what is in a moniker?. *Neuromodulation*. 2009;12:1–4.

21. Goldner JL, Nashold Jr BS, Hendrix PC. Peripheral nerve electrical stimulation. *Clin Orthop Relat Res*. 1982;163:33–41.

22. Trentman TL, Slavin KV, Freeman JA, Zimmerman RS. Occipital nerve stimulator placement via a retromastoid to infraclavicular approach: a technical report. *Stereotact Funct Neurosurg*. 2010;88:121–125.

23. Slavin KV, Nersesyan H, Wess C. Peripheral neurostimulation for treatment of intractable occipital neuralgia. *Neurosurgery*. 2006;58:128–132.

24. Popeney CA, Aló KM. Peripheral neurostimulation for the treatment of chronic, disabling transformed migraine. *Headache*. 2003;43:369–375.

25. Rogers LL, Swidan S. Stimulation of the occipital nerve for the treatment of migraine: current state and future prospects. *Acta Neurochir Suppl*. 2007;97:121–128.

26. Hagen JE, Bennett DS. Occipital nerve stimulation for treatment of migraine. *Pract Pain Manage*. 2007;7:43–45 56.

27. Schwedt TJ, Dodick DW, Trentman TL, Zimmerman RS. Occipital nerve stimulation for chronic cluster headache and hemicrania continua: pain relief and persistence of autonomic features. *Cephalalgia*. 2006;26:1025–1027.

28. Burns B, Watkins L, Goadsby PJ. Treatment of medically intractable cluster headache by occipital nerve stimulation: Long-term follow-up of eight patients. *Lancet*. 2007;369:1099–1106.

29. Magis D, Allena M, Bolla M, De Pasqua V, Remacle JM, Schoenen J. Occipital nerve stimulation for drug-resistant chronic cluster headache: A prospective pilot study. *Lancet Neurol*. 2007;6:314–321.

30. Burns B, Watkins L, Goadsby PJ. Treatment of intractable chronic cluster headache by occipital nerve stimulation in 14 patients. *Neurology*. 2009;72:341–345.

31. Leone M, Franzini A, Cecchini AP, Broggi G, Bussone G. Stimulation of occipital nerve for drug-resistant chronic cluster headache. *Lancet Neurol*. 2007;6:289–291.

32. Johnson MD, Burchie KJ. Peripheral stimulation for treatment of trigeminal postherpetic neuralgia and trigeminal posttraumatic neuropathic pain: a pilot study. *Neurosurgery*. 2004;55:135–142.

33. Slavin KV, Wess C. Trigeminal branch stimulation for intractable neuropathic pain: technical note. *Neuromodulation*. 2005;8:7–13.

34. Narouze SN, Kapural L. Supraorbital nerve electric stimulation for the treatment of intractable chronic cluster headache: a case report. *Headache*. 2007;47:1100–1102.

35. Reed KL, Black SB, Banta II7 CJ, Will KR. Combined occipital and supraorbital neurostimulation for the treatment of chronic migraine headaches: initial experience. *Cephalalgia*. 2010;30:260–271.

36. Krutsch JP, McCeney MH, Barolat G, Al Tamimi M, Smolenski A. A case report of subcutaneous peripheral nerve stimulation for the treatment of axial back pain associated with postlaminectomy syndrome. *Neuromodulation*. 2008;11:112–115.

37. Falco FJE, Berger J, Vrable A, Onyewu O, Zhu J. Cross talk: a new method for peripheral nerve stimulation. An observational report with cadaveric verification. *Pain Physician*. 2009;12:965–983.

38. Bernstein CA, Paicius RM, Barkow SH, Lempert-Cohen C. Spinal cord stimulation in conjunction with peripheral nerve field stimulation for the treatment of low back and leg pain: a case series. *Neuromodulation*. 2008;11:117–123.

39. Lipov EG, Joshi JR, Sanders S, Slavin KV. Use of Peripheral subcutaneous field simulation for the treatment of axial neck pain: a case report. Neuromodulation. 2009;12:292–295.

40. Yakovlev AE, Peterson AT. Peripheral nerve stimulation in treatment of intractable postherpetic neuralgia. *Neuromodulation*. 2007;10:373–375.

41. Stinson Jr LW, Roderer GT, Cross NE, Davi BE. Peripheral subcutaneous electrostimulation for control of intractable post-operative inguinal pain: a case report series. *Neuromodulation*. 2001;4:99–104.

42. Paicius RM, Bernstein CA, Lempert-Cohen C. Peripheral nerve field stimulation in chronic abdominal pain. *Pain Physician*. 2006;9:261–266.

43. Johnson RD, Green AL, Aziz TZ. Implantation of an intercostal nerve stimulator for chronic abdominal pain. *Ann R Coll Surg Engl*. 2010;92:W1–3.

44. Carayannopoulos A, Beasley R, Sites B. Facilitation of percutaneous trial lead placement with ultrasound guidance for peripheral nerve stimulation trial of ilioinguinal neuralgia: A technical note. *Neuromodulation*. 2009;12:296–301.

45. Monti E. Peripheral nerve stimulation: a percutaneous minimally invasive approach. *Neuromodulation*. 2004;7:193–196.

46. Zabara J. Inhibition of experimental seizures in canines by repetitive vagal stimulation. *Epilepsia*. 1992;33:1005–1012.

47. Penry JK, Dean JC. Prevention of intractable partial seizures by intermittent vagal stimulation in humans: Preliminary results (abstract). *Epilepsy*. 1990;31(suppl.):S40–S43.

48. Amar AP, Heck CN, Levy ML, et al. An institutional experience with cervical vagus nerve trunk stimulation for medically refractory epilepsy: rationale, technique, and outcome. *Neurosurgery*. 1998;43:1265–1276 discussion 1276-1280.

49. Tecoma ES, Iragui VJ. Vagus nerve stimulation use and effect in epilepsy: what have we learned?. *Epilepsy Behav*. 2006;8:127–136.

50. Sackeim HA, Rush AJ, George MS, et al. Vagus nerve stimulation (VNS) for treatment-resistant depression: efficacy, side effects, and predictors of outcome. *Neuropsychopharmacology*. 2001;25:713–728.

51. Sackeim HA, Brannan SK, Rush AJ, George MS, Marangell LB, Allen J. Durability of antidepressant response to vagus nerve stimulation (VNS). *Int J Neuropsychopharmacol*. 2007;10:817–826.

52. Mauskop A. Vagus nerve stimulation relieves chronic refractory migraine and cluster headaches. *Cephalalgia*. 2005;25:82–86.

53. Merrill CA, Jonsson MA, Minthon L, et al. Vagus nerve stimulation in patients with Alzheimer's disease: Additional follow-up results of a pilot study through 1 year. *J Clin Psychiatry*. 2006;67:1171–1178.

Commentary on Peripheral Nerve

Jeffery E. Arle, MD, PhD

Lahey Clinic, Director Functional Neurosurgery and Research, Department of Neurosurgery, Burlington, MA Associate Professor of Neurosurgery Tufts University School of Medicine Boston, MA

Dr Slavin has provided a well-balanced survey of PNS, including the applications of both field-type stimulation and direct stimulation of cranial nerves such as the trigeminal and vagus nerves. As he notes, this area of neuromodulation may be the fastest growing, primarily due to its use in treating many common disorders, efficaciously or not, including migraine, depression, obesity, and regional low back pain. These are indications that have notable benefit already or realistic potential for efficacy with other treatment approaches (e.g. bariatric surgery for obesity, deep brain stimulation (DBS) for depression), but which may still see benefit from PNS in a less invasive manner. So further development in these directions is vital and, fortunately, likely to continue.

Such newer developments, however, raise one of the important aspects of PNS that Dr Slavin mentions, but which deserves further elaboration. PNS is only marginally approved and reimbursed at present. Beyond the fact that refinements may be made in devices used for PNS, obtaining approvals from third-party payors to perform PNS can be a hindrance to further development of these techniques. Largely anecdotal successes, but successes nonetheless, have been reported. In particular, the ability to treat migraine and refractory low back pain has a wide potential population that can find help for what are typically very difficult and disabling disorders.

As Dr Slavin alluded to, as well, the mechanisms of analgesia in PNS are understudied and may allow for refinement of IPG parameter spaces if even small breakthroughs could be made in this regard. For example, is there an optimal frequency or pattern of stimulation when an electrode is placed at different distances from the dorsal root ganglion or are there different electrode designs or configurations that might optimize field stimulation over direct nerve stimulation? In addition, very little work has been done modeling VNS stimulation. Such work might lead to refinements of treatment for depression over epilepsy, or better use of VNS for failed bariatric patients where it had been abandoned previously, for example.

In sum, there are three important areas that can be explored with PNS, both for hopes in advancing the field and, as a reader, gaining a better understanding of the PNS field currently. These include:

1. reimbursement and approval
2. better controlled outcome studies
3. refinement of PNS analgesic mechanisms that can drive device refinement.

STUDY QUESTIONS

1. Define a decision algorithm for placing PNS, with and without failed SCS or other treatments, for various locations.
2. Consider the programming parameter spaces used in stimulating a peripheral nerve currently available, including cycling on and off, and how each might contribute or detract from the ability to obtain satisfactory pain relief for the patient.

Part III

The Neuromodulation Therapy Interface

Chapter 6

The Electrode – Materials and Configurations

Daniel R. Merrill PhD
Vice President, Technical Affairs, Alfred Mann Foundation, Santa Clarita, CA, USA

PHYSICAL BASIS OF THE ELECTRODE/ELECTROLYTE INTERFACE

When a metal electrode is placed inside a physiological medium, such as extracellular fluid (ECF), an interface is formed between the two phases. In the metal electrode phase and in attached electrical circuits, charge is carried by electrons. In the physiological medium or, in more general terms, the electrolyte, charge is carried by ions, including sodium, potassium, and chloride in the ECF. The central process that occurs at the electrode–electrolyte interface is a transduction of charge carriers from electrons in the metal electrode to ions in the electrolyte.

In the simplest system, two electrodes are placed in an electrolyte, and electrical current may pass between the electrodes through the electrolyte. One of the two electrodes is termed a working electrode (WE), and the second is termed a counter electrode (CE). The working electrode is defined as the electrode that one is interested in studying, with the counter electrode being necessary to complete the circuit for charge conduction.

There are two primary mechanisms of charge transfer at the electrode–electrolyte interface, illustrated in Figure 6.1. One is a non-faradaic reaction, where no electrons are transferred between the electrode and electrolyte. Non-faradaic reactions include redistribution of charged chemical species in the electrolyte. The second mechanism is a faradaic reaction, in which electrons are transferred between the electrode and electrolyte, resulting in reduction or oxidation of chemical species in the electrolyte.

Capacitive/Non-Faradaic Charge Transfer

If only non-faradaic redistribution of charge occurs, the electrode/electrolyte interface may be modeled as a simple electrical capacitor called the double layer capacitor C_{dl}. This capacitor is formed due to several physical

Essential Neuromodulation. DOI: 10.1016/B978-0-12-381409-8.00006-1

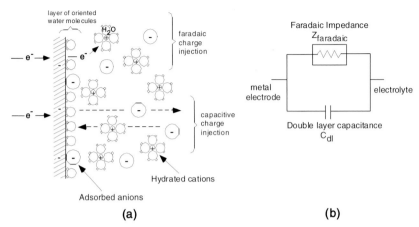

FIGURE 6.1   The electrode/electrolyte interface. Faradaic charge transfer (top) and capacitive redistribution of charge (bottom) is shown as the electrode is driven negative. (A) Physical representation; (B) two element electrical circuit model for mechanisms of charge transfer at the interface.

phenomena [1–6]. When a metal electrode is placed in an electrolyte, charge redistribution occurs as metal ions in the electrolyte combine with the electrode. This involves a transient transfer of electrons between the two phases, resulting in a plane of charge at the surface of the metal electrode, opposed by a plane of opposite charge, as counterions, in the electrolyte.

If the net charge on the metal electrode is forced to vary (as occurs with charge injection during stimulation), a redistribution of charge occurs in the solution. Suppose that two metal electrodes are immersed in an electrolytic salt solution. Next, a voltage source is applied across the two electrodes so that one electrode is driven to a relatively negative potential and the other to a relatively positive potential. At the interface that is driven negative, the metal electrode has an excess of negative charge (see Fig. 6.1). This will attract positive charge (cations) in solution towards the electrode and repel negative charge (anions). In the interfacial region, there will be net electroneutrality, because the negative charge excess on the electrode surface will equal the positive charge in solution near the interface. The bulk solution will also have net electroneutrality. At the second electrode, the opposite processes occur, i.e. the repulsion of anions by the negative electrode is countered by attraction of anions at the positive electrode.

If the total amount of charge delivered is sufficiently small, only charge redistribution occurs, there is no transfer of electrons across the interface, and the interface is well modeled as a simple capacitor. If the polarity of the applied voltage source is then reversed, the direction of current is reversed, the charge redistribution is reversed, and charge that was injected from the electrode into the electrolyte and stored by the capacitor may be recovered.

Faradaic Charge Transfer and the Electrical Model of the Electrode–Electrolyte Interface

Charge may also be injected from the electrode to the electrolyte by faradaic processes of reduction and oxidation, whereby electrons are transferred between the two phases. Reduction, which requires the addition of an electron, occurs at the electrode that is driven negative, while oxidation, requiring the removal of an electron, occurs at the electrode that is driven positive. Unlike the capacitive mechanism, faradaic charge injection forms products in solution that cannot be recovered upon reversing the direction of current if the products diffuse away from the electrode. Figure 6.1b illustrates a simple electrical circuit model of the electrode–electrolyte interface, consisting of two elements [7–9]. C_{dl} is the double layer capacitance, representing the ability of the electrode to cause charge flow in the electrolyte without electron transfer. $Z_{faradaic}$ is the faradaic impedance, representing the faradaic processes of reduction and oxidation where electron transfer occurs between the electrode and electrolyte. One may generally think of the capacitance as representing charge storage, and the faradaic impedance as representing charge dissipation.

The following are examples of faradaic electrode reactions. Cathodic processes, defined as those where reduction of species in the electrolyte occur as electrons are transferred from the electrode to the electrolyte, include such reactions as:

$$2\,H_2O + 2\,e^- \rightarrow H_2\!\uparrow + 2\,OH^- \quad \text{reduction of water} \tag{6.1}$$

$$PtO + 2\,H^+ + 2\,e^- \rightleftharpoons Pt + H_2O \quad \text{oxide formation and reduction} \tag{6.2}$$

$$IrO + 2\,H^+ + 2\,e^- \rightleftharpoons Ir + H_2O \quad \text{oxide formation and reduction} \tag{6.3a}$$

$$IrO_2 + 4\,H^+ + 4\,e^- \rightleftharpoons Ir + 2\,H_2O \quad \text{oxide formation and reduction} \tag{6.3b}$$

$$2\,IrO_2 + 2\,H^+ + 2\,e^- \rightleftharpoons Ir_2O_3 + H_2O \quad \text{oxide formation and reduction} \tag{6.3c}$$

$$Pt + H^+ + e^- \rightleftharpoons Pt - H \quad \text{hydrogen atom plating} \tag{6.4}$$

Anodic processes, defined as those where oxidation of species in the electrolyte occur as electrons are transferred to the electrode, include:

$$2\,H_2O \rightarrow O_2\!\uparrow + 4\,H^+ + 4\,e^- \quad \text{oxidation of water} \tag{6.5}$$

$$Pt + 4\,Cl^- \rightarrow [PtCl_4]^{2-} + 2\,e^- \quad \text{corrosion} \tag{6.6}$$

Reaction 6.1 is the irreversible reduction of water forming hydrogen gas and hydroxyl ions. The formation of hydroxyl raises the solution pH. Reversible reactions, where species remain bound or close to the electrode surface, are demonstrated by reactions 6.2 through 6.4. Reactions 6.2 and 6.3a, 6.3b, and

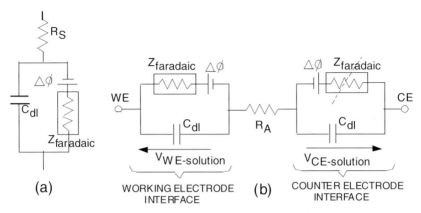

FIGURE 6.2 Electrical circuit models. (A) Single electrode–electrolyte interface; (B) two-electrode system. External access to the system is at two points labeled 'WE' and 'CE'. If the counter electrode has a large surface area, it may be considered as strictly a capacitance as shown.

6.3c are the reversible formation and subsequent reduction of an oxide layer on platinum and iridium, respectively. Reaction 6.4 is reversible adsorption of hydrogen onto a platinum surface, responsible for the so-called pseudocapacity of platinum. In reaction 6.5, water molecules are irreversibly oxidized, forming oxygen gas and hydrogen ions, thus lowering the pH. Reaction 6.6 is the corrosion of a platinum electrode in a chloride-containing medium.

The electrode interface model of Figure 6.1b demonstrates the mechanisms of charge injection from an electrode; however it neglects the equilibrium interfacial potential $\Delta\phi$ that exists across the interface at equilibrium. This is modeled as shown in Figure 6.2a, along with the solution resistance R_S (also known as the access resistance R_A or the ohmic resistance R_Ω) that exists between two electrodes in solution.

If one begins with a system that is in equilibrium and then forces the potential of an electrode away from its equilibrium value, for example by connecting a voltage or current source between the working and counter electrodes, the electrode is said to become polarized. Polarization is measured by the overpotential η, which is the difference between an electrode's potential and its equilibrium potential (both measured with respect to some third reference electrode):

$$\eta \equiv E - E_{eq} \qquad (6.7)$$

The net current density across an electrode/electrolyte interface due to a faradaic reaction is proportional to an exponential function of the overpotential, fully described by the current–overpotential equation below [9].

$$i_{net} = i_0 \left\{ \frac{[O](0,t)}{[O]\infty} \exp\left(-\alpha_c n f \eta\right) - \frac{[R](0,t)}{[R]\infty} \exp\left(+(1-\alpha_c) n f \eta\right) \right\} \qquad (6.8)$$

where i_{net} is the net faradaic current density across the electrode–electrolyte interface, i_0 is the exchange current density, $[O](0,t)$ and $[R](0,t)$ are concentrations at the electrode surface (x = 0) as a function of time, $[O]_\infty$ and $[R]_\infty$ are bulk concentrations, α_c is the cathodic transfer coefficient and equals ≈ 0.5, n is the number of moles of electrons per mole of reactant oxidized, $f \equiv F/R\ T$, F is Faraday's constant $\approx 96\,485\,C$/mole of electrons, R is the gas constant $\approx 8.314\,Joules$/mole-°K, and T is the absolute temperature.

For a sufficiently small overpotential (a small potential excursion away from equilibrium), there is little faradaic current, and current flows primarily through the capacitive branch of Figure 6.1, charging the electrode capacitance, not through the faradaic branch. As more charge is delivered through an electrode interface, the electrode capacitance continues to charge, the overpotential increases, and the faradaic current (proportional to $\exp(\eta)$) begins to be a significant fraction of the total injected current. For substantial cathodic overpotentials, the left term of equation 6.8 dominates; for substantial anodic overpotentials the right term dominates.

Reversible and Irreversible Faradaic Reactions

Faradaic reactions are divided into reversible and irreversible reactions [9]. A reversible process is one where the reactants are reformed from the products upon reversing the direction of current. The degree of reversibility depends on the relative rates of kinetics (electron transfer at the interface) and mass transport of reactants to the electrode surface. A faradaic reaction with very fast kinetics relative to the rate of mass transport is reversible. With fast kinetics, large currents occur with small potential excursions away from equilibrium. Since the electrochemical product does not move away from the surface extremely fast (relative to the kinetic rate), there is an effective storage of charge near the electrode surface and, if the direction of current is reversed, then some product that has been recently formed may be reversed back into its initial (reactant) form.

In a faradaic reaction with slow kinetics, large potential excursions away from equilibrium are required for significant currents to flow. In such a reaction, the potential must be forced very far from equilibrium before the mass transport rate limits the net reaction rate. In the lengthy time frame imposed by the slow electron transfer kinetics, chemical reactant is able to diffuse to the surface to support the kinetic rate, and product diffuses away quickly relative to the kinetic rate. Because product diffuses away, there is no effective storage of charge near the electrode surface, in contrast to reversible reactions. If the direction of current is reversed, product will not be reversed back into its initial (reactant) form, since it has diffused away within the slow time frame of the reaction kinetics. Irreversible products may include species that are soluble in the electrolyte, precipitate in the electrolyte, or evolve as a gas (e.g. reactions 6.1 and 6.5). Irreversible faradaic reactions result in a net change in the chemical environment, potentially creating chemical species that are damaging to tissue or the electrode. As a general principle, an objective of electrical stimulation design is to avoid irreversible faradaic reactions.

CHARGE INJECTION ACROSS THE ELECTRODE–ELECTROLYTE INTERFACE DURING ELECTRICAL STIMULATION

Charge Injection During Pulsing: Interaction of Capacitive and Faradaic Mechanisms

As illustrated in Figure 6.1, there are two primary mechanisms of charge injection from a metal electrode into an electrolyte. The first consists of charging and discharging the double layer capacitance, causing a redistribution of charge in the electrolyte but no electron transfer from the electrode to the electrolyte. C_{dl} for a metal in aqueous solution has values on the order of 10–$20\,\mu F/cm^2$ of real area (geometric area multiplied by the roughness factor). For a small enough total injected charge, all charge transfer occurs by charging and discharging of the double layer. Above some injected charge density, charge transfer commences via faradaic reactions where electrons are transferred between the electrode and electrolyte, thus changing the chemical composition in the electrolyte by reduction or oxidation reactions. Figure 6.1 illustrates a single faradaic impedance representing the electron transfer reaction $O + n\,e^- \rightleftharpoons R$. Generally, there may be more than one faradaic reaction possible, which is modeled by several branches of $Z_{faradaic}$ (one for each reaction), all in parallel with the double layer capacitance.

As current is passed between a working electrode and counter electrode through an electrolyte, both the working and counter electrodes' potentials move away from their equilibrium values, with one moving positive of its equilibrium value and the other moving negative of its equilibrium value. Total capacitance is proportional to area, with capacitance $C_{dl} = (\text{capacitance/area}) \times \text{area}$. Capacitance/area is an intrinsic material property. Capacitance is defined as the ability to store charge, and is given by:

$$C_{dl} \equiv \frac{dq}{dv} \tag{6.9}$$

where q = charge and v = the electrode potential with respect to some reference electrode.

An electrode with a large area and total capacity (as is often the case for a counter electrode) can store a large amount of charge (dq) with a small overpotential (dv). By maintaining the potential of the counter electrode fairly constant during charge injection (near its equilibrium value), there is little faradaic current. It is common to neglect the counter electrode in analysis and, while this is often a fair assumption, it is not always the case. Significant overpotentials may be realized at a working electrode with small surface area, as may be required to achieve high spatial resolution during stimulation.

In addition to the double layer capacitance, some metals have the property of pseudocapacity [8], where a faradaic electron transfer occurs but, because the product remains bound to the electrode surface, the reactant may be recovered (the reaction may be reversed) if the direction of current is reversed. Although electron transfer occurs, in terms of the electrical model of Figure 6.1, the pseudocapacitance is better modeled as a capacitor, since it is a charge storage (not dissipative) process. Platinum is commonly used for stimulating electrodes as it has a pseudocapacity (by reaction 6.4) of $210 \, \mu C/cm^2$ real area [10], or equivalently $294 \, \mu C/cm^2$ geometric area using a roughness factor of 1.4.[1]

It is a general principle when designing electrical stimulation systems that one should avoid the onset of irreversible faradaic processes which may potentially create damaging chemical species, and keep the injected charge at a low enough level where it may be accommodated strictly by reversible charge injection processes. Unfortunately, this is not always possible because a larger injected charge may be required to cause the desired effect (e.g. initiating action potentials). Reversible processes include charging and discharging the double layer capacitance, reversible faradaic processes involving products that remain bound to the surface, such as the reversible formation and reduction of a surface oxide (reactions 6.2, 6.3) or plating of hydrogen atoms on platinum (reaction 6.4), and reversible faradaic processes where the solution phase product remains near the electrode due to mass diffusion limitations.

The net current passed by an electrode is the sum of currents through the two parallel branches shown in Figure 6.1, given by:

$$i_{total} = i_C + i_f \qquad (6.10)$$

where i_C is the current through the capacitance and i_f is the current through the faradaic element.

The current through the Faradaic element is given by the current–overpotential equation 6.8. The current through the capacitance is given by equation 6.11 below:

$$i_C = C_{dl} \, dv/dt = C_{dl} \, d\eta/dt \qquad (6.11)$$

The capacitive current depends upon the rate of potential change, but not the absolute value of the potential. The faradaic current, however, is exponentially dependent upon the overpotential. As an electrode is driven away from its equilibrium potential, essentially all charge initially flows through the capacitive branch since the overpotential is small. As the overpotential increases, the faradaic branch begins to conduct a relatively larger fraction of the injected current. When the overpotential becomes great enough, the faradaic impedance becomes sufficiently small that the faradaic current equals the injected current.

Methods of Controlling Charge Delivery During Pulsing

Charge injection from an electrode into an electrolyte (e.g. extracellular fluid) is commonly controlled by one of three methods. In the current controlled or

---

1. The relationship between capacitance and stored charge is given by equation 6.9. A one volt potential excursion applied to a double layer capacitance of $20 \, \mu F/cm^2$ yields $20 \, \mu C/cm^2$ stored charge, which is an order of magnitude lower than the total charge storage available from platinum pseudocapacitance.

galvanostatic method, a current source is attached between the working and counter electrodes and a user-defined current is passed. In the voltage controlled or potentiostatic method, current is driven between the working electrode and counter electrode as required to maintain the working electrode potential with respect to a third (reference) electrode. This method is generally not used for stimulation, and is not discussed further here. In the third method, V_{WE-CE} control, a voltage source is applied between the working and counter electrodes. While this is the simplest method to implement, neither the potential of the WE nor the CE (with respect to a third reference electrode) is controlled; only the net potential between the working and counter electrodes is controlled.

The current controlled method is commonly used for electrical stimulation of excitable tissue. In monophasic pulsing, a constant current is passed for a period of time (on the order of tens to hundreds of microseconds), then the external stimulator circuit is open-circuited until the next pulse. In biphasic pulsing, a constant current is passed in one direction, then the direction of current is reversed, and then the circuit is open-circuited until the next pulse. In biphasic pulsing, the first or stimulating phase is used to elicit the desired physiological effect, such as initiation of an action potential, then the second or reversal phase is used to reverse electrochemical processes occurring during the stimulating phase. It is common to use a cathodic pulse as the stimulating phase followed by an anodic reversal phase, although anodic pulsing may also be used for stimulation (see below, Charge injection for extracellular stimulation of excitable tissue).

Figure 6.3 illustrates key pulsing parameters. The frequency of stimulation is the inverse of the period, or time between the start of pulses. The interpulse

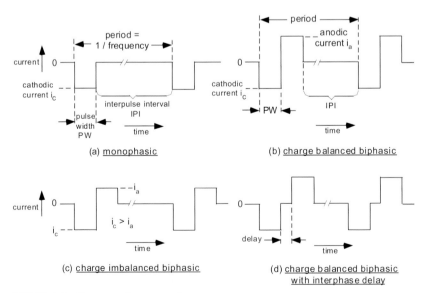

FIGURE 6.3   Common pulse types and parameters.

interval (IPI) is the period of time between pulses. Figure 6.3b illustrates charge balanced biphasic pulsing, where the charge in the stimulation phase equals the charge in the reversal phase. Figure 6.3c illustrates charge imbalanced biphasic pulsing where there are two phases, but the reversal phase has less charge than the stimulating phase. Figure 6.3d illustrates the use of an interphase delay, where an open-circuit is introduced between the stimulating and reversal phases.

Based on the simple electrical model of Figure 6.1, the relative properties of monophasic pulsing, charge balanced biphasic pulsing, and charge imbalanced biphasic pulsing during current control may be predicted. A complete description of mechanisms is given elsewhere [6]. The steady-state response to a train of monophasic pulses results in all injected charge per pulse going into irreversible faradaic reactions that occur during either the pulse or during the open-circuit interpulse interval period. The steady-state response to a train of charge balanced biphasic pulses occurs when one of the two following conditions is met:

1. there are no irreversible faradaic reactions during either the cathodic or anodic phases, and the electrode simply charges and then discharges the double layer, or
2. the same amount of charge is lost irreversibly during the cathodic phase and the anodic phase. The irreversible reactions in the two phases are distinct; for example oxygen reduction occurs during the cathodic phase and electrode corrosion occurs during the anodic phase.

Of the three pulsing protocols (monophasic, charge balanced biphasic, and charge imbalanced biphasic), monophasic pulsing causes the greatest accumulation of unrecoverable charge (corresponding to products of irreversible faradaic reactions) for a given injected charge.

An alternative form of charge injection involves the direct connection of a voltage source between the working and counter electrodes. Upon applying a voltage pulse between the working electrode and counter electrode in V_{WE-CE} control, the current is maximum at the beginning of the pulse as the double layer capacitances of the two electrodes charge and the current is predominantly capacitive. Given a long duration pulse, the current will asymptotically approach a value where the applied voltage maintains a steady-state faradaic current, with current density given by equation 6.8. An exhausting circuit [11,12] shorts together the WE and CE at the end of the monophasic voltage pulse, causing the charge on the working electrode capacitance to discharge rapidly, and the working electrode potential to attain the counter electrode potential. If the counter electrode is sufficiently large, its potential will not be notably perturbed away from its equilibrium potential during pulsing and, upon shorting the WE and CE, the working electrode potential will be brought back to the counter electrode equilibrium potential. The discharge of the working electrode is relatively rapid during V_{WE-CE} control with an exhausting circuit, as the working electrode is directly shorted to the counter electrode. This is contrasted by the relatively slow discharge using monophasic current control with an open circuit during the interpulse interval. During the open circuit period, the WE capacitance discharges through faradaic reactions at the working electrode interface. This leads to a greater accumulation of unrecoverable charge during the open circuit interpulse interval (current control) than with the short circuit interpulse interval (V_{WE-CE} control). However, in current control, appropriate biphasic pulsing waveforms can promote rapid electrode discharge.

Advantages of the V_{WE-CE} control scheme with an exhausting circuit over the current control scheme include:

1. the circuitry is simpler (it may be a battery and an electronic switch) and
2. unrecoverable charge accumulation is lower during the interpulse interval than it would be with monophasic current control.

Disadvantages of the V_{WE-CE} control scheme include:

1. maximum stimulation of excitable tissue occurs only at the beginning of the pulse when current is maximum, and stimulation efficiency decreases throughout the pulse as current decreases; whereas with current control the current is constant throughout the pulse
2. an increase in resistance anywhere in the electrical conduction path will cause an additional voltage drop, decreasing the current and potentially causing it to be insufficient for stimulation, whereas with current control the current is constant (assuming the required voltage is within the range of the stimulator); and
3. neither the current driven nor the charge injected are under direct control using voltage control [13].

Because the level of neuronal membrane depolarization is related to the applied current, these factors result in a reduction in reproducibility between stimulation sessions during V_{WE-CE} control. Moreover, because tissue properties can change over time, stimulation efficacy may change when using V_{WE-CE} control.

Electrochemical Reversal

The purpose of the reversal phase during biphasic stimulation is to reverse the direction of electrochemical processes that occur during the stimulating phase, minimizing unrecoverable charge. Upon delivering current in the stimulation phase and then reversing the direction of current, charge on the electrode capacitance will discharge, returning the electrode potential towards its pre-pulse value. If only double layer charging occurred, then upon passing an amount of charge in the reversal phase equal to the charge delivered in the stimulation phase (a charge balanced protocol), the electrode potential will return precisely to its pre-pulse potential by the end of the reversal phase and the potential curve will be a simple sawtooth as shown in Figure 6.4a. If reversible faradaic reactions occur during the stimulation phase, then charge in the reversal or secondary

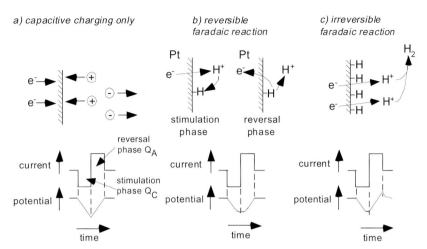

FIGURE 6.4 Electrochemical processes and potential waveforms during charge balanced stimulation. (a) Capacitive charging only; (b) reversible hydrogen plating; (c) irreversible hydrogen evolution.

phase may go into reversing these reactions. Figure 6.4b illustrates an example reversible faradaic process, in this case charging of the pseudocapacitance (reduction of protons and plating of monatomic hydrogen onto the metal electrode surface) as may occur on platinum. During reversal, the plated hydrogen is oxidized back to protons. Because the electrochemical process occurring during the reversal phase is the exact opposite of that occurring during the stimulation phase, there is zero net accumulation of electrochemical species. Reversible faradaic reactions include adsorption processes as in Figure 6.4b, as well as processes where the solution phase product remains near the electrode due to mass diffusion limitations. If irreversible faradaic reactions occur, upon passing current in the reverse direction, reversal of electrochemical product does not occur as the product is no longer available for reversal (it has diffused away). An example shown in Figure 6.4c is the formation of hydrogen gas after a mono-layer of hydrogen atoms has been adsorbed onto the platinum surface.

The use of biphasic stimulation (either charge balanced or charge imbalanced) moves the electrode potential out of the most negative ranges immediately after stimulation. In comparison, the cathodic-first monophasic stimulation protocol allows the electrode potential to remain relatively negative during the interpulse interval and, during this time, faradaic reduction reactions may continue. In the presence of oxygen, these reactions may include reduction of oxygen and formation of reactive oxygen species, which have been implicated in tissue damage [14–18]. The charge imbalanced waveform has the added advantage that the electrode potential at the end of the anodic pulse is less positive than with charge balanced biphasic pulsing, thus less charge goes into irreversible oxidation reactions, such as corrosion, when using the charge imbalanced protocol.

CHARGE INJECTION FOR EXTRACELLULAR STIMULATION OF EXCITABLE TISSUE

The goal of electrical stimulation of excitable tissue is often the triggering of action potentials in axons, which requires the artificial depolarization of some portion of the axon membrane to threshold. In the process of extracellular stimulation, the extracellular region is driven to relatively more negative potentials, equivalent to driving the intracellular compartment of a cell to relatively more positive potentials. Charge is transferred across the membrane due to both passive (capacitive and resistive) membrane properties as well as through active ion channels [19]. The process of physiological action potential generation is well reviewed in the literature (in particular see Principles of Neural Science by Kandel et al, 2000) [20]. The mechanisms underlying electrical excitation of nerve have been reviewed elsewhere [21–25]. In the simplest case of stimulation, a monopolar electrode (a single current carrying conductor) is placed in the vicinity of excitable tissue. Current passes from the electrode, through the extracellular fluid surrounding the tissue of interest and, ultimately, to a distant counter electrode.

During cathodic stimulation, the negative charge of a working electrode causes redistribution of charge on an axon membrane, with negative charge collecting on the outside of the membrane underneath the cathode (depolarizing the membrane). Associated with the depolarization of the membrane under the cathode is movement of positive charge intracellularly from the distant axon to the region under the electrode, and hyperpolarization of the membrane at a distance away from the electrode. If the electrode is instead driven as an anode (to more positive potentials), hyperpolarization occurs under the anode and depolarization occurs at a distance away from the anode. During such anodic stimulation, action potentials may be initiated at the regions distant from the electrode where depolarization occurs, known as virtual cathodes. The depolarization that occurs with anodic stimulation is roughly one seventh to one third that of the depolarization with cathodic stimulation; thus cathodic stimulation requires less current to bring an axon to threshold. During cathodic stimulation, anodic surround block may occur at sufficiently high current levels, where the hyperpolarized regions of the axon distant from the cathode may suppress an action potential that has been initiated near the electrode. This effect is observed at higher current levels than the threshold values required for initiation of action potentials with cathodic stimulation.

Selectivity is the ability to activate one population of neurons without activating a neighboring population. Spatial selectivity is the ability to activate a localized group of neurons, such as restricting activation to a certain fascicle or fascicles within a nerve trunk. Changes in the transmembrane potential due to electrical excitation are greatest in fibers closest to the stimulating electrode because the induced extracellular potential decreases in amplitude with distance from the stimulation electrode. Thus, activation of neurons closest to the electrode requires the least current. As the distance between the electrode and desired population of neurons for activation increases, larger currents are required, which generally means neurons between the electrode and desired population are also activated. Fiber diameter selectivity is the ability to activate fibers within a certain range of diameters only. Fibers with greater internodal distance and larger diameter experience greater changes in the transmembrane potential due to electrical excitation [26]. Using conventional electrical stimulation waveforms with relatively narrow pulses, the largest diameter fibers are activated at the lowest stimulus amplitude. In motor nerves, activating large diameter fibers first corresponds to activating the largest motor units first. This recruitment order is opposite of the physiological case where the smallest motor units are recruited first. Fang and Mortimer [27] have demonstrated a waveform that allows a propagated action potential in small diameter fibers but not large diameter fibers. Hyperpolarizing pulses have a greater effect on larger fibers than smaller, just as for depolarizing pulses. This means that sustained hyperpolarization can be used to block action potential initiation selectively in the large fibers, so that the corresponding depolarizing stimuli can selectively activate small fibers. Electrical stimulation protocols have also been developed [28] for triggering of action potentials in specific cell types (e.g. interneurons) and structures (e.g. nerve terminals).

The relationship between the strength (current) of an applied constant current pulse required to initiate an action potential and the duration of the pulse, known as the strength–duration curve, is shown in Figure 6.5a. The threshold current I_{th} decreases with increasing pulsewidth W. At very long pulsewidths, the current is a minimum, called the rheobase current I_{rh}. The qualitative nature of the strength–duration curve shown is representative of typical excitable tissue. The quantitative aspects, e.g. the rheobase current, depend upon factors such as the distance between the neuron population of interest and the electrode, and are determined empirically. Figure 6.5b illustrates the charge–duration curve, which plots the threshold charge $Q_{th} = I_{th}W$ versus pulsewidth. At longer pulsewidths, the required charge to elicit an action potential increases, due to two phenomena. First, over a period of tens to hundreds of microseconds, charge is redistributed

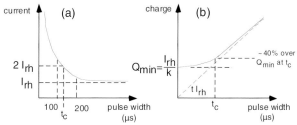

FIGURE 6.5   Strength–duration and charge–duration curves for initiation of an action potential. Rheobase current I_{rh} is the current required when using an infinitely long pulsewidth. Chronaxie time t_c is the pulse width corresponding to two times the rheobase current.

through the length of the axon, and does not all participate in changing the trans-membrane potential at the site of injection [29,30]. Second, over a period of several milliseconds, accommodation (increased sodium channel inactivation) occurs. The minimum charge Q_{min} occurs as the pulsewidth approaches zero. In practice, the Q_{th} is near Q_{min} when narrow pulses are used (tens of microseconds).

MECHANISMS OF DAMAGE

An improperly designed electrical stimulation system may cause damage to the tissue being stimulated or damage to the electrode itself. Damage to an electrode can occur in the form of corrosion if the electrode is driven anodically such that the electrode potential exceeds a value where significant metal oxidation occurs. Corrosion is an irreversible faradaic process. It may be due to dissolution where the electrochemical product goes into solution, or the product may form an outer solid layer on a passivation film that cannot be recovered. Charge balanced waveforms (see Fig. 6.3b) are more likely to reach potentials where corrosion may occur during the anodic reversal phase and the open circuit interpulse interval than are monophasic waveforms. The charge imbalanced waveform (see Fig. 6.3c) has advantages both in preventing tissue damage due to sustained negative potentials during the interpulse interval, and in preventing corrosion by reducing the maximum positive potential during the anodic reversal phase.

The mechanisms for stimulation-induced tissue damage are not well understood. Two major classes of mechanisms have been proposed. The first is that tissue damage is caused by intrinsic biological processes as excitable tissue is overstimulated. This is called the mass action theory, and proposes that damage occurs from the induced hyperactivity of many neurons firing, or neurons firing for an extended period of time, thus changing the local environment. Proposed mass action mechanisms include depletion of oxygen or glucose, or changes in ionic concentrations, both intracellularly and extracellularly, e.g. an increase in extracellular potassium. In the CNS, excessive release of excitatory neurotransmitters, such as glutamate, may cause excitotoxicity. The second proposed mechanism for tissue damage is the creation of toxic electrochemical reaction products at the electrode surface during cathodic stimulation at a rate greater than that which can be tolerated by the physiological system.

McCreery et al [31] have shown that both charge per phase and charge density are important factors in determining neuronal damage to cat cerebral cortex. The McCreery data show that as the charge per phase increases the charge density for safe stimulation decreases. When the total charge is small (as with a microelectrode), a relatively large charge density may safely be used. Shannon [32] reprocessed the McCreery data and developed an expression for the maximum safe level for stimulation, given by:

$$\log (Q/A) = k - \log (Q) \qquad (6.12)$$

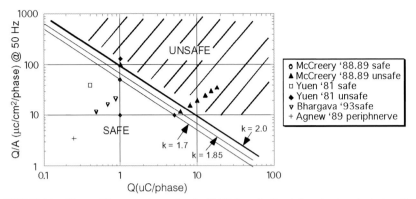

FIGURE 6.6 Charge (Q) versus charge density (Q/A) for safe stimulation. A microelectrode with relatively small total charge per pulse might safely stimulate using a large charge density, whereas a large surface area electrode (with greater total charge per pulse) must use a lower charge density.

where Q is charge per phase (μC/phase), Q/A is charge density per phase (μC/cm^2/phase), and 2.0>k>1.5, fit to the empirical data.

Figure 6.6 illustrates the charge versus charge density relationship of equation 6.12 using k values of 1.7, 1.85 and 2.0, with histological data from the 1990 McCreery study using cat parietal cortex, as well as data from Yuen et al [33] on cat parietal cortex, Agnew et al [34] on cat peroneal nerve, and Bhargava [35] on cat sacral anterior roots. Above the threshold for damage, experimental data demonstrates tissue damage, and below the threshold line, experimental data indicate no damage.

Supporting the concept that damage is due to electrochemical reaction products is the work by Lilly et al [36], which demonstrated that loss of electrical excitability and tissue damage occurs when the cerebral cortex of monkey is stimulated using monophasic current pulses. Later, Lilly et al [37] showed that biphasic stimulation caused no loss of excitability or tissue damage after 15 weeks of stimulation for 4–5 hours per day. The concept that monophasic is a more damaging form of stimulation than charge balanced biphasic was confirmed by Mortimer et al [38], who reported that breakdown of the blood–brain barrier during stimulation of the surface of cat cerebral cortex occurs when monophasic pulses were used at power densities greater than 0.003 W/in^2 (0.5 mW/cm^2), but does not occur with charge balanced biphasic pulses until a power density of 0.05 W/in^2 (8 mW/cm^2) is exceeded. Pudenz et al [39,40] further showed that monophasic stimulation of the cat cerebral cortex causes vasoconstriction, thrombosis in venules and arterioles and blood–brain barrier breakdown within 30 seconds of stimulation when used at levels required for a sensorimotor response, however, charge balanced biphasic stimulation could be used for up to 36 hours continuously without tissue damage if the charge per phase was below 0.45 μC (4.5 μC/cm^2). Also supporting the hypothesis that damage is due to electrochemical products are observations of cat muscle that suggest some non-zero level of reaction product can be tolerated [41,42]. Scheiner and Mortimer [41] studied the utility of charge imbalanced biphasic stimulation, demonstrating that this waveform allows greater cathodic charge densities than monophasic prior to the onset of tissue damage as reactions occurring during the cathodic phase are reversed by the anodic phase, and also that greater cathodic charge densities can be used than with the charge balanced waveform prior to electrode corrosion since the anodic phase is no longer constrained to be equal to the cathodic phase, thus the electrode potential reaches less positive values during the anodic phase and interpulse interval.

In 1975, Brummer and Turner [43] proposed that two principles should be followed to achieve electrochemically safe conditions during tissue stimulation:

1. perfect symmetry of the electrochemical processes in the two half-waves of the pulses should be sought. This implies that we do not generate any electrolysis products in solution. One approach to achieve this would appear to involve the use of perfectly charge-balanced waveforms of controlled magnitude.
2. The aim should be to inject charge via non-faradaic or surface-faradaic processes, to avoid injecting any possibly toxic materials into the body.

Their model for safe stimulation interprets the charge balanced waveform in electrochemical terms. Any process occurring during the first (stimulating) phase, whether it is charging of the electrode or a reversible faradaic process, is reversed during the second (reversal) phase, with no net charge delivered. The observation that monophasic stimulation causes greater tissue damage than biphasic stimulation at the same amplitude, pulsewidth and frequency is explained by the fact that, during monophasic stimulation, all injected charge results in generation of electrochemical reaction products.

Reversible processes include charging and discharging of the double layer capacitance, as well as surface bound reversible faradaic processes such as reactions 6.2 through 6.4. Reversible reactions often involve the production or consumption of hydrogen or hydroxyl ions as the charge counterion. This causes a change in the pH of the solution immediately adjacent to the electrode surface. Ballestrasse et al [44] gave a mathematical description of these pH changes, and determined that the pH may range from 4 to 10 near a 1 μm diameter electrode during biphasic current pulses, but this change extended for only a few microns. Irreversible processes include faradaic reactions where the product does not remain near the electrode surface, such as reactions 6.1, 6.5 and 6.6.

Free radicals are known to cause damage to myelin, the lipid cell membrane and DNA of cells. A likely candidate for a mechanism of neural tissue damage due to electrochemical products is peroxidation of the myelin by free radicals produced on the electrode surface. Several researchers [45–50] have demonstrated the great susceptibility of myelin to free radical damage.

Morton et al [51] have shown that oxygen reduction occurs on a gold electrode in phosphate buffered saline under typical neural stimulating conditions.

Oxygen reduction reactions that may occur during the cathodic stimulating phase include reactions that generate free radicals, such as superoxide and hydroxyl, and hydrogen peroxide, collectively known as reactive oxygen species. These species may have multiple deleterious effects on tissue [14–18]. As free radicals are produced they may interfere with chemical signaling pathways that maintain proper perfusion of nervous tissue. Nitric oxide has been identified as the endothelium-derived relaxing factor, the primary vasodilator [52–54]. Nitric oxide is also known to prevent platelet aggregation and adhesion [55–57]. Beckman et al. [58] have shown that the superoxide radical reacts with nitric oxide to form the peroxynitrite radical. Oxygen-derived free radicals from the electrode may reduce the nitric oxide concentration and diminish its ability as the principal vasodilator and as an inhibitor of platelet aggregation. Superoxide depresses vascular smooth muscle relaxation by inactivating nitric oxide, as reviewed by Rubanyi [59].

An electrochemical product may accumulate to detrimental concentrations if the rate of faradaic reaction, given by the current–overpotential relationship of equation 6.8, exceeds the rate for which the physiological system can tolerate the product. For most reaction products of interest there is some sufficiently low concentration near the electrode that can be tolerated over the long term. This level for a tolerable reaction may be determined by the capacity of an intrinsic buffering system. For example, changes in pH are buffered by several systems including the bicarbonate buffer system, the phosphate buffer system, and intracellular proteins. The superoxide radical, a product of the reduction of oxygen, is converted by superoxide dismutase and cytochrome c to hydrogen peroxide and oxygen. The diffusion rate of a toxic product must be considered, as it may be the case that high concentrations only exist very near the site of generation (the electrode surface).

DESIGN COMPROMISES FOR EFFICACIOUS AND SAFE ELECTRICAL STIMULATION

A stimulating system must be both efficacious and safe. Efficacy of stimulation generally means the ability to elicit the desired physiological response, which can include initiation or suppression of action potentials. Safety has two primary aspects. First, the tissue being stimulated must not be damaged and, second, the stimulating electrode itself must not be damaged, as in corrosion. An electrode implanted into a human as a prosthesis may need to meet these requirements for decades.

Efficacy requires that the charge injected must exceed some threshold (see Fig. 6.5). However, as the charge per pulse increases, the overpotential of the electrode increases, as does the fraction of the current going into faradaic reactions (which may be damaging to tissue or the electrode). Judicious design of stimulation protocols involves acceptable compromises between stimulation efficacy, requiring a sufficiently high charge per pulse, and safety, requiring a sufficiently low charge per pulse, thus preventing the electrode from reaching potentials where deleterious faradaic reactions occur at an intolerable rate. The overpotential an electrode reaches, and thus faradaic reactions that can occur, depend on several factors in addition to the charge per pulse, including:

1. waveform type (see Fig. 6.3)
2. stimulation frequency
3. electrode material (a high charge storage capacity allows relatively large charge storage prior to reaching overpotentials where irreversible faradaic reactions occur)
4. electrode geometric area and roughness (determining real area) and therefore total capacitance
5. train effects.

Increasing either the stimulus phase pulsewidth or the reversal phase pulsewidth of a charge balanced stimulation protocol has the effect of increasing unrecoverable charge into irreversible reactions. Any factor which either drives the electrode potential into a range where irreversible reactions occur (such as a long stimulus phase pulsewidth) or fails to reverse quickly the electrode potential out of this range (such as a long reversal phase pulsewidth) will allow accumulation of unrecoverable charge.

The fundamental design criteria for an electrochemically safe stimulation protocol can be stated:

the electrode potential must be kept within a potential window where irreversible faradaic reactions do not occur at levels that are intolerable to the physiological system or the electrode.

If irreversible faradaic reactions do occur, one must ensure that they can be tolerated (e.g. that physiological buffering systems can accommodate any toxic products) or that their detrimental effects are low in magnitude (e.g. that corrosion occurs at a very slow rate, and the electrode will last for longer than its design lifetime).

The charge–duration curve shown in Figure 6.5 demonstrates that to minimize the total charge injected in an efficacious stimulation protocol, one should use short duration pulses. In practice, pulses on the order of tens of microseconds approach the minimum charge, and are often reasonable design solutions. During this relatively short duration one may be able to avoid faradaic reactions that would occur at higher levels of total charge with longer pulses. While it is desirable to use short duration pulses on the order of tens of microseconds, there are applications for which biological constraints require longer duration pulses. The time constants of several key ion channels in the membranes of excitable tissue are measured in hundreds of microseconds to milliseconds. By using stimulating pulses with comparable durations one can selectively manipulate the opening and closing of these ion channels. Grill and Mortimer [60] have reviewed stimulus waveforms used for spatial and fiber diameter selective neural stimulation, illustrating the response of the neural membrane to different waveforms. Selective waveforms often require stimulation or reversal phases with long pulsewidths relative to conventional stimulus waveforms; thus wave-forms optimized for physiological responses may not be efficient for reversing electrochemical processes. Researchers have demonstrated the ability to inactivate selectively the larger neurons in a nerve trunk [61], selectively inactivate the superficial fibers in a nerve by pre-conditioning [62], and prevent anodic break. Lastly, there are applications where tonic polarization mandates the use of very long (>1 s) monophasic pulses; for example tonic hyperpolarization of the soma to control epileptic activity [63,64]. The use of these various waveforms with long pulsewidths allows greater accumulation of any electro-chemical product, thus requiring additional diligence by a neurophysiologist or prosthesis designer to prevent electrochemical damage.

In addition to biological constraints on the pulse durations, the required current for a short pulsewidth may also be a limitation. In order to inject the minimum charge required for effect, a large current is required (see Fig. 6.5). This is not always possible, as may be the case with a battery-powered stimulator with limited current output.

Figure 6.7 summarizes key features of various stimulation waveform types. The cathodic monophasic waveform, illustrated in Figure 6.7a, consists of

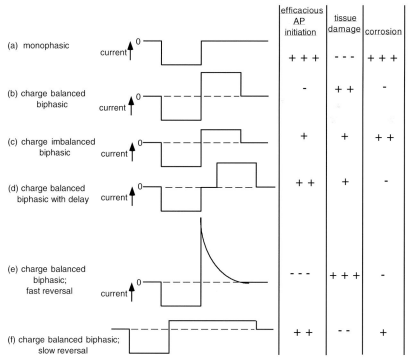

FIGURE 6.7 Comparison of stimulating waveforms. Six prototypical waveforms are rated for relative merit in efficacy and safety. +++=best (most efficacious, least damaging to tissue or the electrode), ---=worst.

pulses of current passed in one direction, with an open-circuit during the inter-pulse interval. At no time does current pass in the opposite direction. Commonly, the working electrode is pulsed cathodically for stimulation of tissue (as shown), although anodic stimulation may also be used. Of the waveforms illustrated in Figure 6.7, the monophasic is the most efficacious for stimulation. However, monophasic pulses are not used in long-term stimulation where tissue damage is to be avoided. Greater overpotentials are reached during monophasic puls-ing than with biphasic pulsing. Furthermore, the electrode potential during the interpulse interval of cathodic monophasic pulsing remains relatively negative as the charged electrode capacitance slowly discharges through faradaic reac-tions, allowing reduction reactions which may be deleterious to tissue to proceed throughout the entire period of stimulation. Biphasic waveforms are illustrated in Figure 6.7b–f. The first (stimulating) phase elicits the desired physiological effect, such as initiation of an action potential, and the second (reversal) phase is used to reverse the direction of electrochemical processes occurring during the stimulating phase. If all processes of charge injection during the stimulat-ing phase are reversible, then the reversal phase will prevent net changes in the chemical environment of the electrode, as desired. The charge balanced bipha-sic waveform (Fig. 6.7b) is widely used to prevent tissue damage. It should be noted that charge balance does not necessarily equate to electrochemical balance. During certain instances of stimulation there are irreversible faradaic reactions during the cathodic phase (e.g. oxygen reduction), and then differ-ent irreversible reactions during the anodic phase (e.g. electrode corrosion) that are not the reverse of the cathodic faradaic reactions. Such electrochemi-cal imbalance leads to a waveform where the potential at the end of the anodic phase is positive of the pre-pulse potential, allowing irreversible reactions such as electrode corrosion to occur. The charge imbalanced waveform, illustrated in Figure 6.7c, may be used to reduce the most positive potentials during the anodic phase with respect to the charge balanced waveform, and prevent elec-trode corrosion [41].

In addition to electrode corrosion, a second concern with the charge balanced biphasic waveform is that the reversal phase not only reverses electrochemical processes of the stimulation phase, but may also reverse some of the desired phys-iological effect of the stimulation phase, i.e. it may suppress an action potential that would otherwise be induced by a monophasic waveform. This effect causes an increased threshold for biphasic stimulation relative to monophasic. Gorman and Mortimer [65] have shown that by introducing an open-circuit interphase delay between the stimulating and reversal phases, the threshold for biphasic stimulation is similar to that for monophasic. This is illustrated in Figure 6.7d. Although the introduction of an interphase delay improves threshold, it also allows the electrode potential to remain relatively negative during the delay period. A delay of $100\,\mu s$ is typically sufficient to prevent the suppressing effect of the reversal phase, and may be a short enough period that deleterious faradaic reaction products do not accumulate to an unacceptable level.

As illustrated in Figure 6.7e and f, the more rapidly charge is injected during the anodic reversal phase, the more quickly the electrode potential is brought out of the most negative range, and thus the less likely that tissue damage will occur. A high current reversal phase, however, means more of a suppressing effect on action potential initiation, and also means the electrode potential will move positive during the reversal phase, thus risking electrode corrosion.

When evaluating the electrochemistry of a stimulating electrode system, both the working electrode and counter electrode should be considered. If the area, and thus total capacitance, of a counter electrode is relatively large, there is a small potential change for a given amount of injected charge. Such an electrode will not be perturbed away from its resting potential as readily as a small electrode, and all charge injection across this large counter electrode is assumed to be by capacitive charging, not faradaic processes.

MATERIALS USED AS ELECTRODES FOR CHARGE INJECTION AND REVERSIBLE CHARGE STORAGE CAPACITY

General Requirements, Biocompatibility and Reversible Charge Storage Capacity

The ideal material for use as a stimulating electrode satisfies the following six requirements:

1. the passive (unstimulated) material must be biocompatible, so it should not induce a toxic or necrotic response in the adjacent tissue, nor an excessive foreign body or immune response
2. the material must be mechanically acceptable for the application. It must maintain mechanical integrity given the intended tissue, surgical procedure and duration of use. The material must not buckle if it is to pass through the meninges. If a device is to be used chronically, it must be flexible enough to withstand any small movement between the device and tissue following implantation
3. the complete device must be efficacious. This requires that sufficient charge can be injected with the chosen material and electrode area to elicit action potentials. The required charge is quantified by the charge–duration curve
4. during electrical stimulation, faradaic reactions should not occur at levels that are toxic to the surrounding tissue. The level of reaction product that is tolerated may be significantly higher for acute stimulation than chronic stimulation
5. during electrical stimulation, faradaic corrosion reactions should not occur at levels that will cause premature failure of the electrode. This again depends greatly on the intended duration of use. During acute stimulation corrosion is rarely a concern, whereas a device that is intended for a 30-year implant must have a very low corrosion rate

6. the material characteristics must be acceptably stable for the duration of the implant. For a chronic electrode, the device electrical impedance must be stable. The conducting and insulating properties of all materials must remain intact.

Dymond et al [66] tested the toxicity of several metals implanted into the cat cerebral cortex for 2 months. Materials were deemed toxic if the reaction to the implanted metal was significantly greater than the reaction to a puncture made from the same metal that was immediately withdrawn (Table 6.1). Stensaas and Stensaas [67] reported on the biocompatibility of several materials implanted passively into the rabbit cerebral cortex (Table 6.1). Materials were classified into one of three categories depending upon changes occurring at the implant/cortex interface:

1. Non-reactive. For these materials, little or no gliosis occurred, and normal CNS tissue with synapses was observed within 5 μm of the interface
2. Reactive. Multinucleate giant cells and a thin layer (10 μm) of connective tissue surrounded the implant. Outside of this was a zone of astrocytosis. Normal CNS tissue was observed within 50 μm of the implant
3. Toxic. These materials are separated from the cortical tissue by a capsule of cellular connective tissue and a surrounding zone of astrocytosis.

TABLE 6.1 Classification of Material Biocompatibility

	Classification by Dymond [66]	Classification by Stensaas and Stensaas [67]	Other References
Conductors:			
Aluminum		Non-reactive	
Cobalt		Toxic	
Copper		Toxic	Toxic [69–71]
Gold	Non-toxic	Non-reactive	
Gold–nickel–chromium	Non-toxic		
Gold–palladium–rhodium	Non-toxic		
Iron		Toxic	
Molybdenum		Reactive	
Nickel–chromium (Nichrome)		Reactive	Non-toxic [69]

	Classification by Dymond [66]	Classification by Stensaas and Stensaas [67]	Other References
Nickel–chromium–molybdenum	Non-toxic		
Nickel–titanium (Nitinol)			Biocompatible [72,73]
Platinum	Non-toxic	Non-reactive	Biocompatible [74,75]
Platinum–iridium	Non-toxic		Biocompatible [76]
Platinum–nickel	Non-toxic		
Platinum–rhodium	Non-toxic		
Platinum–tungsten	Non-toxic		
Platinized platinum (Pt black)	Non-toxic		
Rhenium	Non-toxic		
Silver	Toxic	Toxic	Toxic [69–71]
Stainless steel	Non-toxic	Non-toxic [69]	
Tantalum		Reactive	
Titanium			Biocompatible [75]
Tungsten		Non-reactive	
Insulators:			
Alumina ceramic		Non-reactive	Biocompatible [75]
Araldite (epoxy plastic resin)		Reactive	
Polyethylene		Non-reactive	
Polyimide			Biocompatible [77]
Polypropylene		Non-reactive	
Silastic RTV	Toxic		
Silicon dioxide (Pyrex)		Reactive	

Continued

	Classification by Dymond [66]	Classification by Stensaas and Stensaas [67]	Other References
Teflon TFE (high purity)		Non-reactive	
Teflon TFE (shrinkable)		Reactive	
Titanium dioxide		Reactive	
Semiconductors:			
Germanium		Toxic	
Silicon		Non-reactive	Biocompatible [78–80]
Assemblies:			
Gold–silicon di-oxide passivated microcircuit		Reactive	

Loeb [68] studied the histological response to materials used by the micro-electronics industry implanted chronically in the subdural space of cats, and found reactions to be quite dependent on specific material formulations and surface preparations.

Platinum has been demonstrated as biocompatible for use in an epiretinal array [74] and in cochlear implants [75]. Both titanium and ceramic [75] and platinum–iridium wire [76] have been shown as biocompatible in cochlear implants. Babb and Kupfer [69] have shown stainless steel and nickel–chromium (Nichrome) to be non-toxic. Copper and silver are unacceptable as stimulating electrodes, as these metals cause tissue necrosis even in the absence of current [66–71]. Nickel–titanium shape memory alloys have good biocompatibility response [72], up to a nickel content of 50% [73].

The first intracortical electrodes consisted of single site conductive micro-electrodes made of material stiff enough to penetrate the meninges, as either an insulated metallic wire or a glass pipette filled with conductive electrolyte. Advances in materials science and microelectronics technology have allowed the development of multiple site electrodes built onto a single substrate, using planar photolithographic and silicon micromachining technologies. Such devices have been made from silicon [78,81], and polyimide [82]. In further advancements, bioactive components have been added to the electrode to direct neurite growth toward the electrode, minimizing the distance between the electrode and stimulated tissue [83–85].

Chronic implantation of any device into the central nervous system, even those materials considered biocompatible, elicits a common response consisting of encapsulation by macrophages, microglia, astrocytes, fibroblasts, endothelia and meningeal cells [86]. The early response to material implantation is inflammation [67,86,87]. The chronic response is characterized by a hypertrophy of the surrounding astrocytes [67] which display elevated expression of intermediate filament proteins such as glial fibrillary acidic protein (GFAP) and vimentin [88], an infiltration of microglia and foreign body giant cells [67] and a thickening of the surrounding tissue that forms a capsule around the device [78,87].

The reversible charge storage capacity (CSC) of an electrode, also known as the reversible charge injection limit [89], is the total amount of charge that may be stored reversibly, including storage in the double layer capacitance, pseudocapacitance, or any reversible faradaic reaction. In electrical stimulation of excitable tissue, it is desirable to have a large reversible charge storage capacity so that a relatively large amount of charge may be injected (thus being efficacious for stimulation) prior to the onset of irreversible faradaic reactions (which may be deleterious to the tissue being stimulated or to the electrode itself). The reversible charge storage capacity depends upon the material used for the electrode, the size and shape of the electrode, the electrolyte composition, and parameters of the electrical stimulation waveform.

In many studies, the CSC has been defined as the maximum charge density that can be applied without the electrode potential exceeding the water window (the potential range over which there is no reduction or oxidation of water) during pulsing. In fact, irreversible processes may occur at potentials within the water window, including such reactions as irreversible oxygen reduction [90,91].

Noble Metals, Stainless Steel and Capacitor Electrodes

The noble metals, including platinum, gold, iridium, palladium, and rhodium, have been commonly used for electrical stimulation, largely due to their relative resistance to corrosion [66,92,93]. These noble metals do exhibit some corrosion during electrical stimulation, as shown by dissolution [94–98] and the presence of metal in the neighboring tissue [99,100]. In addition to corrosion of the electrode, there is evidence of long-term toxic effects on the tissue from dissolution [101–103].

Platinum and platinum–iridium alloys are common materials used for electrical stimulation of excitable tissue. Brummer and Turner [43,104–106] have reported on the electrochemical processes of charge injection using a platinum electrode. They reported that three processes could store charge reversibly, including charging of the double layer capacitance, hydrogen atom plating and oxidation (pseudocapacity, reaction 6.4) and reversible oxide formation and reduction on the electrode surface, and that 300–$350\,\mu C/cm^2$ (real area) could theoretically be stored reversibly by these processes in artificial cerebrospinal fluid (equivalently 420–$490\,\mu C/cm^2$ (geometric area)). This is a maximum reversible charge storage capacity under optimum conditions, including relatively long pulsewidths

(>0.6 ms). Rose and Robblee [107] reported on the charge injection limits for a platinum electrode using 200 μs charge balanced biphasic pulses. The reversible charge injection limit was defined as the maximum charge density that could be applied without the electrode potential exceeding the water window during pulsing. The authors determined the charge injection limit to be 50–100 μC/cm^2 (geometric) using anodic first pulses, and 100–150 μC/cm^2 (geometric) using cathodic first pulses. These values are considerably lower than the theoretical values determined by Brummer and Turner [106], since the electrode potential at the beginning of a pulse begins somewhere intermediate to oxygen and hydrogen evolution and not all of the three reversible processes accommodate charge during the stimulating pulse. Dissolution of platinum in saline increases linearly with the injected charge during biphasic stimulation [96]. Anodic first pulses cause more dissolution than cathodic first pulses, as the electrode potential attains more positive values during the stimulating (first) phase. Robblee et al [97] have shown that in the presence of protein such as serum albumin, the dissolution rate of platinum decreases by an order of magnitude.

Platinum is a relatively soft material and may not be mechanically acceptable for all stimulation applications. Platinum is often alloyed with iridium to increase the mechanical strength. Alloys of platinum with 10–30% iridium have similar charge storage capacity to pure platinum [98]. Iridium is a much harder metal than platinum, with mechanical properties that make it suitable as an intracortical electrode. The reversible charge storage capacities of bare iridium or rhodium are similar to that of platinum. However, when a surface oxide is present on either of these materials, they have greatly increased charge storage capacity over platinum. These electrodes inject charge using valency changes between two oxide states, without a complete reduction of the oxide layer.

Iridium oxide is a popular material for stimulation and recording, using reversible conversion between Ir^{3+} and Ir^{4+} states within an oxide to achieve high reversible charge storage capacity. Iridium oxide is commonly formed from iridium metal in aqueous electrolyte by electrochemical activation (known as anodic iridium oxide films on bulk iridium metal, or AIROF), which consists of repetitive potential cycling of iridium to produce a multilayered oxide [89,98, 108–110]. Such activated iridium oxide films have been used for intracortical stimulation and recording using iridium wire [111–115] or with micromachined silicon electrodes using sputtered iridium on the electrode sites [116,117]. The maximum charge density that can be applied without the electrode potential exceeding the water window was reported for activated iridium oxide using 200 μs charge balanced pulses as ±2 mC/cm^2 (geometric) for anodic first pulsing and ±1 mC/cm^2 for cathodic first [118,119]. By using an anodic bias, cathodic charge densities of 3.5 mC/cm^2 (geometric) have been demonstrated both in vitro [118,119] and in vivo [120]. Iridium oxide films can also be formed by thermal decomposition of an iridium salt onto a metal substrate (known as thermally prepared iridium oxide films, or TIROF) [121], or by reactive sputtering of iridium onto a metal substrate (known as sputtered iridium oxide films, or SIROF) [122].

Meyer and Cogan [115] reported on a method to electrodeposit iridium oxide films onto substrates of gold, platinum, platinum–iridium and 316LVM stainless steel, achieving reversible charge storage capacities of $>25\,mC/cm^2$.

The stainless steels (types 303, 316 and 316LVM) as well as the cobalt–nickel–chromium–molybdenum alloy MP35N are protected from corrosion by a thin passivation layer that develops when exposed to atmospheric oxygen and which forms a barrier to further reaction. In the case of stainless steel, this layer consists of iron oxides, iron hydroxides and chromium oxides. These metals inject charge by reversible oxidation and reduction of the passivation layers. A possible problem with these metals is that if the electrode potential becomes too positive (the transpassive region), breakdown of the passivation layer and irreversible metal dissolution may occur at an unacceptable rate [92,123,124], potentially leading to failure of the electrode. A cathodic charge imbalance has been shown to allow significantly increased charge injection without electrode corrosion [41,125]. Titanium and cobalt–chromium alloys are also protected from corrosion by a surface oxide passivation layer, and demonstrate better corrosion resistance than does stainless steel [126]. 316LVM stainless steel has good mechanical properties and has been used for intramuscular electrodes. The charge storage capacity of 316LVM is only 40–$50\,\mu C/cm^2$ (geometric), potentially necessitating large surface area electrodes.

Capacitor electrodes inject charge strictly by capacitive action, as a dielectric material separates the metal electrode from the electrolyte, preventing Faradaic reactions at the interface [127–129]. The tantalum/tantalum pentoxide (Ta/Ta_2O_5) electrode has a high charge storage capacity, achieved by using sintered tantalum or electrolytically etched tantalum wire to increase the surface area [130]. Guyton and Hambrecht [127,128] have demonstrated a sintered Ta/Ta_2O_5 electrode with a charge storage capacity of $700\,\mu C/cm^2$ (geometric). The Ta/Ta_2O_5 electrodes have sufficient charge storage capacity for electrodes in the range of $0.05\,cm^2$ and charge densities up to $200\,\mu C/cm^2$ (geometric), however, they may not be acceptable for microelectrode applications where the required charge densities may exceed $1\,mC/cm^2$ [129]. Tantalum capacitor electrodes must operate at a relatively positive potential to prevent electron transfer across the oxide. If pulsed cathodically, a positive bias must be used on the electrode.

Table 6.2 lists several parameters of interest for materials commonly used for stimulation.

Inherently Conducting Polymers

Although platinum and iridium oxide possess several desirable characteristics, including low impedance, high charge injection capacity and high corrosion resistance, their interface with neural tissue may be suboptimal. As an alternative interface for neural stimulation, inherently conducting polymers (ICPs) have been developed including polypyrrole (PPy), polyaniline (PANi), and

TABLE 6.2 Reversible Charge Storage Capacity and Other Parameters in Electrode Material Selection

	Reversible Charge Storage Capacity ($\mu C/cm^2$)	Reversible Charge Injection Processes	Corrosion Characteristics	Mechanical Characteristics
Platinum AF, 200 μs: CF, 200 μs:	300–350 r [106] 50–100 g [107] 100–150 g [107]	double layer charging, hydrogen atom plating, oxide formation and reduction	relatively resistant; and greatly increased resistance with protein	relatively soft
Platinum/iridium alloys	Similar CSC to Pt			stronger than Pt
Iridium	Similar CSC to Pt			stronger than Pt
Iridium oxide	AF: ± 2200 g [118,119] CF: ± 1200 g [118,119] AB: ± 3500 g [118–120]	oxide valency changes	highly resistant [98,120]	
316LVM Stainless steel	40–50 g	passive film formation and reduction	resistant in passive region; rapid breakdown in transpassive region	strong and flexible
Tantalum/Tantalum pentoxide	700 g [127,128] 200 g [129]	capacitive only	corrosion resistant [130–133]	

r = real area; g = geometric area; AF = anodic first, charge balanced; CF = cathodic first, charge balanced; AB = cathodic first, charge balanced, with anodic bias.

poly(3, 4-ethylenedioxythiophene) (PEDOT). Advantages of conducting polymers over metal electrodes include improved biocompatibility, higher charge injection capacity and lower electrode impedance.

Conducting polymers are often electrochemically deposited onto metal seed layers such as platinum and gold. The adhesion of the polymer to the metal is critical for stability of an implant. Thicker films display a lower impedance and larger charge capacity, however, as the thickness increases there is a greater likelihood of cracking and delamination.

Polypyrrole, polythiophene, and their derivatives can be electrochemically polymerized from aqueous solution and deposited on neural electrodes [134–140]. Bioactive molecules, such as cell adhesion molecules, extracellular matrix proteins, and growth factors, can be incorporated into the polymer to promote neuronal growth and binding to the electrode [134–137,141,142]. The rough surface area of the polymer may induce a smaller inflammatory response with respect to a smooth metal implant, thus offering superior implant integration with the surrounding tissue. Polymers can be modified with peptides and proteins [141,143], polysaccharides, and living cells [144,145]. A conducting polymer's surface and structure may be modified to improve charge injection capacity and improve biocompatibility.

The properties of PPy can be modified by various dopants and preparation methods. A common dopant is polystyrenesulfonate (PSS). PEDOT along with dopant PSS is stable after hundreds of cyclic voltammetric scans, and PEDOT/PSS coating decreases electrode impedance by almost two orders of magnitude [136,137]. PEDOT is more electrochemically stable than PPy. Cui and Zhou [146] reported on PEDOT electrochemically deposited onto thin-film platinum stimulating electrodes. These coated electrodes displayed much lower impedance than the thin-film platinum due to the high surface area and high ion conductivity of the film, and had a charge injection limit of $2.3\,mC/cm^2$, similar to iridium oxide and much higher than thin-film platinum.

SIZE AND SHAPE CONSIDERATIONS

As previously detailed, a central tenet in stimulation design is a compromise between efficacy, requiring a sufficiently high charge per pulse, and safety, requiring a sufficiently low charge per pulse. As an additional factor in evaluating efficacy and safety, one must consider not only charge but also charge density (charge divided by electrode area). In terms of efficacy, the total charge per pulse determines the volume within which neurons are excited, and the charge density determines the proportion of neurons close to an electrode that are excited.

The McCreery data [31] shown in Figure 6.6 demonstrate an inverse relationship: as the charge per pulse increases the allowable charge density for safe stimulation decreases. When the total charge is small (as with a microelectrode), a relatively large charge density may safely be used. Conversely, for a given injected charge, increasing the electrode area decreases the charge density and moves the protocol towards the safe region. However, increasing electrode area decreases selectivity, recruiting neurons over a larger volume. Often, the maximum allowable electrode area is determined according to the volume of tissue to be stimulated. If an electrode area is constrained to some small value due to anatomical reasons, achieving both efficacy and safety may require use of a material with high CSC.

Stimulation of muscle, peripheral nerve or cortical surface requires relatively high charge per pulse (ca. 0.2–5 μC), thus platinum or stainless steel electrodes must be of fairly large surface area (on the order of 1 mm^2) to stay within their reversible charge storage capacity. Such electrodes are often fabricated as flat or slightly curved surfaces to lie against a tissue plane (e.g. epimysial, epineurial, epidural or subdural surfaces). Intracortical stimulation requires much less total charge per pulse, however, in order to achieve selective stimulation, the electrode size must be very small, resulting in high charge density requirements. A penetrating microelectrode with a geometric surface area of 2000 μm^2 may have a charge per pulse on the order of 0.008–0.064 μC, yielding a charge density of 400–3200 μC/cm^2 [120,147]. Such high charge densities may be achieved using iridium oxide electrodes with anodic pulses, or cathodic pulses with an anodic bias. Microelectrode designs typically yield electrode surface areas of several hundred to a few thousand μm^2. Example technologies include the Utah electrode array [148–150], the Michigan array [151,152], and microwires [153].

The reversible charge storage capacity is dependent upon the electrode real surface area and geometry. The geometric area of an electrode is usually easily calculated, but the real area (equal to the geometric area multiplied by the roughness factor) is the value that determines the total charge capacity. The real area of an electrode may change during the course of stimulation.

The current density emanating from a spherical (ball) electrode to a distant counter electrode is uniform. Any deviation from a spherical geometry causes a non-uniform current density [154], with increased local current density at a discontinuity or edge. Since current density is directly proportional to electrochemical reaction rate normalized to an area, these locations of increased current density may be prone to localized electrode corrosion [155]. Increased current density at an electrode edge is also associated with localized tissue burns [156].

One method of directing current from an electrode to target tissue is by selectively covering the conducting electrode surface with an insulator, such as silicone or polyimide, and creating apertures for current flow. If such an electrode is connected to a voltage source, the effect of decreasing the aperture size (thus making stimulation more selective) is to increase resistance presented to the voltage source and decrease the net current flowing from the electrode. If the electrode is connected to a current source, the effect of decreasing the aperture size is to increase the current density through the aperture, with increased local stimulation efficacy, but at the risk of increased electrode corrosion, tissue heating, and tissue damage.

Another method of selectively directing current is simply to bend a flat electrode to match a tissue contour, for example by forming a semi-cylindrical shape onto the spinal cord in the epidural space. A simple advantage of this method is to mitigate the distancing that necessarily occurs between a flat electrode and a curved target tissue surface. Also, if the curved electrode is judiciously designed without sharp edges (thus preventing localized increases in current density on the electrode surface), one can sidestep an increase in reaction rate driving electrode corrosion; yet from the perspective of tissue the effective 'rain' of current density increases into the volume, improving stimulation efficacy.

If multiple contacts are used, for example to steer currents, one must exercise caution that electrode selection does not inadvertently result in high current density and possible electrode corrosion, tissue heating or tissue damage. Assume a particular selection of electrodes is balanced in the sense that there is no electrode or tissue damage over the long term. Later, an adjustment is made so that electrode A no longer carries current, increasing the flow of current from electrodes B and C. If B and C were previously just under a safe limit for net charge delivery to prevent corrosion over a specified lifetime, the removal of electrode A may unwittingly move electrodes B or C into an unsafe region.

REFERENCES

1. Helmholtz von HLF. Ueber einige gesetze der vertheilung elektrischer strome in korperlichen leitern mit anwendung auf die thierisch-elektrischen versuche. *Ann Physik*. 1853;89:211–233.
2. Guoy G. Constitution of the electric charge at the surface of an electrolyte. *J Physique*. 1910;9:457–467.
3. Chapman DL. A contribution to the theory of electrocapillarity. *Philos Mag*. 1913;25:475–481.
4. Stern O. Zur theorie der elektrolytischen doppelschicht. *Z Elektrochem*. 1924;30:508–516.
5. Grahame DC. The electrical double layer and the theory of electrocapillarity. *Chem Rev*. 1947;41:441–501.
6. Merrill DR. The electrochemistry of charge injection at the electrode/tissue interface. *Implantable neural prostheses: techniques and applications*. Springer; 2009:85–138.
7. Randles JEB. Rapid electrode reactions. *Disc Faraday Soc*. 1947;1:11–19.
8. Gileadi E, Kirowa-Eisner E, Penciner J. *Interfacial electrochemistry: an experimental approach*. Reading: Addison-Wesley; 1975.
9. Bard AJ, Faulkner LR. *Electrochemical methods*. New York: John Wiley and Sons; 1980.
10. Biegler T, Rand DAJ, Woods R. Limiting oxygen coverage on platinized platinum: relevance to determination of real platinum area by hydrogen adsorption. *J Electroanal Chem*. 1971;29:269–277.
11. Donaldson NdN, Donaldson PEK. When are actively balanced biphasic ('Lilly') stimulating pulses necessary in a neural prosthesis? I. Historical background, Pt resting potential, dQ studies. *Med and Biol Eng and Comput*. 1986;24:41–49.
12. Donaldson NdN, Donaldson PEK. When are actively balanced biphasic ('Lilly') stimulating pulses necessary in a neural prosthesis? II. pH changes, noxious products, electrode corrosion, discussion. *Med and Biol Eng and Comput*. 1986;24:50–56.
13. Weinman J, Mahler J. An analysis of electrical properties of metal electrodes. *Med Electron Biol Eng*. 1964;2:229–310.
14. Halliwell B. Reactive oxygen species and the central nervous system. *J Neurochem*. 1992;595:1609–1623.
15. Stohs SJ. The role of free radicals in toxicity and disease. *J Basic Clin Physiol Pharmacol*. 1995;63-4:205–228.

16. Hemnani T, Parihar MS. Reactive oxygen species and oxidative DNA damage. *Indian J Physiol Pharmacol*. 1998;424:440–452.

17. Imlay JA. Pathways of oxidative damage. *Annu Rev Microbiol*. 2003;57:395–418.

18. Bergamini CM, Gambetti S, Dondi A, Cervellati C. Oxygen, reactive oxygen species and tissue damage. *Curr Pharm Des*. 2004;1014:1611–1626.

19. Hille B. *Ionic channels of excitable membranes*. Sunderland: Sinauer Associates; 1984.

20. Kandel ER, Schwartz JH, Jessell TM. *Principles of neural science*. 4th edn. New York: McGraw Hill; 2000.

21. Merrill DR, Bikson M, Jefferys JGR. Electrical stimulation of excitable tissue: design of efficacious and safe protocols. *J Neurosci Methods*. 2005;1412:171–198.

22. McNeal DR. Analysis of a model for excitation of myelinated nerve. *IEEE Trans Biomed Eng*. 1976;234:329–337.

23. Ranck JB. Extracellular Stimulation In: Patterson MM, Kesner RP, eds. *Electrical stimulation research techniques*. New York: Academic Press; 1981.

24. Mortimer JT. Electrical Excitation of Nerve In: Agnew WF, McCreery DB, eds. *Neural prostheses: fundamental studies*. Englewood Cliffs: Prentice-Hall; 1990.

25. Durand D. Electrical Stimulation of Excitable System In: Bronzino JD, ed. *Biomedical engineering handbook*. Boca Raton: CRC Press; 1995.

26. Rattay F. Analysis of models for extracellular fiber stimulation. *IEEE Trans Biomed Eng*. 1989;36:676–682.

27. Fang Z, Mortimer JT. Selective activation of small motor axons by quasitrapezoidal current pulses. *IEEE Trans Biomed Eng*. 1991;382:168–174.

28. McIntyre CC, Grill WM. Extracellular stimulation of central neurons: influence of stimulus waveform and frequency on neuronal output. *J Neurophysiol*. 2002;884:1592–1604.

29. Warman EN, Grill WM, Durand D. Modeling the effects of electric fields on nerve fibers: determination of excitation thresholds. *IEEE Trans Biomed Eng*. 1992;3912:1244–1254.

30. Plonsey R, Barr RC. *Bioelectricity: a quantitative approach*. New York: Plenum Press; 1988.

31. McCreery DB, Agnew WF, Yuen TGH, Bullara LA. Charge density and charge per phase as cofactors in neural injury induced by electrical stimulation. *IEEE Trans Biomed Eng*. 1990;3710:996–1001.

32. Shannon RV. A model of safe levels for electrical stimulation. *IEEE Trans Biomed Eng*. 1992;394:424–426.

33. Yuen TGH, Agnew WF, Bullara LA, Jacques S, McCreery DB. Histological evaluation of neural damage from electrical stimulation: considerations for the selection of parameters for clinical application. *Neurosurg*. 1981;93:292–299.

34. Agnew WF, McCreery DB, Yuen TGH, Bullara LA. Histologic and physiologic evaluation of electrically stimulated peripheral nerve: considerations for the selection of parameters. *Ann Biomed Eng*. 1989;17:39–60.

35. Bhargava, A. Long-term effects of quasi-trapezoidal pulses on the structure and function of sacral anterior roots. MS thesis, Case Western Reserve University, Department of Biomedical Engineering, Cleveland, OH. 1993.

36. Lilly JC, Austin GM, Chambers WW. Threshold movements produced by excitation of cerebral cortex and efferent fibers with some parametric regions of rectangular current pulses (cats and monkeys). *J Neurophysiol*. 1952;15:319–341.

37. Lilly JC, Hughes JR, Alvord EC, Garkin TW. Brief noninjurious electric waveforms for stimulation of the brain. *Science*. 1955;121:468–469.

38. Mortimer JT, Shealy CN, Wheeler C. Experimental nondestructive electrical stimulation of the brain and spinal cord. *J Neurosurg*. 1970;325:553–559.

39. Pudenz RH, Bullar LA, Dru D, Talalla A. Electrical stimulation of the brain II: effects on the blood-brain barrier. *Surg Neurol*. 1975;4:265–270.

40. Pudenz RH, Bullara LA, Jacques P, Hambrecht FT. Electrical stimulation of the brain III: the neural damage model. *Surg Neurol*. 1975;4:389–400.

41. Scheiner A, Mortimer JT. Imbalanced biphasic electrical stimulation: muscle tissue damage. *Ann Biomed Eng*. 1990;18:407–425.

42. Mortimer JT, Kaufman D, Roessmann U. Intramuscular electrical stimulation: tissue damage. *Ann Biomed Eng*. 1980;8:235–244.

43. Brummer SB, Turner MJ. Electrical stimulation of the nervous system: the principle of safe charge injection with noble metal electrodes. *Bioelectrochem Bioenerg*. 1975;2:13–25.

44. Ballestrasse CL, Ruggeri RT, Beck TR. Calculations of the pH changes produced in body tissue by a spherical stimulation electrode. *Ann Biomed Eng*. 1985;13:405–424.

45. Chan PH, Yurko M, Fishman R. Phospholipid degradation and cellular edema induced by free radicals in brain slice cortical slices. *J Neurochem*. 1982;38:525–531.

46. Chia LS, Thompson JE, Moscarello MA. Disorder in human myelin induced by superoxide radical: an in vitro investigation. *Biochem and Biophys Res Commun*. 1983;1171:141–146.

47. Konat G, Wiggins RC. Effect of reactive oxygen species on myelin membrane proteins. *J Neurochem*. 1985;45:1113–1118.

48. Sevanian, A. In Lipid peroxidation, membrane damage, and phospholipase A2 action. CRC Reviews, Cellular Antioxidant Defense Mechanisms, Vol.II, 77–95. 1988.

49. Griot C, Vandevelde RA, Peterhans E, Stocker R. Selective degeneration of oligiodendrocytes mediated by reactive oxygen species. *Free Radic Res Commun*. 1990;114:181–193.

50. Buettner GR. The pecking order of free radicals and antioxidants: lipid peroxidation, alpha-tocopherol, and ascorbate. *Arch Biochem Biophys*. 1993;3002:535–543.

51. Morton SL, Daroux ML, Mortimer JT. The role of oxygen reduction in electrical stimulation of neural tissue. *J Electrochem Soc*. 1994;141:122–130.

52. Furchgott RF. Studies on relaxation of rabbit aorta by sodium nitrite: the basis for the proposal that the acid-activatable inhibitory factor from retractor penis is inorganic nitrite and the endothelium-derived relaxing factor is nitric oxide, vasodilatation. In: vanhoutte PM, ed, *Vascular smooth muscle, peptides, autonomic nerves and endothelium*. New York: Raven Press; 1988:401–414.

53. Ignarro LJ, Byrns RE, Wood KS. Biochemical and pharmacological properties of endothelium-derived relaxing factor and its similarity to nitric oxide radical. In: vanhoutte PM, ed, *Vascular smooth muscle, peptides, autonomic nerves and endothelium*. New York: Raven Press; 1988:427–436.

54. Umans J, Levi R. Nitric oxide in the regulation of blood flow and arterial pressure. *Ann Rev Physiol*. 1995;57:771–790.

55. Azuma H, Ishikawa M, Sekizaki S. Endothelium-dependent inhibition of platelet aggregation. *British J Pharm*. 1986;88:411–415.

56. Radomski MW, Palmer RMJ, Moncada S. Endogenous nitric oxide inhibits human platelet adhesion to vascular endothelium. *Lancet*. 1987;2:1057–1058.

57. Moncada S, Palmer RMJ, Higgs EA. Nitric oxide: physiology, pathology and pharmacology. *Pharmacol Rev*. 1991;432:109–142.

58. Beckman JS, Beckman TW, Chen J, Marshall PA, Freeman BA. Apparent hydroxyl radical production by peroxynitrite: implications for endothelial injury from nitric oxide and superoxide. *Proc Natl Acad Sci USA*. 1990;87:1620–1624.

59. Rubanyi GM. Vascular effects of oxygen derived free radicals. *Free Radic Biol and Med*. 1988;4:107–120.

60. Grill WM, Mortimer JT. Stimulus waveforms for selective neural stimulation. *IEEE Eng Med Biol*. 1995;14:375–385.
61. Fang Z, Mortimer JT. A method to effect physiological recruitment order in electrically activated muscle. *IEEE Trans Biomed Eng*. 1991;382:175–179.
62. Grill WM, Mortimer JT. Inversion of the current-distance relationship by transient depolarization. *IEEE Trans Biomed Eng*. 1997;441:001–009.
63. Gluckman BJ, Neel EJ, Netoff TI, Ditto WL, Spano ML, Schiff SJ. Electric field suppression of epileptiform activity in hippocampal slices. *J Neurophysiol*. 1996;766:4202–4205.
64. Ghai RS, Bikson M, Durand DM. Effects of applied electric fields on low-calcium epileptiform activity in the CA1 region of rat hippocampal slices. *J Neurophysiol*. 2000; 841:274–280.
65. Gorman PH, Mortimer JT. The effect of stimulus parameters on the recruitment characteristics of direct nerve stimulation. *IEEE Trans Biomed Eng BME*. 1983;30:407–414.
66. Dymond AM, Kaechele LE, Jurist JM, Crandall PH. Brain tissue reaction to some chronically implanted metals. *J Neurosurg*. 1970;33:574–580.
67. Stensaas SS, 7 Stensaas LJ. Histopathological evaluation of materials implanted in the cerebral cortex. *Acta Neuropathol*. 1978;41:145–155.
68. Loeb GE, Walker AE, Vematsu S, Konigsmark BW. Histological reaction to various conductive and dielectric films chronically implanted in the subdural space. *J Biomed Mater Res*. 1977;112:195–210.
69. Babb TL, Kupfer W. Phagocytic and metabolic reactions to chronically implanted metal brain electrodes. *Exp Neurol*. 1984;862:171–182.
70. Fisher G, Sayre GP, Bickford RC. Histological changes in the cat's brain after introduction of metallic and plastic-coated wire. In: Sheer DE, ed. *Electrical stimulation of the brain*. Austin: University of Texas Press; 1961:55–59.
71. Sawyer, P.N. & Srinivasan, S. In Medical Engineering (Ray, C.D., ed.) 1099-1110. Chicago: Year Book Medical Publishers; 1974.
72. Ryhanen J, Kallioinen M, Tuukkanen J, et al. In vivo biocompatibility evaluation of nickel-titanium shape memory alloy: muscle and perineural tissue responses and encapsule membrane thickness. *J Biomed Mater Res*. 1998;413:481–488.
73. Bogdanski D, Koller M, Muller D, et al. Easy assessment of the biocompatibility of Ni-Ti alloys by in vitro cell culture experiments on a functionally graded Ni-NiTi-Ti material. *Biomaterials*. 2002;2223:4549–4555.
74. Majji AB, Humayun MS, Weiland JD, Suzuki S, D'Anna SA, deJuan Jr E. Long-term histological and electrophysiological results of an inactive epiretinal electrode array implantation in dogs. *Invest Ophthalmol Vis Sci*. 1999;409:2073–2081.
75. Chouard CH, Pialoux P. Biocompatibility of cochlear implants. *Bull Acad Natl Med*. 1995;1793:549–555.
76. Niparko JK, Altschuler RA, Xue XL, Wiler JA, Anderson DJ. Surgical implantation and biocompatibility of central nervous system auditory prostheses. *Ann Otol Rhinol Laryngol*. 1989;9812:965–970.
77. Stieglitz T, Meyer JU. Implantable microsystems. Polyimide-based neuroprostheses for interfacing nerves. *Med Dev Technol*. 1999;106:28–30.
78. Hoogerwerf AC, Wise KD. A three-dimensional microelectrode array for chronic neural recording. *IEEE Trans Biomed Eng*. 1994;41:1136–1146.
79. Schmidt S, Horch K, Normann R. Biocompatibility of silicon-based electrode arrays implanted into feline cortical tissue. *J Biomed Mater Res*. 1993;2711:1393–1399.

80. Kristensen BW, Noraberg J, Thiebaud P, Koudelka-Hep M, Zimmer J. Biocompatibility of silicon-based arrays of electrodes coupled to organotypic hippocampal brain slice cultures. *Brain Res.* 2001;896:1–17.

81. Jones KE, Campbell PK, Normann RA. A glass/silicon composite intracortical electrode array. *Ann Biomed Eng.* 1992;204:423–437.

82. Rousche PJ, Pellinen DS, Pivin DP, Williams JC, Vetter RJ, Kipke DR. Flexible polyimide-based intracortical electrode arrays with bioactive capability. *IEEE Trans Biomed Eng.* 2001;481:361–370.

83. Kennedy PR. The cone electrode: A long-term electrode that records from neurites grown onto its recording surface. *J Neurosci Methods.* 1989;293:181–193.

84. Kennedy PR, Bakay RA. Restoration of neural output from a paralyzed patient by a direct brain connection. *Neuroreport.* 1998;98:1707–1711.

85. Kennedy PR, Bakay RA, Moore M, Adams K, Montgomery G. Neural activity during acquisition of cursor control in a locked-in patient. *Soc Neurosci Abstr.* 1999;251:894.

86. Rudge JS, Smith GM, Silver J. An in vitro model of wound healing in the central nervous system:analysis of cell reaction and interaction at different times. *Exp Neurol.* 1989;103:1–16.

87. Turner JN, Shain W, Szarowski DH, et al. Cerebral astrocyte response to micromachined silicon implants. *Exp Neurol.* 1999;156:33–49.

88. Bignami A, Dahl D. The astroglial response to stabbing: immunoflourescence studies with antibodies to astrocyte-specific protein (GFA) in mammalian and sub-mammalian vertebrate. *Neuropathol Appl Neurobiol.* 1976;251:23–43.

89. Robblee LS, Rose TL. Electrochemical guidelines for selection of protocols and electrode materials for neural stimulation. In: Agnew WF, McCreery DB, eds. *Neural prostheses: fundamental studies.* Englewood Cliffs: Prentice-Hall; 1990:25–66.

90. Merrill DR. *Electrochemical processes occurring on gold in sulfuric acid under neural stimulation conditions, PhD thesis.* Cleveland, OH: Case Western Reserve University, Department of Biomedical Engineering; 2002.

91. Merrill DR, Stefan IC, Scherson DA, Mortimer JT. Electrochemistry of gold in aqueous sulfuric acid solutions under neural stimulation conditions. *J Electrochem Soc.* 2005;1527:E212–E221.

92. White RL, Gross TJ. An evaluation of the resistance to electrolysis of metals for use in bio-stimulation probes. *IEEE Trans Biomed Eng BME.* 1974;21:487–490.

93. Johnson PF, Hench LL. An in vitro analysis of metal electrodes for use in the neural environment. *Brain Behav Evol.* 1977;14:23–45.

94. Brummer SB, McHardy J, Turner MJ. Electrical stimulation with Pt electrodes: Trace analysis for dissolved platinum and other dissolved electrochemical products. *Brain Behav Evol.* 1977;14:10–22.

95. Black RD, Hannaker P. Dissolution of smooth platinum electrodes in biological fluids. *Appl Neurophysiol.* 1979;42:366–374.

96. McHardy J, Robblee RS, Marsten M, Brummer SB. Electrical stimulation with platinum electrodes. IV. Factors influencing platinum dissolution in inorganic saline. *Biomater B1.* 1980;129–134.

97. Robblee RS, McHardy J, Marsten M, Brummer SB. Electrical stimulation with platinum electrodes. V. The effects of protein on platinum dissolution. *Biomater B1.* 1980; 135–139.

98. Robblee RS, Lefko JL, Brummer SB. Activated iridium: An electrode suitable for reversible charge injection in saline solution. *J Electrochem Soc.* 1983;130:731–733.

99. Robblee RS, McHardy J, Agnew WF, Bullara LA. Electrical stimulation with Pt electrodes. VII. Dissolution of Pt electrodes during electrical stimulation of the cat cerebral cortex. *J Neurosci Meth.* 1983;9:301–308.

100. Tivol WF, Agnew WF, Alvarez RB, Yuen TGH. Characterization of electrode dissolution products on the high voltage electrode microscope. *J Neurosci Meth.* 1987;19:323–337.

101. Rosenberg B, VanCamp L, Krigas T. Inhibition of cell division in Escherichia coli by electrolysis products from a platinum electrode. *Nature.* 1965;205:698–699.

102. Rosenberg B. Some biological effects of platinum compounds: New agents for the control of tumours. *Platin Met Rev.* 1971;15:42–51.

103. Macquet JP, Theophanides T. DNA-platinum interactions. Characterization of solid DNA-K$_2$PtCl$_4$ complexes. *Inorg Chim Acta.* 1976;18:189–194.

104. Brummer SB, Turner MJ. Electrochemical considerations for safe electrical stimulation of the nervous system with platinum electrodes. *IEEE Trans Biomed Eng BME.* 1977;24:59–63.

105. Brummer SB, Turner MJ. Electrical stimulation with Pt electrodes. I. A method for determination of 'real' electrode areas. *IEEE Trans Biomed Eng BME.* 1977;24:436–439.

106. Brummer SB, Turner MJ. Electrical stimulation with Pt electrodes. II. Estimation of maximum surface redox (theoretical non-gassing) limits. *IEEE Trans Biomed Eng BME.* 1977;24:440–443.

107. Rose TL, Robblee LS. Electrical stimulation with Pt electrodes. VIII. Electrochemically safe charge injection limits with 0.2 ms pulses. *IEEE Trans Biomed Eng.* 1990;37:1118–1120.

108. Rand DAJ, Woods R. Cyclic voltammetric studies on iridium electrodes in sulfuric acid solutions. Nature of oxygen layer and metal dissolution. *J Electroanal Chem Interfacial Electrochem.* 1974;55:375–381.

109. Zerbino JO, Tacconi NR, Arvia AJ. The activation and deactivation of iridium in acid electrolytes. *J Electrochem Soc.* 1978;125:1266–1276.

110. Mozota J, Conway BE. Surface and bulk processes at oxidized iridium electrodes-I: monolayer stage and transition to reversible multilayer oxide film behavior. *Electrochim Acta.* 1983;28:1–8.

111. Bak MK, Girvin JP, Hambrecht FT, Kufta CV, Loeb GE, Schmidt EM. Visual sensations produced by intracortical microstimulation of the human occipital cortex. *Med Biol Eng Comput.* 1990;28:257–259.

112. McCreery DB, Yuen TGH, Agnew WF, Bullara LA. Stimulation with chronically implanted microelectrodes in the cochlear nucleus of the cat: histologic and physiologic effects. *Hear Res.* 1992;62:42–56.

113. Loeb GE, Peck RA, Martyniuk J. Toward the ultimate metal microelectrode. *J Neurosci Meth.* 1995;63:175–183.

114. Liu X, McCreery DB, Carter RR, Bullara LA, Yuen TGH, Agnew WF. Stability of the interface between neural tissue and chronically implanted intracortical microelectrodes. *IEEE Trans Rehabil Eng.* 1999;73:315–326.

115. Meyer RD, Cogan SF. Electrodeposited iridium oxide for neural stimulation and recording electrodes. *IEEE Trans Neural Syst Rehabil Eng.* 2001;91:2–10.

116. Anderson DJ, Najafi K, Tanghe SJ, et al. Batch-fabricated thin-film electrodes for stimulation of the central auditory system. *IEEE Trans Biomed Eng.* 1989;36:693–704.

117. Weiland JD, Anderson DJ. Chronic neural stimulation with thin-film, iridium oxide electrodes. *IEEE Trans Biomed Eng.* 2000;477:911–918.

118. Beebe X, Rose TL. Charge injection limits of activated iridium oxide electrodes with 0.2 msec pulses in bicarbonate buffered saline. *IEEE Trans Biomed Eng BME.* 1988;35:494–495.

119. Kelliher EM, Rose TL. Evaluation of charge injection properties of thin film redox materials for use as neural stimulation electrodes. *Mater Res Soc Symp Proc.* 1989;110:23–27.

120. Agnew WF, Yuen TGH, McCreery DB, Bullara LA. Histopathologic evaluation of prolonged intracortical electrical stimulation. *Exp Neurol*. 1986;92:162–185.

121. Robblee LS, Mangaudis MM, Lasinsky ED, Kimball AG, Brummer SB. Charge injection properties of thermally-prepared iridium oxide films. *Mater Res Soc Symp Proc*. 1986;55:303–310.

122. Klein JD, Clauson SL, Cogan SF. Morphology and charge capacity of sputtered iridium oxide films. *J Vac Sci Technol*. 1989;A7:3043–3047.

123. Loucks RB, Weinberg H, Smith M. The erosion of electrodes by small currents. *Electroenceph Clin Neurophysiol*. 1959;11:823–826.

124. Greatbatch W, Chardack WM. Myocardial and endocardiac electrodes for chronic implantation. *Ann NY Acad Sci*. 1968;148:234–251.

125. McHardy J, Geller D, Brummer SB. An approach to corrosion control during electrical stimulation. *Ann Biomed Eng*. 1977;5:144–149.

126. Gotman I. Characteristics of metals used in implants. *J Endourol*. 1997;116:383–389.

127. Guyton DL, Hambrecht FT. Capacitor electrode stimulates nerve or muscle without oxidation-reduction reactions. *Science*. 1973;181:74–76.

128. Guyton DL, Hambrecht FT. Theory and design of capacitor electrodes for chronic stimulation. *Med and Biol Eng*. 1974;7:613–620.

129. Rose TL, Kelliher EM, Robblee LS. Assessment of capacitor electrodes for intracortical neural stimulation. *J Neurosci Meth*. 1985;12:181–193.

130. Johnson PF, Bernstein JJ, Hunter G, Dawson WW, Hench LL. An in vitro and in vivo analysis of anodized tantalum capacitive electrodes: corrosion response, physiology and histology. *J Biomed Mater Res*. 1977;11:637–656.

131. Bernstein JJ, Hench LL, Johnson PF, Dawson WW, Hunter G. Electrical stimulation of the cortex with Ta_2O_5 capacitive electrodes. In: Hambrecht FT, Reswick JB, eds. *Functional electrical stimulation*. New York: Marcel-Dekker; 1977:465–477.

132. Donaldson PEK. The stability of tantalum-pentoxide films in vivo. *Med Biol Eng*. 1974;12:131–135.

133. Lagow CH, Sladek KJ, Richardson PC. Anodic insulated tantalum oxide electrocardiograph electrodes. *IEEE Trans Biomed Eng*. 1971;18:162–164.

134. Cui XY, Hetke JF, Wiler JA, et al. Electrochemical deposition and characterization of conducting polymer polypyrrole/PSS on multichannel neural probes. *Sens Actuators*. 2001;A93:8–18.

135. Cui X, Lee VA, Raphael Y, et al. Surface modification of neural recording electrodes with conducting polymer/biomolecule blends. *J Biomed Mater Res*. 2001;56:261–272.

136. Cui XY, Martin DC. Electrochemical deposition and characterization of poly(3, 4-ethylenedioxythiophene) on neural microelectrode arrays. *Sens Actuators B Chem*. 2003;89: 92–102.

137. Cui XY, Martin DC. Fuzzy gold electrodes for lowering impedance and improving adhesion with electrodeposited conducting polymer films. *Sens Actuators A Phys*. 2003;103:384–394.

138. Xiao YH, Cui XY, Martin DC. Electrochemical polymerization and properties of PEDOT/S-EDOT on neural microelectrode arrays. *J Electroanal Chem*. 2004;573:43–48.

139. Yang JY, Martin DC. Microporous conducting polymers on neural microelectrode arrays II. Physical characterization. *Sens Actuators A Phys*. 2004;113A:204–211.

140. Kim DH, Abidian M, Marti DC. Conducting polymers grown in hydrogel scaffolds coated on neural prosthetic devices. *J Biomed Mater Res*. 2004;71A:577–585.

141. Cui XY, Wiler J, Dzamann M, et al. In vivo studies of polypyrrole/peptide coated neural probes. *Biomaterials*. 2003;24:777–787.

142. Kim DH, Sequerah C, Hendricks JL, et al. Effect of immobilized nerve growth factor (NGF) on conductive polymers electrical properties and cellular response. *Adv Funct Mater.* 2007;17:79–86.

143. Stauffer WR, Cui XT. Polypyrrole doped with 2 peptide sequences from laminin. *Biomaterials.* 2006;27:2405–2413.

144. Richardson-Burns SM, Hendricks JL, Foster B, et al. Polymerization of the conducting polymer poly(3, 4-ethylenedioxythiophene) (PEDOT) around living neural cells. *Biomaterials.* 2007;28:1539–1552.

145. Richardson-Burns SM, Hendricks JL, Martin DC. Electrochemical polymerization of conducting polymers in living neural tissue. *J Neural Eng.* 2007;4:L6–L13.

146. Cui X, Zhou D. Poly (3,4-ethylenedioxythiophene) for chronic neural stimulation. *IEEE Trans Neural Syst Rehabil Eng.* 2007;15:502–508.

147. McCreery DB, Bullara LA, Agnew WF. Neuronal activity evoked by chronically implanted intracortical microelectrodes. *Exp Neurol.* 1986;92:147–161.

148. Campbell PK, Jones KE, Huber RJ, Horch KW, Normann RA. A silicon-based, three dimensional neural interface: manufacturing processes for an intracortical electrode array. *IEEE Trans Biomed Eng.* 1991;38:758–768.

149. Jones KE, Campbell PK, Normann RA. A glass/silicon composite intracortical electrode array. *Ann Biomed Eng.* 1992;20:423–437.

150. Normann RA, Maynard EM, Rousche PJ, Warren DJ. A neural interface for a cortical vision prosthesis. *Vision Res.* 1999;39:2577–2587.

151. Wise KD, Angell JB, Starr A. An integrated-circuit approach to extracellular microelectrodes. *IEEE Trans Biomed Eng.* 1970;17:238–247.

152. Drake KL, Wise KD, Farraye J, Anderson DJ, BeMent SL. Performance of planar multisite microprobes in recording extracellular single-unit intracortical activity. *IEEE Trans Biomed Eng.* 1988;35:719–732.

153. McCreery D, Lossinsky A, Pikov V, Liu X. Microelectrode array for chronic deep-brain microstimulation and recording. *IEEE Trans Biomed Eng.* 2006;53:726–737.

154. Bruckenstein S, Miller B. An experimental study of non-uniform current distribution at rotating disk electrodes. *J Electrochem Soc.* 1970;117:1044–1048.

155. Shepherd RK, Murray MT, Houghton ME, Clark GM. Scanning electron microscopy of chronically stimulated platinum intracochlear electrodes. *Biomater.* 1985;6:237–242.

156. Wiley JD, Webster JJ. Analysis and control of the current distribution under circular dispersive electrodes. *IEEE Trans Biomed Eng BME.* 1982;29:381–385.

Commentary on The Electrode – Materials and Configurations

Mark Stecker, PhD, MD,
Marshall University Medical Center, Huntington, WV, USA

Although neural activity may be modulated by many means, electrical neuromodulation requires the use of electrodes. This technique has substantial advantages over the older technique of producing lesions that destroy neural structures, but the intricacy of the electrode–tissue interface makes the delivery of electrical energy a complex process that must be understood by both practitioners and developers in the field. There are three major issues associated with electrical neuromodulation. The first is how to deliver the electrical stimulation to the correct anatomic location. The second is how to choose the stimulation waveform to optimize activation of target neural structures. The third issue is the prevention of injury during the delivery of the stimulus.

DELIVERY OF ELECTRICAL ENERGY

Although it is possible to focus high frequency electromagnetic radiation so that it has a peak intensity at a point remote from the generator, it is impossible to create a low frequency electric fields with this property in a homogeneous medium. The transition between these two regimes is determined by the wavelength of the electromagnetic radiation and the size of the target. In order to focus electromagnetic energy on a millimeter size target would require radiation with a wavelength smaller than this. This translates to electromagnetic radiation with a frequency of a least 300 GHz which is in the far infrared part of the electromagnetic spectrum. In comparison, 300 Hz electromagnetic radiation, like that used in clinical neuromodulation, would have a wavelength on the order of 1000 km and could not be focused on any clinically relevant targets. Practically, this implies that for the low frequency stimulation used in clinical situations, the maximum intensity of the electric field is at the surface of one of the electrodes. Thus, either the electrode must be placed near the structure being stimulated, such as with deep brain stimulation (DBS), or a very high amplitude stimulus must be delivered to electrodes that are far from the target, as with transcranial motor evoked potentials. What then is the optimal shape and size for a stimulating electrode? The critical design parameters are that the electrode must be small enough not to cause significant mechanical damage to the structure to be stimulated but must be large enough so that when stimulation of the entire target area is achieved, the electric field near the electrode must not be so large that it causes actual damage to the target tissue. In order to understand this principle, a simple illustration will be helpful. Consider the case in which a spherical electrode of diameter a is placed in a nucleus to be stimulated in such a way that the distance between the center of the electrode and the nearest neuron is b ($b \geq a$), while the distance to the most distant neuron is c. If the medium is electrically homogeneous, then Ohm's law states that the current density produced by the electrode is proportional to the electrical field which is inversely proportional to the square of the distance from the center of the electrode. Since the current density is directly related to both the toxic and the therapeutic effects of the stimulation, the conditions for safe and effective stimulation are:

$$\frac{I_0}{a^2} \leq J_{max} \text{ (to avoid toxicity)}$$

$$\frac{I_0}{b^2} \geq J_{min} \text{ (to be therapeutically effective)}$$

and so the range of useful stimulation currents is:

$$J_{max}\, a^2 \geq I_0 \geq J_{min}\, b^2$$

where J_{max} is the current density above which tissue damage occurs, J_{min} is the minimum current density necessary to a therapeutic effect and I_0 is the total stimulating current. This illustrates clearly that the larger the nucleus, the larger is the required stimulating current, but the closer the electrode is to neurons the lower the current must be to avoid injury.

Of course, stimulation is not limited to placing a single electrode in the nucleus of interest with a distant reference electrode (monopolar configuration) and the choice of the electrode configuration is also critical to delivering the electrical stimulus to the target nucleus. For example, a bipolar configuration in which both the anode and the cathode are placed in or near the target nucleus will produce a very different spatial distribution of stimulation than the monopolar configuration. Near the electrodes the stimulation will be similar to that produced by an individual monopolar electrode but there will be a more localized distribution of current than an equivalent monopolar configuration. This may be helpful when the nucleus to be stimulated is small or there are nearby structures that if stimulated would produce harmful side effects. However, the electric field generated is more complex and less spatially uniform than that of the monopolar configuration and so uniform stimulation of the target nucleus may be less likely. In addition, cathodal and anodal stimulation do not have fully equivalent effects on neural structures so that the variation in stimulation effect over the nucleus in question will be further increased. On the other hand, if multiple monopolar electrodes are placed in a nucleus with a remote reference electrode, more uniform stimulation of the nucleus can be produced. For the sake of illustration, consider the case of multiple spherical electrodes of diameter a placed in a nucleus of radius c so that the maximum distance between the electrodes is d and the minimum distance between each electrode and the closest neural structure is b. If I_0 is the current flowing through each electrode, then the (approximate) criterion for avoiding toxicity is:

$$\frac{I_0}{a^2} \leq J_{max}$$

while the criterion for effective stimulation has the form:

$$\frac{I_0}{d^2} = \geq J_{min}$$

(Of course the exact formula accounting for the effects of all electrodes will have more complex dependence on d with a constant other than 1 multiplying the left hand term in the above inequality) and so the range of useful stimulation current is:

$$J_{max} a^2 \geq I_0 \geq J_{min} \frac{C^2}{N^{\frac{2}{3}}}, \text{ since } d \sim \frac{C}{N^{\frac{1}{3}}}$$

where N is the number of electrodes placed into the nucleus. This means that the stimulation of the entire nucleus can be achieved with a lower level of current passing through each electrode which minimizes the risk of electrical injury to the tissue. Practically, this improvement in electrical safety may be mitigated by the increased risk of mechanical injury during placement of this large number of electrodes.

The term spatial homogeneity was mentioned in the above discussions. This refers to a state in which the conductivity and the dielectric constant of the material are essentially the same at all locations in the nucleus to be stimulated. Since neural structures are never homogeneous, it is important to consider what effects this may have on the ability to deliver localized currents. First, there may be anatomically defined low current pathways within the nucleus being stimulated. This may occur for instance because conduction along a saline solution will be greater than the conduction across a poorly conducting structure as a lipid bilayer. Current will then be shunted through these low resistance pathways and away from high resistance pathways producing regions that are both more susceptible to damage and to therapeutic stimulation. Second, although the maximum current density will still occur near the electrodes, as discussed above, there may be local maxima and minima in the electric current produced by these homogeneities which also reduces the possibility of a uniform therapeutic effect of electrical stimulation on the nucleus. It is important to recognize that the reciprocity principle demands that the neural structures most readily stimulated during stimulation with two electrodes are those that also produce the greatest evoked potentials when neural activity is recorded from those same two electrodes. This provides an alternative means to understand and make use of the effects of spatial inhomogeneities.

OPTIMIZING NEURAL ACTIVATION

Directing electrical energy to the correct target is space is important but the temporal characteristics of the stimulus are also critical in determining its effect. One general principle for short pulses is embodied in the strength–duration curve which illustrates the fact that higher currents are required to produce an equivalent stimulus with short duration pulses than with long duration currents. However, even with very long stimuli, there is a minimum stimulation current (called the rheobase) required to have a therapeutic effect. The other descriptor of this curve is the chronaxie which is the duration at which the necessary stimulation current is twice that of the rheobase. The simplicity that comes with stimulation of single axons disappears when entire neural networks are stimulated. In this case, such things as the time between pulses and the pattern of the pulses become critical. This is seen simply in the situation where transcranial motor evoked potentials are to be elicited under anesthesia. In this case, the no muscle motor response can be obtained with single pulse stimulation no matter what current is used. However, responses can be easily obtained if the stimulus is a train of pulses with the optimal inter-stimulus interval. In general, temporal coding is one means that can be used to transmit information in the nervous system. Stimulation that reproduces certain patterns may be used to send specific instructions to the neural network to produce defined output patterns. Although, at the present time, our knowledge of these temporal codes is limited, research in this area will improve our ability to control the effects of stimulation.

INJURY PRODUCED BY ELECTRICAL STIMULATION

Damage to tissue can easily be caused by electrical stimulation through many mechanisms. These mechanisms fall into two categories: those related to the flow of current itself and those related to specific chemical reactions occurring at the surface of the electrode. In the first category are the changes in temperature that occur as a result of current flowing through the resistive medium. The degree of temperature elevation depends on many factors other than the current and the resistivity of the medium including the rate of heat conduction out of the region stimulated. This possibility of thermal injury during stimulation should not be underestimated, since even stimulation of platinum electrodes with 20 V for a few minutes can produce temperatures near the boiling point of water. In living tissue, temperature elevations more than 5 °C above physiologic can quickly cause protein denaturation and cellular death and so much smaller degrees of stimulation for shorter periods of time could cause significant thermal injury.

There are many ways in which reactions at the electrode surface can be classified. One is whether the reaction is reversible or irreversible. Clearly, the smallest probability of damage to the nucleus in which the electrode is placed occurs when the stimulus is such that only reversible reactions occur. Another useful classification is whether the material from which the electrode is made is changed during the reaction. If the electrode material undergoes chemical change during stimulation, there is a high probability of damage to the electrode or deposition of the electrode material into the target nucleus both of which could produce significant problems with stimulation.

Stimulation-induced chemical reactions that do not change the electrode material can have very significant effects on local neuronal function. In fact, a few seconds of DC stimulation at 20 V can produce regions on the order of a centimeter or more in which the pH is greater than 8 or less than 5. Not only can these changes in pH cause changes in important chemical equilibria but also can cause changes in protein and enzymatic function that can lead to cellular injury. It is also possible that toxic substances (such as bleach, free radicals, and chlorine gas) may be produced during stimulation. The negative effects of these reactions are accentuated by low frequency and DC stimulation and are mitigated by high frequency stimulation with short pulses. This is for two reasons: first, any toxic products produced during a very short period of stimulation will necessarily remain near the electrode. When stimulation is terminated, there is a probability that the reverse reaction will occur and the toxic substances will decline in concentration. In addition, when short pulse stimuli are used, any toxic substances that have not had a chance to undergo a reverse reaction may have the opportunity to diffuse away from the electrode and reduce the concentration that would otherwise be present.

It is important to know what reactions will occur at different stimulation voltages. Each reaction is associated with a change in the composition of the materials and hence a change in the energy of the system. The fundamental equation of electrochemistry which relates the concentration of reduced and oxidized chemical species (oxidation–reduction or redox reactions involve the transfer of electrons from one chemical species to another) to the electrical potential is the Nernst equation:

$$E = E^0 - \frac{RT}{zF} \ln \frac{C_{Red}}{C_{Ox}}$$

$$\frac{RT}{zF} \approx 0.059 \, V (T = 37°C)$$

where C_{Red} and C_{Ox} and the concentration of the reduced and oxidized species, T is the temperature, z is the number of electrons transferred during the reaction, R is the ideal gas constant and F is the Faraday constant. E is the change in electrical potential as the ion comes in contact the stimulating electrode and E^0 is the difference in energy between the reduced and oxidized species. The values of E^0, the standard electrode potential, for many different reactions are well known and are typically in the range of −3 V to 3 V and are strongly dependent on pH. This means that most oxidation–reduction reactions can be driven from a situation where one species is dominant to a situation in which another species is dominant by a voltage change of only a few Volts or less. Although the Nernst equation provides information on the reactions that will happen in equilibrium, it does not provide information about the rates of these reactions. This is especially important because of the complexity of the electrode–solution interface. The prime example of this is the stainless steel electrode. Although iron rusts in a saline solution relatively quickly, stainless steel includes nickel and chromium as well as iron. The chromium reacts with oxygen to form a surface (passivation) layer which prevents the oxidation reaction of rusting. However, during electrical stimulation, especially with pulses longer than 1 ms and more than 5 V, this passivation layer breaks down and allows the iron in the anode to oxidize. This process becomes more significant as the duration and voltage of stimulation increases so much so that a stainless steel wire anode will vaporize in seconds with 20 V DC stimulation. These phenomena are highly material specific and, for example, tungsten anodes decompose rapidly under pulse stimulation and not DC stimulation, while platinum anodes and cathodes do not decompose significantly even under the extremes of stimulation possible with current equipment.

Choosing the correct electrode material, electrode shape, configuration and stimulation paradigm for neuromodulation involves the understanding of a large number of physical, chemical and biologic phenomena associated with electrical stimulation.

STUDY QUESTIONS

1. Review details of charge transfer at the electrode–tissue interface. Considering the duration of each pulse in most stimulation systems currently available, how do stimulation of constant current and constant voltage differ ultimately, and how might this manifest clinically?
2. Review the purpose of each phase in a biphasic pulse. How might biphasic pulses be modified to achieve other effects of stimulation, or eliminate adverse effects?

Chapter 7

The Electrode – Principles of the Neural Interface: Axons and Cell Bodies

Cameron C. McIntyre, PhD

Cleveland Clinic Foundation, Department of Biomedical Engineering, Cleveland, OH, USA

INTRODUCTION

Therapeutic interventions with electrical stimulation of the nervous system have a long and successful history. The fundamental feature of clinical applications of neurostimulation is the interaction between the stimulating electrode, its resulting electric field, and the surrounding neural tissue. Decades of research have elucidated many of the basic principles of these interactions, however, ongoing research efforts continue to be necessary to refine the details. The general goals of such studies are to expand scientific understanding on the effects of electric fields on the brain and enable more efficacious stimulation delivery for medical devices. The goal of this chapter is to provide an overview of current knowledge on the interactions between permanently implanted stimulating electrodes and the nervous system. While many of the presented details are from deep brain stimulation (DBS) examples, the general concepts are applicable to all forms of clinical neurostimulation (spinal cord stimulation (SCS), functional electrical stimulation (FES), cochlear stimulation, etc.).

ELECTRODE–BRAIN INTERFACE

The interface between a permanently implanted electrode and the brain is a complex entity that can dramatically affect the functioning of the clinical device. An anatomically and electrically stable interface is necessary to allow for the determination of stimulation parameters that can provide consistent therapeutic benefit. This interface is made up of many different components. The basis of this interface is the transfer of charge from the metal electrode to the ionic medium of the brain; thereby establishing a potential gradient from the cathode to the anode [1]. This potential gradient passes through the extracellular space whose electrical conductivity is inhomogeneous, meaning it changes as a function of distance from the electrode, and anisotropic, meaning it changes as a function of

Essential Neuromodulation. DOI: 10.1016/B978-0-12-381409-8.00007-3

direction. Very close to the electrode (within a few hundred microns) the tissue medium is typically made up of glia cells and extracellular matrix proteins that form a low conductivity encapsulation layer around the implanted foreign body. The periphery of this encapsulation layer transitions into the normal brain tissue. These biological features create a complex volume conductor for the generated electric field. This field is then indiscriminately applied to all of the neural tissue surrounding the implanted electrode. The spatial change in the electric field, or the second spatial derivative of the extracellular voltage distribution, along a neural process (e.g. axon or dendrite) directly influences its subsequent polarization [2,3]. Some neural elements surrounding the electrode will generate action potentials in response to the applied stimulus. Most of these action potentials will be transmitted from their initiation zone to their axon terminals, resulting in synaptic action on their target. This interaction with an end organ, or a larger neural network, underlies the basic therapeutic function of the device. From this perspective, the fundamental goal of a neurostimulation device is to control the release of neurotransmitters.

As neurostimulation device users and engineers, our goal is to tap into the existing nervous system circuitry and rekindle its lost function or modulate its abnormal activity. There exist two basic features we can manipulate to achieve our goal. First is the design/placement of the electrode and, second, is the stimulation parameter settings we choose to apply. The details of selecting/choosing these features are covered elsewhere, but they both dictate the first step in characterizing the effects of neurostimulation – quantifying the electric field.

ELECTRIC FIELD

The electric field generated by an implanted electrode is a three-dimensionally complex phenomenon that is distributed throughout the brain. While the fundamental purpose of neurostimulation technology is to modulate neural activity with applied electric fields, historically, much of the device design work and clinical protocols were primarily based on anatomical considerations (i.e. stimulation of a specific brain nucleus). This approach was taken because logical hypotheses could be generated to relate the effects of selectively stimulating a given nucleus to a behavioral outcome. However, without considering the complete system of electrode placement in that nucleus, stimulation parameter settings, electrical characteristics of the electrode, and electrical properties of the surrounding tissue medium, it is impossible to determine if the stimulation effects will be contained in that nucleus or if they will extend to surrounding brain regions [4]. Therefore, the first step in predicting the effects of neurostimulation is to characterize the voltage distribution generated in the brain.

Recently, Miocinovic et al [5] experimentally measured the spatial and temporal characteristics of the voltage distribution generated by DBS electrodes implanted in the brain of a non-human primate. Recordings were made during voltage-controlled and current-controlled stimulation. We found that three factors directly affected the voltage measurements:

1. a voltage drop at the electrode–electrolyte interface
2. inhomogeneity and anisotropy of the tissue medium
3. capacitive modulation of the stimulus waveform.

A relatively dramatic drop in voltage occurs in the transition from the polarization of the electrode contact to the ionic medium. This voltage drop ($\approx 40\%$ for DBS electrodes [5,6]) is rooted in the complex non-linear and frequency-dependent reactions that actually take place at the electrode–electrolyte interface [1]. Therefore, it is as if for a 1 V stimulus only ≈ 0.6 V makes it through the transition from an electron-mediated to ion-mediated polarization.

Once the polarization has been transmitted to the ionic medium, it must transverse the brain's inflammatory response to the implanted electrode. This collection of cellular infiltrate, protein deposits, and collagen matrices increases the electrode impedance [7]. As a result, the effective strength of voltage-controlled stimulation can be further reduced by encapsulation tissue because the injected current is inversely proportional to electrode impedance [8]. Fortunately, this issue can be alleviated by using current-controlled stimulation [9].

Whether using voltage-controlled or current-controlled stimulation, the spatial distribution of voltage in the tissue medium will be dictated by the anisotropy and inhomogeneity of the brain. These factors can be estimated from diffusion tensor imaging [10], and have been used extensively in DBS electric field models [11]. The effects of inhomogeneity are most apparent at and around the ventricles where the relatively low conductivity brain tissue is juxtaposed to the relatively high conductivity fluid cavities. In such regions, the voltage spread is enhanced in the directions of high conductivity. The effects of anisotropy are most apparent at and around large fiber tracts, such as the internal capsule. In such regions, the voltage spread is enhanced parallel to and hindered transverse to the fiber tract.

The above paragraphs provide an overview of the factors that impact the spatial distribution of voltage in the tissue medium, however, temporal aspects of stimulus waveform are also modified by the electrode–brain interface [5,12]. The implanted pulse generator is designed to generate a stimulus pulse with a specific shape (typically a rectangular pulse of ≈ 0.1 ms). However, when using voltage-controlled stimulation, the capacitance of the electrode–electrolyte interface results in a non-linear decay of the potential during a voltage-controlled stimulus pulse. On the other hand, a current-controlled waveform is not affected by the interface capacitance, but instead exhibits a non-linear rising of the voltage measured in the brain due to the capacitance of the bulk tissue medium. These modifications to the stimulus waveform decrease the relative excitability of the pulse to the surrounding neural tissue [6,12].

STIMULATED NEURAL ELEMENTS

Once the electrode is implanted, stimulus pulse applied, and electric field generated in the tissue medium, the next step is to characterize the neural response. In general, three classes of neurons can be directly affected by electrical stimulation: local cells, afferent inputs, and fibers of passage. Local cells represent neurons that have their cell body in close proximity to the electrode and an axon that projects locally and/or to a different brain region. Afferent inputs represent neurons that project to the region near the electrode and whose axon terminals make synaptic connections with local cells. Fibers of passage represent neurons where both the cell body and axon terminals are far from the electrode, but the axonal process of the neuron traces a path that comes in close proximity to the electrode. Experimental measurements indicate that local cells, afferent inputs, and fibers of passage have similar thresholds for activation [13]. And, local cells can be directly excited by the stimulus (see below) and/or have their excitability indirectly altered via activation of afferent inputs that make synaptic connections on their dendritic arbor (see below).

CABLE EQUATION AND ACTIVATING FUNCTION

Much of our understanding of how a neuron responds to extracellular stimulation has come from computational models. The modeling techniques typically used to predict the neural response to extracellular stimulation date back to McNeal [2], who was the first to integrate an electric field model and a neuron model to simulate action potential generation. Modern extracellular stimulation modeling still relies on those same two fundamental components:

1. a model of the voltage distribution generated by the stimulating electrode(s), and
2. a model of the neuron(s) being stimulated.

Voltage distribution models range from simple (i.e. theoretical point source electrode in an infinite homogeneous isotropic medium) to complex (i.e. finite element volume conductor with explicit representation of electrode geometry, time dependence, and tissue inhomogeneity/anisotropy) [6]. Irrespective of the voltage distribution model selected, the simulated extracellular potentials (Ve[n]) at the location of individual compartments of neurons in the surrounding tissue medium can be predicted by defining a common coordinate system between the stimulating electrode and the neuron. The neural response to the stimulation can then be simulated with electrical circuits of conductances (Gm[n]) and capacitors (Cm[n]) in parallel, defining the transmembrane voltage [14]. The individual compartments of a single neuron are then connected in series by resistors representing the intracellular resistance (Gi[n]) [15]. Neuron models of this type are commonly referred to as multicompartment cable models. When an extracellular stimulus is applied to the neuron model, the membrane current at compartment n is equal to the sum of the incoming axial currents and the sum of the capacitive and ionic currents through the membrane:

$$Cm[n](dVm[n]/dt) + Ii[n] =$$
$$Gi[n-1](Vi[n-1] - Vi[n] + Ve[n-1] - Ve[n]) + Gi[n](Vi[n+1] - Vi[n]$$
$$+ Ve[n+1] - Ve[n])$$

where the transmembrane voltage at each compartment (Vm[n]) is defined by difference between the intracellular (Vi[n]) and extracellular (Ve[n]) potentials [2].

The response of an individual neuron to the applied field is related to the second derivative of the extracellular potential distribution along each neural process (commonly referred to as the 'activating function') [3]. Each neuron (or neural process) surrounding the electrode will be subject to both depolarizing and hyperpolarizing effects from the stimulation [16]. Therefore, a neuron can be either activated or suppressed in response to extracellular stimulation in different ways and in different neural processes depending on its positioning with respect to the electrode and the stimulation parameters.

DIRECT NEURAL ACTIVATION

The measurable response of a neuron to a sufficiently strong stimulus is the generation of an action potential. The coupled analysis of modeling and experimental results over the last four decades provides four general conclusions on the effects of stimulation [17]:

1. when stimulating local cells with extracellular sources, action potential initiation (API) typically takes place in a node of Ranvier of the axon of that cell. This site of API is not necessarily the closest point of the neuron to the electrode, but the axon is the most excitable part of the neuron and hence the most susceptible to extracellular activation. This axonally generated action potential then propagates in both directions (orthodromically and antidromically) along the fiber
2. when stimulating axon terminals and fibers of passage with extracellular sources, API takes place in a node of Ranvier relatively close to the electrode. Once again, this action potential propagates in both directions along the fiber
3. for most neuron–electrode orientations, anodic stimuli are more effective in activating local cells than cathodic stimuli
4. cathodic stimuli are more effective in activating fibers of passage than anodic stimuli.

However, it should always be noted that activation of any neuron with extracellular electric fields is dependent on four main factors [17]:

1. electrode geometry and the electrical conductivity of the tissue medium. The response of the neuron to stimulation is dependent on the electric field generated by the electrode, which is dependent on the size and shape of

the electrode. In addition, the inhomogeneous and anisotropic electrical properties of the CNS tissue medium affect the shape of the electric field. Therefore, both the type of electrode used and the region of the nervous system where it is inserted will affect the neural response to stimulation

2. stimulation parameters. Changes in stimulation parameters can affect the types of neurons activated, and the volume of tissue over which activation will occur. The four primary stimulation parameters are the polarity, duration, and amplitude of the stimulus pulse, as well as the stimulation frequency. In general, alterations in the stimulus pulse duration and amplitude will affect the volume of tissue activated by the stimulus, and alterations in the stimulus polarity and frequency will affect the types of neurons activated by the stimulus

3. geometry of the neuron and its position with respect to the electrode. In general, the closer the neuron is to the electrode the lower the stimulation current necessary for activation. However, complex neural geometries such as dendritic trees and branching axons result in a large degree of variability in current–distance relationships (threshold current as a function of electrode-to-neuron distance). Therefore, the orientation of the neural structures with respect to the electrode is of similar importance as the geometric distance between them, especially for small electrode-to-neuron distances

4. ion channel distribution on the neuron. Axonal elements of a neuron consist of a relatively high density of action-potential-producing sodium channels compared to cell bodies and dendrites. As a result, the axonal elements of a neuron are the most excitable and regulate the neural output that results from application of extracellular electric fields. However, while the cell body and dendrites may not be directly responsible for the action potential spiking that results from the stimulus, they do contain several types of calcium and potassium channels that can affect neuronal excitability on long time scales when trains of stimuli are used.

INDIRECT SYNAPTIC MODULATION

Previous experimental and modeling results have shown that the threshold for indirect, or trans-synaptically evoked excitation or inhibition of local cells stimulated with extracellular sources is similar to (in some cases dependent on electrode location less than) the threshold for direct excitation of local cells [18]. Indirect excitation or inhibition of local cells is the result of stimulation-induced release of neurotransmitters that result from the activation of axon terminals activated by the stimulus. In general, axon terminals are activated at low stimulus amplitudes relative to local cells. Therefore, when considering the effect of the stimulus on local cells near the electrode, it is probable that a large numbers of axon terminals are activated resulting in high levels of synaptic activity on the dendritic tree of the local cell. This stimulation induced trans-synaptic activity can be predominantly excitatory, predominantly inhibitory, or any relative mix of excitation and inhibition depending on the types and numbers of synaptic receptors activated.

Therefore, the interpretation of the effects the stimulation on the neuronal output of local cells is made up of two components:

1. the direct effect of the extracellular electric field on the local cell
2. the indirect effect of the stimulation-induced trans-synaptic excitation and/ or inhibition.

As a result, an action potential can be generated either by the stimulus pulse itself or by indirect synaptic activation. However, activation of axon terminals by extracellular stimuli is non-selective to excitatory or inhibitory neurotransmitter release. In general, the indirect effects of extracellular stimulation of local cells result in a biphasic response of a short period of depolarization followed by a longer period of hyperpolarization. This biphasic response is the result of the interplay between the time courses of the traditionally fast excitatory synaptic action and the traditionally slow inhibitory synaptic action. The role of indirect effects on the output of local cells can be enhanced with high frequency stimulation. If the inter-stimulus interval is shorter than the time course of the synaptic conductance, the indirect effects will summate. Because inhibitory synaptic action traditionally has a longer time course than excitatory synaptic action, the effect of this summation is hyperpolarization of the cell body and dendritic arbor of the local cell. This hyperpolarization can limit the neuronal output when stimulating at high frequencies. However, because API from direct activation takes place in the axon of local cells, the efferent output of local cells is typically an action potential in response to each stimulus pulse, given that the stimulus amplitude is strong enough for direct activation of that cell's axon.

STIMULATION-INDUCED NETWORK ACTIVITY

The fundamental purpose of a neurostimulation device is to generate a desired clinical outcome (i.e. improvement in symptoms) without induction of stimulation-induced side effects. The stimulation effects at the site of the electrode are only the first step in that process. Those stimulation-induced action potentials are transmitted to the axon terminals of the activated neurons, resulting in stimulation-induced synaptic action on the neural network. Prevailing hypotheses suggest that most forms of central nervous system neurostimulation rely on this stimulation-induced network activity to generate the desired clinical outcome [19].

Defining relationships between the anatomical placements of the electrode, the stimulation parameter settings, the relative proportion of neurons directly stimulated, the stimulation-induced network activity, and the resulting behavioral outcomes, represent the state-of-the-art process for deciphering the therapeutic mechanisms of neurostimulation therapies. However, integration of such systems is so complex that it typically requires computational models and numerous simplifying assumptions to analyze appropriately. In turn, numerous scientific questions remain unanswered on the stimulation-induced network activity generated by therapies like DBS. Nonetheless, as new experimental data become available, and modeling technology evolves, it will be possible to integrate synergistically the results of systems neurophysiology with large-scale neural network models to create a realistic representation of the brain circuits being modulated by neurostimulation. Such advances will enable the development of novel stimulation technology (electrodes, pulsing paradigms, pulse generators, etc.) that can be optimized to achieve specific clinical goals; thereby improving patient outcomes through better understanding of the fundamentals of neurostimulation.

REFERENCES

1. Merrill DR, Bikson M, Jefferys JG. Electrical stimulation of excitable tissue: design of efficacious and safe protocols. *J Neurosci Meth*. 2005;141:171–198.
2. McNeal DR. Analysis of a model for excitation of myelinated nerve. *IEEE Trans Biomed Eng*. 1976;23:329–337.
3. Rattay F. Analysis of models for external stimulation of axons. *IEEE Trans Biomed Eng*. 1986;33:974–977.
4. Maks CB, Butson CR, Walter BL, Vitek JL, McIntyre CC. Deep brain stimulation activation volumes and their association with neurophysiological mapping and therapeutic outcomes. *J Neurol Neurosurg Psychiatry*. 2009;80:659–666.
5. Miocinovic S, Lempka SF, Russo GS, et al. Experimental and theoretical characterization of the voltage distribution generated by deep brain stimulation. *Exp Neurol*. 2009;216:166–176.
6. Chaturvedi A, Butson CR, Lempka SF, Cooper SE, McIntyre CC. Patient-specific models of deep brain stimulation: influence of field model complexity on neural activation predictions. *Brain Stimul*. 2010;3:65–77.
7. Lempka SF, Miocinovic S, Johnson MD, Vitek JL, McIntyre CC. In vivo impedance spectroscopy of deep brain stimulation electrodes. *J Neural Eng*. 2009;6:046001.
8. Butson CR, Maks CB, McIntyre CC. Sources and effects of electrode impedance during deep brain stimulation. *Clin Neurophysiol*. 2006;117:447–454.
9. Lempka SF, Johnson MD, Miocinovic S, Vitek JL, McIntyre CC. Current-controlled deep brain stimulation reduces in vivo voltage fluctuations observed during voltage-controlled stimulation. *Clin Neurophysiol*. 2010; (in press).
10. Tuch DS, Wedeen VJ, Dale AM, George JS, Belliveau JW. Conductivity tensor mapping of the human brain using diffusion tensor MRI. *Proc Natl Acad Sci USA*. 2001;98:11697–11701.
11. McIntyre CC, Mori S, Sherman DL, Thakor NV, Vitek JL. Electric field and stimulating influence generated by deep brain stimulation of the subthalamic nucleus. *Clin Neurophysiol*. 2004;115:589–595.
12. Butson CR, McIntyre CC. Tissue and electrode capacitance reduce neural activation volumes during deep brain stimulation. *Clin Neurophysiol*. 2005;116:2490–2500.
13. Ranck JB. Which elements are excited in electrical stimulation of mammalian central nervous system: a review. *Brain Res*. 1975;98:417–440.
14. Hodgkin AL, Huxley AF. A quantitative description of membrane current and its application to conduction and excitation in nerve. *J Physiol*. 1952;177:500–544.
15. Rall W, Burke RE, Holmes WR, Jack JJ, Redman SJ, Segev I. Matching dendritic neuron models to experimental data. *Physiol Rev*. 1992;72(Suppl.):S159–186.

16. McIntyre CC, Grill WM. Excitation of central nervous system neurons by non-uniform electric fields. *Biophys. J*. 1999;76:878–888.

17. Lee DC, McIntyre CC, Grill WM. Extracellular electrical stimulation of central neurons: quantitative studies. In: Finn, Lopresti, eds. *Handbook of Neuroprosthetic Reseach Methods*. CRC Press; 2003.

18. Gustafsson B, Jankowska E. Direct and indirect activation of nerve cells by electrical pulses applied extracellularly. *J Physiol*. 1976;258:33–61.

19. McIntyre CC, Hahn PJ. Network perspectives on the mechanisms of deep brain stimulation. *Neurobiol Dis*. 2010;38:329–337.

Considerations for Quantitative Modeling of Excitation and Modulation of CNS neurons[1]

Warren M. Grill, PhD

Department of Biomedical Engineering, Duke University, Durham, NC

INTRODUCTION

Electrical stimulation has a long history as a tool to study the form and function of the nervous system and, in the past several decades, has emerged as an effective means to restore function following neurological disease or injury. Two persistent and long-standing challenges of using electrical stimulation in the central nervous system (CNS) are the ambiguity as to which neural elements are activated under different conditions [1] and how selective activation of targeted elements can be accomplished through, for example, selection of electrode geometry and stimulation parameters [2].

Quantitative approaches using first principles, concepts like the activating function, cable modeling of neurons, and experimental measurements have generated substantial insight into these questions. In this commentary, I focus on several fundamental issues whose importance in understanding and controlling electrical activation of the central nervous system has been highlighted by recent advances in the field. In several instances, the issues are presented in the context of determining and controlling the effects of deep brain stimulation (DBS), although the principles carry over to many applications of CNS stimulation.

Electrical activation of the nervous system has traditionally been thought of and analyzed as a two-part problem. The first part is determining, through measurement or calculation, the electrical potentials (voltages) generated in the tissue by the application of stimulation pulses. The second part is determining, again through measurement or calculation, and now, through imaging, the response of neurons to the stimulation pulses (i.e. to the voltages imposed in the tissue). However, recent progress highlights the need to add a third part to this problem – the network effects of stimulation. That is, given the changes in the pattern of activity in the neurons directly affected by stimulation, what changes occur either downstream from the point of stimulation or even further

---

1. Preparation of this material was supported in part by the US National Institutes of Health (R01 NS040894).

distant within interconnected networks of neurons. Again, interplay between experimental and modeling approaches is beginning to decipher the network effects of CNS stimulation.

PART ONE: GENERATION OF POTENTIALS IN TISSUE

Early quantitative models of electrical stimulation relied on simple representations of the electrode and surrounding volume conductor. While approximating the electrode as a point source is a good representation of a sharp microelectrode [3], it was not readily apparent that such an approximation was adequate for the large cylindrical and disk macroelectrodes used for clinical applications of electrical stimulation. In a recent analysis, we found that, under a broad range of conditions, the point source is an adequate representation of a cylindrical DBS electrode, especially if the intent is to calculate the activation of a population of axons [4]. When considering activation of a population, as compared to a particular neuron, positive and negative errors in the estimation of threshold of individual axons cancelled one another, and led to small errors in both the number and spatial extent of stimulated neurons.

A second consideration is the representation of the volume conductor representing the tissue(s) surrounding the electrode. Again, the volume conductor was initially most often represented as an infinite homogeneous, isotropic, medium. However, the nervous system is indeed inhomogeneous (meaning the electric properties depend on position) and strongly anisotropic (meaning that the electrical properties depend on the direction that the current is moving through the tissue), and these properties of the volume conductor can influence the thresholds and spatial distribution of neuronal activation [5]. Measurements in vivo suggest that the potentials in the immediate vicinity of DBS electrodes are similar to those predicted in a homogeneous isotropic volume conductor [6]. However, a more recent analysis highlighted the importance of including tissue electrical properties to make accurate predictions of the thresholds to activate fibers within an adjoining white matter tract [7]. Such highly anisotropic regions are poorly represented by a homogeneous isotropic medium, and failure to include anisotropic electrical conductivity led to an underestimation of thresholds or, equivalently, an overestimation of the volume of tissue stimulated with a given intensity. Collectively, these results suggest that simple volume conductor models will suffice to predict activation within a local gray matter region around the electrode, but prediction of activation of white matter regions, especially further from the electrode, requires more realistic depictions of the electrical properties of the tissue.

A third consideration is the representation of the electrical properties of the tissue. The electrode–tissue interface has capacitive properties (i.e. the double layer capacitance) that can strongly influence the potentials in the tissue and neural excitation when using regulated voltage pulses [8]. As well, the tissue has a capacitance associated with it, but the reactive component of the conductivity is small relative to the real component [9], and thus only small changes in conductivity are expected over the range of signal power frequencies relevant to neural stimulation. This expectation is consistent with measurements of neural tissue conductivity over a range of frequencies [10–12] and recent theoretical and experimental results support the validity of the quasistatic approximation [13] – i.e. treating the volume conductor as purely resistive – when calculating the potentials in problems of neural stimulation. Measurements in vivo indicate that the impedance of cortical tissue is frequency-independent between 10 Hz and 5 kHz [14], the frequency band that contains the bulk of the energy in typical neural stimulating pulses [15]. Thus, the tissue does not alter the temporal properties of the stimulation pulse (i.e. the tissue does not act like a filter), or in other words, the capacitive properties of the tissue can be ignored.

PART TWO: EFFECTS OF EXTRACELLULAR POTENTIAL ON NEURONS

Subsequent to determining the distribution of electrical potentials in the tissue, the second challenge is determining the response of neurons to stimulation. I first consider direct effects, i.e. those mediated by electrical effects on the transmembrane potential of surrounding neurons. Subsequently, I consider indirect effects, i.e. those mediated by synaptic transmission following direct excitation of presynaptic axons or terminals.

Direct excitation of post-synaptic cells

In addition to the familiar factors that determine the effects of extracellular stimulation on neurons – stimulation parameters, electrode geometry, electrode location [1] – more recent results have revealed that the direct effects of stimulation are also strongly dependent on the rate and pattern of intrinsic activity present in the cell in the absence of stimulation, as well as the temporal pattern of the stimulation train [16].

As part of an effort to determine the mechanisms of action of DBS, we quantified the effect of DBS at different amplitudes and frequencies on the output of intrinsically active model neurons. DBS produced frequency-dependent modulation of the output firing and, above a critical frequency, all intrinsic activity was masked and replaced by firing at the stimulation frequency [17,18]. Changes in the coefficient of variation of the interspike interval of the bursting neurons as a function of the stimulation frequency and stimulation intensity matched remarkably well the shape of tremor amplitude as a function of DBS frequency measured in persons with DBS of the ventral intermediate thalamic nucleus (Vim) for essential tremor [18]. These findings highlight the importance of considering ongoing intrinsic neural activity when determining the effects of stimulation and are consistent with DBS masking pathological burst activity and with regularization of neuronal firing pattern as one of the mechanisms underlying the effectiveness of DBS.

Indirect effects on post-synaptic cells

The thresholds for excitation of presynpatic terminals and subsequent indirect effects on local neurons (mediated by synaptic transmission) are similar to thresholds for direct effects (mediated by stimulus current) during extracellular stimulation [19,20]. Thus, it is important to consider indirect effects of extracellular stimulation on neurons [21]. Further, recent evidence suggests that indirect effects can propagate quite far from the stimulating electrodes, and robust antidromic and re-orthodromic propagation of action potentials in branched axons generates widespread effects across the brain [22]. It is important, however, to consider that extended, especially high frequency, activation can lead to synaptic depression, which will mute downstream effects of stimulation, as well as potentially mask intrinsic activity passing through the same synapses [23,24].

A recent study using calcium imaging to infer activation of neurons within $\approx 200\,\mu m$ of the electrode by low amplitude microstimulation – where increase in calcium-mediated fluorescence was an indicator of action potential firing – demonstrated that microstimulation resulted in a widespread but sparse pattern of neuronal excitation [25]. Interestingly, the number and distribution of activated neurons was largely similar following blockade of excitatory synaptic transmission, thus demonstrating that activation resulted from direct stimulation, presumably of groups of axons, which then resulted in antidromic activation of the imaged cells [26].

Indeed, previous electrophysiological studies demonstrated long-range effects of microstimulation, including both inhibition and excitation, several millimeters from the electrode [27]. This extensive spread of activation was corroborated by studies using functional magnetic resonance imaging (fMRI) to map changes in hemodynamics during microstimulation of the visual cortex in anesthetized monkeys [28]. Presumably, the extensive spread arose in part due to the horizontal connections within the cortex.

These data in no way contradict the well established current–distance relationship – that the threshold current to activate a neuron is proportional to the square of the distance between the electrode and the neuron [1,21]. Rather they demonstrate that the relationship does not hold true when considering a population of neurons (although this appears to be the case for a population of axons [29]). That is, the radius of a sphere bounding all activated neurons within a volume does not necessarily grow as the square root of the current, nor will all neurons within the bounding sphere be activated.

PART THREE: NETWORK EFFECTS OF CNS STIMULATION

As described above, the cellular effects of CNS stimulation are becoming increasing clear, but it remains to be determined what the effects of local stimulation are on downstream neural elements, as well as how these local changes in activity are propagated within a network of highly interconnected neurons, including feedback pathways.

In an effort to understand further the network effects of CNS stimulation, investigators have begun building network models of the basal ganglia and/or thalamus [30–32]. In the first effort in this direction, Rubin and Terman [32] assembled a small network of interconnected Hodgkin–Huxley style neurons to quantify the effects of deep brain stimulation of the subthalamic nucleus (STN) on firing in downstream nuclei (globus pallidus), as well as the thalamus. The model suggested that STN DBS tended to regularize the firing in the neurons in globus pallidus pars interna (GPi), which were bursty in the parkinsonian condition, a result consistent with experimental measurements of GPi neuronal activity during STN DBS [33,34]. Further, the model predicted that the more regular, tonic firing of GPi neurons generated persistent inhibition of thalamus, thereby improving thalamic fidelity. Indeed, analysis of thalamic neuronal activity during STN DBS demonstrated that high frequency symptom relieving DBS made thalamic neurons much less likely to burst, while low frequency ineffective DBS made thalamic neurons more likely to burst [34]. Further, we recently found that the changes in thalamic fidelity in a modified version of the Rubin and Terman network model in response to different temporal patterns of DBS were strikingly well correlated with changes in bradykinesia in human subjects during STN DBS with the same temporal patterns of stimulation [35]. Collectively, these results demonstrate the potential value of network models of the effects of CNS stimulation, as well as the interplay between model and experiment.

Another tool to evaluate the network effects of CNS stimulation is the use of functional neural imaging, including positron emission tomography (PET) and functional magnetic resonance imaging (fMRI). Recent functional imaging studies support the notion that the effect of DBS is to change the temporal patterns of activity in a spatially distributed brain network. DBS reduced glucose metabolism in the pallidum and pons, areas believed to be hyperactive in Parkinson's disease (PD), and increased glucose metabolism in cortical areas believed to be underactive in PD [36]. Importantly, the changes in network activity were correlated with the reduction in parkinsonian symptoms, and this suggests that these changes are strongly linked to the mechanisms of symptom amelioration [36]. Similarly, effective DBS increased rCBF in areas that were underactive in PD relative to controls, including supplementary motor cortex, and decreased rCBF in areas that were overactive in PD as compared to controls, including premotor cortex and cerebellum [37]. Together these studies suggest that DBS normalizes neuronal activity, not just in the vicinity of the electrode, but across highly interconnected brain networks.

STUDY QUESTIONS

1. Consider that the fidelity of synaptic transmission from an incoming action potential is variable and well below 100% – how might such a detail affect information transfer?
2. Review the main components of neurons affected by DBS in the short term (< a few seconds). How might longer-term changes (> hours) from DBS be manifested within cells?

REFERENCES

1. Ranck JB Jr. Which elements are excited in electrical stimulation of mammalian central nervous system: a review. *Brain Res.* 1975;98:417–440.
2. Kuncel AM, Grill WM. Selection of stimulus parameters for deep brain stimulation. *Clin Neurophysiol.* 2004;115:2431–2441.
3. McIntyre CC, Grill WM. Finite element analysis of the current-density and electric field generated by met al microelectrodes. *Ann Biomed Eng.* 2001;29:227–235.
4. Zhang T, Grill WM. *Effect of electrode geometry on deep brain stimulation: monopolar point source vs. Medtronic 3389 lead.* Miami, FL: 25th Southern Biomedical Engineering Conference; 2009.
5. Grill WM. Modeling the effects of electric fields on nerve fibers: influence of tissue electrical properties. *IEEE Trans Biomed Eng.* 1999;46:918–928.
6. Miocinovic S, Lempka SF, Russo GS, et al. Experimental and theoretical characterization of the voltage distribution generated by deep brain stimulation. *Exp Neurol.* 2009;216:166–176.
7. Chaturvedi A, Butson CR, Lempka SF, Cooper SE, McIntyre CC. Patient-specific models of deep brain stimulation: influence of field model complexity on neural activation predictions. *Brain Stimul.* 2010;3:65–77.
8. Butson CR, McIntyre CC. Tissue and electrode capacitance reduce neural activation volumes during deep brain stimulation. *Clin Neurophysiol.* 2005;116:2490–2500.
9. Ackmann JJ, Seitz MA. Methods of complex impedance measurements in biologic tissue. *Crit Rev Biomed Eng.* 1984;11:281–311.
10. Ranck JB Jr. Specific impedance of rabbit cerebral cortex. *Exp Neurol.* 1963;7:144–152.
11. Nicholson PW. Specific impedance of cerebral white matter. *Exp Neurol.* 1965;13:386–401.
12. Ranck JB Jr, BeMent SL. The specific impedance of the dorsal columns of the cat: an anisotropic medium. *Exp Neurol.* 1965;11:451–463.
13. Plonsey R, Heppner D B. Considerations of quasi-stationarity in electrophysiological systems. *Bull Math Biophys.* 1967;29:657–664.
14. Logothetis NK, Kayser C, Oeltermann A. In vivo measurement of cortical impedance spectrum in monkeys: implications for signal propagation. *Neuron.* 2007;55:809–823.
15. Bossetti CA, Birdno MJ, Grill WM. Analysis of the quasi-static approximation for calculating potentials generated by neural stimulation. *J Neural Eng.* 2008;5:44–53.
16. Birdno MJ, Grill WM. Mechanisms of deep brain stimulation in movement disorders as revealed by changes in stimulus frequency. *Neurotherapeutics.* 2008;5:4–25.
17. Grill WM, Snyder AN, Miocinovic S. Deep brain stimulation creates an informational lesion of the stimulated nucleus. *Neuroreport.* 2004;15:1137–1140.

18. Kuncel AM, Cooper SE, Wolgamuth BR, Grill WM. Amplitude- and frequency-dependent changes in neuronal regularity parallel changes in tremor with thalamic deep brain stimulation. *IEEE Trans Neural Syst Rehabil Eng.* 2007;15:190–197.

19. Baldissera F, Lundberg A, Udo M. Stimulation of pre- and postsynaptic elements in the red nucleus. Exp. *Brain Res.* 1972;15:151–167.

20. Gustafsson B, Jankowska E. Direct and indirect activation of nerve cells by electrical pulses applied extracellularly. *J Physiol.* 1976;258:33–61.

21. Tehovnik EJ, Tolias AS, Sultan F, Slocum WM, Logothetis NK. Direct and indirect activation of cortical neurons by electrical microstimulation. *J Neurophysiol.* 2006;96:512–521.

22. Grill WM, Cantrell MB, Robertson MS. Antidromic propagation of action potentials in branched axons: implications for the mechanisms of action of deep brain stimulation. *J Comput Neurosci.* 2008;24:81–93.

23. Urbano FJ, Leznik E, Llinas RR. Cortical activation patterns evoked by afferent axons stimuli at different frequencies: an in vitro voltage-sensitive dye imaging study. *Thal Rel Sys.* 2002;1:371–378.

24. Anderson TR, Hu B, Iremonger K, Kiss ZH. Selective attenuation of afferent synaptic transmission as a mechanism of thalamic deep brain stimulation-induced tremor arrest. *J Neurosci.* 2006;26:841–850.

25. Histed MH, Bonin V, Reid RC. Direct activation of sparse, distributed populations of cortical neurons by electrical microstimulation. *Neuron.* 2009;63:508–522.

26. Lipski J. Antidromic activation of neurones as an analytic tool in the study of the central nervous system. *J Neurosci Meth.* 1981;4:1–32.

27. Butovas S, Schwarz C. Spatiotemporal effects of microstimulation in rat neocortex: a parametric study using multielectrode recordings. *J Neurophysiol.* 2003;90:3024–3039.

28. Tolias AS, Sultan F, Augath M, et al. Mapping cortical activity elicited with electrical microstimulation using FMRI in the macaque. *Neuron.* 2005;48:901–911.

29. Mahnam A, Hashemi SM, Grill WM. Measurement of the current-distance relationship using a novel refractory interaction technique. *J Neural Eng.* 2009;6:036005.

30. Rubin JE, Terman D. High frequency stimulation of the subthalamic nucleus eliminates pathological thalamic rhythmicity in a computational model. *J Comput Neurosci.* 2004;16:211–235.

31. Arle JE, Mei LZ, Shils JL. Modeling parkinsonian circuitry and the DBS electrode. I. Biophysical background and software. *Stereotact Funct Neurosurg.* 2008;86:1–15.

32. Shils JL, Mei LZ, Arle JE. Modeling parkinsonian circuitry and the DBS electrode. II. Evaluation of a computer simulation model of the basal ganglia with and without subthalamic nucleus stimulation. *Stereotact Funct Neurosurg.* 2008;86:16–29.

33. Hashimoto T, Elder CM, Okun MS, Patrick SK, Vitek JL. Stimulation of the subthalamic nucleus changes the firing pattern of pallidal neurons. *J Neurosci.* 2003;23:1916–1923.

34. Dorval AD, Russo GS, Hashimoto T, Xu W, Grill WM, Vitek JL. Deep brain stimulation reduces neuronal entropy in the MPTP-primate model of Parkinson's disease. *J Neurophysiol.* 2008;100:2807–2818.

35. Dorval 2nd AD, Kuncel AM, Birdno MJ, Turner DA, Grill WM. Deep brain stimulation alleviates parkinsonian bradykinesia by regularizing pallidal activity. *J Neurophysiol.* 2010;104:911–921.

36. Grafton ST, Turner RS, Desmurget M, et al. Normalizing motor-related brain activity: subthalamic nucleus stimulation in Parkinson disease. *Neurology.* 2006;66:1192–1199.

37. Trost M, Su S, Su P, et al. Network modulation by the subthalamic nucleus in the treatment of Parkinson's disease. *Neuroimage.* 2006;31:301–307.

Chapter 8

The Electrode – Principles of the Neural Interface: Circuits

Erwin B. Montgomery Jr, MD

Dr Sigmund Rosen Scholar in Neurology, Professor of Neurology, University of Alabama at Birmingham, Birmingham, AL

It is an empiric fact that deep brain stimulation (DBS) directly and proximately (antidromicaly and oligosynapticaly) activates wide regions of the brain at time scales comparable to the immediate clinical effect. Further, evidence and reason support the notion that the widespread and simultaneous activations of brain systems are causal to the therapeutic effect of DBS. Combined, these facts and reasoned inferences challenge the most basic tenets of today's conception of physiology and pathophysiology.

It will be argued that the DBS effects of interest are the result of activating distributed systems, *the DBS systems effect hypothesis*, rather than specific structures uniquely and in isolation. The DBS systems effect is not the trivial case, in the philosophical sense, that DBS affects wide regions of the brain, eventually. Ultimately, DBS must affect the muscle because motor performance improves, for example in patients with Parkinson's disease, presumably by engaging the motor cortex (MC), spinal cord and possibly brainstem. The claim of the DBS systems effect hypothesis will be more fully explicated subsequently, but it stands in contrast to claims that the effects of interest, usually those associated with some clinically meaningful effect, are due to actions restricted to the stimulated target; what will be termed *the DBS local effect hypothesis*. Adjudicating between these conflicting claims first presupposes some notions, implicit or explicit, of causality and how such causality is established. Indeed, it will be argued that the claims of DBS local effect hypothesis were a default position because of the presuppositions current in more general notions of basal ganglia physiology and pathophysiology. Alternatives, such as the DBS systems effect hypothesis, have (had) a more difficult time earning its respect because it opposes not only the DBS local effect hypothesis but also the antecedent notions of causality in basal ganglia physiology and pathophysiology. Thus, the implications of a DBS systems effect hypothesis are wide and deep.

Essential Neuromodulation. DOI: 10.1016/B978-0-08-089064-7.00004-1

One might argue that the very elegant work by Eidelberg and colleagues demonstrating an effect of DBS on the Parkinson network of neurometabolic changes would argue for a systems effect [1]. However, because of the time scales or temporal resolution of neurometabolic imaging, the network or systems wide effect could be only in the trivial sense as described above. Thus, stimulation of the subthalamic nucleus (STN) or globus pallidus interna (GPi) could result in eventually activating all the nuclei and cortical structures associated with the basal ganglia but at a time scale not relevant to the very rapid dynamics necessary for the effects of DBS or motor control which is over the course of milliseconds. Note, for example, the time difference in the inter-stimulus interval for an effective 130 pulses per second (pps) and an ineffective 100 pps DBS is approximately 3 ms. Yet, this 3 ms difference is sufficient to result in a benefit or none. The issue is raised primarily to demonstrate the importance of understanding the effects of DBS (and basal ganglia physiology in general) on a time scale on the order of milliseconds. Consequently, any analytic or scientific method that does not have a temporal resolution on the order of milliseconds, at best will provide little insight, and at worst, be misleading resulting in complacency.

EVIDENCE IN SUPPORT OF A DBS SYSTEMS EFFECT HYPOTHESIS

The following reviews the empiric evidence that DBS activates directly and proximately wide areas of the brain. DBS of the basal ganglia structures will be taken as the exemplar. Note that the argument for the DBS systems effect hypothesis does not discount the observations established by investigators favoring the DBS local effect hypothesis; rather, the criticism is the extrapolation from those observations to inferences of neuronal physiology and pathophysiology sufficient to explain behavior.

First, a definition of what is meant by direct and proximate is offered. Direct implies an effect in a neural element consequent to the electromagnetic fields induced by the DBS pulse. Proximate means that these direct effects have an immediacy, that are operating in a narrow time frame, in this case within one to two synapses from the direct effect. For example, as will be further discussed, the direct effect of the DBS pulse on the efferent axon of the GPi neuron causes an affect in neurons of the ventrolateral thalamus (VL) with the assumption that this affect is important to the clinically important DBS effects. This is to distinguish the proximate effect from the more trivial case of eventual activation of the motor system all the way through to the muscles that is necessary to manifest the clinical DBS effect.

Antidromic activation of axons projecting to and in the vicinity of the DBS

Microelectrode recordings of neurons throughout the basal ganglia–thalamic–cortical (BG–TH–CTX) system during subthalamic nucleus (STN)

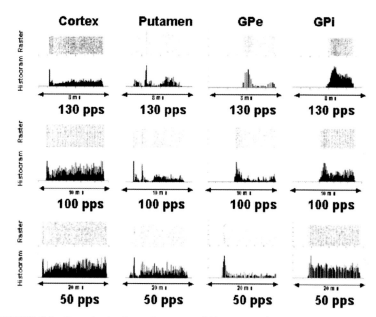

FIGURE 8.1   Post-stimulus interval rasters and histograms of neuronal responses to the DBS pulse. The time interval represented is the time between two successive DBS pulses with the first pulse occurring at the left margin of the time line. Note that the time line varies depending on the DBS frequency. For 130 pps, the time line is 8 ms; for 100 pps, 10 ms; and for 50 pps, 20 ms. The top figure of each pair is the raster where each dot represents the time of a neuronal action potential. Each row in the raster represents a single DBS pulse. The neuronal activities are collapsed across the rows to produce the histogram of neuronal activities in response to the DBS pulse. Rasters and histograms are shown for representative neurons in the motor cortex (cortex), putamen, globus pallidus externa (GPe) and globus pallidus interna (GPi). In each case, the DBS produces a robust response in each neuron. Thus, DBS produces neuronal activations throughout the basal ganglia–thalamic–cortical system. Also interesting, the qualitative characteristics of the response are the same regardless of the DBS frequency. In the motor cortex, one can see a robust, temporally consistent response at 1.2 ms consistent with an antidromic response. Similar antidromic responses were seen in the sensory cortex and GPe (not shown). With permission from [2].

DBS demonstrate findings consistent with antidromic activation of neurons in the MC, somatosensory cortex (SM), and globus pallidus externa (GPe) [2]. Thus, axons from the MC, SM, and GPe projecting to or in the vicinity of the STN are activated directly by the DBS pulse and that activation is conducted retrograde to cause an action potential in neurons of the MC, SM, and GPe. Figure 8.1 shows peri-event rasters and histograms of neuronal activities following the DBS pulse in the STN of a non-human primate. As can be seen, there is a robust short latency action potential response that is temporally consistent, meaning very little variation in the time of occurrence unlike the more broad and varying subsequent responses. The reason is that most of the variability in communication of information between neurons lies in the diffusion of neurotransmitters across the synaptic cleft which is not a factor in antidromic activation. Further, this response attributed to antidromic activation is seen with different frequencies of DBS stimulation. Regular orthodromic activation requires driving responses in the postsynaptic neuron. Higher frequencies take advantage of temporal summation in the postsynaptic neuron to increase the probability of action potential generation. The robust, short latency, temporally consistent response seen over multiple frequencies are three out of four criteria for an antidromic response. The fourth criterion is collision. To demonstrate collision, one looks for episodes where a spontaneous action potential occurs in the neuron of interest just prior to the DBS pulse. This spontaneous action potential then induces a refractory period that blocks the ascending antidromic action potential and consequently, no response is seen in the neuron.

Unfortunately, in the experiments of which Figure 8.1 is an example, it was not possible to demonstrate collision because insufficient lengths of data were recorded; thereby resulting in rare occurrence of a spontaneous action potential just prior to the DBS pulse. At the time of the experiments, the possibility of antidromic activations was unanticipated.

Antidromic activation of the MC in non-human primates undergoing STN DBS is consistent with the finding of robust cortical electroencephalographic (EEG) potentials at short latencies in response to STN DBS (Fig. 8.2) consistent with antidromic activation. Further evidence of axonal activation came from the EEG evoked potentials in humans using paired-pulse DBS experiments [3]. Pairs of DBS pulses were delivered through the STN and the evoked potentials measured from scalp electrodes. If the time interval between the pair of pulses is more than the refractory period of the stimulated neuronal element, the axon, for example, the effect of the pair of pulses would be relatively doubled compared to a pair of pulses whose interval between pulses were less than the refractory period. There was a marked change in the magnitude of the STN DBS EEG evoked potential between an interval between pulse pairs of 0.5 and 0.75 ms

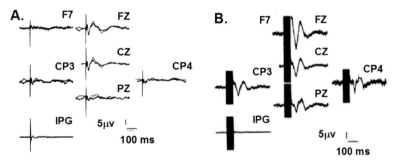

FIGURE 8.2   Electroencephalographic (EEG) evoked potentials from left STN DBS in a human. (A) The evoked potentials associated with a single pulse; (B) the evoked potentials associated with a brief train of high frequency DBS. As can be seen the evoked potentials are primarily frontal midline and over the left motor area with the single pulse. This distribution is consistent with the predominant projections of the basal ganglia to the cortex. Also note that the response is more robust and more widespread with brief trains of DBS pulses. With permission from [3].

suggesting that the refractory period of the neuronal element responsible for the EEG evoked potential is between 0.5 and 0.7 ms. This refractory period is more consistent with axons than it is with neuronal cell bodies (somas) or dendrites. This is evidence that the EEG evoked potential to STN DBS is mediated by axons and most likely axons between the MC and the STN.

In retrospect, it is understandable that STN DBS would result in antidromic activations of the MC, SC, and GPe as these structures send axons to the STN. Further, biophysical principles demonstrate that axon terminals have the lowest threshold to electrical stimulation.

What could not be anticipated from the anatomy is the antidromic activation of other structures such as the contralateral STN. Microelectrode recordings in the STN contralateral to STN DBS demonstrate clear evidence of increased activity including antidromic activation with collision (unpublished observations). These data clearly demonstrate that axons from the contralateral STN pass close enough in the vicinity of the DBS-stimulated STN to be antidromically activated. The precise terminations of these axons from the contralateral STN running in the vicinity of the STN DBS are unknown.

Alluding to the problematic notion of causation, the question is whether the antidromic activation of the STN neurons contralateral to the DBS are causal to the improvement of symptoms ipsilateral to the STN DBS [4]. If one were to make the reasonable assertion that ipsilateral improvement is related in some manner to neuronal activity changes in the contralateral STN, then the conclusion would be that antidromic activation and increased neuronal activity is related to improvement. At the very least, contralateral STN activation does not worsen ipsilateral symptoms and this contradicts current concepts of basal ganglia pathophysiology where overactivity of the STN is causally related to the bradykinesia of Parkinson's disease. This will be discussed further subsequently.

Another counterintuitive observation is that GPi DBS produces antidromic activation of VL neurons (Fig. 8.3) [5] Again, axons from the VL pass in the vicinity of the GPi but their precise terminations are unknown. The question arises: how many other axons to and from structures pass in the vicinity of the DBS targets? Current neuroanatomy knowledge is insufficient to answer this question because generally neuroanatomical studies have been interested primarily in the origin and termination of pathways and not the course of the axons. Consequently, there may be many unanticipated and undetected mechanisms evoked by DBS and their causal relations to the clinical effects of DBS are unknown.

Every antidromic activation is associated with corresponding orthodromic activations. For example, antidromic activation of VL neurons in turn causes orthodromic monosynaptic activation of neurons that are the target of VL axons such as the MC and possibly others. Consider antidromic activation of MC neurons. Antidromic action potentials conducted up the MC axon from the STN could invade and be conducted down axon collaterals to many other structures enervated by MC such as other neurons in the ipsi- and contralateral cortex, cerebellum,

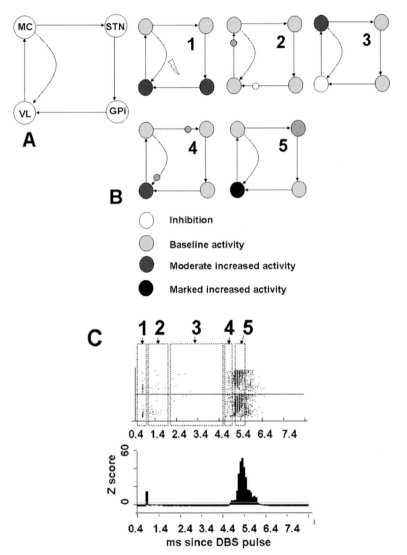

FIGURE 8.3   C shows the post-stimulus raster and histogram (see Figure 8.1 for explanation) for a ventrolateral thalamic (VL) neuron in response to a globus pallidus interna (GPi) DBS pulse delivered at time 0. The horizontal line demarcates the two sets of DBS. The lowest row in each set represents the first DBS pulse. As can be seen, the response can be divided into five zones. The response in the first zone demonstrates robust, short latency and temporally consistent neuronal activity consistent with antidromic activation. Note that the antidromic activation decreases with continued DBS. The activity in zone 2 represents normal VL activity for those DBS pulses that did not elicit an antidromic activation. Zone 3 shows inhibition of VL neuronal activity consequent to activation of the GPi efferent to the VL. Zone 4 shows a modest increase in neuronal activity probably related to post-inhibitory rebound increased excitability. Zone 5 shows a robust increase of neuronal activity, which increases with continued DBS. One explanation for these changes is represented in (B). (B1) shows the time of the DBS

putamen (PT), STN, red nucleus and others. Thus, there is a myriad of neurons in many structures and systems within one synapse of the STN DBS effect.

Orthodromic activation of neurons remote from the vicinity of the DBS

Figure 8.1 demonstrates robust responses in representative neurons throughout the BG–TH–CTX system in response to STN DBS at latencies and temporal consistencies that would not be consistent with antidromic activations and, consequently, are likely to be orthodromic activations. Further, STN DBS produces orthodromic activation of the contralateral STN [6,7]. Also, GPi DBS produces orthodromic responses in the basal ganglia receiving neurons of VL [5]. The precise mechanisms underlying the orthodromic activations are not clear. These activations could be a consequence of antidromic action potentials reaching an axon collateral and then orthodromic conduction down the axon collateral to a subsequent neuron. Alternatively, orthodromic action potentials could be initiated in neurons at the axon hillock or first internode in neurons located in the vicinity of the stimulated target.

EVIDENCE RELATED TO DBS SYSTEMS VERSUS LOCAL EFFECTS HYPOTHESES

The empiric evidence that a great many structures are within oligosynaptic distance of the DBS effect is clear. The issue is whether the widespread effects or the local effects are causal to the clinically meaningful DBS effects. Given that there is some level of organization of structures into systems linked by sequential or parallel monosynaptic connections, some of the widespread effects can be organized into systems effects. For example, the antidromic activation of MC in turn activates the striatum (STR) and STN. The STR in turn activates (inhibition and/or post-inhibition rebound excitation) the GPe and GPi while the STN activates the GPe and GPi. Organization of some of the DBS effects into a set of systems effects does not exclude the incidental activation of axons in passage from systems extraneous to the BG–TH–CTX system. The question now becomes whether activation of a system rather than a specific DBS target nuclei or structure is causal to the clinically meaningful DBS effects.

Establishing cause and effect in any domain is problematic [8]. The large majority of scientific research is based on correlational analyses and it is difficult to distinguish a necessary connection, to use David Hume's terms [8] from a

---

pulse, which produces an antidromic activation of the VL neuron which in turn travels orthodromically to excite the motor cortex (MC) neuron in (B2) and (B3). In addition, there is orthodromic activation of the GPi efferents to VL as in (B2) resulting in inhibition of the VL neuron as in (B3). Following inhibition there is a rebound increased activity (B4). At the same time, an orthodromic action potential is generated in MC and travels to VL (B4). The orthodromic pulse from MC combines with the post-inhibitory rebound to produce a marked increase in VL neuronal activity.

constant conjunction; hence epiphenomenal. Further, if multiple, different and contrary inferential claims are based on the same observed correlations, it becomes difficult to adjudicate between the conflicting claims. Typically, the decisions required by an adjudicating necessity, such as awarding of grant support, are influenced by non-scientific factors such as which competing school of thought has greater representation or constituency on review panels [9]. There is a great advantage for the incumbent. Consequently, one school of thought may accept the observational correlations established by the contrary school of thought and still dismiss them as epiphenomenal or, more often, just ignore them.

The first key observation is that there are multiple effective targets for DBS for any condition. For example, DBS of the GPi [10], STN [10], VL [11], MC [12] and GPe [13] are effective for the treatment of Parkinson's disease. The only reason the STR is not included is that STR DBS has not been attempted, to the knowledge of this author*. Added to this list of DBS targets for Parkinson's disease is spinal cord stimulation as demonstrated in the rodent model of Parkinson's disease [14] and vestibular nerve stimulation at least for postural stability [15].

Rationally, there are two options for explanation. First, there are as many different mechanisms as there are effective DBS targets, in the case of Parkinson's disease there would have to be at least seven different mechanisms; or there is one (or a few) mechanism(s) in common. If one accepts Occam's Razor, that it is vain to do with more that which can be done with fewer, the reasonable conclusion is that it is likely that there is one (or a few) mechanism(s). This argues for the DBS systems effect hypothesis and against the DBS local effect hypothesis. The key would then be to determine what DBS of GPi, STN, VL, MC, GPe, vestibular nerve and spinal cord (and perhaps STR) have in common.

The observations of multiple effective DBS targets extend to other disorders. For example, DBS of the intralaminar nuclei of the thalamus, GPi, and STN are effective for treatments of dystonia [16] and Tourrette's syndrome [17,18]. DBS of the subgenu cingulum [19] and anterior limb of the internal capsule is effective for treating depression [20]. DBS of the anterior limb of the internal capsule [21] and the STN [22] are effective for the treatment of obsessive–compulsive disorder.

The DBS local effect hypothesis only plausibility is its resonance with past and, unfortunately, current theories of basal ganglia physiology and pathophysiology. The GPi Rate theory posits the GPi to be overactive in Parkinson's disease due to dysinhibition as well as due to increased drive from a dysinhibited STN. There is overwhelming evidence that overactivity of the GPi is not a necessary or sufficient condition to produce parkinsonism [23]. Further, the GPi Rate theory follows from a notion of motor function as a sequential, hierarchical and modular process. The net effect is a perception of the GPi as a sole or dominant actor in basal ganglia physiology and pathophysiology. However, this notion is incorrect and the alternative is to view the basal ganglia as a system or more specifically the BG–TH–CTX system where physiological function is diffusely represented

---

*Recently, DBS of the putamen has been demonstrated to improve bradykinesia (unpublished observations).

in a parallel and distributed manner throughout the BG–TH–CTX system [23,24]. Consequently, it makes no sense to talk about the GPi as having a specific or unique function independent of the rest of the BG–TH–CTX system. This illogic is an example of the Mereological fallacy where the properties of a whole are ascribed to a part or parts. Consequently, it makes little sense to attribute the effects of DBS for movement disorders as though they were mediated solely by the GPi.

REGARDING EVIDENCE FOR THE DBS LOCAL EFFECT HYPOTHESIS

An extensive review and critique of the evidence offered in support of the DBS local effect hypothesis is beyond the charge of this chapter. However, a few points are warranted. First, any number of experiments confined to the study of a single structure does not and cannot provide evidentiary support for the DBS local effect hypothesis. This is even true when the results appear self-consistent and seem to have external validity according to the supporting theory invoked. Doing so would be falling victim to the Fallacy of Confirming the Consequence, which is of the form 'if a implies b is true and b is true then a is true; b may be true for any number of reasons other than a such as © and/or d.' The inference that a is true given the above is only possible to the extent that there is no c or d that also is true or has a reasonable probability of being true. In this case, the argument is 'if reducing GPi neuronal activity by GPi DBS implies improved parkinsonism and improved parkinsonism is true, then it must be true that reducing GPi neuronal activity by GPi DBS is true'. While reducing GPi neuronal activity by GPi DBS may be true, there is no evidence that this is causally related to the improved parkinsonism because it may be epiphenomenal. Other experiments would be necessary to demonstrate that other mechanisms, particularly those outside of the DBS target, are not causal. Unfortunately, the large majority of research on the neuronal mechanisms have studied only the stimulated target and these are not sufficient to make any claims in favor of the DBS local effect hypothesis, even if they were replicated a million times.

EVIDENCE FOR THE DBS SYSTEMS EFFECT HYPOTHESIS

There are at least two notions of a systems effect. In one case to be argued subsequently is that stimulation of the BG–TH–CTX system ultimately results in activation of the MC and that injecting activity anywhere in the system drives the MC. This would be an argument for a weak (in the philosophical sense of weak and strong) systems effect. It is weak in the sense that it may be possible to improve disorders, such as parkinsonism, by stimulation of the MC alone by injecting a pulse into any structure that sends axons to or received from the MC. A stronger notion of a systems effect is that activation of the MC, while a necessary condition, it is not a sufficient condition. The weak notion of a systems effect would allow activation of only the MC as sufficient. The strong sense implies that activations of other structures within the BG–TH–CTX system are necessary, though not sufficient.

The weak notion of the DBS systems effect hypothesis

STN DBS causes antidromic activation of the MC as described above. GPi DBS antidromically activates VL output neurons, which necessarily cause monosynaptic activation of the MC. Also, it is highly likely that VL DBS also activates MC. It is not known whether GPe DBS also activates directly (via MC projection axons) or indirectly (via activation of VL to MC axons passing near the GPe DBS). This possibility cannot be discounted given the surprising and unanticipated findings of VL activation with GPi DBS. Consequently, one mechanism in common to most, if not all, effective DBS targets for Parkinson's disease might involve short latency activation of MC. One study in favor of an MC activation as the causal mechanism are the findings of specific activation or inhibition of STN neurons do not reverse parkinsonism in the rodent model. However, activation of MC axons does [25].

A brief side note, neurometabolic imaging demonstrates reduced cerebral blood flow [26,27] (though see [28] to the contrary) in regions demonstrating STN EEG evoked potentials and microelectrode recordings demonstrate increased neuronal activities. Thus, there is a disconnect between neurometabolic changes and changes in actual neuronal activity. At the least, there raises serious concerns, and skepticism, in attempting to infer neuronal activities from metabolic changes. Yet, the seductiveness of the neurometabolic images [29] seem to trump direct neurophysiological recordings (see [30] and [31] in reply). One would expect that direct recordings of electrophysiological measures neuronal activities would be a more direct and more meaningful measure.

The strong notion of the DBS systems effect hypothesis

As the effects of DBS on motor function necessarily involve changes in motor unit recruitment, which likely involve changes in MC neuronal behavior, DBS clearly must have an effect on the MC. However, the effect of DBS on MC neurons is very different in the DBS systems effect hypothesis, where the effect is direct (oligosynaptic), compared to the DBS local effect hypothesis (polysynaptic). The weak notion of the DBS systems effect hypothesis posits that the direct activation of the MC, either directly or from anywhere within the BG–TH–CTX system, is both necessary and sufficient.

The strong notion of the DBS systems effect hypothesis is that while direct activation of the MC is necessary, it is not sufficient. Other activities are necessary within the BG–TH–CTX system so that the MC activation becomes sufficient. Further, the response to DBS has to build over time to reach sufficiency. Evidence for this comes from the demonstrations of varying latencies to clinical responses [32]. For example, the reduction of tremor occurs on the order of a few seconds, improvement in bradykinesia in Parkinson's disease occurs over tens of seconds and changes in gait may take tens of minutes. (Note the very long-term latencies of changes in dystonia and other disorders are beyond the scope of the chapter.) The DBS local effect theory has difficulty accounting for the latencies with the exception of the possibility of neurotransmitter depletion [33,34], while other studies discount the significance of any depletion [35], or the accumulation of other chemical agents that suppress activity such as adenosine [36]. While depletion of neurotransmitters or the accumulation of other agents may take time and hence, explain the latency to clinical effects, it is not at all clear how such changes would explain the differences in the latencies to different clinical effects. The argument would have to be that different levels of depletion and/or accumulations are necessary for the resolution of the different symptoms. However, this risks an unnecessary profusion of causes against which Occam's admonition, against doing with more that which can be done with fewer, is of good counsel.

The question becomes: is there an evolution of neuronal activity that would correspond to the latencies to clinical effect? Unfortunately, there is a paucity of data and the most appropriate data are from studies not directly intended to examine this question. One would think that further more definitive studies would be considered important and of some priority. One example is the change in VL thalamic neuronal activities in response to GPi DBS shown in Figure 8.3. There are several features in the post-stimulus raster. There is an initial short latency temporally consistent and robust response indicative of antidromic activation. This is followed by an inhibition beginning at approximately 3 ms and lasting approximately 3 ms probably related to the postsynaptic inhibitory potentials consequent to γ-aminobutyric acid (GABA) release by activated afferents from the GPi. This is followed by a relatively modest increase in VL neuronal activity consistent with post-inhibitory rebound excitability probably mediated by I_h conductance channels. There is a much more robust increase in VL neuronal activity following the modest rebound. There are two additional features. It appears that the antidromic activation diminishes over the time course of the DBS train. At the same time, there appears to be a build up of the late robust increase in activity.

There are at least two possible explanations for the decreasing antidromic activation. At the same time as the antidromic action potential is ascending the VL neuron's axon there is either hyperpolarization or shunting inhibition of the neuronal cell body, perhaps from stimulation of presynaptic elements releasing inhibitory neurotransmitters. However, antidromic activation is very robust as demonstrated by its relative independence on stimulation frequencies. In other words, while high frequency stimulation improves the probability of an orthodromic activation through temporal summation, this is usually not necessary for antidromic activations and, consequently, the probability of a given DBS pulse producing an antidromic activation is relatively independent of stimulation frequency. Also against the possibility of hyperpolarization or shunting inhibition is the relative lack of inhibition just after the DBS pulse and occurring at times when the neuron did not have an antidromic activation of its axon [5].

Another possibility is the reduction of antidromic activations consequent to an increased probability of 'collision'. This occurs when there is an orthodromic generation of an action potential in a neuron just as an antidromic activation is initiated in the neuron's axon. The orthodromic action potential transmitted down the axon is followed by a refractory period in the axon, which prevents the ascending antidromic action potential from reaching the neuronal cell body to manifest a recorded extracellular action potential. The probability of 'collision' increases as the discharge rate of the VL neuron increases and particularly if there is a temporal correlation between the generation of the orthodromic action potential generated in the VL neuron and the stimulation pulse producing the antidromic action potential in the axon of the same VL neuron. Thus, the late increase in VL neuronal activity could be the source of the orthodromic action potentials and their refractory periods in the axon that block the ascending antidromic action potentials and prevent their appearance as a recorded extracellular action potential.

Another, albeit indirect, evidence for a temporally evolving neuronal response is seen from the STN EEG evoked potentials (see Fig. 8.2). As can be seen, the waveform associated with a single pulse is different in morphology and distribution compared to a brief train of DBS pulses. This suggests that the underlying neuronal responses are different between the two. The interactions between the sequences of pulses in the DBS train affect the evoked potential. These observations warrant further investigation and substantiation.

Resonance effects of DBS

The question then becomes what are the mechanisms underlying the temporal evolution of the DBS response. One possible mechanism would be the progressive depletion of neurotransmitters or accumulation of other agents, however, this is unlikely given the discussion above. An alternative would be resonance effects due to re-entrant activities traversing the BG–TH–CTX system that builds an increasing response [2,32]. This presupposes the BG–TH–CTX system as interconnected, nested, non-linear, re-entrant polysynaptic oscillators, what is called the Systems Oscillators theory. Evidence for the systems oscillators theory is presented elsewhere [23,24,32].

There are several lines of evidence supporting the Systems Oscillators theory and its extension that DBS causes resonance amplification within the BG–TH–CTX system. Relatively short latency responses would have to be found in most, if not all, of the structures of the BG–TH–CTX system. This has been demonstrated as shown in Figure 8.1. STN DBS in the non-human primate produces increased neuronal activity in the MC, sensory cortex, PT, GPi, GPe, and VL (not shown) within 6 ms of the DBS pulse. If the DBS effect is to generate activities that repetitively traverse the various oscillators within the BG–TH–CTX system, then one would expect to see increased activity in many, if not all, the various nodes (cortex and nuclei) of the oscillators. This is the case as shown in Figure 8.1. However, such demonstration is not proof of re-entrant activity.

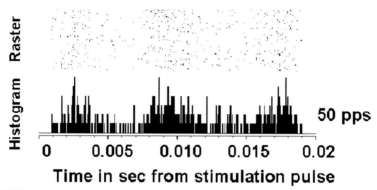

FIGURE 8.4 Post-stimulus raster of an MC neuronal activity in response to 50 pps STN DBS in a non-human primate (see Figure 8.1 for explanation). As can be seen, there is a periodic increase and decrease of neuronal activity between DBS pulses.

Some neurons within the BG–TH–CTX system display repetitive behaviors in response to STN DBS in the non-human primate. An example is shown in Figure 8.4. In the example shown in Figure 8.4, there are repetitive increases and decreases in the MC neuron's response to 50 pps STN DBS. In this case, there are three cycles within the 20 ms inter-stimulus interval suggesting a frequency of approximately 150 Hz. This is significant because the Systems Oscillators theory predicts that the fundamental frequency within the VL–MC feedback loop is approximately 147 Hz and further [32,37], high frequency DBS at these frequencies is most efficacious for improving upper extremity function in patients with Parkinson's disease.

Additional supportive evidence comes from paired pulse STN DBS experiments in non-human primates. The argument is that if a DBS pulse initiates activations that traverse a closed feedback loop, then a second pulse given at exactly the right time would summate with the returning activation from the first pulse to result in a greater response to the second pulse (Figure 8.5). A schematic of the data analysis is shown in Figure 8.6 and a representative example of analyzed data is shown in Figure 8.7.

As seen in Figure 8.6, there are increased responses to the second of the paired pulses over pre-stimulation baseline with different inter-stimulation pulse intervals. The increase at 1 and 2 ms probably represents temporal summation at the site of stimulation. This is followed by an absence of a response at 3 ms consistent with a refractory period due to activations with the 1 and 2 ms paired pulses as demonstrated in the autocorrelogram shown in Figure 8.6. There is an increased response with the 4 ms inter-stimulation, which may reflect the post-refractory period rebound increased excitability. However, the increased effects with longer inter-stimulation pulse intervals are not associated with any changes intrinsic to the neuron recorded, as evidenced by no corresponding changes in the autocorrelogram. Rather, the effects are most consistent with resonance effects as schematically represented in Figure 8.4. Also note that, with

(A)   **Resonance Effect**

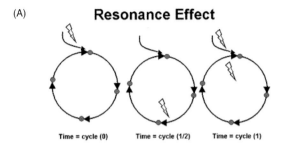

Time = cycle (0)   Time = cycle (1/2)   Time = cycle (1)

(B)   **Stimulation sequence**

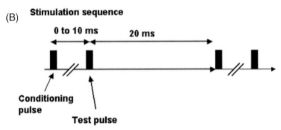

0 to 10 ms   **20 ms**

Conditioning pulse

Test pulse

FIGURE 8.5 (A) Schematic representation of the resonance effect. The first stimulation (conditioning pulse) causes an excitation to traverse the closed loop. If the second stimulation (test pulse) is delivered just as the excitation effect from the first or conditioning pulse returns to the original site, the temporal summation on the neuronal cell membrane will amplify the response. (B) Schematic representation of the paired-pulse stimulus trains. The inter-stimulus-interval represents a specific frequency (1/interval). This study examined the frequencies represented by the intervals from 1 to 10 ms (1000 to 100 Hz) at 1-ms increments.

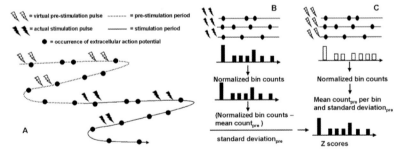

FIGURE 8.6 Schematic representation of the analysis methods for detecting a resonance effect for paired-pulse stimulation. A set of virtual stimulus pulse pairs were created during the pre-stimulation period by translating the timing of the actual stimulation pulse pairs into the pre-stimulation period (A). Post-stimulus rasters and histograms were constructed indexed to the second pulse of the actual (B) and virtual (C) stimulation pulses. The rasters were collapsed across rows into the time bins (0.4 ms) of the histograms resulting in counts of extracellular action potentials. This was normalized by dividing by the number of sets of paired pulse stimuli resulting in probabilities of neuronal discharge in each time interval following the second of the stimulus pair. The mean probabilities per bin and the standard deviation were calculated for the virtual stimulation histograms (C). The mean was then subtracted from each time bin probability during the actual stimulation and divided by the standard deviation resulting in a z score (B).

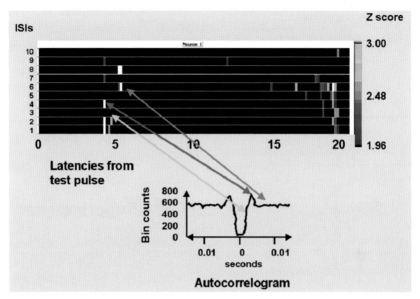

FIGURE 8.7 Resonance effects for different inter-stimulus intervals (ISI) in a GPe neuron. Resonance effects with z scores greater than 1.96 are color-coded. The colored bar represents the ISI and the latency to the resonance effect. Also shown is the autocorrelogram for a motor cortex neuron, which shows a refractory period up to 3 ms, followed by a post-refractory rebound of increased excitability at approximately 4 ms. The refractory period is associated with no significant resonance effects at an ISI of 3 ms, whereas the rebound excitability is associated with a resonance effect with an ISI of 4 ms. The significant resonance effects at longer ISIs are not associated with increased membrane excitability, suggesting that the resonance effects at longer ISIs are less likely to neuronal specific mechanisms and may be more related to re-entrant circuit mechanisms.

longer time intervals from the second pulse, there is a progressive increase in the increases of neuronal activities consistent with multiple re-entrant cycles.

Another example of possible resonance effect would be negative resonance. If the second pulse is given just before the effects of the first pulse reach the stimulation site, the second pulse could induce a refractory period that would prevent further cycling of the re-entrant activity. Unfortunately, research proposals to study further the phenomena of positive and negative resonance were not favorably received and not supported. However, there is one example from STN DBS EEG evoked potentials as shown in Figure 8.8. As can be seen, there is a robust STN DBS EEG evoked potential to single pulses (given at 2 pps) that lasts approximately 50 ms. However, a second pulse given 25 ms after the first aborts the STN DBS EEG evoked potential. This was from a single patient; consequently, further study is warranted.

The next question is: how could a systems resonance effect induced by DBS posited by the Systems Oscillators theory cause amplification of the signal thereby improving signal-to-noise ratio? The improvement in motor control consequent to DBS can be considered as an improvement in the signal that

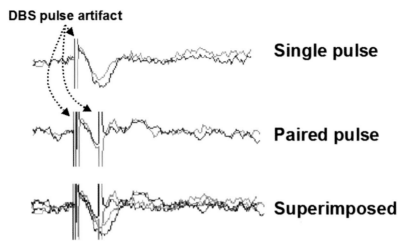

FIGURE 8.8   Left STN DBS EEG evoked potential recorded from the right parietal region in response to a single or paired pulse. The paired pulse inter-pulse interval was 25 ms.

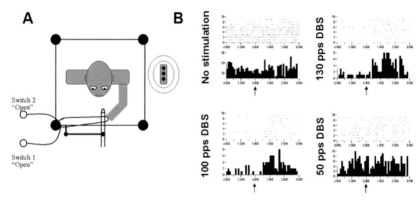

FIGURE 8.9   A non-human primate was trained to make an arm pulling movement in response to an auditory go signal (A). (B) shows peri-event rasters and histograms of a PT neuron under conditions of no DBS and DBS at 130 pps, 100 pps, and 50 pps. The upward arrow shows the time of the go signal onset. As can be seen, there is little modulation of the neuronal activity correlated with the behavior. However, with 130 pps DBS, there is a robust modulation of neuronal activities related to the behavior. This seen to a lesser extent with 100 pps DBS and not with 50 pps DBS. Modified with permission from [2].

improves information for the precise orchestration of muscle activities [37]. An example of such an effect is shown in Figure 8.9. In these experiments, a non-human primate is trained to make an arm movement following an auditory 'go' signal during different STN DBS conditions [2]. As seen in the no DBS condition, there is little modulation of the neuronal activity associated with the behavior. However, at 130 pps DBS, there is a change in the neuronal activities and, specifically, there is a modulation of neuronal activities correlated with the behavior. This modulation is seen, but to a lesser extent, with 100 pps DBS and not at all at 50 pps DBS. One possible interpretation is that the signal-to-noise ratio related to the modulation of the neuronal activity was too low in the no DBS condition but improved with the 130 pps DBS condition (examples of the converse also were noted).

The resonance effect described above is analogous to the operations of an AM radio. A radio station sends out an electromagnetic wave, known as a carrier wave, at a certain frequency. The information transmitted is encoded in the amplitude of the waves. A radio receives this information as well as information from many other radio stations but these are at different frequencies. In order to select the specific radio station transmission, the radio has an oscillator that can be tuned to the frequency of the carrier wave of the specific radio station. This amplifies the signal above those of competing radio stations.

IMPLICATIONS OF DBS FOR THE PHYSIOLOGY AND PATHOPHYSIOLOGY OF THE BG–TH–CTX SYSTEM

DBS is a remarkably effective therapy and, in fact, is better than the best pharmacological therapy as evident by the improvement with DBS when all manner of pharmacological therapies (which include dopamine neuron replacement strategies) have failed. STN and GPi DBS are better than pharmacological therapies for Parkinson's disease [38,39]. The question becomes whether the DBS mechanisms are more synonymous with the physiology and pathophysiology compared to the presumed mechanisms of pharmacological therapies. Specifically, if the DBS mechanisms of action relate to re-entrant activity within the multiple interconnected, nested, non-linear, re-entrant oscillators of the BG–TH–CTX system, then perhaps the normal physiology also is based on interactions among these oscillators and diseases are the result of disturbances in these systems of oscillators. In Parkinson's disease, dopamine replacement is not able to normalize the BG–TH–CTX system physiology compared to DBS. This raises considerable problems for the use of neurochemical modulation as a means to understand neuronal physiology (what is called the Pharmacology [Neurochemistry]-as-Physiology paradigm [23]). There is a long history that posits that pharmacology replicates and therefore allows inferences to neuronal physiology [40] and that long history is suspect.

There may be striking parallels between the DBS local and systems effects hypotheses and local and systems physiology/pathophysiology hypotheses. The local physiology/pathophysiology hypothesis is represented by the GPi Rate and Action Selection/Focused Attention theories. The GPi Rate theory posits that the central mechanism is the relative activities of the GPi and hence, is a strictly local theory of pathophysiology. Similarly, the Action Selection/Focused Attention theory poses the GPi as central to the generation of movement. The function of the GPi is to suppress unwanted movements while inhibition of the GPi allows desired movements; again a local theory of physiology. In contrast, the Systems Oscillators theory holds that the disorders of the basal ganglia are disorders of the system; hence a systems physiology/pathophysiology hypothesis. Just as DBS of most, if not all, of the structures within the BG–TH–CTX system can improve Parkinson's disease, so can lesions of most, if not all, of the structures within the BG–TH–CTX system produce parkinsonism consistent with a systems pathophysiology [23,24,32,37].

Another parallel between the systems wide effects of DBS and a systems based notion of physiology is reflected in the time course of neuronal modulation. As shown in Figure 8.1, STN DBS produces changes in neuronal activities that share the same millisecond time scales. Similarly, recordings in the MC and PT demonstrate that changes in behaviorally related neuronal activities are nearly simultaneous in the MC and PT [41].

INTELLECTUAL ANTECEDENTS TO THE DBS LOCAL EFFECT

The concept of physiology and pathophysiology based on the dynamics and interactions of many interconnected, nested, non-linear re-entrant oscillators comprising the BG–TH–CTX system is a radical departure from current concepts such as those instantiated in the GPi Rate and the Action Selection/ Focused Attention theories. But it is more fundamental than that. Theories like the Systems Oscillators theory are contrary to approaches and concepts perhaps hundreds of years old. These older concepts are based on a hierarchical, modular and sequential organization [24] and processing within the motor systems pioneered by such eminent neuroscientists as Sir Charles Sherrington. Indeed, the hierarchical, sequential and modular organization concepts necessitate a local physiology/pathophysiology hypothesis. Consider the predecessor theory to the GPi Rate theory, which was the Dopaminergic/Cholinergic Striatal Imbalance theory. This theory argued that a lack of dopamine relative to acetylcholine in the STR resulted in parkinsonism; clearly a local physiology/pathophysiology hypothesis. It will be interesting whether the community of scientists will be open minded to the possibilities of theories that counter such a long tradition and perspective [9].

Central to local physiology/pathophysiology hypotheses are one-dimensional push–pull dynamics. In the GPi Rate theory, the GPi is either overactive, in the case of hypokinetic disorders such as parkinsonism, or underactive, in cases of hyperkinetic disorders. In the Action Selection/Focused Attention theory, either the GPi is inhibited to facilitate movements or activated to prevent movements. In the Dopaminergic/Cholinergic Striatal Imbalance theory, there was either an excess or deficiency of dopamine relative to acetylcholine. Similarly, in the DBS local effect hypothesis, there is either increased or decreased activity in the neurons in the stimulated target.

The dynamics inherent in a system physiology/pathophysiology hypothesis is far more complex which would be necessary even to begin to explain the complexities of behavior. The dynamics involved in the Systems Oscillators theory include positive and negative resonance, coherence in frequency, phase or amplitude, beat interactions between oscillators, information interactions based on commensurate or non-commensurate frequencies, holographic memory, self-organization, transitions between metastable states, among others [24,32].

IMPLICATIONS OF THE DBS SYSTEMS EFFECT HYPOTHESIS ON FUTURE SURGICAL THERAPIES

The DBS systems versus local effect hypotheses have implications for target selection of current and emerging DBS therapies. The DBS local effect hypothesis holds that the physiological or clinical effect is unique to the stimulated target. Consequently, the physiological or clinical effects due to DBS of other targets cannot be predicted nor would there be any rational justification for their consideration. An example is the selection of the subgenu cingulum for DBS in the treatment of medically refractory depression [19]. Positron emission tomography (PET) scanning demonstrated areas of decreased metabolism and other areas with increased metabolism, such as the subgenu cingulum. Based on the now faulty premise that DBS inhibits (as well as the potential incorrect inference that increased or decreased metabolism actually reflects changes in neuronal activity) the metabolic overactivity, the subgenu cingulum was chosen as the target. And it was successful. However, what is not and cannot be known at this time is whether DBS of any other area in the systems served by the subgenu cingulum would not be equally effective.

The DBS systems effect hypothesis, particularly the Systems Oscillators theory, argues that stimulation anywhere in the system may be effective. In the case above, what if DBS of the subgenu cingulum had greater surgical risks compared to DBS of another structure within the systems? It could be just as effective and safer to target one of the other structures.

The DBS systems effect hypothesis greatly expands the range of potential interventions. The DBS local effect hypothesis follows from the same mode of thinking as surgical ablative therapies and, consequently, risks unnecessarily limiting the possibilities of DBS. This risk is particularly acute if the DBS local effect hypothesis posits DBS inhibition of neuronal activities in the DBS target. An example counter to the mind set involved in selecting the subgenu cingulum DBS for depression is use of DBS of the intralaminar nuclei of the thalamus for minimally conscious state patients [42]. The development of DBS for minimally conscious state patients was not based on the presupposition that DBS inhibits the stimulated target or that the effects were local.

FINAL NOTE

The most accurate and honest statement is that the therapeutic mechanisms of action of DBS are unknown. However, research has identified a large number of responses to DBS and which of these, if any, will be the one or few that are responsible for the therapeutic effects is unknown. But what is clear is that the responses to DBS go well beyond the stimulated target. Unfortunately, research to investigate the systems wide effects of DBS has not been expeditious for many reasons. But the most significant reason likely is the very narrow view of how the basal ganglia functions and how its physiology is altered at the neuronal level by disease that have prejudiced grant reviewers. Old notions such as the GPi Rate and the Action Selection/Focused Attention theories have long outlasted any potential heuristic value and now these theories are an impediment [23,37]. These theories had the seductive quality of simplicity and it is human nature to give far more explanatory power to theories than is warranted [43]. These theories have a resistance to dismissal even in the face of significant contravening data because of their simplicity that leads to a relatively effortless intuitive appreciation, also an all-too-human trait [43].

In some ways, the GPi Rate and Action Selection/Focused Attention theories are really symptoms of a more fundamental and pervasive problem that is a notion and perspective depending on and requiring a hierarchical, sequential, and modular organization of the motor nervous system. This latter perspective has long and deep historical roots and overcoming them likely will be difficult. The old adage applies 'when all one has is a hammer, the whole world seems like a nail'. When experimental reasoning and methods follow wholly from hierarchical, sequential, and modular organization of the motor nervous system perspective, the DBS local effect and the local physiology/pathophysiology hypotheses are difficult to escape. It is hard to convince someone of alternatives if all they see is one thing; to them there simply are no alternatives and attempts to find them are nonsense and not fundable.

The one clear conclusion is that, while DBS-related research has not proven how DBS works, it has provided remarkable opportunities to understand how the BG–TH–CTX system and, more generally, the brain, work. It is likely that the therapeutic mechanisms of DBS may never be known unless there is a revolution in our understanding of the physiology and pathophysiology of the BG–TH–CTX system and that understanding will have to be orders of magnitude more complex and sophisticated than current theories. Any new theory will have to be based on dynamics comparable to the complexities of the behaviors to be explained. This is going to require a great deal of re-thinking.

REFERENCES

1. Asanuma K, Tang C, Ma Y, et al. Network modulation in the treatment of Parkinson's disease. *Brain*. 2006;129:2667–2678.
2. Montgomery Jr EB, Gale JT. Mechanisms of action of deep brain stimulation (DBS). *Neurosci Biobehav Rev*. 2008;32:388–407.
3. Baker K, Montgomery Jr EB, Rezai AR, et al. Subthalamic nucleus deep brain stimulus evoked potentials: physiology and therapeutic implications. *Mov Disord*. 2002;17:969–983.
4. Walker HC, Watts RL, Guthrie S, et al. Bilateral effects of unilateral subthalamic deep brain stimulation on Parkinson's disease at 1 year. *Neurosurgery*. 2009;65:302–309.

5. Montgomery Jr EB. Effects of GPi stimulation on human thalamic neuronal activity. *Clin Neurophysiol.* 2006;117:2691–2702.

6. Walker, H.C., Huang, H., Guthrie, S.L., et al. (2008). Subthalamic neuronal activity is altered by contralateral subthalamic deep brain stimulation in Parkinson disease. Movement Disorders Society International Congress.

7. Novak P, Klemp JA, Ridings LW, et al. Effect of deep brain stimulation of the subthalamic nucleus upon the contralateral subthalamic nucleus in Parkinson disease. *Neurosci Lett.* 2009;463:12–16.

8. Hume D. *enquiries concerning human understanding and concerning the principle of morals.* 3rd edn.. Oxford: Clarendon Press; 1975.

9. Kuhn TS. *The structure of scientific revolutions.* Chicago: The University of Chicago Press; 1996.

10. Deep-Brain Stimulation for Parkinson's Disease Study GroupDeep-brain stimulation of the subthalamic nucleus or the pars interna of the globus pallidus in Parkinson's disease. *N Engl J Med.* 2001;345:956–963.

11. Koller W, Pahwa R, Busenbark K, et al. High-frequency unilateral thalamic stimulation in the treatment of essential and parkinsonian tremor. *Ann Neurol.* 1997;42:292–299.

12. Pagni CA, Altibrandi MG, Bentivoglio A, et al. Extradural motor cortex stimulation (EMCS) for Parkinson's disease. History and first results by the study group of the Italian neurosurgical society. *Acta Neurochir Suppl.* 2005;93:113–119.

13. Vitek JL, Hashimoto T, Peoples J, et al. Acute stimulation in the external segment of the globus pallidus improves Parkinsonian motor signs. *Mov Disord.* 2004;19:907–915.

14. Fuentes R, Petersson P, Siesser WB, et al. Spinal cord stimulation restores locomotion in animal models of Parkinson's disease. *Science.* 2009;323:1578–1582.

15. Yamamoto Y, Struzik Z, Soma R, et al. Noisy vestibular stimulation improves autonomic and motor responsiveness in central neurodegenerative disorders. *Ann Neurol.* 2005;58:175–181.

16. Kupsch A, Benecke R, Muller J, et al. Pallidal deep-brain stimulation in primary generalized or segmental dystonia. *N Engl J Med.* 2006;355:1978–1990.

17. Ackermans L, Temel Y, Visser-Vandewalle V. Deep brain stimulation in Tourette's syndrome. *Neurotherapeutics.* 2008;5:339–344.

18. Welterm ML, Malletm L, Houetom JL, et al. Internal pallidal and thalamic stimulation in patients with Tourette syndrome. *Arch Neurol.* 2008;65:952–957.

19. Lozano AM, Mayberg HS, Giacobbe P, et al. Subcallosal cingulate gyrus deep brain stimulation for treatment-resistant depression. *Biol Psychiatry.* 2008;64:461–467.

20. Malone DAJ, Dougherty DD, Rezai AR, et al. Deep brain stimulation of the ventral capsule/ventral striatum for treatment-resistant depression. *Biol Psychiatry.* 2009;65:267–275.

21. Greenberg BD, Malone DA, Friehs GM, et al. Three-year outcomes in deep brain stimulation for highly resistant obsessive-compulsive disorder. *Neuropsychopharmacology.* 2006;31:2384–2393.

22. Mallet L, Polosan M, Jaafari N, et al. Subthalamic nucleus stimulation in severe obsessive-compulsive disorder. *N Engl J Med.* 2008;359:2121–2134 Erratum in N Engl J Med. (2009) 2123, 2361.

23. Montgomery Jr EB. Basal ganglia physiology and pathophysiology: a reappraisal. *Parkinsonism Relat Disord.* 2007;13:455–465.

24. Montgomery Jr EB. Dynamically coupled, high-frequency reentrant, non-linear oscillators embedded in scale-free basal ganglia-thalamic-cortical networks mediating function and deep brain stimulation effects. *Nonlinear Stud.* 2004;11:385–421.

25. Gradinaru V, Mogri M, Thompson KR, et al. Optical deconstruction of parkinsonian neural circuitry. *Science.* 2009;324:354–359.

26. Karimi M, Golchin N, Tabbal SD, et al. Subthalamic nucleus stimulation-induced regional blood flow responses correlate with improvement of motor signs in Parkinson disease. *Brain*. 2008;131:2710–2719.

27. Tanei T, Kajita Y, Nihashi T, et al. Changes in regional blood flow induced by unilateral subthalamic nucleus stimulation in patients with Parkinson's disease. *Neurol Med Chir (Tokyo)*. 2009;49:507–513.

28. Sestini S, Scotto di Luzio A, Ammannati F, et al. Changes in regional cerebral blood flow caused by deep-brain stimulation of the subthalamic nucleus in Parkinson's disease. *J Nucl Med*. 2002;43:725–732.

29. McCabea DP, Castelb AD. Seeing is believing: The effect of brain images on judgments of scientific reasoning. *Cognition*. 2008;107:343–352.

30. Montgomery Jr EB. Basal ganglia pathophysiology in Parkinson's disease. *Ann Neurol*. 2009;65:618.

31. Obeso JA, Olanow CW. Basal ganglia pathophysiology in Parkinson's disease reply to Montgomery. *Ann Neurol*. 2009;65:618–619.

32. Montgomery Jr EB. *Deep brain stimulation: principles and practice*. New York: Oxford University Press; 2010.

33. Lozano AM, Eltahawy H. How does DBS work?. *Suppl Clin Neurophysiol*. 2004;57:733–736.

34. Iremonger KA, Anderson TR, Hu B, Kiss ZH. Cellular mechanisms preventing sustained activation of cortex during subcortical high-frequency stimulation. *J Neurophysiol*. 2006;96:613–621.

35. McIntyre CC, Savasta M, Walter BL, et al. How does deep brain stimulation work? Present understanding and future questions. *J Clin Neurophysiol*. 2004;21:40–50.

36. Bekar L, Libionka W, Tian G, et al. Adenosine is crucial for deep brain stimulation–mediated attenuation of tremor. *Nat Med*. 2008;14:75–80.

37. Montgomery Jr EB. Theorizing about the role of the basal ganglia in speech and language: the epidemic of miss-reasoning and an alternative. *Commun Disord Rev*. 2008;2:1–15.

38. Schupbach WMM, Maltete D, Houeto JL, et al. Neurosurgery at an earlier stage of Parkinson disease: a randomized, controlled trial. *Neurology*. 2007;68:267–271.

39. Weaver FM, Follett K, Stern MB, et al. Bilateral deep brain stimulation vs best medical therapy for patients with advanced Parkinson disease: a randomized controlled trial. *J Am Med Assoc*. 2009;301:63–73 2009.

40. Valenstein ES. *The war of the soups and sparks: the discovery of neurotransmitters and the dispute over how nerves communicate*. New York: Columbia University Press; 2005.

41. Montgomery Jr EB, Buchholz SR. The striatum and motor cortex in motor initiation and execution. *Brain Res*. 1991;549:222–229.

42. Schiff ND, Giacino JT, Kalmar K, et al. Behavioural improvements with thalamic stimulation after severe traumatic brain injury. *Nature*. 2007;448:600–603.

43. Johnson-Laird PN. *How we reason*. New York: Oxford University Press; 2006.

STUDY QUESTIONS

1. Consider the current language used to describe neuronal circuitry dynamics (e.g. firing rates, variability and regularity, power spectra). How might descriptions of dynamic activity appear if they involved characterizing groups of coupled oscillators, multiplexers, holographic storage, and phased attractor states? Consider the role of neuromodulation in such a description.

2. What is considered information within the nervous system? Is information between groups of neurons transferred through a high-pass or low-pass filtering effect?

Commentary on The Electrode – Principles of the Neural Interface: Circuits

Mark Stecker PhD, MD,

Marshall University Medical Center, Huntington, WV, USA

The essential feature of deep brain stimulation (DBS) is that electrical stimulation of a small neural structure can produce changes in a larger part of the nervous system resulting in significant clinical changes. Thus, by necessity, DBS must work by affecting large systems of neurons. However, the mechanisms underlying this effect are unknown. Even the question as to whether DBS produces its system effect through localized stimulation or depression of neural activity is not fully understood as there is evidence for both effects.

Since the detailed function of the actual systems underlying DBS are poorly understood, it is helpful to consider the effects that DBS might have on a simple control system (Figure 8.1.1) in different circumstances. In this model system, if there is no non-linear element, the output is related through the input by the relation:

$$O(\omega) = \frac{G(\omega)}{1-(f_+-f_-)G(\omega)} I(\omega) \qquad 8.1$$

where $O(\omega)$ is the frequency dependent output, $I(\omega)$ is the frequency dependent input, $G(\omega)$ is the gain and is the angular frequency. There are two possible ways in which such a system can produce output oscillations. If there is an oscillatory input, i.e. $I(\omega)$ is non-zero for at least some frequencies, so will the output demonstrate oscillations. Oscillations can occur even if the input is zero $I(\omega)=0$ for those frequencies for which1 $(f_+-f_-)G(\omega)=0$. These would be designated "feedback oscillations" since they occur only when there is positive feedback. If, for the moment, only the problem of treating

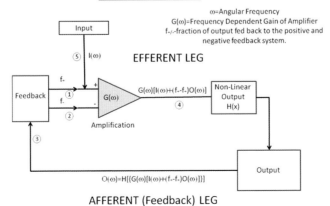

General Control System

ω=Angular Frequency
G(ω)=Frequency Dependent Gain of Amplifier
f-/-fraction of output fed back to the positive and negative feedback system.

Input

⑤ I(ω)   EFFERENT LEG

f-
Feedback   +
①   G(ω)[I(ω)+(f--f-)O(ω)]   Non-Linear Output H(x)
f-   G(ω)
②   -   ④
Amplification

③

Output

O(ω)=H[{G(ω)[I(ω)+(f--f-)O(ω)]}]

AFFERENT (Feedback) LEG

movement disorders with DBS is considered, it is useful to see how oscillations in each of these cases responds to DBS directed to the various physiologic locations in Figure 8.1.1

If DBS works by blocking impulses traveling through the stimulated region and the oscillatory drive comes from outside this feedback system, then DBS directed to locations 5 and 4 will stop the oscillations. Stimulating at location 1 will reduce the amplitude of the oscillations by reducing positive feedback but not eliminate them. Stimulating at location 2 will increase the amplitude of oscillations and stimulating at 3 can either increase or decrease the amplitude of the oscillations depending on the values of f_+, f_-. In each of these situations, stimulation may change the shape of the output spectrum but, if the input is highly peaked around a single frequency, stimulation will not change that frequency greatly. Changing the intensity of the stimulus has a large effect on the amplitude of the output but only a smaller effect on the frequency spectrum. In the case where the output oscillations occur primarily as the result of feedback, then stimulating at locations 1, 3 and 4 will stop the oscillations. Stimulating at 5 will have no effect and stimulating at location 2 will eliminate or change the frequency of the oscillations as it changes the relative balance of positive and negative feedback. Slight changes in the amplitude of the stimulus (either more or less than a critical value) will either stop the oscillations or change their frequency depending on the frequency dependence of the amplifier gain.

Other observations can be used to clarify the physiologic function of the specific locations in Figure 8.3. Spontaneous recording during passive movements will generate activity at 1, 2, 3 and maybe at 4 but not 5. Mechanically stopping the movement will eliminate activity at 1, 2 and 3 but will not change activity at 5. If the feedback hypothesis is true, then activity at 4 will also stop in this case.

The above results were based on the assumption that DBS blocks pathways in an amplitude dependent manner. What if DBS really functions by injecting a signal at the DBS frequency into the control system that was discussed earlier. In the case of a purely linear system, this would result in output frequencies at the frequency of the DBS stimulation. This is not what is observed in actuality. However, the situation changes dramatically if there are non-linear elements in the system. In this case, injecting a signal at one frequency at any point can change the response not only at the DBS frequency but also all other frequencies. Consider the simple case in which the non-linear output restricts the total power output of the system to a specific maximum and minimum value. If the DBS stimulator injects a high amplitude signal into the system at one frequency so that it pushes the output to a point near its maximum, then signals at other frequencies cannot pass through the non-linear element without being reduced in amplitude or else the total power output would exceed the specified limits. This results in an effective reduction in the amplifier gain at frequencies other than that of the DBS stimulator and changes the gain of the feedback system at the clinical tremor frequencies. As described above, this can have dramatic effects on both feedback oscillations and the output resulting from an oscillatory input.

This above discussion illustrates some of the general principles that are involved in analyzing the effects of introducing stimulation into a general control system and the importance of the concepts of both positive and negative feedback as well as the critical role of non-linear responses.

This model applies not only to the case of movement disorders. It also applies to the case in which the output is a behavior, such as when treating psychiatric problems with DBS, and to the case where the output is the overall level of activity in groups of neurons, as in treatment of epilepsy. The model describing the modulation and perception of pain would be similar as well.

Chapter 9

Device Materials, Handling, and Upgradability

John Kast, BSME, Gabi Molnar, MS and Mark Lent, BSME, MSMOT
Medtronic Inc., Minneapolis, MN USA

INTRODUCTION

A neuromodulation system may consist of two or more components, including the neurostimulator, a lead which contains the electrodes for stimulation and, in some cases, an extension that bridges the connection between the neurostimulator and the lead. It is important to understand both the therapies as well as the technology and engineering trade-offs when designing implantable neuromodulation systems (INS). This chapter will focus on device physical design and materials, as well as device safety considerations, with specific focus on deep brain stimulation (DBS) and spinal cord stimulation (SCS) systems.

NEUROSTIMULATOR FORM FACTOR AND MATERIALS

Device Construction

An implantable neurostimulation system, or INS, consists of six main components: the device enclosure, a power source, electronics circuitry, communication and/or recharge antenna, feedthroughs, and the device connector/header (Figure 9.1). These systems may be classified into three types based on their power source: rechargeable, primary power source, and transcutaneous inductively-coupled devices. Most of the systems implanted today for DBS and SCS are rechargeable or primary power source devices.

The device enclosure is a hermetic package. The power source and electronics circuitry are contained within this hermetic device enclosure to ensure safe containment and prevention of body fluids from contacting these components and, conversely, to prevent potentially harmful substances present in the internal components from reaching the body. In rechargeable devices, typically a magnetic or radiofrequency (RF) energy is transferred from an external instrument with a transmitting antenna to a receiving antenna located on or inside the INS.

Communication to the electronics circuitry is done with wireless telemetry to an antenna located within the electronics circuitry. An external instrument, such as a patient programmer or a clinician programmer, may be used to establish

Essential Neuromodulation. DOI: 10.1016/B978-0-12-381409-8.00009-7

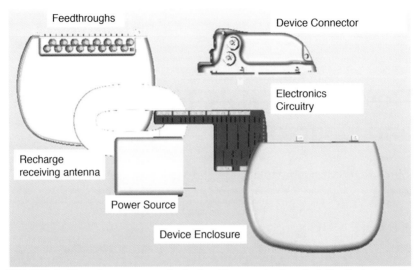

FIGURE 9.1   The main components of an implantable neurostimulator, including the device enclosure, power source, recharge receiving antenna, feedthroughs, electronics circuitry, and device connector/header.

communication with the device in order to obtain device status updates or adjust stimulation parameters. A clinician programmer may provide a greater range of adjustment than a patient programmer, as the patient may be restricted to certain ranges of use by the clinician, depending on the specific therapy and system.

The device header houses the electrical contacts that mate with the proximal extension or lead connectors. The number of these contacts matches the number of independently programmable electrodes on the distal end of the lead. The extension or lead is inserted, positioned, and secured inside the device header. Each electrical contact in the device header connects to a feedthrough conductor. The feedthroughs are electrical conductors that carry the stimulation current from the electronics circuitry through the hermetic enclosure. These feedthroughs are made of an insulative material, such as glass or ceramic, to maintain the electrical isolation between each other and from the device conductive enclosure material. The material of the device header electrically insulates the connection of each feedthrough conductor. The header configuration determines how many different leads can be utilized by the device. The header typically mates with single lead or dual lead systems with four to eight electrodes per lead.

Implant Locations and form Factor

The physical form factor of the INS is derived by balancing the clinical therapy needs with the technology and engineering attributes of the device. These attributes include: number of leads and electrodes, power source type and capacity, and the stimulation output capabilities of the electronics.

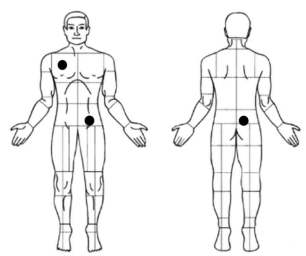

FIGURE 9.2 Typical device implant locations are indicated by black dots and include the subclavicular region, the lower abdomen, and the upper buttocks.

Significant technological advances in power sources, electronics circuitry, and connector interconnects have allowed reductions in the size of the INS. The sizes of implantable neurostimulation devices, as measured by total volume, range from very small at 0.2 mL to relatively large at 50 mL. The size of the INS can restrict where the device is implanted, with the larger devices (greater than 20 mL) typically implanted in a subcutaneous or submuscular pocket located below the subclavicular region, lower abdomen or upper buttocks (Figure 9.2).

The size of the implanted device is primarily driven by the number of therapy electrodes that may be activated for stimulation and the type and capacity of the power source. In general, rechargeable devices or transcutaneous inductively coupled devices are smaller. Neuromodulation therapies that require less stimulation energy or use a lower number of electrical contacts are significantly smaller in size. These smaller devices are implanted in different areas of the body and typically are implanted closer to the target therapy site. For example, the cochlear transcutaneous inductively coupled powered implants are small enough for implantation in the mastoid process in the skull. Implanting the device closer to the target therapy site also has advantages in minimizing lead migration and conductor fracture failures.

The device shapes and form factors are also designed to minimize tissue and skin erosion using rounded edges and large edge radii. The form factor and especially the thickness are key considerations for implant location. Lower abdomen implant locations can tolerate thicker and larger devices; for example, cardiac defibrillators with a volume greater than 200 mL and a thickness of 20 mm have been implanted in this region. In contrast, the subclavicular region requires thinner and smaller devices to prevent skin erosion. Smaller devices also provide an improved

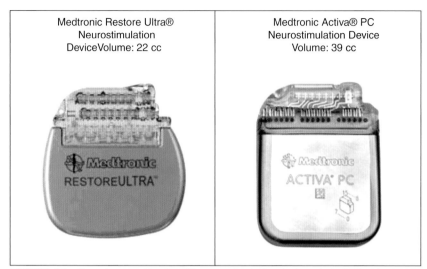

| Medtronic Restore Ultra® Neurostimulation DeviceVolume: 22 cc | Medtronic Activa® PC Neurostimulation Device Volume: 39 cc |

FIGURE 9.3 Form factors of implantable neurostimulation devices. Left: SCS rechargeable device with control of up to 16 electrodes. Right: DBS device containing a primary power source.

cosmesis effect for the patient to better conceal the system within subcutaneous tissues. Device form factors for SCS and DBS are shown in Figure 9.3.

All devices, extensions, leads, and accessories are supplied packaged and sterilized. However, infection at the incision sites is a potential complication for patients who receive implanted neurostimulation systems. In one study, they found that most infections with implanted DBS devices were associated with the pocket location and that the infection agents were those most commonly associated with skin-based infections [1]. With infections that involve the areas of the INS or extension, partial hardware removal sparing the lead accompanied by a course of postoperative intravenously administered antibiotics was successful in treating the infection in most cases [1,2]. Similarly, infections involving the incision site of SCS systems also tend to involve the pocket location [3]. The use of perioperative prophylactic antibiotics has been suggested for infection control [3,4].

Implant Considerations for Recharge and Telemetry

Implanting depth is an important consideration for recharge and telemetry. Implanting the device too shallow can result in skin erosion, while implanting the device too deep can result in poor telemetry communication and recharge coupling. The recharge power transmission is significantly reduced with deeper implant depths. Implant depths of less than 1.5–2 cm are often recommended for rechargeable systems. The receiving recharge antenna is sometimes located toward or external to one side of the implanted device. Keeping the recommended recharge receiving antenna orientation is important for consistent recharge coupling performance. These devices are often sutured in place to prevent device flipping or migration.

Device Materials and Biocompatibility

Materials that are implanted in the human body are biocompatible and biostable for the designed application. Device manufacturers ensure that materials implanted in the body meet all required standards for implantable medical devices. International standards for biological evaluations of medical device materials include ISO 10993-1 which outline a set of comprehensive tests and protocols required for medical devices. The materials are categorized based on the type of body contact that the medical device has with the human body. The standard outlines four types of body contact: non-contacting medical devices, surface-contacting devices, external communicating devices, and implanted devices.

Implanted medical devices are further defined based on the specific application sites of the device. These sites are categorized into two different groups: direct tissue/bone or blood contact. In addition to the site of the implant, the required tests also vary depending on the duration of the implant or contact. The ISO 10993-1 standard categorizes three different periods of exposure: limited exposure, prolonged exposure or permanent exposure. Implanted neurostimulation devices have been evaluated by these biological tests including cytotoxicity, sensitization, irritation or intracutaneous reactivity, systemic toxicity, subacute and subchronic toxicity, genotoxicity, and hemocompatibility to ensure biocompatibility. In addition to biocompatibility, the biostability of the materials and designs that are implanted are evaluated. This biostability evaluation includes detailed mechanical, electrical, and chemical characterization of the material properties after being subjected to the human body for the defined exposure. Materials that are resistant to degradation and corrosion are key characteristics that the device manufacturers consider when selecting materials for chronic, implantable systems such as neurostimulation devices.

The materials of the implanted device that have direct tissue contact for permanent exposure periods include the device enclosure, device header and, with some designs, the recharge receiving antenna. Titanium is the most common material used for the hermetic package. It exhibits high levels of corrosion resistance, is non-magnetic, lightweight, non-toxic and biologically compatible with human tissue and bone. Titanium also has excellent mechanical strength and durability characteristics. It is often formed into thin-walled shield halves that are laser welded together to create the hermetic enclosure. For device systems that implement monopolar stimulation, this titanium hermetic package is utilized as the return common electrode. Titanium has several implantable grades. Commercially pure titanium, such as Grade 1 or 2, is most commonly used. Commercially pure titanium has excellent formability and elongation which allows cold working and forming of custom device shapes and a relatively tight bend radius. Other titanium alloys, such as Grade 9 or 23, include other alloys such as aluminum or vanadium which increase the electrical resistivity of the material. This increase in electrical resistivity results in lower magnetic eddy current loss during inductive power transfer used with rechargeable systems. These alloys do not have the formability or elongation properties of commercially pure titanium which translates into larger bend radius constraints for the device form factor. However, the improved recharge performance allows the device to be implanted deeper.

The receiving recharge coil is sometimes located external to the hermetic titanium package. This also improves the efficiency of the power transmission. When the receiving coil is located external to the hermetic titanium enclosure, it is packaged in magnetically transparent materials such as polyurethane, silicone rubber, polysulfone, ceramic, glass or biocompatible epoxy.

The electrical contacts in the device header are typically made from titanium, platinum, or iridium alloy materials. Commonly used insulating materials include polyurethane, silicone rubber, polysulfone, and biocompatible epoxy.

LEAD SYSTEM CONFIGURATION AND MATERIALS

The portion of the neurostimulation system that connects proximally to the neurostimulator and contains the electrodes distally is referred to as the lead. The lead provides an electrical pathway from the neurostimulator to the electrodes via the conductors that is isolated from the environment of the body. The lead must be designed to conform to the surrounding anatomy, to enable adequate modulation of neural tissue, to be biocompatible, and to be reliable throughout the lifetime of the device. Materials of the lead in contact with body tissues must be selected to minimize the inflammatory response due to the insertion of a foreign object into the body. Additionally, the lead design should minimize the invasiveness of the procedure and should consider the potential of lead removal from the body without damage or disruption of neural tissue.

The lead may be connected directly to the neurostimulator and, in some cases, may be connected to an extension which bridges the connections between the neurostimulator and the lead. The lead or extension is secured to the neurostimulator connector using set screws or spring-lock mechanisms. It is important to establish a secure electrical connection between the lead and device header as improper connections may lead to increases in system impedance that may affect the therapy delivered. In addition, non-ionic fluids should be used for wiping the lead, and connections should be dried since fluid in the connection may result in short circuit. A short circuit may cause stimulation at the connection site, intermittent, or loss of stimulation. Extensions are typically used between a DBS lead and the neurostimulator. If the neurostimulator needs to be replaced due to infection, battery replacement, or other reasons, the extension may be disconnected from the neurostimulator without having to handle the previously implanted lead. For SCS leads, several types of extensions are available that allow flexibility in the number and type of leads that may be connected to the neurostimulator for programming. For example, either one lead may be connected to a single extension, or two leads may be connected to a single bifurcated extension. Thus, using combinations of extensions, it is possible to place up to four leads in multiple locations connected to a single SCS device. Lead insulators are typically made out of a robust, biocompatible, and flexible material such as polyurethane or silicone rubbers. Percutaneous, cylindrical leads such as those used in SCS and DBS are designed to have blunt tips that reduce the likelihood of tissue damage during insertion and, in some cases, also help steer the lead into place.

The conductors, typically wires, within the lead may be arranged in a variety of different ways, including multilumen, where the conductors are placed side-by-side running parallel to each other (some SCS leads) and helical, multifilar, where the conductors are coiled into a long helix (DBS leads). The advantage of using coiled conductors is reduced stress and torsion during tension, bending and twisting, which reduces the likelihood of conductor fracture under these conditions. The lead conductors are typically made of corrosion-resistant materials, such as MP35N (a nickel alloy) or platinum. The conductors themselves may be individually insulated to prevent shorting. Materials for coating these conductor wires may include polytetrafluoroethylene (PTFE) or ethylene tetrafluoroehtylene (ETFE) which are both corrosion-resistant.

The electrodes deliver electrical stimulation and are the interface between the implanted system and the excitable tissue. The function of the electrodes is to provide sufficient current to activate or inactivate the target neural tissue, without causing significant damage to the electrode or surrounding tissue. Metals such as platinum or platinum–iridium are typically used for electrodes. The shape of the electrode is designed to be appropriate to the target anatomy and achieve the desired spatial activation. Typical shapes include cylindrical electrodes (as in SCS percutaneous leads and DBS leads), or flattened electrodes which could be circular, oval, or rectangular with rounded edges such as those in SCS paddle leads. The number and spacing of electrodes on a lead is also related to the size of the target anatomy or the resolution required for targeting. Leads with small spacing between electrodes may be used to target anatomical sites with finer resolution, compared to leads with larger spacing between electrodes which may be selected for covering a larger area.

SCS Lead Complications

Implanted leads are subject to a variety of adverse environmental conditions including corrosion caused by bodily fluids and mechanical stresses caused by body movement, discussed further below. Therefore, mechanical failures are a common cause of re-operation in patients with implanted SCS systems, as is lead migration [5–8]. Thus, the lead must be designed to avoid, to the extent possible within technical capabilities, both migration and breakage. Vertical or horizontal lead migration may result in a loss of proper paresthesia coverage and reduced therapy outcomes.

Various types of lead anchors may also be utilized to help reduce or minimize lead migration. The use of a soft silicone anchor versus a rigid anchor attached to the lead with silicone medical adhesive to prevent slippage is recommended. Solid anchors may result in fractured conductors at lower cycles of a bending fatigue test compared to the soft anchors which caused no failures at 1 million cycles [9]. The anchor should be attached to the lumbodorsal fascia using a figure-of-8 non-absorbable suture to minimize tissue trauma; 2-0 non-absorbable suture is recommended, and ligatures should not be overtightened on the anchor or connector boot. If the anchor has a tail, the tip of the anchor should be pushed through the fascia to maximize the bend radius of the lead (Figure 9.4A). Pushing the end of the anchor through the fascia may prevent fractures observed distal to the point of anchor where the lead exits from the deep fascia caused by increased stresses by the repeated bending motion of the spine [4,9]. In some cases, such as when using the Medtronic Titan™ anchor, biomechanical testing has shown that it is not necessary to push the anchor through the fascia to obtain appropriate lead retention (Figure 9.4B). The anchor should be placed close to the midline near the spinous process to prevent lead movement caused by muscle contractions. However, it is important to note that there is no existing technology to anchor leads at the specific

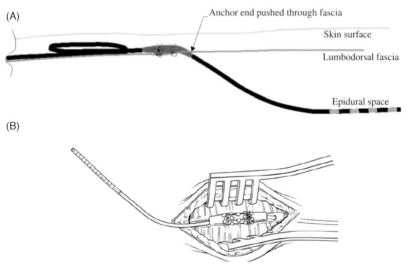

(A)

Anchor end pushed through fascia

Skin surface

Lumbodorsal fascia

Epidural space

(B)

FIGURE 9.4 Anchor placements. (A) With some anchors, the end must be pushed through the deep fascia before securing it to the deep fascia to maximize the bend radius of the lead with flexion and extension of the spine (from [4]). (B) With other anchors, suturing the anchor to the fascia is sufficient to provide adequate anchoring of the lead.

point of therapy delivery within the body. In current practice, anchors that are somewhat distant from the precise point of therapy delivery are utilized, meaning that the potential for migration of the lead at the delivery locus remains.

Leads are exposed to cyclic loads as a result of biomechanical motions in the spine. The magnitude of these loads is dependent on several factors, primary of which is whether a strain-relief loop is applied between the sutured extension connector and the entry point into the epidural space. When using percutaneous leads, a strain relief loop should be considered after anchoring the lead, before connecting the lead to an extension (if used) [4]. When using surgical paddle leads, biomechanical testing showed improved performance when using a strain-relief loop and no anchor compared with using an anchor and no strain-relief loop [9]. If an extension is used, the connector should be placed near the lead or near the pulse generator to prevent the formation of a third point of fixation within the system.

Multichannel devices have reduced the need for re-operation as a result of lead migration. Leads containing up to 16 independently programmable electrodes have a significantly greater reliability than single-channel systems [10].

DBS Lead Complications

Reported DBS lead complications have included lead fractures, lead migration, and infections and/or erosions [1,11–14].

Lead fractures typically occur when the connector between the extension and the lead was located below the mastoid, likely due to movements of the neck that increase stresses at the connector site [11]. Thus, lead fractures have been reported in patients with essential tremor, cervical dystonia, and dyskinesia [11,14]. If a patient complains of stimulation at a connector site, intermittent, or loss of stimulation, it is possible that a lead fracture has occurred. Fractures may be detected by conducting an impedance test and confirmed by x-ray [15]. Often, lead fractures are treated by replacing the fractured lead with a new lead [14]. Lead migration in the upward direction has been reported, mainly in movement disorder patients. Migration of the lead-extension connector from the parietal to the cervical area with slippage from the anchoring system may have been the cause of the lead movement [11,14].

Bow-stringing near the DBS extension wires has also been reported [16]. This is caused by the formation of scar tissue around the extension that over time becomes overtightened, noticeably protruding beneath the skin, leading to limitation of movement and patient discomfort. This is more often associated when two extension wires are tunneled along the same side and often necessitates surgical revision.

Skin erosion may be related to scalp thickness and device size, as erosion typically occurs at the lead-extension connector site [14]. The design of a lower profile connector has helped reduce skin erosions at the connector site [11]. Another consideration is to avoid placing bulky components of the implanted system underneath skin incisions [11]. No evidence supports the use of a curvilinear versus a straight incision to reduce erosions [2]. Nonetheless, curvilinear incisions have been suggested to help prevent erosions [11] likely because the incision is not placed directly over the burr hole. Recessing of components into a drilled bone trough in patients with thin skin may be useful to lower the profile of the component [1].

SAFETY CONSIDERATIONS

MRI Interactions

Magnetic resonance imaging (MRI) procedures may be unsafe for individuals with implanted neuromodulation systems. Three types of magnetic fields are used in MRI [17], and each of these three fields may interact with implanted medical devices.

The first type of field is a static magnetic field (typically 0.2–3.0 T). This magnetic field may interact with ferromagnetic objects, such as neurostimulation devices, by producing force and torque on the device. Strong enough forces may move or dislodge a device from its existing position. Injury would result if the lead or device is located in an area of the body that contains vital structures.

The second type of field used in MRI is gradient, or 'time varying' magnetic fields. These fields may induce electrical fields and currents in patients which may cause stimulation of nerves or muscles. At sufficient exposure levels, peripheral nerve stimulation may be perceived as 'tingling' or 'tapping' sensations [18]. Increasing the strength of the gradient field 50–70% above the perceived peripheral nerve stimulation threshold may lead to uncomfortable or painful sensations [19]. At extremely high levels, cardiac stimulation may occur, but this is highly unlikely using commercial MRI systems [18,19]. The presence of a metallic implant with conducting components will tend to concentrate the currents induced by the gradient fields. The concentration of currents is expected to occur if the implanted device has the shape of a long wire or forms a closed loop of sufficient size, such as a lead used for a neuromodulation system [17]. With sufficient magnitude and concentration of induced current, nerve stimulation may occur near the electrodes. A second effect on implanted devices is the induction of current in the metallic object, which will induce a magnetic moment and therefore torque on the object by the static MRI field. The implant will then exhibit high frequency vibration which may be uncomfortable in the case of large, highly conductive implants [17]. A third effect on implanted devices is that the induced eddy currents in the object may cause heating of the device. This heating may cause patient discomfort and may result in tissue damage.

The third type of magnetic field is radiofrequency (RF) fields. The MRI RF fields induce a current in the lead, resulting in a scattered electric field within the tissue, with the largest intensity near the ends of the lead [17]. As with field gradients, currents are primarily induced in components that are shaped as a long wire or that form a closed loop of sufficient size, such as neurostimulation leads. The lead acts as an antenna that picks up the RF energy resulting in induced currents, which may cause excessive heating at the electrodes. Increased heating may result in thermal lesions which can lead to coma, paralysis, or death. There are many variables that may impact the resulting heating. For example, the position of the implant in the patient relative to the transmit RF coil, the length and dimensions of the implant relative to the wavelength of the RF field in the patient, and the connection of an implant to another device (i.e. lead to neurostimulator). In contrast to leads, large metallic implants with smooth edges, such as an INS, have demonstrated minimal RF-heating [18].

The measure of the rate at which energy is absorbed by the body when exposed to an RF field is known as specific absorption rate, or SAR. SAR is measured in units of watts per kilogram (W/kg) and it may be averaged over the whole body, or over a small sample volume, such as the head.

Interactions between MRI machines and implanted systems are not predictable, due to variations in types of MRI scanners and protocols available, system configurations including lead paths and device placements, and lead and device construction. Therefore, it is important to follow the labeling for each specific manufacturer's devices in order to maximize patient safety. Applicable imaging guidelines concerning patient screening and potential device interactions should also be observed.

The presence of metallic components in an implantable system may cause an artifact in an MRI image near the system components which are easily recognizable. This artifact may be observed as a local distortion of the image and/or as a void in the image. The size of the artifact depends on several factors, including device shape, orientation, quantity, position, magnetic susceptibility, pulse sequences, and image processing method [18].

Device-related MRI considerations for DBS systems

Current, FDA-approved DBS systems manufactured by Medtronic, Inc have specific labeling to ensure patient safety (Medtronic, 2010). The guidelines apply to combinations of the following components and are summarized in Box 9.1:

- Neurostimulator models Itrel^{TM} II 7424, Soletra^{TM} 7426, Kinetra^{TM} 7428, Activa^{TM} PC 37601, Activa^{TM} RC 37612
- Lead extension models 7495, 7482, 7482A, 37085
- Lead models DBS 3387, 3389
- Pocket adaptor models: 64001 (1×4 pocket adaptor), 64002 (2×4 pocket adaptor)

It is important to follow the complete set of guidelines from the device manufacturer for the specific implanted product to ensure safety when using MRI, otherwise serious injury of patients may occur with inappropriate MRI conditions [21,22].

BOX 9.1 Summary of MRI Guidelines for DBS

MRIs may be performed only using a 1.5T horizontal bore MRI (not an open sided or other field strength MRI system) and a transmit/receive head coil. The MR parameters should be limited to those that produced a head SAR of 0.1 W/kg or less for all RF pulse sequences. Make sure the SAR value is for the head, and not whole body SAR or local body SAR as some machines may display. Careful consideration should be given to patients with tremor, as tremor may return when the stimulation is turned off and cause artifact in the MRI images. If the neurostimulator is reset during an MRI examination, it can be reprogrammed.

An impedance test measuring impedances and battery currents should be performed to test for open circuits. If an open circuit is suspected and an x-ray confirms that it was caused by a broken lead wire, an MRI should not be performed. Increased heating may occur at the site of the fracture or at the electrodes which may result in thermal lesions.

Device-Related MRI Consideration for SCS Systems

Several different products for spinal cord stimulation are commercially available. However, only devices from a single manufacturer (Medtronic, Inc, Minneapolis, MN) have FDA-approval to allow certain types of MRI procedures in patients. The MRI guidelines are not applicable for RF neurostimulation systems and are summarized in Box 9.2.

As with DBS, it is important to follow the complete set of guidelines from the device manufacturer for the specific implanted product to ensure safety when using MRI, otherwise serious injury of patients may occur with inappropriate MRI conditions.

BOX 9.2 Summary of MRI Guidelines for SCS

MRI examinations are restricted to those of the head only (no other body part) using an RF transmit/receive head coil at 1.5T horizontal bore MRI, and the head SAR must be limited to 1.5 W/kg for all RF pulse sequences [23]. MRI examinations of any other part of the body are not recommended as these require the use of an RF transmit body coil which may cause hazardous temperatures at the lead electrodes. If the neurostimulator is reset during an MRI examination, it can be reprogrammed.

The head coil must not cover any implanted system component. In addition, MRI procedures should not be prescribed for patients undergoing trial stimulation and that have systems that are not fully implanted. As with the DBS guidelines, an MRI should not be performed in patients with broken lead wires, since increased heating may occur at the site of the fracture or at the electrodes.

Diathermy

Diathermy is a treatment that uses RF energy to accelerate tissue healing by local heating. These treatments are typically used to relieve pain, stiffness and muscle spasms, to reduce joint contractures, to reduce swelling and pain after surgery, or to promote wound healing. Use of diathermy is contraindicated in patients implanted with DBS or SCS systems as a current may be induced within the lead conductors and cause subsequent heating at the electrodes. This may lead to severe injury or death [24,25].

Environmental Problems

Environmental influences, such as electromagnetic interference (EMI), may affect implanted neurostimulation devices [26,27]. Enough interference may be generated to change the parameters of the INS, turn an INS on or off, or cause a neurostimulator to shock or jolt the patient. In addition, it is possible for the extension, lead, or both to pick up EMI and deliver excess voltage which may in turn cause excessive heating at the electrodes.

Routine diagnostic procedures such as fluoroscopy and x-rays are not expected to affect the system operation. However, other medical devices may interfere with neurostimulation systems. The following guidelines may be used as being generally applicable to neurostimulation systems [27], but it is important to follow the specific requirements as provided in individual device system labeling.

- Other medical devices. The neurostimulator may affect the performance of other implanted devices, such as cardiac pacemakers and implantable defibrillators. Careful programming may be needed to optimize the patient's benefit from each device.
- External defibrillation. To minimize current flowing through the system, the defibrillation pads should be as far away from the INS as possible, the defibrillation pads should be positioned perpendicular to the implanted INS-lead system, and use the lowest clinically appropriate energy output. Some have also suggested turning the INS stimulation amplitude to zero [26].
- Electrocautery. The current path (ground plate) should be kept as far away from the neurostimulation components as possible. Bipolar cautery is recommended.
- High radiation sources. If high radiation sources such as cobalt 60 or gamma radiation are required, they should not be directed at the neurostimulator. Lead shielding should be placed over the device to prevent radiation damage if the therapy will be delivered near the neurostimulator.
- Lithotripsy. Devices with high output ultrasonic frequencies may damage the neurostimulator circuitry and, if they must be used, the beam should not be focused near the device.

- Psychotherapeutic procedures. The safety of equipment that generates EMI such as electroshock therapy and transcranial magnetic stimulation has not been established.

Home appliances generally do not produce enough EMI to interfere with the device. However, items with magnets, such as stereo speakers, refrigerators, and freezers, may cause the stimulator to switch on or off, but do not change the programmed parameters. The use of radiofrequency sources including cell phones, AM/FM radios, cordless phones and wired telephones may contain permanent magnets and should be kept at least 10 cm away from the neurostimulator.

Commercial equipment (arc welders, induction furnaces, resistance welders), communication equipment (microwave transmitters, linear power amplifiers, high-power amateur transmitters), and high voltage power lines may interfere with the neurostimulator if approached too closely.

Theft detectors and security screening devices may cause the stimulator to be turned on or off. If patients must pass through the security device, they should pass through the middle if two security gates are present or as far from the gate as possible if only one gate is present.

If a patient suspects that a device is interfering with their neurostimulator, he/she should move away from it or turn the device off. The patient programmer may be used to set the neurostimulator to the desired on or off state, either by the patient or by a trained family member or clinician. In patients implanted with DBS devices, identification of unintended deactivation of the INS is easier for patients with tremor-dominant disease and typically these patients would turn on the INS on their own [26]. Some patients may not be aware that their neurostimulator is inadvertently turned off and, in some cases, this may constitute a medical emergency [28]. In patients implanted with SCS devices, a loss of paresthesia may be detected if a device is inadvertently turned off. All patients should be educated on the possibility of EMI interfering with the functioning of their device and what to do in cases of device deactivation. Patients should also be educated on the importance of carrying their programmer with them at all times.

CONCLUSIONS

The design of implantable medical devices involves knowledge of interactions of devices and the human body, as well as interactions of devices with external environmental influences. Device materials, form factor, and component design aim to maximize patient safety and comfort while simultaneously delivering effective stimulation therapy.

REFERENCES

1. Sillay KA, Chen JC, Montgomery EB. Long-term measurement of therapeutic electrode impedance in deep brain stimulation. *Neuromodulation*. 2010;13(3):195–200.

2. Rezai AR, Kopell BH, Gross RE, et al. Deep brain stimulation for Parkinson's disease: surgical issues. *Mov Disord*. 2006;21(S14):S197–S218.

3. Follett KA, Boortz-Marx RL, Drake JM, et al. Prevention and management of intrathecal drug delivery and spinal cord stimulation system infections. *Anesthesiology*. 2004;100:1582–1594.

4. Kumar K, Buchser E, Linderoth B, et al. Avoiding complications from spinal cord stimulation: practical recommendations from an international panel of experts. *Neuromodulation*. 2007;10:24–33.

5. Cameron T. Safety and efficacy of spinal cord stimulation for the treatment of chronic pain: a 20-year literature review. *J Neurosurg (Spine 3)*. 2004;100:254–267.

6. Kumar K, Hunter G, Demeria D, et al. Spinal cord stimulation in treatment of chronic benign pain: challenges in treatment planning and present status, a 22-year experience. *Neurosurgery*. 2006;58:481–496.

7. Quigley DG, Arnold J, Eldridge PR, et al. Long-term outcome of spinal cord stimulation and hardware complications. *Stereotact Funct Neurosurg*. 2003;81:50–56.

8. Rosenow JM, Stanton-Hicks M, Rezai AR, et al. Failure modes of spinal cord stimulation hardware. *J Neurosurg Spine*. 2006;5:183–190.

9. Henderson JM, Schade CM, Sasaki J, et al. Prevention of mechanical failures in implanted spinal cord stimulation systems. *Neuromodulation*. 2006;9:183–191.

10. North RB, Ewend MG, Lawton MT, et al. Spinal cord stimulation for chronic, intractable pain: superiority of "multi-channel" devices. *Pain*. 1991;44:119–130.

11. Blomstedt P, Hariz MI. Hardware-related complications of deep brain stimulation: a ten year experience. *Acta Neurochir (Wien)*. 2005;147:1061–1064.

12. Kenney C, Simpson R, Hunter C, et al. Short-term and long-term safety of deep brain stimulation in the treatment of movement disorders. *J Neurosurg*. 2007;106:621–625.

13. Lyons KE, Wilkinson SB, Overman J, et al. Surgical and hardware complications of subthalamic stimulation: a series of 160 procedures. *Neurology*. 2004;63:612–616.

14. Oh MY, Abosch AA, Kim SH, Lang AE, Lozano AM. Long-term hardware-related complications of deep brain stimulation. *Neurosurgery*. 2002;50(6):1268–1276.

15. Farris S, Vitek J, Giroux ML. Deep brain stimulation hardware complications: the role of electrode impedance and current measurements. *Mov Disord*. 2008;23:755–760.

16. Miller PM, Gross RE. Wire tethering or 'bowstringing' as a long-term hardware-related complication of deep brain stimulation. *Stereotact Funct Neurosurg*. 2009;87:353–359.

17. Nyenhuis JA, Park SM, Kamondetdacha R, et al. MRI and implanted medical devices: basic interactions with an emphasis on heating. *IEEE Trans Device Mater Reliab*. 2005;5:467–480.

18. Shellock FG, MRI safety and neuromodulation systems. Krames ES, Peckham PH, Rezai AR, eds. *Neuromodulation*, vol 1. Amsterdam: Elsevier; 2000:243–281.

19. Schaefer D, Bourland JD, Nyenhui JA. Review of patient safety in time-varying gradient fields. *J Magn Reson Imag*. 2000;12:20–29.

20. Medtronic*MRI guidelines for Medtronic deep brain stimulation systems*. Minneapolis: Medtronic Inc.; 2010.

21. Henderson JM, Tkach J, Phillips M, et al. Permanent neurological deficit related to magnetic resonance imaging in a patient implanted with implanted deep brain stimulation electrodes for Parkinson's disease: case report. *Neurosurgery*. 2005;57:E1063.

22. Spiegel J, Fuss G, Backens M, et al. Transient dystonia following magnetic resonance imaging in a patient with deep brain stimulation electrodes for the treatment of Parkinson's disease. *J Neurosurg*. 2003;99:772–774.

23. Medtronic*Appendix B: MRI and neurostimulation therapy for chronic pain. Medtronic pain therapy: using neurostimulation for chronic pain*. Minneapolis: Medtronic Inc.; 2007.

24. Nutt JG, Anderson VC, Peacock JH, et al. DBS and diathermy interaction induces severe CNS damage. *Neurology*. 2001;56:1384–1386.

25. Roark C, Whicher S, Abosch A. Reversible neurological symptoms caused by diathermy in a patient with deep brain stimulators: case report. *Neurosurgery*. 2008;62:E256.

26. Blomstedt P, Jabre M, Bejjani B, et al. Electromagnetic environmental influences on implanted deep brain stimulators. *Neuromodulation*. 2006;9:262–269.

27. Medtronic*Activa RC and Activa PC: Multi-program neurostimulators*. Minneapolis: Medtronic Inc.; 2007.

28. Hariz MI, Johansson F. Hardware failure in parkinsonian patients with chronic subthalamic nucleus stimulation is a medical emergency. *Mov Disord*. 2001;16:166–168.

Commentary on Device Materials, Handling, and Upgradability

Jay L. Shils, PhD

Director of Intraoperative Monitoring, Department of Neurosurgery, Lahey Clinic, Burlington, MA

This chapter by Kast et al is an excellent description of the key elements in the material designs of present macroelectrode neuromodulation systems. My commentary will address future considerations in materials and handling needs. With the electrode surface much larger than the neuronal elements themselves (i.e. neuronal cell bodies and axons), the physics of the tissue–electrode interface might be further modified by electrode design to enable better modulation control. Current densities at this interface are usually low enough to overpower scar development around the electrode and deliver acceptable energy (or record neuronal activity with a large enough signal-to-noise ratio) over the lifetime of the patient. On the other hand, as future applications of neuromodulation move toward implantation of multiple smaller (surface areas of $100\,\mu m^2$ or smaller) electrodes [1,2] placed directly into neuronal tissues, electrode sizes will likely approach the scale of the neurons they are designed to control or record.

Size reduction has two major issues that then become relevant and contribute to the potential reduction in the usable electrode lifetime. The first is that much higher current densities on these electrodes can either cause corrosion of the electrode surface or, just as dangerous, tissue damage (see chapters 6, 7 and 8). The other issue, currently a subject of intense research [3–5], is that there can be changes in the tissue around the implanted electrode stemming from both the damage caused by pushing the electrode into the tissue, causing penetration wound effects local to the electrode. Research is being done looking at the speed of implant, the force of the implant and the sharpness of the electrode tips [6–8]. Since the electrode stimulation and/or recording surfaces are on the scale of the tissue elements involved in the foreign-body reaction, the damaging effects are much stronger and have greater potential of permanently rendering that electrode useless. Work to solve this issue includes looking at new material such as polyimide [9] electrode coatings, and pulse shapes and patterns [10] that can reduce the failure of the electrode tissue interface.

Flexible material technologies would also be of benefit to future neuromodulation systems. Implants for peripheral nerves (see Chapter 5) that could fit around the nerves for either full stimulation or focal stimulation, complex arrays and grids that may need to be implanted in sulci, or even intraspinal implants should have the flexibility needed to move with and not harm the tissue, conform to the shape of the tissue plane desired and adjust shape from implanting tools to the specific tissue location. Present integrated circuit technology is advancing very fast in this area. Wearable technologies, IC lithography, and IC substrates presently exist; it is only demonstrating their biocompatibility that is needed.

For many of these microarray implants, the electrodes will need to last for many years and will also not be able to be easily replaced. Once the electrodes are in the tissue and homeostasis has been reached, getting the electrode out of the tissue if necessary, and replacing it if needed in the exact same location will be extremely difficult if not impossible. Since a basic goal of these electrodes is to act on a focal area (i.e. a single to a few neural elements), damage from removal or the likelihood of not getting to the 'exact' same location is very high. As discussed in the power chapter (see Chapter 11), RF power transmission or very long-lasting batteries will be needed. RF technologies presently exist, but the depth of implant is inversely related to the efficiency of the transmission. As the microelectronics industry and consumer electronics advance, battery technology will then likely move forward at a faster pace.

REFERENCES

1. Maynard EM, Nordhausen CT, Normann RA. The Utah electrode array: a recording structure for potential brain-computer interfaces. *Electroencephalogr Clin Neurophysiol*. 1997;102:228–239.
2. Bai Q, Wise KD. Single-unit neural recording with active microelectrode arrays. *IEEE Transact Biomed Eng*. 2001;48:911–920.
3. Azemi E, Gobbel GT, Cui XT. Seeding neural progenitor cells on silicon-based neural probes. *J Neurosurg*. 2010;113:673–681.
4. Polikov VS, Tresco PA, Reichert WM. Response of brain tissue to chronically implanted neural probes. *J Neurosci Methods*. 2009;148:1–18.
5. Ludwig KA, Uram JD, Yang J, Martin DC, Kipke DR. Chronic neural recordings using silicon microelectrode arrays electrochemically deposited with a poly(3,4-ethylenedioxythiophene) (PEDOT) film. *J Neural Eng*. 2006;3:59–70.
6. Kozai TDY, Marzullo TC, Hooi F, et al. Reduction of neurovascular damage resulting from microelectrode insertion into the cerebral cortex using in vivo two-photon mapping. *J Neural Eng*. 2010;7:046011.
7. Bjornsson CS, Oh SJ, Al-Kofahi YA, et al. Effects of insertion conditions on tissue strain and vascular damage during neuroprosthetic device insertion. *J Neural Eng*. 2006;3:196–207.
8. Johnson MD, Kao OE, Kipke DR. Spatiotemporal pH dynamics following insertion of neural microelectrode arrays. *J Neurosci Meth*. 2007;160:276–287.

9. Chen Y-Y, Lai H-Y, Lin S-H, et al. Design and fabrication of a polyimide-based microelectrode array: Application in neural recording and repeatable electrolytic lesions in rat brain. *J Neurosci Meth*. 2009;182:6–16.

10. Wagenaar DA, Pine J, Potter SM. Effective parameters for stimulation of dissociated cultures using micro-electrode arrays. *J Neurosci M*. 2004;138:27–37.

STUDY QUESTIONS

1. How might MRI compatability in neuromodulation systems be accomplished? To what degree would such a development improve delivery of this therapy?

2. While biocompatibility plays a vital role at the outset of developing the materials in neuromodulation devices, in what ways later, when devices are being implanted, used, and potentially revised or removed do the device materials and handling characteristics have relevance?

Chapter 10

Electronics

Emarit Ranu, MSEE, MSBS, EMT-B
Boston Scientific Neuromodulation, Valencia, CA

THE HISTORY OF SCS DEVICES

From Fish to Electronics

The history of spinal cord stimulation is part of a larger human story of pain reduction spanning perhaps thousands of years. In the first century CE, Scribonius Largus reported the use of the torpedo fish in treating gout and headache after observing accidental contact with the electrically active fish relieved gout pain [1]. Following the use of various electrostatic friction machines [2], electrochemical based devices [3] and magnetically derived current apparatus [4] from the 17th to 20th centuries for pain mitigation (and some less appropriate symptoms), the Electreat was patented in 1919 [5] (Fig. 10.1). This device was operated by two standard 'D' cell batteries which powered an internal mechanically controlled induction device as a source of pulsing current applied to a roller and/or sponge electrode(s). The Electreat may have been the precursor to the modern TENS unit (transcutaneous electrical nerve stimulation, a term used

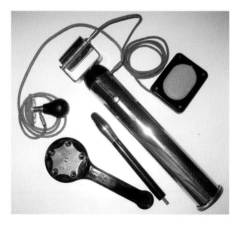

FIGURE 10.1   Electreat. From Emarit Ranu.

Essential Neuromodulation. DOI: 10.1016/B978-0-12-381409-8.00010-3

by Burton and Maurer in 1974 [6]). Such devices apply current directly to the skin via patch electrodes for pain mitigation. In early SCS, TENS units were used to test patient tolerance to stimulation prior to implantation [7].

THE FIRST IMPLANTED DORSAL COLUMN STIMULATORS

The first human dorsal column stimulator was implanted in 1967 by Shealy and designed by Mortimer [8,9], following experimentation in a feline model [10]. The system used a single cathodic electrode sutured through the dura mater while the anodic electrode was placed in the intramuscular space, both electrodes being composed of the material Vitallium® [11]. Subcutaneous, hypodermic needle-accessed jacks permitted connection of the hand-made stimulator device to the electrodes.

Their second stimulator, implanted 7 months later, was designed by Mortimer and based on Medtronic's Angiostat and Barostat carotid sinus nerve stimulators [12–14]. Mortimer's second device used platinum–iridium electrodes of the same shape as the first stimulator but was RF-coupled thus requiring an external coil for the provision of power (Fig. 10.2). A portable box housed the transmitter electronics, connected to the external coil and contained the stimulation parameter controls (e.g. rate, amplitude and, in the case of the variable frequency transmitter, the rate of frequency change). An account of their first, second and subsequent implants can also be found in [15].

First Commercial SCS Devices

The technology behind SCS initially was derived from cardiac pacemakers [16], which are single source devices. An incomplete look at the SCS business field finds that the first commercially produced SCS device was the Myelostat from

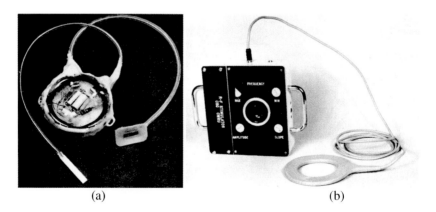

(a)   (b)

FIGURE 10.2 (A) Mortimer's second stimulator with electronics exposed, implanted by Shealy and (B) the variable frequency transmitter box with coupling coil.

Medtronic, based on the Angiostat and Barostat carotid sinus stimulators, made available in 1968 [12,17]. Avery Laboratories, originally founded to develop phrenic nerve pacing applications, offered their own device in 1972 [11].

Cordis (purchased by Johnson & Johnson in 1996) introduced the 199A in 1976 [11]. This represented the first totally implantable device, being entirely self-contained in epoxy and powered with a mercury battery. It was a modified cardiac pacemaker with the ability to change externally amplitude and rate, though was first used with movement disorders [18]. In 1980, they developed the 900X–MKI which was the first SCS device to gain The Food and Drug Administration (FDA) approval for pain relief [19]. It is important to note that, prior to 1976, the FDA did not have the complete legal mandate to regulate and require safety and efficacy testing of medical devices [20].

Medtronic received approval to market their Itrel device in late 1984 [21]. The design of this device continued to leverage the company's cardiac devices [22]. It was a totally implantable, primary cell, SCS device. Neuromed developed their Quattrode RF system in 1980 [23–25]. The company was acquired by Quest Medical in 1995, which then changed its name to become Advanced Neuromodulation Systems (ANS) in 1998 which, in turn, was purchased by St Jude Medical (STJ) in 2005. Advanced Bionics (AB), initially a cochlear implant company, whose Pain Division was purchased by Boston Scientific (BSC) in 2004, introduced the first rechargeable, multisource, fully-implantable SCS device with their Precision system [26]. The system has technology similar to the AB Clarion Multi-Strategy cochlear implant with 16 simultaneously active channels (released in 1995) [27]. At the time of this writing, all three companies currently manufacturing SCS systems offer rechargeable devices. Only MDT and STJ offer primary cell devices.

From Bipolar to Multipolar: the Evolution of Commercial Leads

Connected to the SCS device is an integral part of the therapy delivery system: the lead(s). With their electrode(s) placed over the proper portion(s) of dorsal column, leads are responsible for permitting the application of stimulation to accessible fibers associated with the patient's pain, as discussed later. The first leads, again borrowed from endocardial pacing technology, eventually addressed the application specific design needs of SCS [22]. As time moved on, leads were designed with more electrodes and varying geometries to mitigate effects of lead migration [22].

The first leads were of course from those of Mortimer's first device, a pair of which is shown in Figure 10.3. They consisted of a single cathode and a single anode, where the cathode was placed endodurally (see [28] for a similar description). During the early stages of the treatment modality, leads were surgically placed. Percutaneous leads were initially used as a minimally invasive screening tool, prior to the surgical implantation of permanent leads [29–33].

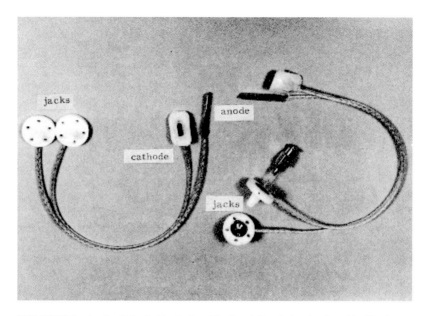

FIGURE 10.3   A pair of identical leads from Mortimer's first device, implanted by Shealy.

Moving to a partial look at the commercial realm, in 1978, Medtronic intro-duced a percutaneously inserted electrode for permanent use. In 1981, the com-pany released a four-electrode percutaneous lead, the PISCES (percutaneously inserted spinal cord electrical stimulation) model 3484 [34]. In 1986, Neuromed made available an eight-electrode RF system, based on their Octrode lead [35]. The work of J. Law, published in 1987, suggested the advantage of multiple rows of electrodes for low-back pain patients [36]. In late 1994, Neuromed received approval to market their Dual Octrode device [37], a dual lead sys-tem representing the first 16-electrode system, just 2 months before Medtronic gained approval for their eight-electrode Mattrix system [38].

At the time of this writing, all three companies currently manufacturing SCS systems (MDT, STJ and BSC) offer 16-electrode systems with varying lead and electrode geometries in both percutaneous and surgically implanted paddle leads. Currently, a greater number of electrodes is not available. The various leads include electrodes as small as 3 mm to as large as 6 mm with spacing as tight as 1 mm to as wide as 12 mm. Currently, only MDT and STJ offer tripole paddle leads: the Specify 5-6-5 (model 39565) and Lamitrode (models 8, 8C and 16C) series, respectively. A tripole configuration may penetrate deeper into the dorsal columns [39,40], possibly dependent on the electrical capabilities of the system [41].

The number of electrodes is but one means of assessing the targeting abil-ity of a lead or how well it mitigates lead migration effects. For example, the number of simultaneously active electrodes and the number of sources both provide field superposition in independent ways. Having more than one channel allows for independent sets of stimulation parameters to be applied rapidly in sequence, which can provide the cognitive perception of simultaneity. It is also a means to emulate simultaneously active electrodes. All of these features are part of a suite of tools used to recruit the specific fibers associated with the patient's pain, each discussed later.

Targeting Fibers Spatially

The need for placing SCS leads over the area(s) of the dorsal columns that will place paresthesia in the patient's pain area has partially driven lead design to move from Mortimer's first lead having a two electrode approach (a single cathode and a single anode) to today's 16-electrode leads. Additional electrodes can also mitigate loss of paresthesia due to migration. From a biomedical engineering design standpoint, coupled with the electrical capabilities of the system, the following characteristics of the lead(s) all work together to provide the clinician a level of control and selectivity in placing paresthesia:

1. the number of electrodes
2. the electrode geometry
3. the relative positions of the electrodes (electrode spacing).

Addressing the first point, for a given electrode geometry and spacing, as the number of electrodes increases, the available population of neurons that can be recruited into paresthesia increases. This is simply due to there being more accessible points of stimulation delivery over the dorsal columns. However, both the geometry and relative positions of the electrodes play an important factor in determining the recruitable population. As noted above, these factors may not be to the exclusion of the electrical capabilities of the pulse generator. In fact, as will be discussed later, therapeutic potential of a lead can be dependent on the ability of the pulse generator.

Next, looking at the second point, as the size of an electrode increases its contact area and depth with tissue increases accordingly. Thus, a larger electrode can permit access to a larger neuron population than a smaller electrode. However, when considering electrode size one must consider both charge density and spatial resolution. As the contact area with the tissue increases, the charge density on the electrode decreases (when the stimulation current is kept constant). With a larger electrode then, the stimulation current may need to be quite high to ensure that the charge density is adequate to provide any useful therapy. Such a case would require a high output device with the requisite design limitations. Conversely, a smaller electrode has a larger charge density on its surface. This limits the available stimulation current for therapy as exposure to excessive charge density is well understood to cause physiological damage [42].

Spatially, a smaller electrode permits a more confined and therefore selective population of neurons to be recruited thus allowing paresthesia to be more defined. However, a small electrode may not provide access to the entirety of

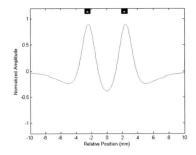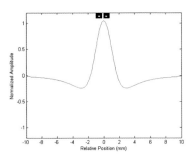

FIGURE 10.4 Overall activating functions from two equal and simultaneous point-source mono-pole cathodes 2.2 cm from an unmyelinated axon (A) 5 mm apart and (B) 1 mm apart illustrating their interaction in space (modeling from [43]).

the population of neurons necessary to recruit for complete paresthesia coverage. This suggests that a larger number of smaller electrodes is favorable to either fewer electrodes, larger electrodes or some combination of the two.

Finally, relative electrode position is an important design factor. If electrodes are spaced far apart, then their activating functions may not overlap to provide a cumulative effect which would effectively fill in the gaps in recruitable neurons between electrodes. This is demonstrated in Figure 10.4.

Thus, an electrode with a size that does not challenge charge density safety or stimulation efficacy, placed in an array of similar electrodes utilizing a spacing permitting activating functions to overlap constructively, can give the clinician access to both large swaths of fibers for broad paresthesia coverage and provide spatial resolution to target small fiber populations for paresthesia selectivity.

ELECTRONICS FOR SPINAL CORD STIMULATION

Next, the various aspects of the design of spinal cord stimulator pulse generators will be presented. These include the pulse delivery type (constant voltage or constant current), the pulsewidth, the pulse rate, the number of pulse sources, the power source (RF, primary cell or rechargeable), implantable pulse generator (IPG) efficiency concerns, telemetry design needs, clinically specific application design and reliability issues. All of these aspects work in concert with each other and the clinician to provide clinical results.

Pulse Delivery Types

The pulses delivered by an IPG can either be constant voltage or constant current. The implementation of each has both design implications and physiological implications. However, prior to the discussion of pulse types, it is important to recapitulate how the membrane potential of a neuron is determined.

Equation 10.1 shows the well-known Goldman-Hodgkin-Katz equation [44]. It illustrates how the concentration gradients of important ions with respect to the inside and outside of a cell dictate the membrane potential. When the local membrane potential meets or exceeds threshold, the voltage-gated sodium channels open. If enough of these channels open and remain open long enough, an action potential is generated which may be conducted via saltatory conduction along the axon.

$$V_{rest} = \frac{RT}{F} ln \frac{PK[K^+]_{out} + PNa[Na^+]_{out} + PCl[Cl^-]_{in}}{PK[K^+]_{in} + PNa[Na^+]_{in} + PCl[Cl^-]_{out}} \qquad (10.1)$$

Ions in solution can be driven to move by three mechanisms: the concentration gradient, relative charge and externally applied potentials. The first is the osmotic basis for charge movement and is the mechanism primarily responsible for sodium influx when the sodium channels open. The second is the electrical basis of ion movement: opposite charges attract and like charges repulse. This mechanism can either oppose or work with osmotic forces. The third exploits the electrical basis of charge movement artificially: an externally applied potential will move ions according to their polarities, the field orientation and osmotic forces. This in turn changes the ion concentrations at the membrane and therefore alters the local membrane potential.

Constant Voltage

The simplified model in Figure 10.5 shows an electrolyte solution represented by a resistance and two electrodes each with their own associated resistance and capacitance. The solution resistance models how well ions flow in the medium[1].

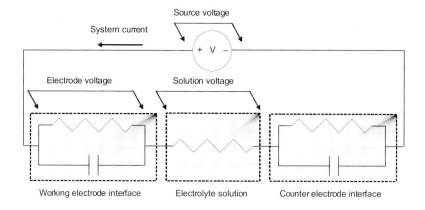

FIGURE 10.5   Electrode circuit constituents in solution and specific measurement points. Electrode resistance, capacitance and solution resistances are shown.

1. The solution resistance depends on the electrical properties of the ionic constituents in the medium (e.g. the molarity of the different species in solution) and the contact area with the electrodes. The electrical, or ohmic, resistance (in series with the source and not shown) depends on the material characteristics of the electrodes and connecting wires (e.g. the length of the wires and the composition of the metals).

At the electrode, the resistance models actual electron transfer into solution via reduction–oxidation (redox). This is an undesirable process, resulting in a local change in chemistry and electrode corrosion, so materials are selected to keep this resistance very large. The capacitance models the charge redistribution at the electrode–solution interface. The electrical resistance of the electrodes and wires are not included in this model since they merely model energy loss intrinsic to the system. In practice, a significant amount of energy can be spent to overcome high resistance lead wires, while intrinsic lead capacitance can change the pulse shape.

If the counter electrode is very large and does not transfer electrons by redox reactions, its resistance is very small and capacitance very large. Thus, the counter electrode becomes a reference electrode at near zero potential. When a constant voltage is applied between the two electrodes, there is a static voltage drop in solution due to the solution resistance. However, the capacitive component of the working electrode becomes polarized due to the storing of charge. As a result, it begins to develop an electrode voltage, thereby decreasing the solution voltage, since the voltage between the working electrode and solution is conserved and therefore must equal the source voltage.

Over the course of the constant voltage pulse, the voltage in solution (where excitable neurons are located) decreases and in the limit for a long pulse becomes zero. Mathematically, this is expressed as:

$$v(t) = V_0(e^{-t/RC} - 1) \qquad (10.2)$$

where V_0 is the initial applied voltage of the pulse between, t is time in seconds, R is the resistance in the circuit and C is the capacitance, assuming the electrode resistance is zero. At $t=0$, the start of the constant voltage pulse, the capacitive element is uncharged and has zero voltage across it. However, as time progresses, the capacitive element charges and effectively moves the applied voltage from solution to the electrode. Concurrently, the amount of charge moved in the system decreases. With increasing time, the electric field becomes confined in the volume just at the electrode–solution interface (the electrical double layer or Helmholtz layer) decreasing the electric field in solution. The thickness of the electrical double layer, from the Gouy-Chapman-Stern model, is on the order of nanometers, placing excitable tissue well outside the layer in neurostimulation applications.

The waveforms shown in Figure 10.6 illustrate the time courses of the current and voltages at specific points in the system referenced in Figure 10.5. As mentioned, note how the current decreases over the course of the applied constant voltage pulse. Intuitively this makes sense since, as more charge is attracted to the electrode, it would take more driving force (i.e. more voltage) to continue to attract additional charge of that same polarity (though opposites attract, like charges are driven apart).

As electrical conditions change due to changes in electrode contact area and physiological response to foreign body introduction [45], the solution

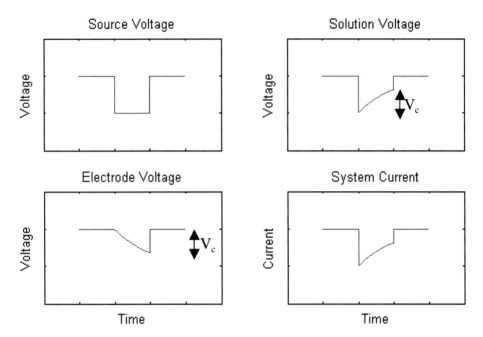

FIGURE 10.6 Illustration of the current and voltage(s) during a constant voltage cathodic pulse. Amplitude and time axes are arbitrary in scale. The effect of the capacitor voltage is shown. Refer to measurement points in Figure 10.5.

resistance and even capacitance will change accordingly causing the voltage in solution to change as well. Thus, for several reasons, the constant voltage approach does not control the injected charge and therefore the applied electric field in solution, leaving both to vary. The physiological consequences of this may include unreliable stimulus repeatability, and decreased stimulation efficacy over the course of the pulse and can result in higher stimulation thresholds [46,47].

Constant Current

In the constant current approach, as illustrated in Figure 10.7, the voltage applied to the entire system is increased over the course of the pulse to ensure that the charge injection rate is continuous. The increasing voltage serves to counter the polarization of the working electrode as its capacitive element charges. This increasing driving force then continues to move charge at the desired rate. This ensures that the electric field in solution is constant while the electric field at the electrode–solution interface increases over the course of the pulse. Unlike the constant voltage approach, since the electric field in solution remains unchanged, the excitable tissue is exposed to an unvarying rate of change of ion movement. As discussed previously, this may help to maintain stimulus repeatability and efficacy over the course of the pulse.

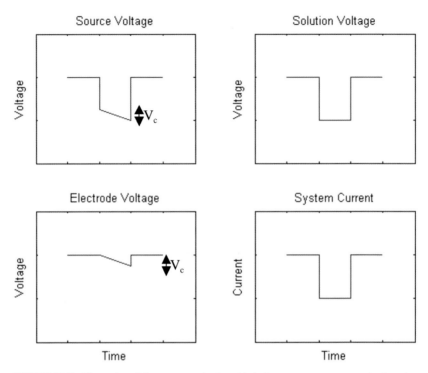

FIGURE 10.7 Illustration of the current and voltage(s) during a constant current cathodic pulse. Amplitude and time axes are arbitrary in scale. The effect of the capacitor voltage is shown. Refer to measurement points in Figure 10.5.

Derivation for how the voltage across a capacitor should vary to achieve a constant current i_0 when the initial voltage is zero is found from the current–voltage relationship of a capacitor (Equation 10.3) by integrating the voltage over the current (Equation 10.4). The result is found in Equation 10.5. It can be seen that to keep a constant current (and therefore a constant electric field) in solution, the voltage on the source must increase to counter the increasing voltage accumulated on the electrode due to the capacitance. Much like a battery (a constant voltage source) can be thought of as varying the current to a load to maintain a constant voltage, a constant current source can be thought of as varying the voltage on a load to maintain a constant current.

$$i = C \frac{dv}{dt} \tag{10.3}$$

$$v(t) = \frac{1}{C} \int_0^t i_0 \, d\tau \tag{10.4}$$

$$v(t) = \frac{i_0 t}{C} \tag{10.5}$$

Advantages and Disadvantages of Each

The constant voltage approach only controls the net voltage between the electrodes, not in solution. The electrode–solution boundary acts as a capacitor
which stores charge when the voltage is applied. Over the course of the pulse,
that stored charge reduces the electric field in solution. As the electric field
in solution decreases, there is less of a driving force to move ions to affect
the membrane potential, as the Goldman–Hodgkin–Katz equation describes.
However, the constant voltage method is simpler to realize electronically.

The primary advantage of the constant current approach is its ability to control
the electric field in solution. It is this field that drives ion movement in the tissue.
This approach compensates for the polarization effect at the electrode–solution
interface by increasing the system voltage over the course of the pulse to ensure
the rate of charge injected is constant which, in turn, keeps the electric field in
solution constant, preserving the driving force to move ions affecting membrane
potential. However, the electronic realization of this approach is more complex.

Electronics Design Differences

As shown in Figure 10.8a, a battery with an electronic switch is the basic means
to provide constant voltage stimulation. In practice, the source is never directly
connected to the tissue. This is to isolate the source from the tissue to prevent
DC stimulation in case of circuit failure. The switch first connects the capacitor to the source, permitting it to charge. When the switch is thrown to the
other position, the capacitor discharges into the tissue, providing stimulation.
This topology by its very nature does not deliver a constant voltage pulse at the
electrodes since, as charge from the isolating capacitor is released, the voltage
must drop. However, if C is relatively large, it can come close to delivering a
constant voltage pulse (i.e. if C stores much more charge than delivered) to the
electrodes, though of course not to the solution for reasons already discussed. In
practice, additional electronics would be placed between the switch and battery
to regulate the output voltage to the desired value independent of variations in
battery voltage. Figure 10.8b shows the addition of a switch network to select
the polarity (cathode or anode) of the output branches.

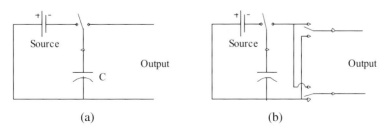

(a)   (b)

FIGURE 10.8 Fundamental electrical implementation of a constant voltage pulse generator with an isolation capacitor. (a) The cathode is the top branch and the anode is the bottom.
(b) Addition of a switch network to select either output branch as an anode or cathode.

The maximum voltage in this example is limited directly by the maximum available voltage of the source (e.g. the battery). Voltage can be increased beyond the source voltage with switching regulators known as boost or step-up converters. These converters increase the output voltage available at the electrodes to deliver stimulation, which is discussed later. Because power (the mathematical product of current and voltage) must be conserved, the current pulled from the battery is higher than the current delivered by the converter. However, the ability to increase the voltage beyond that of the source comes at the expense of reduced efficiency, increased cost and, in the case of battery-controlled devices, reduced battery life. The maximum current is limited by the voltage source resistance (causing the output voltage to decrease as current increases) and design tolerances of the electronics. In the case of battery-controlled devices, the battery chemistry reaction may limit the maximum output current, though tends not to be a factor in SCS applications.

The constant current approach is more complicated to implement electrically as shown by one constant current source example circuit in Figure 10.9a with identical transistors. I_{REF} is set by a reference such as an active resistor, R_{SET}. Because Q_2 shares the same V_{GS} as Q_1, I_{OUT} will be the same as I_{REF}. However, when R_{LOAD} gets too big, $V_D < V_{GS}$ and the transistor leaves the saturation region and enters the triode region where I_{OUT} is linearly related to R_L and is no longer pinned to I_{REF}. At this point the output current decreases below I_{REF} and is not constant, unless V_{Source} is increased so that $V_D > V_{GS}$ and Q_2 returns to the saturation region. The voltage across R_{LOAD} required to obtain I_{OUT} is the 'compliance voltage' and is $V_{Source} - V_D$. The maximum available compliance voltage in this case is $V_{Source} - V_{GS}$. The reduction by V_{GS} accounts for the drop across Q_2 since the drain cannot be at ground potential and must be above V_{GS} to maintain operation in the saturation region.

To add programmability to the current, a cascade of additional transistors is added as shown in Figure 10.9b. Switching on additional transistors allows the application of more current. To mitigate DC stimulation in case of component

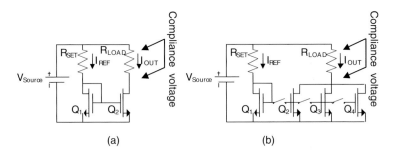

(a)   (b)

FIGURE 10.9 A monophasic constant current source implementation. (a) Simple CMOS current mirror and (b) switched topology for current programmability. In this example, the anode is on the + side of V_{Source} and the cathode is on the drain(s) of the transistor(s). Compliance voltage measurement point is shown.

failure, a capacitor can be added in series with the load for isolation. Because the circuit ensures constant current through the load over the course of the pulse, the capacitor would not reduce the voltage in the load. However, at the completion of each pulse, the capacitor must have a means to discharge. Polarity of the output can be selected by adding a switch network over R_{LOAD} to connect either side of the load to the transistor drain(s) or V_{Source}.

The maximum current in this example is limited by the maximum available source voltage, source resistance, tolerances of the electronics and, in the case of a battery-controlled device, the reaction rate (again, typically not a factor in SCS).

A device can have independent limits on both its output current and available voltage at the load. For example device A may have a 10 mA current limit and a 10 V output limit whilst device B may have a 7.5 mA current limit and a 15 V output limit. Ohm's Law shows that device B can deliver more current above a load resistance of 2000 ohms. The current provided by device A is limited to 10 mA at resistances below 1000 ohms and drops below 10 mA as the load resistance increases above 1000. Similar boundaries can be found for device B. If the devices are constant voltage systems, when the capacitance of the electrode(s) is included, the maximum current is only applicable at the start of the pulse, as we know the current will decrease over the course of the pulse and the voltage in solution will decrease too.

Finally, it's important to note that the demonstrative examples presented here are not specifically intended to deliver biphasic pulses. Such circuits, though conceptually related to those presented, are more complicated.

Single Versus Multisource

The number of sources available in a system, especially with respect to the number of electrodes on the lead, is a major factor in controlling the amount of current on any single electrode. As shown in Figure 10.10, when a single source system (either current or voltage) has more than one active electrode, the current in each will be identical when the load impedances are the same. However, when the load impedances change, more current will flow on the electrode with the lower impedance, as is evident from Ohm's Law.

In a multisource system, each electrode can be programmed to provide the specified output independent of the load impedances on all other active electrodes. Independent control of each electrode may permit real-time and dynamic movement of paresthesia [48]. In the case of a multisource constant current device, once programmed, the current at each electrode is maintained regardless of changes in impedances anywhere in the system (unless the compliance voltage is inadequate, as discussed in the previous section). This is illustrated in Figure 10.11. In the case of a multisource constant voltage system, current at each electrode is controlled by electrode load impedance, though each voltage can be adjusted independently at each patient programming visit[2]. It is

---

2. No such system has been or currently is available on the market at the time of this writing.

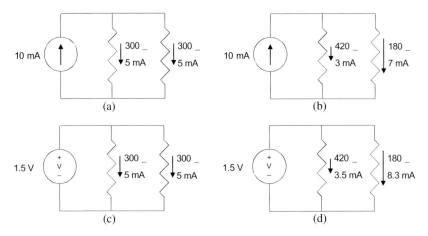

FIGURE 10.10 Single source constant current system with two electrodes and load impedances (a) same and (b) different. Single source constant voltage system with two electrodes and load impedances (c) same and (d) different.

interesting to note, that given two otherwise identical constant current systems (lead geometry, electrical specifications, etc.), a multisource system is indistinguishable from a single source system only when two electrodes are active (one cathode and one anode). This is because the load current is not dependent on the inactive electrodes in a single source constant current system.

Targeting Fibers Electrically

The need to direct stimulation at specific points along the spinal cord is accomplished both by the location and geometry of the lead and its electrodes (discussed previously), as well as the ability of the system to provide controlled electrical output on any active electrode [49]. A single source system is restricted to defining any combination of electrodes as a cathode, anode or off and the total system amplitude. Those electrodes that are active will share the source output. Thus, targeting may be ineffective in a single source system due to the inability to specify the output current on each of the active electrodes, despite the lead design.

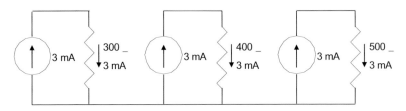

FIGURE 10.11 Constant current multisource system with different load impedances.

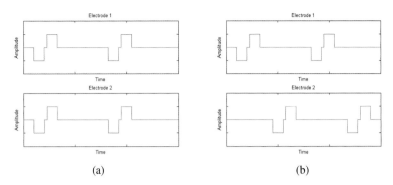

FIGURE 10.12 (a) Interleaved versus (b) simultaneous stimulation timing of adjacent electrodes. Pulse shape, amplitude and time axes are arbitrary in scale.

Interleaving Versus Multisource

One method that attempts to expand the ability of a single source system to emulate a multisource system is by rapidly switching two adjacent electrodes on and off in succession each at a specified amplitude. This approach is clearly different than that of simultaneous pulses on different electrodes as shown in Figure 10.12. The idea behind interleaving is to permit controlled output at more than one electrode (since only one electrode is on at a time) and possibly exploit summation of the individual activating functions in the time domain. This might be possible since the neurons may still be somewhat depolarized following the first pulse: the second pulse in the series may drive the already partially depolarized neurons to threshold. This mechanism is not the same as that of simultaneously active electrodes where each electrode contributes to the depolarization of the neurons separately and cumulatively.

While there are no comparative clinical data currently published between the two methods, modeling with thick fibers (those with the longest time constant) suggests that this method is unable to approximate the same effect as having simultaneously active electrodes [50]. This is illustrated in Figure 10.13 which

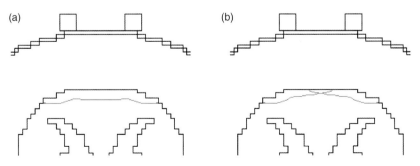

FIGURE 10.13 (a) Simultaneous versus (b) interleaved areas of activation in the spinal cord when using adjacent electrodes. From [50].

shows the area of the dorsal columns stimulated when two adjacent electrodes are (a) on simultaneously and (b) interleaved. The area of neurons unable to be recruited in the interleaved approach is apparent.

However, there is clinical utility in this interleaving approach by applying this technique to generate paresthesia in specific and isolated anatomic areas. This is done by applying different stimulation settings (active electrodes, pulsewidth, rates, amplitudes) rapidly in succession, even with concurrently active electrodes. For example, one setting could be applied for foot pain then another setting could be applied for upper leg pain. These independent settings may be repeatedly applied quickly enough such that the patient is unable consciously to resolve the system alternating between the two settings. This would cause the perception that both areas of isolated paresthesia are occurring simultaneously. Though both single and multisource systems can use this interleaving approach, only the multisource system is able to define simultaneously precise outputs on each electrode allowing the associated stimulation fields to overlap in time and space permitting access to a greater volume of tissue, as Figure 10.13a shows in comparison to Figure 10.13b.

Advantages and Disadvantages of Each

The single source approach has the advantage of being much simpler to design and is therefore less expensive. It may also have the benefit of being easier to program from the clinician's standpoint since electrodes can only be programmed as a cathode, anode or off. When using the interleaved approach to create the effect of multiple areas of stimulation, however, programming complexity increases as the relative amplitudes (and other variables as appropriate) must be specified for each setting. Software algorithms in the programmer relieve some of that burden.

The multisource approach, in addition to being more expensive, could have a higher probability of failure given the additional complexity. However, it provides greater flexibility in directing stimulation output to specific electrodes. The multisource current topology can both compensate for electrode polarization and maintain the programmed output as resistances in the system change. From a programming perspective, this approach is the most complicated as each active electrode must have its output deliberately specified and typically requires software algorithms to manage the programming.

Electronics Design Differences

A single source system will require an electronic switch network to apply the single source to any number of electrodes on the lead. There must be as many switches as there are electrodes, each with three positions: (1) off, (2) connect to positive side of source to function as an anode and (3) connect to negative side to function as a cathode. When a switch is closed, the source shares its output with the associated electrode. The source amplitude is variable and controlled by the programmer. When the polarity of an electrode is changed, all active electrodes must return to zero amplitude to avoid overstimulation.

For a multisource system, each electrode must have its own dedicated source. Each source in turn must be individually adjustable between zero and maximum output selectable as either a cathode or anode. Because the amplitude of each electrode can be defined continuously from positive to negative or vice versa, any single electrode can transition between polarities without forcing the amplitude to zero on all active electrodes.

Pulsewidth Clinical Utility

As is illustrated in Figure 10.14, the length of time a single stimulation pulse is applied is called the pulsewidth. The time between the cathodic pulse and anodic pulse (the putative charge injection and recovery phases) is named the interphase interval. The time between the same part of successive series of pulses is the interpulse interval, the inverse of which is the rate.

Studies have shown both clinically and theoretically that when all other factors remain constant, the pulsewidth can determine the lower limit on the axonal diameter size of those fibers that can be stimulated [51–53]. This inverse relationship between pulsewidth duration and recruitable fiber size suggests that a longer pulsewidth may allow a larger population of fibers to be recruited [54] possibly permitting paresthesia to be more broadly placed across the body. Additionally, histological study of the superficial dorsal columns has shown that nearly 85% of the fibers are smaller than $7\,\mu m$, and only 1% of the fibers are larger than $10\,\mu m$ [55], whereas modeling suggests that a pulsewidth of $210\,\mu s$ will activate fibers only as small as $9\,\mu m$ before possibly uncomfortable dorsal root stimulation occurs [56]. Currently available SCS systems provide a maximum pulsewidth between 450 and $1000\,\mu s$.

Strength–Duration Curve and Targeting Fibers Based on Diameter

The strength–duration curve illustrates which combinations of amplitude and pulsewidth result in recruitment of a fiber with known diameter, at a specific

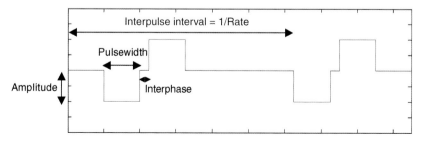

FIGURE 10.14   Stimulation pulse characteristics.

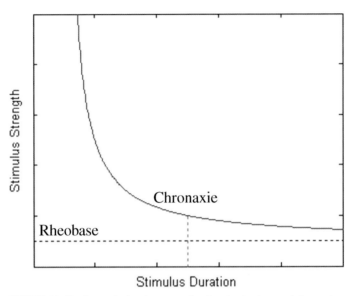

FIGURE 10.15   Strength–duration curve showing the rheobase and chronaxie.

distance from an electrode as illustrated in Figure 10.15. The perception threshold is believed to be the boundary between sub- and super-threshold fiber activation states. The rheobase is defined as the smallest stimulus amplitude, when applied for an infinite duration, that would result in fiber activation. It is an asymptote of the perception threshold boundary and is therefore measured practically at a point where the rate of charge injected is just short of rate of charge loss due to membrane resistance, usually in the hundreds of milliseconds. The chronaxie is the time it takes to activate a fiber when the amplitude is twice the rheobase. For two neurons with similar rheobases, the chronaxie gives a measure of the relative excitability of the neurons: shorter chronaxies apply to neurons that are more excitable.

Smaller diameter fibers have a strength–duration curve that is pushed to the right, indicating that, for the same amplitude, a longer pulsewidth is required for recruitment when compared with a larger diameter fiber. Intuitively, this may make sense since smaller diameter fibers, owing to less myelination, have more current leakage through and along the membrane. The longer pulsewidth may be necessary to compensate for the loss in injected current. Additionally, a short interphase can limit the recruitment of fibers as the subsequent anodic pulse can hyperpolarize the membrane before a regenerative action potential is initiated by the initial cathodic pulse [57].

Figure 10.16 shows actual clinical data from a patient subjected to three different pulsewidths. The amplitude was set at the perception threshold plus 80% of the difference between the perception threshold and the maximum tolerable threshold; i.e. at a 'strong but comfortable' stimulation intensity level.

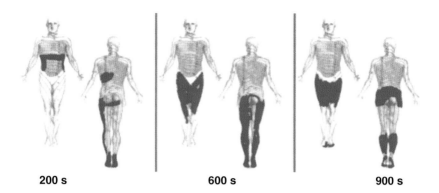

200 s **600 s** **900 s**

FIGURE 10.16 Clinical data from a patient showing change in paresthesia coverage when only the pulsewidth is varying [58].

The pulsewidth was then set at three different values. At each value, the patient drew a paresthesia diagram. It is clearly evident that the change in pulsewidth is responsible for activating different populations of fibers in the dorsal columns.

Electronics Design Issues

The larger challenge when designing for an extended pulsewidth is ensuring that the system has enough output power available over the course of the pulse. Recall that any system will have a step-up, or boost converter, onboard both to regulate the internal power source for the control circuitry and for providing a high enough output to offer effective stimulation.

An example step-up converter uses a switching topology shown in Figure 10.17. When the switch is closed, the inductor stores energy from the source. The capacitor provides energy to the load (accumulated from the previous switching event), while the diode prevents the capacitor from being shorted by the switch closure. When the switch is opened, the inductor current is forced into the load[3] and charges the capacitor. The inductor voltage is related only to its output current and not to the source voltage, thus providing for the voltage boost. By introducing a duty-cycle into the switch, either the output voltage or current can be precisely controlled by modulating the duty-cycle based on measuring the output. However, when the stimulation pulsewidth is very high, the demand for a constant output by the load can exceed the output available by the converter due to the inductor and/or capacitor being unable to deliver the necessary energy over the course of the entire pulse. Though designing those lumped elements with larger values may seem the solution, it would in turn require a source capable of providing a larger output as well. In both cases, additional cost and system size would be necessary.

---

3. In this case, the load is not exclusively the tissue, but includes all of the circuits to where power is being delivered.

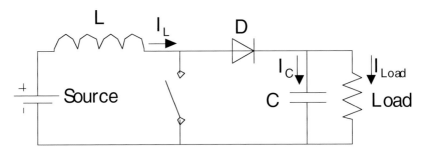

FIGURE 10.17   Typical step-up converter circuit topology.

The Clinical Significance of Pulse Rate

Typically, the pulse rate provides for the qualitative aspect of the stimulation. Very slow pulses can be perceived as 'thumpy' and sometimes irritating, while very fast pulses can be perceived as uncomfortable. Very fast pulses have been implicated in neuronal blocking [59,60], preventing action potentials from propagating. No such technology utilizing very high frequency effects is currently available for use in spinal cord stimulation, though its use on the anterolateral system to mitigate pain may be interesting.

Commensurate with the interleaved approach discussed previously, as the stimulation rate increases there may be a slight integrative effect if pulses are delivered closely. However, one study has indicated that the perception threshold does not change as rate is increased up to nearly 250 Hz, suggesting that axon membrane temporal summation does not occur [61]. Additionally, patients do report that paresthesia can become more intense with increasing rate, though the reason for this is not clear.

As the rate increases, the power necessary to provide the pulses increases linearly. For example, a 120 Hz rate will require twice as much power as a 60 Hz rate, assuming identical pulse shapes. This is derived simply from power being the average energy over time. Power usage is a very important concern affecting the useable lifetime of non-rechargeable/primary cell devices, a moderate concern in rechargeable devices (affecting the recharge interval) and is of little concern in an RF device, as discussed in the next section.

RF Versus Primary Cell Versus Rechargeable Devices

All stimulation systems require a power source to effect their functions. There are four options for providing power:
  1. direct, percutaneous connection
  2. indirect transcutaneous power coupling by the use of RF coils
  3. non-rechargeable primary cell battery integrated into the implant
  4. a rechargeable battery integrated into the implant.

Option (1) has been deprecated due to advances in technology. The remaining options are implemented in currently marketed devices.

Advantages and Disadvantages of Each

An RF coupled system can provide relatively high stimulation output compared to a battery-based system since there are no associated output limitations (beyond the electrical design of the implant). The output is effectively limited only by the amount of energy that can be coupled into the device. This is controlled by the output power of the transmitting coil and, perhaps more importantly, the ability of the patient (both via skill and overall anatomy of the implant area) to keep the transmitter coil well aligned with the receiver coil in the implant. Factors such as body shape, biomechanics of the implant location and physical activity level can all contribute to coil alignment. If the alignment is not within the proper tolerances, the system will be unable to deliver therapy due to poor power coupling. Additionally, such a system is free from battery limitations so related revision surgeries would not occur. Because the part of the system providing power is not implanted, cost for such a system can be the lowest of the three discussed.

Primary cell devices could be ideal for patients who would be unable to charge a rechargeable device (due to physical or psychological deficits), are known very low power users or even have a short life expectancy. Since the battery, now a part of the implant, must be well-designed, the cost for a device could be higher than an RF system, though likely still lower than a rechargeable device.

Rechargeable devices may be appropriate for users that do not have extraordinary power requirements and intend to keep their implants for a protracted time period. Some rechargeable batteries can be damaged by depletion, requiring explant of the system for replacement [62,63]. One battery manufacturer offers a battery for implantable medical devices with an internal chemistry specifically designed to permit depletion without damage [64] and is currently used in the SCS system offered by Boston Scientific.

Electronics Design Differences

From a design perspective, RF devices must have a means of rectifying the sinusoidal input waveform and regulating the energy coupled into the device. The input energy level will be varying due to the factors contributing to non-constant coil alignment. Because the device is only powered when the coil is properly applied, the internal microcomputer must tolerate spontaneous abrupt loss and re-establishment of power. This is done both electronically as well as with software. Capacitive energy storage is used to provide enough power to properly run the software shutdown routines when adequate coupling is lost. Heat is generated due to eddy currents on the implant casing when it is made of met al. The system can also be designed to modulate its transmitter output based on the coupling efficiency to compensate for poor coil alignment and can extend the external battery life. This would require a communications pathway between the implant and transmitter. Such a topology is discussed in the next section.

Rechargeable systems must have a charging circuit that regulates charge current. This is to ensure damage to the battery does not occur by excessive heat generation due to high charge current or continued application of current after the battery is recharged. Greater overall system efficiency and less heat generation can be had by providing a means to shut down the charger when the system is either recharged or due to some safety concern. As in an RF system, heat is generated due to eddy currents from the charger on the implant case when it is made of met al.

Primary cell systems do not require the complexities associated with an onboard charging circuit or dealing with the challenges of an RF approach.

In all systems, it is prudent to have a temperature sensor that will shut down the system beyond some threshold to mitigate patient discomfort and injury due either to malfunction or patient misuse. In addition, as with an RF system, any battery-based system requires internal power regulation to ensure that a constant output is available to power the internal electronics.

Recharging and Energy Transfer Methods

Similar to the challenges with an RF device, a rechargeable device necessitates a method to transmit energy to the implant and the implant requires a means to rectify and regulate that energy. In addition, there should be a means to communicate to the charger that the battery is recharged, allowing the charger to be shut off.

One such charging topology is shown in Figure 10.18. The charger can either be plugged into the mains or, for greater patient freedom, have its own battery for power (that battery itself can be rechargeable as well). The charger circuit includes a power amplifier to provide energy to the output coil, which serves as an antenna to transmit a high density electromagnetic field in the immediate vicinity of the charger. The implant has a receiver coil, acting as an antenna to couple the transmitted energy into the device. That energy is rectified, changing it from sinusoidal to DC. This energy is then presented to the battery in a controlled manner, ensuring that the recharge current is appropriate given the present charge state of the battery. When the battery is full, or if the implant temperature is excessive, the implant coil load impedance can be modulated (a technique called LSK, for load shift keying). This 'back telemetry' is detected by the charger as an extreme change in the coupling coefficient and can be used as a means to communicate that the charger should shut down.

Telemetry and Communication Challenges

The use of RF methods to communicate to implants and recharge them can present unique issues, especially in the presence of other technologies that either communicate with RF or generate RF noise of their own.

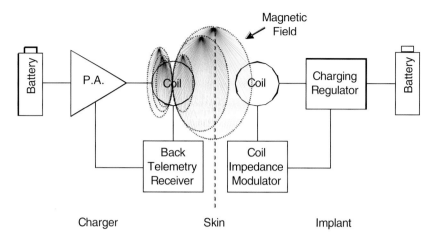

FIGURE 10.18   Transcutaneous implant recharging diagram.

The communications frequency to the implant must be low enough to permit passage through the body as higher frequencies are absorbed more easily. Lower communications frequencies have larger wavelengths, as the relation $c = f\lambda$ illustrates, where c is the speed of light in meters per second, f is the frequency in Hertz and λ is the wavelength in meters. Larger wavelength frequencies require a larger radiating element meaning that the antenna must be correspondingly larger to radiate efficiently. Typical communications frequencies for implants are in the hundreds of kHz range with wavelengths on the order of 3000 meters. A practical implementation of an antenna for this application would be a wirewound coil. Such an antenna is very inefficient, radiates poorly and has a radiation pattern in the shape of a donut that approaches a null as the axis of the donut is approached. This reduces the available communications range to a region close to the implant and, as a result, requires attention to orientation by the patient. Effective communications at an arm's length away (a few feet) is a reasonable design goal given expected usage.

In the case of an RF-powered device, and for rechargeables, the energy coupling field should be well contained in a volume very close to the transmitter. This ensures less energy is wasted as radiation and mitigates causing interference to nearby devices. In this case, a lower radiation frequency is advantageous as the body does not absorb it well and a wirewound antenna keeps the field highly localized. In the case of a rechargeable device, because both the communications and charging frequencies need to be low and thus close together, communications with the implant can be difficult if not impossible during recharging due to the interference with the implant receiver.

Electronic Design Issues

Implants and their associated external parts should be designed to decrease sensitivity to interference. This can be addressed by filtering signals outside the center bandwidth of the communications frequency. Additionally, the communications protocol should use error checking (such as CRC or checksums) and a means to ensure that the stimulator controller of one patient cannot adjust the settings on the implant of another patient. The latter is realized by providing each implant with a unique identifier and specifying that identifier on each communication between the remote and implant.

Rechargeable devices will also require filtering on the implant telemetry receiver to ensure that energy from the recharger is poorly coupled into the receiver since overloading the receiver with such high energy may cause damage. Regardless, all devices should be designed with proper filtering on the receiver input so that signals from other unrelated devices (such as cell phones, Bluetooth transmitters, CB radios, etc.) do not impede the implant from carrying out legitimate communications.

Efficiency and Power Usage Concerns

How efficiently an implant uses its available power source is important for a couple of reasons. All inefficiencies create heat which can irritate or even damage physiology if too hot. Also a less efficient system requires more power. In the case of a primary cell device, this would lower operational lifetime. In the case of a rechargeable device, this increases the number of times a battery needs to recharged which also decreases the operational lifetime of the device. Inefficiencies in RF systems would have less of an effect on operational lifetime. Since no system can be 100% efficient, there are a number of methods to minimize the effects of inefficiencies. Also, it is important to use just enough energy to provide the patient satisfactory relief such that the burden of energy restoration to the system is minimized. For example, recharging in a rechargeable device, replacement surgery in a primary cell device and less available tolerance in coil alignment with RF devices.

The resistance in the lead wires contributes to wasted energy. This resistance can be minimized by using wires with a lower resistance and ensuring that contact resistance between lead and implant connection point is low. By maximizing the contact area between the lead tail and header connectors the resistance can be minimized.

Receiver Cycling

All implants have a radio receiver onboard to listen for commands from the programmer and patient controller or remote. Practically, over the course of a day, a very small fraction of time is used to communicate with the implant. Thus, operating the onboard receiver continuously is a poor use of energy. The receiver can be cycled on and off to conserve power usage. For example, the receiver can be powered on every second for only a millisecond, representing a 0.1% duty cycle with the associated energy savings. If the receiver identifies an attempt to communicate with the implant in that 1 millisecond window, the receiver will stay on longer to process the command. To cope with this small window of opportunity for communication to the implant, the programmer or patient controller might send out its initial request for the attention of the implant for one second. This ensures that the receiver will be on at some point during that request.

Therapy Parameters

It is important to use only the lowest settings necessary to provide the patient with satisfactory relief. The energy usage per pulse depends on the therapy settings. That energy is proportional to the square of the amplitude and only linearly proportional to both the pulsewidth and rate, described by the relation $E \approx A^2 \times pw \times rate$. Therefore, a doubling in amplitude causes a four times increase in energy usage, while a doubling in either pulsewidth or rate causes only a doubling of energy usage. Doubling of all three parameters results in a sixteen times energy increase.

The pain a patient experiences may vary in its quality over the course of a day or other time period. As such, the patient may not require paresthesia for the entire day and/or may need to adjust the intensity or coverage area depending on their activities and pain profile. Providing the patient with amplitude (and even pulsewidth and rate) control and several settings from which to choose helps to ensure efficient use of the system. When using the interleaved method to provide paresthesia in isolated areas, each area can be adjustable independently by the patient providing both comfort and efficient use of implant energy. Also, the system could be programmed by a clinician automatically to cycle therapy settings on and off as defined by a schedule. Finally, an upper bound on amplitude can be programmed by the clinician to mitigate accidental overstimulation.

Reliability

Implantation of medical devices in humans necessitates that the probability of failure be very low for ethical and cost reasons. It is impossible to achieve a 0% failure rate, however, by addressing known failure modes, the probability of failure can otherwise be reduced. A survey of how the physician and clinician handle the devices in the operating room during implantation, storage and shipping means, human biomechanics and expected patient activities, all provide insight into the physical design needs of the system.

The system should be able to tolerate the vibration of being shipped and even dropped from a reasonable height both inside and outside of the packaging. Leads are subjected to repeated flexing during patient movement and should be designed to tolerate a reasonable number of bend cycles over a reasonable diameter bend. Leads must tolerate varying suture techniques, including suturing with a direct purse-string. Such a technique can introduce significant compressive force on the lead causing kinking, friction between lead wires if not isolated internally (leading to wire fracture) or even a tear in the external insulation (causing current leakage into the surrounding physiology). Leads must also tolerate repeated handling and insertion into the IPG in the operating room setting.

Susceptibility to the use of electrocautery must be minimized by designing appropriate means of protecting implant electronics from associated discharges. Exposure to x-ray and fluoroscopy must not affect usage. The use of external defibrillators may cause device failure, though any such failure should not cause significant injury to the patient. The system could be subjected to magnetic resonance imaging (MRI) and ultrasound (therapeutic and/or diagnostic). The use of non-ferromagnetic metals may help mitigate risk if MRI is used. The use of non-ceramic components may help reduce risk of failure due to ultrasound exposure.

Though they may be rare, especially active patients may be involved in physical exercise such as sit-ups or SCUBA diving, the latter of which would require the design to address operating pressures. Although the human body is very reliable in temperature regulation, the system may be subjected to temperature extremes during shipping and storage so its specifications must be guaranteed within these scenarios. Also, the system must have a means to minimize and even dissipate the heat it generates during usage so as not to damage the surrounding physiology.

All implanted materials must be inert, biocompatible, non-biodegradable and not react with commonly used operating room solutions such as antimicrobial agents. Electrical components, either sourced from vendors or specifically designed, must meet all physical, temperature and electrical specifications. Welds and solder joints should be well applied and free from defects.

FUTURE

Though the treatment modality has existed now for more than 40 years, the technology continues to progress and there are a number of directions that spinal cord stimulation can explore moving forward.

Pulse Shape

In present and past systems, the depolarization pulse shape has been designed to be square or near square. This may be because of its relatively simple electronic implementation. However, not only are there no known square electrical pulses in human physiology, neurons may respond differently to pulses with different shapes [65,66]. Different shapes may cause a different quality to the paresthesia [67,68] and/or affect the efficiency of stimulation or recruitment profile.

Automated Quality of Life Monitoring

Implanted medical devices are in a prime location to monitor other human functions. For example, movement could be recorded by an accelerometer perhaps as a means to measure changes in quality of life. Cardiac function indications might be available similar to EKG measurement. These are currently being investigated for use in pacemaker applications.

Fiber Selection Via Pre-Pulsing

Modeling has shown that a subthreshold depolarizing pulse can desensitize fibers near the cathode, causing a relative increase in threshold current for a subsequent pulse [69]. With an increase in threshold current large enough, the flanking hyperpolarization will prevent the action potential from propagating. Since fibers closest to the electrode would be affected by this prepulse, this may allow fibers further from the electrode to be stimulated to the exclusion of the closer fibers. Since larger fibers are more affected by this technique, this could also allow selectivity of smaller fibers. This may be useful for more selective paresthesia and even preventing stimulation of the roots, whose population includes the largest proprioceptive fibers.

Auto-Adjusting for Lead Migration

One system currently in the market can determine the relative lead positions and indicate the amount of rostrocaudal skew between adjacent leads [70]. This kind of information can be used to permit the clinician to recapture paresthesia by correcting the settings for migration, freeing the clinician from having to explore entirely new settings for the patient. Currently, this can be done manually by the clinician, but it may be expanded to be automatically compensated for. In some cases, this would permit the patient to avoid radiation exposure from fluoroscopy or x-ray to determine lead location. Methods to determine the absolute position of the lead are not currently available, however, if developed would provide extraordinary utility to both the clinician and patient.

Auto-Adjusting for Postural Changes

When the distance between the leads and spinal cord change as a result of postural changes and pulmonary activity, the distribution of injected charge can change. This may cause a change in paresthesia intensity and even coverage. Strategies may be developed to determine lead location and adjust stimulation parameters in response to those changes. For example, accelerometer- or electrical-based techniques may be used.

Battery and Charging Technologies

While current battery technology is based on chemical storage of energy, very high capacity capacitors exist that can quickly store energy in a dense field. Capacitors have a very low internal electrical resistance and do not have the concerns of chemical reaction heating found in conventional batteries. This would permit extremely fast recharging times, on the order of seconds. Furthermore, energy storage capacity would not be affected by patient usage profile or number of discharge cycles. As it matures, this technology is being investigated for other unrelated applications.

REFERENCES

1. Compositiones Medicae of 46 AD.
2. Stillings D. A survey of the history of electrical stimulation for pain to 1900. *Med Instrum.* 1975;9:255–259.
3. Cambridge NA. Electrical apparatus used in medicine before 1900. *Proc R Soc Med.* 1977;70:635–641.
4. Turrell WJ. The landmarks of electrotherapy. *Arch Phys Med Rehabil.* 1969;50:157–160.
5. Kent, C.W. (1919). Electric massage machine. US Patent number 1,305,725, filed July 5, 1918 and issued June 3, 1919.
6. Burton C, Maurer DD. Pain suppression by transcutaneous electrical nerve stimulation. *IEEE Trans. Biomed. Eng.* 1974;21:81–88.
7. Shealy, C.N., Liss, S. & Liss, B.S. Evolution of electrotherapy: From TENS to cyberpharmacology. In Bioelectromagnetic medicine (Rosch, P.J. & Markov, M.S., eds), Marcel Dekker, Inc., New York; 2004:87–108.
8. Mortimer JT. *Pain suppression in man by dorsal column electroanalgesia.* Cleveland, OH: PhD dissertation, Case Western Reserve University; 1968.
9. Shealy CN, Mortimer JT, Reswick JB. Electrical inhibition of pain by stimulation of the dorsal columns. Preliminary clinical report. *Anesth Anal Cur Res.* 1967;46:469–491.
10. Shealy CN, Taslitz N, Mortimer JT, Becker DP. Electrical inhibition of pain: experimental evaluation. *Anesth Analg.* 1967;46:299–305.
11. Rossi U, The history of electrical stimulation of the nervous system for the control of pain. Simpson BA, ed. *Pain research and clinical management,* vol 15. Amsterdam: Elsevier; 2003:5–16.
12. Gildenberg PL. History of electrical neuromodulation for chronic pain. *Pain Med.* 2006;7(Suppl 1):S7–S13.
13. Cavuoto J. The birth of an industry. In: Krames E, Peckham PH, Rezai AR, eds. *Neuromodulation.* London: Elsevier; 2009:41–47.
14. Schwartz SI, Griffith LSC, Neistadt A, Hagfors N. Chronic carotid sinus nerve stimulation in the treatment of essential hypertension. *Am J Surg.* 1967;114:5–15.
15. Shealy CN, Mortimer JT, Hagfors NR. Dorsal column electroanalgesia. *J Neurosurg.* 1970;32:560–564.
16. Lazorthes Y, Morucci J-P, Clemente G. Biotechnical basis of neurostimulation. In: Lazorthes Y, Upton ARM, eds. *Neurostimulation: an overview.* Mt Krisco: Futura Publishing Company; 1985:11–41.
17. Anon. Dial away pain. *Life Magazine.* 1972;73:61–62.
18. Davis R, Gray E. Technical factors important to dorsal column stimulation. *Appl Neurophysiol.* 1980;44:160–170.
19. FDA (1981). PMA number P800040, docket 81M-0136. Decision date April 14, 1981.
20. Medical Device Amendments of 1976.
21. FDA (1984). PMA number P840001, docket 84M-0415. Decision date November 30, 1984.
22. Shatin D, Mullett K, Hults G. Totally implantable spinal cord stimulation for chronic pain: design and efficacy. *Pacing Clin Electrophysiol.* 1986;9:577–583.

23. Waltz JM, Reynolds LO, Riklan M. Multi-lead spinal cord stimulation for control of motor disorders. *Appl Neurophysiol.* 1981;44:244–257.

24. Waltz JM. Computerized percutaneous multi-level spinal cord stimulation in motor disorders. *Appl Neurophysiol.* 1982;45:73–92.

25. Waltz JM. Spinal cord stimulation: a quarter century of development and investigation. A review of its development and effectiveness in 1,336 cases. *Stereotact Funct Neurosurg.* 1997;69:288–299.

26. FDA (2004). PMA number P030017, docket 04M-0256. Decision date April 27, 2004.

27. Kessler DK. The CLARION multi-strategy cochlear implant. *Ann Otol Rhinol Laryngol Suppl.* 1999;177:8–16.

28. Burton C. Seminar on dorsal column stimulation: Summary of proceedings. *Surg Neurol.* 1973;1:285–289.

29. Erickson DL. Percutaneous trial of stimulation for patient selection for implantable stimulating devices. *J Neurosurg.* 1975;43:440–444.

30. Cook A. Percutaneous trial for implantable stimulating devices. *J Neurosurg.* 1976;44:650–651.

31. Zumpano BJ, Saunders RL. Percutaneous epidural dorsal column stimulation. *J Neurosurg.* 1976;45:459–460.

32. Hosobuchi Y, Adams JE, Weinstein PR. Preliminary percutaneous dorsal column stimulation prior to permanent implantation. Technical note. *J Neurosurg.* 1972;37:242–245.

33. Urban BJ, Nashold Jr BS. Percutaneous epidural stimulation of the spinal cord for relief of pain. Long-term results. *J Neurosurg.* 1978;48:323–328.

34. FDA (1981). 510(k) number K812154. Received July 29, 1981. Decision date August 20, 1981.

35. FDA (1986). 510(k) number K860158. Received January 16, 1986. Decision date March 7, 1986.

36. Law JD. Targeting a spinal stimulator to treat the 'failed back surgery syndrome'. *Appl Neurophysiol.* 1987;50:437–438.

37. FDA (1994). 510(k) number K930536. Received January 27, 1991. Decision date December 15, 1994.

38. FDA (1995). 510(k) number K934065. Received August 20, 1993. Decision date February 14, 1995.

39. Struijk JJ, Holsheimer J, Spincemaille GH, Gielen FL, Hoekema R. Theoretical performance and clinical evaluation of transverse tripolar spinal cord stimulation. *IEEE Trans Rehabil Eng.* 1998;6:277–285.

40. Holsheimer J, Nuttin B, King GW, Wesselink WA, Gybels JM, de Sutter P. Clinical evaluation of paresthesia steering with a new system for spinal cord stimulation. *Neurosurgery.* 1998;42:541–547 discussion 547-549.

41. Struijk JJ, Holsheimer J. Transverse tripolar spinal cord stimulation: theoretical performance of a dual channel system. *Med Biol Eng Comput.* 1996;34:273–279.

42. Merrill DR, Bikson M, Jefferys JG. Electrical stimulation of excitable tissue: design of efficacious and safe protocols. *J Neurosci Meth.* 2005;15:171–198.

43. Rattay F. Analysis of models for extracellular fiber stimulation. *IEEE Trans Biomed Eng.* 1989;36:676–682.

44. Hodgkin AL, Katz B. The effect of sodium ions on the electrical activity of the giant axon of the squid. *J Physiol.* 1949;108:37–77.

45. Grill WM, Mortimer JT. Electrical properties of implant encapsulation tissue. *Ann Biomed Eng.* 1994;22:23–33.

46. Mortimer JT, Motor prostheses. Brooks VB, ed. *Handbook of physiology – the nervous system,* vol 3. Bethesda: American Physiological Society; 1981:155–187.

47. Wessale JL, Geddes LA, Ayers GM, Foster KS. Comparison of rectangular and exponential current pulses for evoking sensation. *Ann Biomed Eng.* 1992;20:237–244.

48. Oakley J, Varga C, Krames E, Bradley K. Real-time paresthesia steering using continuous electric field adjustment. Part I: Intraoperative performance. *Neuromodulation.* 2004;7: 157–167.

49. Manola L, Holsheimer J, Veltink PH, Bradley K, Peterson D. Theoretical investigation into longitudinal cathodal field steering in spinal cord stimulation. *Neuromodulation.* 2007;10: 120–132.

50. Manola, L. & Holsheimer, J. (2005). Dual percutaneous leads for SCS: single vs dual mode stimulation. Abstracts of the 7th Congress of the International Neuromodulation Society, June 10-13, 2005, Rome, Italy.

51. McNeal DR. Analysis of a model for excitation of myelinated nerve. *IEEE Trans Biomed Eng.* 1976;23:329–337.

52. West DC, Wolstencroft JH. Strength-duration characteristics of myelinated and non-myelinated bulbospinal axons in the cat spinal cord. *J Physiol.* 1983;337:37–50.

53. Wesselink WA, Holsheimer J, Boom HB. A model of the electrical behaviour of myelinated sensory nerve fibres based on human data. *Med Biol Eng Comput.* 1999;37:228–235.

54. Gorman PH, Mortimer JT. The effect of stimulus parameters on the recruitment characteristics of direct nerve stimulation. *IEEE BME.* 1983;30:407–414.

55. Feirabend HKP, Choufoer H, Ploeger S, Holsheimer J, van Gool JD. Morphometry of human superficial dorsal and dorsolateral column fibres: significance to spinal cord stimulation. *Brain.* 2002;125:1137–1149.

56. Holsheimer J, Wesselink WA. Optimum electrode geometry for spinal cord stimulation: the narrow bipole and tripole. *Med Biol Eng Comput.* 1977;35:493–497.

57. van den Honert C, Mortimer JT. The response of the myelinated nerve fiber to short duration biphasic stimulating currents. *Ann Biomed Eng.* 1979;7:117–125.

58. Yearwood T, Hershey B, Bradley K, Lee DC. Pulse width programming in spinal cord stimulation: A Clinical Study. *Pain physician.* 2010.

59. Tai C, De Groat WC, Roppolo JR. Simulation analysis of conduction block in unmyelinated axons induced by high-frequency biphasic electrical currents. *IEEE Trans Biomed Eng.* 2005;52:1323–1332.

60. Williamson RP, Andrews BJ. Localized electrical nerve blocking. *IEEE Trans Biomed Eng.* 2005;52:362–370.

61. Cameron T, Peng P, Arango P, Lozano A. Nerve blocks: psychophysics of spinal cord stimulation. *J Pain.* 2004;5(Suppl. 3):S54.

62. RestoreUltra Implant Manual, p. 9, 37712, 2007-9.

63. Eon Clinician's Manual, p. 76, 37-06/9-01F, March 2006.

64. Tsukamoto, H., Kishiyama, C., Nagata, M., Nakahara, H. & Piao, T. (2006). Rechargeable lithium battery for tolerating discharge to zero volts. US Patent number 7,101,642, filed September 20, 2002 and issued September 5, 2006.

65. Grill WM, Mortimer JT. Inversion of the current-distance relationship by transient depolarization. *IEEE Trans Biomed Eng.* 1997;44:1–9.

66. Sahin M, Tie Y. Non-rectangular waveforms for neural stimulation with practical electrodes. *J Neural Eng.* 2007;4:227–233.

67. Alo K, Cartwright T. *Patient preferences for constant current and constant voltage stimulation.* Acapulco, Mexico: 11th North American Neuromodulation Society, 2007; 2009.

68. Alo K. *Patient-reported differences in constant current and constant voltage stimulation.* Acapulco, Mexico: 11th North American Neuromodulation Society, 2007; 2007.

69. Deurloo KEI, Holsheimer J, Bergveld P. The effect of subthreshold prepulses on the recruitment order in a nerve trunk analyzed in a simple and a realistic volume conductor model. *Biol Cybern*. 2001;85:281–291.
70. Bradley, K. (2006). Electrical fluoroscopy: a novel method for determining spatial orientation of spinal cord stimulation leads. Presented at the International Spine Intervention Society Annual Meeting, July 13-16, 2006; Salt Lake City, Utah.

Commentary on Electronics

Henricus Louis Journee, MD, PhD

Department of Neurosurgery, University Medical Center Groningen, Groningen, The Netherlands

The chapter written by Ranu describes the electronics of the neuromodulation therapy interface. It commences with a short historic review on treatment of pain by electrical stimulation. This dates back many centuries, when originally electricity from torpedo fish was employed. Later devices became powered electrochemically or by induction currents from varying magnetic fields. These sources are used in present electrical devices. The review also addresses the development of the technology of commercial spinal cord stimulation (SCS) devices and evolution of commercial leads from bipolar to multipolar configurations proceeding into a still expanding complexity of number of electrode contacts and geometry. This process is mutually coupled to the technological development of electronics, advances in research on the electrical–biological interface between electrodes and neurophysiological elements and to the fast increasing variety of clinical applications in neuromodulation.

The commentary on the previous chapter is given from a contemplative view on current developments in neuromodulation and future directions pertaining to engineering of the electrodes and of contemporary progressing technologies in electronics that are amenable for implantable programmable stimulation devices (IPG). After revisiting definitions on pulses and electrical field parameters pertaining to the design of electrodes and stimulators, this commentary addresses a selection of subjects, which are further elucidated.

BASIC DEFINITIONS IN PULSE DELIVERY

A stimulation pulse injects a certain amount of charge into tissue. The amount of charge per pulse is equal to the product of the administered current times the pulse duration of a single or so-called monophasic pulse. The interpulse interval of stimuli (ipi) is defined as the time epoch between the onset of two repetitively administered stimuli. Stimuli can be mono- or biphasic. In biphasic stimulation, two monophasic pulses of opposite polarity are cascaded. This is depicted in Figure 10.14. Each pulse describes a phase. The time lag between cessation of the first pulse and onset of the oppose polarity pulse is called the interphase time. Usually, this time is shorter than the pulsewidth and is used to change the switching states from positive to negative and vice versa.

CURRENT, CHARGE DENSITY AND DISSIPATED ENERGY IN THE EXAMPLE OF SPATIAL TARGETING OF AXON FIBERS

The current density of an electrode is defined as the amount of injected current per phase divided by the surface of the electrode. Similarly, a charge density per phase is defined as the division of the injected charge by the surface.

According to Ohm's Law, the resistance is defined as the division of voltage by current. The stimulation amplitude can either be expressed in a voltage or current. When the size of an electrode increases, the charge density will decrease when the stimulus current is kept constant like in the example of the author. However, when the stimulus voltage is kept constant, then the current density remains unaltered. This is because the electrode impedance is inversely related to the increased surface area. At constant voltage, the stimulation current is implicitly increased with the surface and thus the current density remains constant.

Increasing the electrode size alters the spatial geometry of the electrical field around the electrodes. With increasing electrode size, both area and depth of stimulation are increased defining a volume containing neural structures (axons) that are stimulated. As a result, the stimulated volume around the electrode is also increased. For a given current density for excitation, the stimulation current and also the current threshold must increase, while the stimulation voltage threshold remains about unchanged. The required stimulation energy per phase, being the product of current, voltage and pulsewidth is also increased along with the increased electrode surface.

SPATIAL SELECTIVITY ENGINEERING BY MULTIELECTRODE GRIDS

The development of leads from single to multielectrode provides better tailored solutions for problems that are still encountered in clinical practice. Spatial targeting permits selective stimulation of nerve structures while minimizing side effects from stimulation of other neural tissues. The larger the number of electrodes, the more flexibility one gets. Tripole configurations already offer a means for selectivity for deeper penetration into the dorsal columns [1–4]. Expanding the size of a multielectrode grid to 16-electrode contacts permits compensation for electrode shift by shifting the active electrodes in a grid in the same direction. When the number of electrodes is increased to higher numbers, like 64, by programming, one is able to construct different types of electrodes. A promising example, which is not used yet in humans, is a cylindrical electrode with 64 contacts that is given the same outer dimensions of a deep brain stimulation electrode, which is used for treatment of movement disorder symptoms as in Parkinson's disease [5]. With the electrode contacts arranged in 16 equally spaced rows, covering a total length equivalent to the state-of-the-art DBS electrode arrangement, one is able to define the interconnected electrode contacts as cylinder contacts of a DBS electrode with the same axial symmetric volume of activated tissue (VTA). This gives a stimulation field in all directions around the electrode. This is preferred when the electrode is placed in middle of the targeted nucleus. However, when the target is displaced from the center, side effects from stimulation of other structures, like the internal capsule, may limit the therapeutic range of the stimulation voltage. This problem can be solved by field steering resulting in displacement of the VTA away from the lead's axis. One can define VTAs with a submillimeter precision. This is relevant for patients with suboptimal placed DBS leads. Field steering provides extra compliance to a surgeon and neurologist since it compensates for displacement errors of the DBS electrode in the magnitude of a few millimeters while one would expect to approximate the optimal therapeutic effects at center placement. This expectation is based on recent theoretical and experimental studies. Field steering gives postoperatively additional degrees of freedom to optimize the DBS therapy, which is of benefit to the neurologist in the postoperative trajectory. When compared to a center position, one also can expect that the energy consumption is not affected as much at a dislocated placement. The activation volume will be limited to the volume of the targeted nucleus and does not expand further outside as would be the case with a misplaced conventional electrode without field steering options. This is also in favor of a longer battery life.

TEMPORAL SELECTIVITY ENGINEERING BASED ON INTERLEAVING

In addition to engineering on spatial selectivity, one can define new functionality in time by interleaving techniques. Interleaving is equivalent to multiplexing. One can:

1. define independent processes working together or
2. create new functionality by combining processes.

By interleaving processes, actions are switched on and off in succession. Each action consists of individual selections of electrodes and stimulation paradigms. These selections can be chosen independently of each other. An example of the first possibility is creating a spatial selectivity for selecting two different anatomic areas to generate paresthesia by the interleaving scheme of biphasic stimuli as depicted in Figure 10.13b. Interleaving offers the possibility to switch alternately over to different electrode connections on different locations on the body which otherwise would not be possible during simultaneous stimulation. An example of the second possibility is the creation of a biphasic pulse by taking two subsequent pulses with equal intensity and pulsewidth, while during the interphase time the electrodes that are defined as anode and as cathode are interchanged. In contrast to monophasic stimulation, action potentials will appear at both electrodes. Another application of interleaving is selective stimulation by conditioning stimulation by prepulsing [6,7]. Subsequent pulse sequences are given with stepwise increasing amplitude. This paradigm results in selective stimulation by blocking thicker fibers, which are easier to stimulate by single pulses. A multiplex scheme can be extended by many more steps. This makes it possible to create ramps which suppress hyperpolarization at sharp flanks of square waves or unwanted generation of action potentials due to anodal break effects. Theoretically, all but one parameter of electrode connection matrices and stimulation paradigms of each action element can be chosen independently of each other, such as pulse amplitude, shape, width and also short high frequency pulse trains that are embedded in an interleaved time frame. Only one common stimulation frequency can be chosen for all processes. Since interleaving is bound to time frames, all separate processes and actions are confined to one common repetition rate.

CONSTANT CURRENT AND VOLTAGE STIMULATION

An ideal current stimulator delivers a preset current of a stimulation pulse and is independent of the load impedance. Similarly, a voltage stimulator generates pulses according to the preset voltage. The electrode impedance of single electrodes will not influence the current from a current stimulator. The current on its way to a pure conductive medium is unaffected by the capacitive characteristics of the electrode impedances. The simplified model in Figure 10.7 shows a preservation of the rectangular current pulse of the electrical field potential across the field enclosed by the electrodes. In contrast, rectangular pulses from constant voltage stimulation are affected by capacitive loading and unloading causing field potentials in the solutions between the electrodes to be distorted. Another feature of current stimulation is that electrochemical reactions as ion reduction and oxidation and charge deposition in double layers can be disregarded as well. Similarly, the current transfer is immune to the non-linear relationship between voltage and current of the bioelectrical interface of the electrode. Constant voltage stimulation is sensitive to electrochemical polarization effects resulting in a bias of several hundreds of millivolts. Threshold voltages may even vary in repeated voltage pulse series. Encapsulation may increase electrode impedance and, consequently, influences electrical fields [8]. The VTA of DBS electrodes are affected by increased impedances from encapsulation layers when constant voltage stimulation is used [9]. Since the field potential, from which the activation function is derived, describes the driving force to move ions affecting membrane potentials, one would conclude that a current source would prevail above a constant voltage stimulator.

However, practice is more complex than in this model.

1. The model of Figure 10.5 is given for a pure resistive solution. However, the impedance of biological tissues is complex since these also have capacitive characteristics. In the linear range for small signals, tissues expose a decreasing course of permittivity, which represents the capacitive

characteristics, as well as decreasing conductivity as a function of frequency. These courses depend on the kind of tissue, but capacitive effects are clearly noticed in all tissues in the frequency range of 1–100 kHz of rectangular pulses with widths of 50–1000 ms covering almost all applications in neuromodulation. The voltage over the solution in Figure 10.7 resulting from a rectangular constant current pulse is not rectangular anymore.

2. For higher intensities, as used during stimulation, non-linear effects make tissue impedances dependent on intensity. These are most prominent near the electrodes where gradients of the electrical field are the highest. Both conductivity and permittivity decrease with intensity, while also hysteresis may become evident.

3. A current will be distributed over parallel conducting pathways of several tissue compartments and conducting fluids. When the conductivity of one of the compartments changes like by formation of edema around a DBS electrode or by geometric changes of the CSF cylinder around the spinal cord due to a change in posture, the redistribution of current leads to a change of potential gradients over the targeted neural tissue. This results in changes of stimulation thresholds. Constant voltage stimulation is less sensitive for changes of parallel conducting compartments when compared to constant current stimulators.

These complicating factors make a preferential choice for constant current or constant voltage less obvious and should be considered more in detail in the context of the specific application of neuromodulation.

Programmable Multielectrodes: Current or Voltage Stimulation?

Programming of multielectrodes allows definition of a variety of electrode configurations by interconnecting one or more electrode contacts forming an anode or cathode. Instead of interconnecting, each electrode contact can be assigned to its own current or voltage source as is shown in Figure 10.10 for two electrode contacts. Individual current sources overcome the disadvantage of redistribution of currents over electrode contacts from one common current source when the load impedances are different as shown in the example of Figure 10.11. This electronic solution is principally different from interconnection of electrodes. Interconnected electrodes that are wired to one common source define an equipotential surface. All electrical field models on which designs of electrodes are based are built on the boundary condition of an electrode surface which has essentially an equipotential. This is not the case with a constant current multi-source system with different load impedances. Then, grouped electrodes will not have one common potential, but expose a set of different potentials. This causes a deviation of the potential field in tissues surrounding the electrode. This would not be the case when, instead of multicurrent sources, multivoltage sources are used since these yield equipotentials. However, when deposits on the electrode like oxide layers or fibrin would affect the local electrical field as well, a choice between current or voltage stimulation still remains less obvious to make.

ENGINEERING ASPECTS OF THE ELECTRONIC CIRCUITS

Solid State Switches

Solid state switches are pivotal components in the circuitry of to-date IPGs. They are basically field effect transistors. Their energy consumption can be neglected since the controlling gate has a very high impedance, the on-state resistance is very low and, in the off state, very high. Solid state switches are very fast when compared to mechanical switches like a reed relay. They can be integrated in large numbers on a chip. They are used for switching outputs on and off, to connect to multielectrodes to define anodes and cathodes in mono-phasic pulse delivery, to create biphasic pulses and for interleaving sequences. When a microcontroller is integrated in the same chip and interfaced with the switches, one can design a complex circuitry that copes with the contemporary increasing demands for neuromodulation.

Safety Aspects

Ranu did not address the engineering aspects of the vulnerability of IPG devices for externally applied voltages to the output of IPGs as may occur in MRI and other sources that generate induction currents in the conduction loop of an IPG, extension leads, electrodes and volume conductor of the body. The characteristics of solid state switches are valid when voltages are kept within the range of power supply. When external AC voltages of, for example, 20 Volts are applied to the output, high impedance states may turn into low impedance, while external currents may corrupt proper functioning of the circuits, resulting in reprogramming of stimulator settings. In our own in vitro experiments, we tested some different IPG devices under conditions of induction currents from the leads in MRI. The power consumption of the battery could increase up to three magnitudes, while programmed settings became reset, while, even when switched off, a temperature increase of several degrees Celsius was measured around the electrodes when placed in a phantom with conducting gel. These observations may also explain the sensations reported in a few patients during MRI. Medtronic advises not to perform MRI on patients with implanted IPGs. The discussion on whether or not it is safe to use MRI is still ongoing. In our institute, IPGs are disconnected from the electrode leads before patients are referred for MRI. Similar solutions could theoretically be realized in the electronic circuit of an IPG. Designing an MRI resistant IPG is a relevant engineering topic with a high need.

Constant Current and Voltage Stimulators

There exist several designs for constant current and voltage stimulation. Constant current and constant voltage stimulators can be categorized into feedback and uncontrolled designs. Both categories can be reduced to simple solutions. The circuit in Figure 10.9 is an uncontrolled constant current stimulator. It is based on the characteristics of two identical field effect transistors. Field effect transistors have almost perfect output characteristics for a constant current source. The accuracy of the circuit in Figure 10.9 depends on differences between transistor characteristics and accuracy of the supply voltage source, which should be constant voltage. Feedback systems are based on a comparison between the set-voltage and voltage that is measured across a shunt resistor representing the delivered stimulation current and requires no stabilized voltage supply. Such a design can also be simplified to a few transistors. A disadvantage of the design in Figure 10.9a is that the dissipated power of $V_{source} \times I_{set} + V_{source} \times I_{out}$ is twice as much as delivered for stimulation: $V_{source} \times I_{out}$, since $I_{out} = I_{ref}$. Adding more identical transistors that together deliver the total current will increase the efficiency. The efficiency with one transistor for current delivery in Figure 10.9a is 50% and in Figure 10.9b with three paralleled transistors 75%.

Biphasic Stimulation

Ranu gave demonstrative examples for constant current stimulator circuits generating monopolar pulses. As mentioned earlier, a conceptual possibility to generate biphasic pulses is by interleaving over two consecutive phases and cross-switching of anode and cathode in the interphase time.

Increasing Compliance by Step-Up Converters and Minimizing IPG Energy Loss

Step-up converters increase the compliance voltage above battery voltage, which is necessary for current stimulation. The voltages are needed to overcome impedance increase like that due to encapsulation layers around DBS electrode contacts [8,9]. The example in Figure 10.17 is based on energy transfer where by switching alternately, energy is intermediately stored in an inductor and subsequently transferred to a capacitor. An alternative concept is to use switches instead of diodes in cascaded diode-capacitor charge pumps. Losses that occur in the transient phases of the switching can be minimized using break-before-make switching schemes. Since leakage currents of to-date capacitors are extreme low, the resulting energy losses are lower than energy losses in coils. Diode-capacitor charge pumps consequently may yield the most energy efficient DC-DC converters. In addition, the clinical user can minimize the energy consumption. It is therefore recommended to follow the guidelines for the user to optimize stimulation parameters to minimize the energy consumption as discussed in Ranu's paragraph on therapy parameters. The bottom line is to increase the pulsewidth and decrease the stimulation amplitude since the energy is linear proportional with pulsewidth and quadratic proportional with amplitude.

CONCLUDING REMARKS

The author has given a survey over the engineering aspects on the design of electrodes and electrical circuitry of stimulators in regard to an increasingly rich variety of applications in neuromodulation. It provides a view on the state-of-the-art and gives a brief view of the future. It may serve as an introduction to technically oriented clinical users and neurophysiologists, medical physicists and electronic engineers for orientation in the field of neuromodulation. Although a wide spectrum of subjects is discussed, one important issue on engineering aspects on safety of IPGs in MRI would have been a welcome and appropriate addition in a chapter on engineering of IPG electronics.

REFERENCES

1. Struijk J. J, Holsheimer J. Transverse tripolar spinal cord stimulation: theoretical performance of a dual channel system. *Med Biol Eng Comput.* 1996;34.4:273–279.
2. Struijk JJ, Holsheimer J, Spincemaille GH, Gielen FL, Hoekema R. Theoretical performance and clinical evaluation of transverse tripolar spinal cord stimulation. *IEEE Trans Rehabil Eng.* 1998;6:277–285.
3. Manola L, Holsheimer J, Veltink PH, Bradley K, Peterson D. Theoretical investigation into longitudinal cathodal field steering in spinal cord stimulation. *Neuromodulation.* 2007;10: 120–132.
4. Holsheimer J, Nuttin B, King GW, Wesselink WA, Gybels JM, de Sutter P. Clinical evaluation of paresthesia steering with a new system for spinal cord stimulation. *Neurosurgery.* 1998;42:541–547 discussion 547-549.
5. Martens HCF, et al. Spatial steering of deep brain stimulation volumes using a novel lead design. *Clin Neurophysiology.* 2010;doi: 10.1016/j.clinph.2010.07.026.
6. Grill WM, Mortimer JT, Stimulus waveforms for selective neural stimulation. *IEEE Eng Med Biol.* 1995;14(4):375–385.
7. Deurloo KEI, Holsheimer J, Bergveld P. The effect of subthreshold prepulses on the recruitment order in a nerve trunk analyzed in a simple and a realistic volume conductor model. *Biol Cybern.* 2001;85:281–291.
8. Grill WM, Mortimer JT. Electrical properties of implant encapsulation tissue. *Ann Biomed Eng.* 1994;22:23–33.
9. Butson CR, Maks CB, McIntyre CC. Sources and effects of electrode impedance during deep brain stimulation. *Clin Neurophysiol.* 2006;117(2):447–454. (Epub 2005 Dec).

STUDY QUESTIONS

1. Suppose a development in circuitry or current delivery were made that seemed worthwhile to implement in present-day IPGs – what are the important considerations and trade-offs between manufacturing, marketing, cost/profit, and increased clinical benefit that need to be weighed in deciding if and when to add the new development to a company's platform?
2. Review details related to electrode size – are there any benefits to smaller electrodes? What are the potential disadvantages? How might changes in circuitry design optimize this balance?

Chapter 11

Power

Tracy Cameron PhD, Ben Tranchina MSEE and John Erickson MSEE
St Jude Medical Neuromodulation Division, Plano, Texas, USA

OVERVIEW

Over the past several decades the field of neuromodulation has seen a rapid growth. This growth has been possible in part by the many advances made in technology; in particular the improvements that have been made in power sources for implantable products. The following chapter examines the diverse power requirements necessary to treat the many indications currently targeted for neuromodulation therapy and how improvements in technology have allowed the expansion of neuromodulation into more and more complex disorders. A comparison of the different types of power sources and the advantages and disadvantages of each will be discussed taking into consideration patient compliance and costs. Finally, this chapter will end by briefly looking at some of the new power sources that are currently in the research stage and how these may affect neuromodulation in the future.

POWER REQUIREMENTS

The development of any new neuromodulation device first begins by examining where and how the device is intended to be used. Several questions need to be answered to allow the engineer to design a device capable of supplying the appropriate output to treat the desired disease. These questions include: what are the energy requirements necessary to provide a therapeutic result? Where will the devices be implanted? And what type of interaction will there be between the devices? The answers to these and other questions will help the designer understand the complexity of the device that is needed, the size of the device, and the materials needed to construct the device. Finally, all of these attributes need to be balanced with the overall cost of the device.

Therapeutic Energy Requirements

An important factor that needs to be addressed early in any design is the capacity and type of power source. There is the obvious need to keep the overall size

Essential Neuromodulation. DOI: 10.1016/B978-0-08-089064-7.00004-1

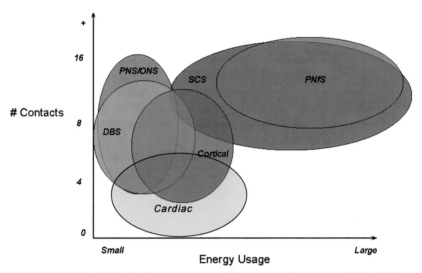

FIGURE 11.1   Graph showing the energy usage and contact requirements needed for different indications.

of any medical device as small as possible, however, this must be balanced with the requirement to provide the necessary complexity and output parameters in order to provide the appropriate therapy to the patient. To date, there have been a number of implantable devices developed to treat a variety of medical conditions with each of these conditions having unique power requirements.

Figure 11.1 demonstrates the energy requirements (energy usage) and the complexity (number of contacts) needed to treat some of the currently approved indications. Energy requirements can be compared using therapeutic average current (ITAVG), which is calculated from:

$$ITAVG = FREQ * PW * AMP$$

where FREQ is the stimulation pulse frequency in Hertz, PW is the stimulation pulsewidth in seconds and AMP is pulse amplitude in amperes.

The highest energy requirements and complexity for currently approved indications are in spinal cord stimulation (SCS). Parameter selection for SCS involves selecting the optimal electrode configuration followed by adjusting the amplitude, width and frequency of electrical pulses [1]. The choice of electrodes depends on the extent of the pain. Systems today can control up to 16 independently programmed electrodes using multiple stimulation programs running sequentially. Stimulation amplitude is measured in either milliamperes or Volts and is a measure of the intensity of stimulation. This is set within a range of 0–25 mA (1–10 V) according to the type of electrode used and the type of nerves stimulated. Lower voltage is chosen for peripheral nerves and paddle type electrodes. Pulsewidth usually varies from 100 to 400 μs. Pulse frequency is measured in Hertz (Hz) and usually delivered between 20 and 120 Hz [2].

Power sources for SCS devices are usually in the range of 2–7 Ah. These high energy requirements are necessary due to the variety of painful conditions and the relatively large distance the current must travel to reach the target nerve fibers. Some of the indications currently being treated with SCS include failed back surgery syndrome, peripheral neuropathy, complex regional pain syndrome, angina pectoris, pain due to peripheral vascular disease, and stump pain [3–6]. Spinal cord stimulation leads are implanted into the epidural space above the spinal cord. They produce their effect by activating fibers below located in the dorsal columns of the spinal cord. Therefore, the electrical field must travel through the epidural tissue, through the dura, through the cerebral spinal fluid, and finally to the dorsal columns. Aside from the actual distance, which can be up to several millimeters, each of these areas has a particular electrical conductance which impedes the flow of current or shunts the field away from the intended target [7]. It has been theorized that only 10% of the current that leaves the electrode actually makes it to the target nerves [8]. On the opposite end of the range are deep brain stimulation (DBS) and cardiac pacemakers. Electrodes implanted into the brain are in direct contact with the target nerve fibers. Therefore, the electrical field needed to activate deep brain targets is much smaller. The therapeutic parameter used for DBS range between 1 and 3 mA amplitudes, 60–210 μs pulsewidths and 100–130 Hz frequencies [9]. Similarly, cardiac pacemakers have a relatively low battery capacity (1–1.5 Ah).

Since neurostimulation is fundamentally delivering small doses of electricity, the system must provide a path through the patient for the electricity to flow and the complex form of Ohm's Law, $V = I * Z$, can be used to provide a quantitative description of the requirements for the power source. The law states that voltage, V, is equal to the product of current, I, and the complex impedance, Z. The power source capacity is frequently given in units of ampere-hours using the symbol Ah. Therefore, given the desired expected life of the power source you can calculate the maximum average current allowed for the neuromodulation device to deliver therapy. We will consider other contributing factors to the maximum average current (IMAVG) in the System Power Requirements section.

SYSTEM POWER REQUIREMENTS

The efficiency of the system being used to deliver the desired therapeutic effect has an impact on the overall power requirements. Figure 11.2 shows a general block diagram for the elements of the system that are in series, including a

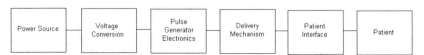

FIGURE 11.2   Block diagram of the elements needed in a power system.

power source, voltage conversion, pulse generator electronics, delivery mechanism, patient interface and patient.

Power Source

The power source, as the name implies, is the source of electrical energy for the system. There are a wide range of voltages, capacities and other characteristics that depend on the type chosen. Details on options for this element of the system are covered later in this chapter.

Voltage Conversion

Now let us consider the contributions of the system elements to the overall power requirements using Ohm's Law, $V = I * Z$, simplification. The electronics and power source must create the voltage (V) necessary to overcome the impedance (Z) presented by the patient, patient interface and delivery mechanism combined in series in order to deliver the therapeutic current (I_{TAVG}). This often requires circuitry in the electronics to multiply the voltage of the power source, resulting in a multiplier (M) times the power source voltage.

Pulse Generator Electronics

In general, the electronics are responsible for converting the raw energy from the power source into the desired stimulation waveforms, monitoring important signals and enabling communication with external patient or clinician devices. The added complexity needed to treat some neuromodulation modalities requires highly advanced electronics. It remains important that electronics used in the design of implantable devices is efficient with respect to its energy consumption and size.

Delivery Mechanism

The delivery mechanism is a mechanical connection that conducts the electrical pulses from the stimulator to electrodes at the patient interface. The electrical resistance of the connection depends upon the construction, materials and the length of the connection. The length is highly dependent upon implant location relative to the neurological target. There are systems that integrate the delivery mechanism with the stimulator, such as the BION® microstimulator, where the length would be less than 1 cm. On the other hand, a stimulator placed in the upper buttock that had a patient interface in the cervical region could be in the range of 90–110 cm.

Patient Interface

Electrical stimulation and recording of excitable tissue is the basis of electrophysiological research and clinical functional electrical stimulation, including deep brain stimulation and stimulation of muscles, peripheral nerves or sensory systems. When a metal electrode is placed inside a physiological medium, such as extracellular fluid (ECF), an interface is formed between the two phases. In the metal electrode phase and in attached electrical circuits, charge is carried by electrons. In the physiological medium, or in more general electrochemical terms the electrolyte, charge is carried by ions, including sodium, potassium, and chloride in the ECF. The central process that occurs at the electrode–electrolyte interface is a transduction of charge carriers from electrons in the metal electrode to ions in the electrolyte. The patient interface is time-varying impedance. Figure 11.3 illustrates a simple electrical circuit model of the electrode–electrolyte interface, consisting of two elements [10–12]. C_{dl} is the double layer capacitance, representing the ability of the electrode to cause charge flow in the electrolyte without electron transfer. $Z_{faradaic}$ is the faradaic impedance, representing the faradaic processes of reduction and oxidation where electron transfer occurs between the electrode and electrolyte. One may generally think of the capacitance as representing charge storage, and the faradaic impedance as representing charge dissipation [13].

It is important that these faradaic reactions are reversible. The electrode material, electrode surface area, maximum charge per phase and charge balance are all considerations for safe pulse delivery. DC blocking capacitors are generally used in the pulse generating electronics to couple pulses to the stimulating electrodes to prevent DC current from generating harmful faradaic reactions.

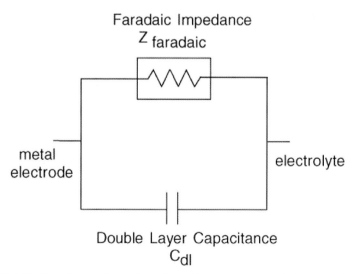

FIGURE 11.3 Two-element electrical circuit model for mechanisms of charge transfer at the interface.

Patient

The patient adds a tissue conductance to the system that varies by implant location. For SCS, when the electrodes are implanted in the cervical region, the typical distance between the electrode surface and the dorsal columns is very close as measured by the threshold of activation [14–16] compared to electrodes implanted in the thoracic region where threshold can be very high. The impedance of the patient is determined by both the conductance of the tissue (ρ) and the geometrical surface area of the stimulating electrodes (a) as given in the following formula:

$$Z = \frac{\rho}{\sqrt{2\pi a}}$$

The smaller the electrode area for a given tissue conductance the higher the impedance [17,18].

Neuromodulation devices exist that provide current, voltage and charge controlled therapy. In order to keep the formula in this chapter more straightforward, we will use the controlled current convention. Conversions may be made to the other conventions using well documented relationships from the physics of electricity.

Now let us consider the contributions of the system elements to the overall Impedance (Z):

$$Z_{Tot} = Z_{Elec} + Z_{DM} + Z_{PI} + Z_P$$

The total impedance Z_{Tot} is equal to the pulse generator electronics impedance Z_{Elec} plus the distribution mechanism impedance Z_{DM} plus the patient interface impedance Z_{PI} plus the patient electrode tissue impedance Z_P. Thus:

$$V = I * Z_{Tot}$$

V is the neuromodulation system voltage requirements to deliver a therapeutic pulse of amplitude I. We now can calculate the required system power source capacity:

$$C = I_{Avg} * T_L$$

The power source capacity C is equal to the average current consumed from the power source times the therapeutic life time in hours.

The formula for average current (I_{AVG}) measured in amperes:

$$I_{AVG} = (M * I_{TAVG}) + I_{EAVG}$$

where

$$M = \frac{V}{V_{PS}} V > V_{PS}$$

The therapeutic voltage V is divided by the power source voltage to obtain M the multiplying factor. Example: if the therapeutic voltage is twice the power source voltage, then it would take twice as much current. Depending on the method of power conversion, M can be a fraction or the next higher integer.

The current required by the electronics (I_{EAVG}) to accomplish the generation of pulses and other functions is highly dependent upon the circuit efficiency and other applications and left for more detailed discussion in other texts.

From the above formula for capacity, a power source can be specified for a given neuromodulation device. When considering a power source, you would want to consider using maximum parameters, treatment duty cycle, manufacturing and product shelf life in determining the final power source capacity.

PATIENT COMPLIANCE AND COST CONSIDERATIONS

Finally, any piece of equipment including an implantable device is only helpful if it is used. Patient compliance is an important feature that needs to be addressed in all aspects of device design. Neurostimulators by design use electricity to produce their effect on nervous tissue. Thus, when in use implantable neurostimulators are constantly depleting their power source. When depleted this can result in the need for surgery (when using a conventional battery) or recharging (when using a rechargeable battery). This burden must be assessed looking at the individual patient and their tolerance. An extreme example of a rechargeable device that required frequent recharging was reported by Trentman et al in a case study that examined the use of a rechargeable stimulator to treat the pain associated with great occipital neuralgia [19]. One patient completed the headache maps approximately 4 months after implant. She subsequently did not appear for her 6-month follow-up and stopped using the Bion before study completion at 12 months. She stated the battery recharging schedule was too demanding, specifically that she was spending 1.5 hours recharging her Bion for every 1.5 hours of use.

Frequent recharge is a disadvantage when using a rechargeable stimulator; however, rechargeable stimulators, when used in the appropriate patient, can often have many advantages, such as decreased patient discomfort and morbidity from procedural complications. Rechargeable stimulators can last over 10 years, significantly reducing the number of replacements due to battery depletion [19,20].

POWER SOURCE OPTIONS

We have established that there are a number of considerations that need to be taken when choosing a power source for an implanted neurostimulator. Today, there are a number of different options that can be chosen. These choices include radiofrequency-powered devices, primary cell devices and, more recently, rechargeable devices. Each of these have their advantages and their limitations, however, all play an important role in providing patients with devices to suit their individual needs.

Early Developments

Some of the early pacemakers (1960s) utilized mercury–zinc batteries and were cast in epoxy, which was porous to allow discharged battery hydrogen to dissipate. These batteries required venting and thus could not be hermetically sealed.

At times, the epoxy allowed fluid leakage into the pacemaker causing premature failures. This technology allowed the growth of the medical device industry
but had many limitations which included size and longevity. The average life of
these pacemakers was still only two years, with some 80% of generator removal
being necessitated by battery failure [21].

Nuclear power sources were tried successfully for some period. These power
sources used plutonium 238 with its half-life of 87 years. The life degrades by
11% over 10 years. These power sources had a very long life but were large
and created problems for patients traveling between states and countries due to
presence of a radioactive fuel. They also needed to be removed upon death of
the patient so they could be properly disposed of. These power sources became
obsolete with the development of lithium batteries.

Introduction of the lithium–iodine battery in 1975 greatly extended the battery life to more than 10 years for some models and replaced the use of mercury–zinc batteries. Lithium is the most active of all the alkali metals, it is easily
handled and, today, is the most common battery used in the implantable medical
device industry [22,23].

The reactions are:

$$Li + Li^+ + e \quad \text{Anode}$$

$$\frac{1}{2}I_2 + e \to I^- \text{Cathode}$$

Giving the combined reaction:

$$Li + \frac{1}{2}I_2 \to LiI$$

The three principal zinc–mercury problems do not exist in the lithium–iodine cell:
no gas is generated, there is no fabricated separator and, since there is no gas, the
cell can be hermetically sealed in a metal can with a glass–metal feed-through
connection. No body moisture can get in and no battery effluent can get out.

Power Source Performance Parameters

A good power source selection is a compromise between various performance parameters to meet the requirements of the specific application. Critical factors are minimum and maximum voltage; minimum, maximum and average discharge current;
continuous or intermittent operation, including size and duration of current pulses;
long shelf life; long service life; high energy density and environmental conditions.

Conventional Batteries

The development of a small, reliable battery that could withstand the hostile
environment of the human body was pivotal in the history of neuromodulation.

In addition to iodine, lithium can also be combined with other materials such
as silver, copper sulfide, bromine chloride, sulfuryl chloride, silver vanadium

TABLE 11.1 Table of Chemistry of Various Primary Cells

Low	Medium	High
Lil	LiSoCl$_2$	LiSVO
	LiCFx	LiSVO-CFx
	LiMnO$_2$	
	LiSVO-CFx	

pentoxide, and lithium-thionyl chloride. Each of these cells has particular advantages and disadvantages.

Primary cells can be categorized as low, medium and high rate. Table 11.1 shows some of the chemistries currently being used in devices.

Pacemakers have generally occupied the low rate category but needs for higher energy telemetry is pushing manufacturers to look at the medium rate chemistries. The high rate category is mostly occupied by defibrillators with their requirement to draw amperes of current when charging their pulse output capacitors. We will concentrate the rest of the discussion on the medium rate group. Table 11.2 shows battery parameters versus the various chemistries.

Primary cells for neuromodulation devices occupy a majority of the device space. In order to minimize size and weight, reduced service life needs to be considered.

Battery reliability has been greatly improved today, resulting in failure due to battery reduced to less than 1%. However, the life of an implanted battery is still limited; lithium batteries used in neuromodulation devices have a useful life of 3–7 years, resulting in the need for battery replacements. The longer years

TABLE 11.2 Battery Parameters Versus the Various Primary cell Chemistries

	SOCl$_2$	CFx	MnO$_2$	SVO-CFx
Operating Voltage (V)	3.6	2.9	3	2.85
Energy Density (W/L)	850–1000	900–1100	500–600	>1050
Internal Resistance (Ω)	<70	<10	<5	<2.5
Self-Discharge (% per year)	<1%	<1%	<1%	<0.25%
Max Pulse Current (mA)	10	25	140	280

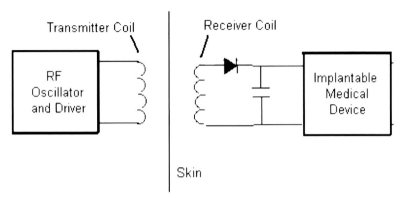

FIGURE 11.4 Schematic of the two main components in a RF powered system.

in some cases require larger capacity batteries or rely on cycle modes (therapy duty cycle) to reduce the average current.

Radiofrequency (RF)

As mentioned above, the early batteries had extremely short battery lives. Because of the high energy demand for SCS devices, manufacturers developed radiofrequency-powered devices. A radiofrequency-powered device has its power source largely external to the body. Power is passed through the skin using a transmitting coil. This coil is coupled to a receiving coil contained within the implanted portion of the device (Fig. 11.4). By using this type of system, there is no power source located inside the body, therefore there is nothing within the body that can be drained. This allows the device to deliver high power demanding parameters without draining a battery. When the externally worn transmitter is drained, the patient simply changes the battery as they would for any other household electronic device. Eliminating the power limitation on the device design allows an RF-powered stimulator to provide a wider range of stimulation parameters. This flexibility allowed expansion of the neuromodulation field into areas such as cochlear implants, pain, etc.

The limitation when using this type of power source is that a patient is required to wear an external transmitter whenever they are required to have the stimulation working. Also, maintaining alignment of the transmitting coil and receiving coil and intercoil distance is problematic. This modality is not a good indication for applications that are life sustaining.

Rechargeable Batteries

Rechargeable batteries use electrochemical reactions that are electrically reversible. Rechargeable batteries come in many different sizes and use different combinations of chemicals. Commonly used secondary cell ('rechargeable battery') chemistries are lead acid, nickel–cadmium (NiCad), nickel–metal hydride (NiMH), lithium ion (Li-ion), and lithium ion polymer (Li-ion polymer). Implantable neuromodulation devices have utilized the lithium ion type chemistries. These batteries can be hermetically sealed and have energy densities among the highest of the rechargeable chemistries. Recharging of all these batteries requires specific charging circuitry to avoid overcharging the batteries which could result in safety and longevity issues. Most of the neuromodulation devices on the market today use batteries in the range of 40–400 mAh. Longevity of these devices has been claimed to last 7–10 years. Some predictions show life times exceeding 10 years.

Construction of these cells is of two types; the jelly roll and stacked plate design. Jelly roll construction is easier to manufacture but is limited to a cylindrical or squashed cylindrical shape. Stacked plate designs require more assembly but can more easily be adapted to the neuromodulation device shape. The useful voltage range of a lithium ion battery is about 4.1–3.4 Volts.

Selection of the rechargeable battery capacity is important. We have calculated the total capacity C required for a power source above. We can either divide that number by the expected recharge cycles or multiply the average current times the minimum time between charges. If the capacity selected is too small then the recharge burden to the patient becomes too high as in the example discussed earlier. If the battery is too large then size, weight and recharge times are more than necessary. Another consideration is the number of recharge cycles. For each recharge cycle, the battery capacity degrades. The number of discharge cycles varies from several hundred to thousands depending on battery chemistry and depth of discharge. The number of cycles and capacity degradation must be taken into account in specifying the initial power source size [24, 25].

Future Power Sources

There are exciting developments in power sources for neuromodulation devices that will change the patient and physician experience dramatically. Research areas include novel approaches to improving the amount of energy stored in a given volume, energy density, and replenishing the energy in rechargeable sources. The resulting paradigm is an array of minimally invasive devices with more physician control over the implant location.

One of the more promising developments for improving energy density and safety is thin-film rechargeable lithium batteries, based on research at the Oak Ridge National Laboratory (ORNL) (http://www.ms.ornl.gov/researchgroups/Functional/BatteryWeb/index.htm). Thin-film batteries have solid state construction using processes normally found in semiconductor manufacturing. The result can be an extremely thin battery that is physically flexible and can be integrated into the structural material of the medical device.

Advances in energy harvesting, the process by which energy is derived from external sources, captured, and stored, could make possible a breakthrough class of self-powered implantable medical devices (IMD) and bio-sensors. A self-powered IMD may have a considerably longer life than those using today's technology. A realization of this vision could obviate the need for replacement surgeries for some patients.

Potential external sources for harvesting include mechanical energy from voluntary and involuntary muscle contraction, body temperature and ambient RF energy. For example, Biophan Technologies' subsidiary TE-Bio has developed a biothermal power source that converts body heat into electricity to power implantable medical devices. NASA has engaged with TE-Bio for advancing high-density, nanoengineered thermoelectric materials for use with implantable medical devices.

Scientists from Princeton University also recently reported that they have developed power-generating rubber film that can harness body movements to power implantable devices. Princeton University engineers said their new material is composed of ceramic nanoribbons embedded onto silicone rubber sheets. They said the material generates electricity when flexed and is highly efficient at converting mechanical energy into electrical energy. They envision placing sheets of the material against the lungs to use breathing motions to power implanted devices. A paper detailing the study, which included postdoctoral researcher Yi Qi and Professor Michael McAlpine, appears in the early online edition of the journal *Nano Letters* (2010).

Most of the of the new energy sources being looked at fall into the low-rate category and are potential sources of power for sensors, but as long as neurostimulation devices remain in the medium-rate category, primary batteries and rechargeable batteries will be the power source of choice for the near future.

REFERENCES

1. Alfano S, Darwin J, Picullel B. *Programming principles in spinal cord stimulation: patient management guidelines for clinicians.* Minneapolis: Medtronics; 2001 pp. 27–33.
2. Kunnumpurath S, Srinivasagopalan R, Vadivelu N. Spinal cord stimulation: principles of past, present, and future practice: a review. *J Clin Monit Comput.* 2009;23:333–339.
3. Burchiel KJ, Anderson VC, Brown FD, et al. Prospective, multicenter study of spinal cord stimulation for relief of chronic back and extremity pain. *Spine.* 1996;21:2786–2794.
4. Kumar K, Toth C, Nath RK, Laing P. Epidural spinal cord stimulation for treatment of chronic pain – some predictors of success. A 15-year experience. *Surg Neurol.* 1998;50:110–121.
5. Kumar K, Taylor R, Jacques L, et al. Spinal cord stimulation versus conventional medical management for neuropathic pain: a multicentre randomised controlled trial in patients with failed back surgery syndrome. *Pain.* 2007;132:179–188.
6. North RB, Kidd DH, Zahurak M, James CS, Long DM. Spinal cord stimulation for chronic, intractable pain: experience over two decades. *Neurosurgery.* 1993;32:384–395.
7. Strujik JJ, Holscheimer J, van Veen BK, Boom HBK. Epidural spinal cord stimulation: calculation of field potentials with special reference to dorsal column nerve fibers. *IEEE Trans Biomed Eng.* 1991;38:104–110.
8. Wesselink WA, Holsheimer J, Boom HBK. Analysis of current density and related parameters in spinal cord stimulation. *IEEE Trans Rehab Eng.* 1998;6:200–207.

9. Volkmann J, Herzog J, Kopper F, Deuschl G. Introduction to the programming of deep brain stimulators. *Mov Disord.* 2002;17:S181–S187.

10. Randles JEB. Rapid electrode reactions. Discuss. Faraday Soc. 1947; 1 11– 9.

11. Gileadi E, Kirowa-Eisner E, Penciner J. Interfacial electrochemistry: an experimental approach. Reading, MA: Addison-Wesley; 1945 Section II.

12. Bard AJ, Faulkner LR. Electrochemical methods. New York: Wiley; 1980 p. 19 and 102 [Chapters 1.3.3, 2, 3.5 and 9.1.3].

13. Merrill DR, Bikson M, Jefferys JGR. Electrical stimulation of excitable tissue: design of efficacious and safe protocols. *J Neurosci Meth.* 2005;141:171–198.

14. Tulgar M, Barolat G, Ketcik B. Analysis of parameters for epidural spinal cord stimulation 1. Perception and tolerance thresholds resulting from 1,100 combinations. *Stereotactic Funct Neurosurg.* 1993;61:129–139.

15. Tulgar M, Barolat G, Ketcik B. Analysis of parameters for epidural spinal cord stimulation 2. Usage ranges resulting from 3,000 combinations. *Stereotactic Funct Neurosurg.* 1993;61: 140–145.

16. Tulgar M, He J, Barolat G, Ketcik B, Struijk H, Holsheimer J. Analysis of parameters for epidural spinal cord stimulation 3. Topographical distribution of paresthesiae – a preliminary analysis of 266 combinations with contacts implanted in the midcervical and midthoracic vertebral levels. *Stereotactic Funct Neurosurg.* 1993;61:146–155.

17. Brown BH, Smallwood RH, Barber DC. *Medical physics and biomedical engineering.* London: Taylor & Francis; 1999.

18. Holsheimer J, Dijkstra EA, Demeulemeester H, Nuttin B. Chronaxie calculated from current-duration and voltage-duration data. *J Neurosci Meth.* 2000;97:45–50.

19. Trentman TL, Rosenfeld DM, Vargas BB, Schwedt TJ, Zimmerman RS, Dodick DW. Greater occipital nerve stimulation via the Bion microstimulator: implantation technique and stimulation parameters. Clinical Trial: NCT00205894. *Pain Phys.* 2009;12:621–628.

20. Hornberger J, Kumar K, Verhulst E, Clark MA, Hernandez J. Rechargeable spinal cord stimulation versus nonrechargeable system for patients with failed back surgery syndrome: a cost-consequences analysis. *Clin J Pain.* 2008;24:244–252.

21. Furman S, Escher DJ. Transtelephone pacemaker monitoring: five years later. *Ann Thorac Surg.* 1975;20:326–338.

22. Moser, J.R. (1972). Solid state lithium-iodine primary battery. US Patent 3,660,163, May 2, 1972.

23. Greatbatch W, Lee JH, Mathias W, Eldridge M, Moser JR, Schneider AA. The solid-state lithium battery: a new improved chemical power source for implantable cardiac pacemakers. *IEEE Trans Biomed Eng.* 1971;18:317–323.

24. Mallela VS, Ilankumaran V, Rao NS. Trends in cardiac pacemaker batteries. *Indian Pacing Elecrophysiol J.* 2004;4:201–212.

25. Wei X, Liu J. Power sources and electrical recharging strategies for implantable medical devices. *Front Energy Power Eng China.* 2008;2:1–13.

26. Qi Y, et al. Piezoelectric ribbons printed onto rubber for flexible energy conversion. *Nano Lett.* 2010;10:524–528.

Commentary on The Electrode – Power

Jay L. Shils, PhD

Director of Intraoperative Monitoring, Dept of Neurosurgery, Lahey Clinic, Burlington, MA

Until recently, and continuing with today's devices, the power source has been one of the key limiting factors in the advancement of neuromodulation devices. They have been hampered by limitations of integrated circuit technology and the basic constraint that no electrical neuromodulation system can run at all without some source of power. As stated here by Cameron, the battery needs to be small enough or compact enough for patient comfort and in daily life, but yet needs to be large enough to hold and/or deliver the appropriate power. The realities of power delivery include:

1. device up-time
2. safe heat dissipation
3. indwelling comfort; and potentially
4. rechargeability ease.

Additionally, one might consider that power could be localized near to where the therapy is delivered, e.g. within or very near the electrode housing itself. Presently, devices generally separate the power and generator from the electrode terminus, requiring wire connections. Device failures then usually require surgical change out and/or manipulation of one or more components. Failure of the implantable pulse generators (IPG) is a rare condition compared to the failure of the lead. Eventually developing small power sources that can be placed at the therapeutic target holds many advantages. Such a device has been developed (consider the Boston Scientific Bion®), but present technology limits its usable life time or energy delivery. When predicting future neuromodulation needs, batteries and power interfaces need to improve. Below are key areas where effort for these improvements could be focused.

Designs can be improved in warning the clinician and patient better before power source depletion in order to minimize loss of therapy. Older power cells would have a slower decline in power over time prior to complete therapy loss thus giving the clinical team and the patient time to plan a battery change. Newer devices offer a much more stable output but have a very steep drop off at the battery end of life which means there is often little or no warning of imminent power loss. In some patients, such as those being treated for psychiatric disorders, this circumstance may be life threatening. In line with this need would be improved power source monitoring and ability to signal immediate recalculations of predicted battery life as therapy changes. Power, therapy current, battery voltage, and electrode impedance should be constantly monitored. This would then allow for potentially even informing the patient and recommending that they call their physician or make attempts to reduce device usage until it can be replaced.

Continuing to reduce overall size of the battery–generator complex is still a worthy goal. IPGs are usually large enough to cause bulges under the skin of most patients and, in smaller patients, these bulges are very noticeable. Microelectronics has already progressed to where it now utilizes less than 5% of the whole implant volume. Reducing the size of the battery further will then make it possible, in some cases, to place the IPG closer to the target, reduce the overall amount of implanted material, and as well likely reduce the potential for infection.

Perhaps in the more distant future, electrodes may be placed where they are not nearly as surgically accessible as they are now. Replacing the power source and/or generator may be impossible without damaging neural elements or other tissue in the process. One such example would be in future therapies that place multicontact, three-dimensional arrays within the spinal cord or brain. Also, the depth of the device would make external power systems, such as RF devices, large and cumbersome, and also the heat generation at the skin surface could be uncomfortable or damaging to the patient. Hybrid systems of internal RF to power circuitry and a stimulator would be needed. Moreover, as the number of implanted electrodes increases, demands on the amplifiers, processing units, and power are also going to need to increase. Smarter processors that can focus or modulate the stimulation to specific areas only at appropriate times may be able to reduce overall power usage.

STUDY QUESTIONS

1. As the size of the electrode decreases, what safety issues need to be considered when designing a power supply?
2. Consider all locations in the body currently used for the placement of IPGs – for DBS, SCS, VNS, and PNS. What are the advantages and disadvantages of each with regard to ease of use for the patient and programmer, ease of placement for the surgeon, likelihood of an erosion or complication, and ability to revise or modify?

Part IV

Placing Neuromodulation in the Human Body

Chapter 12

Surgical Techniques

Jeff Arle, MD, PhD

Director, Functional Neurosurgery and Research, Department of Neurosurgery, Lahey Clinic, Burlington, MA; Associate Professor of Neurosurgery, Tufts University School of Medicine, Boston, MA

INTRODUCTION

Rather than write a 'this is how I do it' chapter, it seems more helpful to get under the skin of surgical techniques used specifically in neuromodulation, so to speak, and try to encapsulate the underlying principles of all surgical aspects in neuro-modulation. This could serve as an underlayer, or foundation, upon which details of various techniques, some of which may be delineated here as well, can be built. Moreover, not only practitioners, but also designers and physiologists, can gain an appreciation for some of the related thoughts and concerns of the surgeon as they engage in neuromodulation, however applied. Other chapters will provide details of patient selection, troubleshooting and revisions, limiting morbidity, and more specifics on indications and applications per se. Further, I will lean heavily on non-percutaneous examples – other than some commentary about implantable pulse generator (IPG) placements and anchoring which apply in all cases. To this end, the chapter here is organized into what I consider to be the three most perti-nent aspects of surgical technique in neuromodulation, as follows:

- emphasizing the physiological target
- being attuned to nuances (tissue, physiology, and the patient)
- assuring that intraoperative resources are adequate for decision making.

EMPHASIZING THE PHYSIOLOGICAL TARGET

Every procedure in neuromodulation involves placement of a device that inter-faces with neural tissue to effect a particular physiological change toward benefiting the patient. The *physiological* aspect of this is the key, as it is an overarching principle helping to guide not only where to place the device, but also in how to place the device, when to place the device, and even in whom to place the device. The physiology available to help in this regard can manifest in various forms: patient feedback, somatosensory evoked potentials (SSEPs), evoked responses, single or multicell microelectrode recordings, electromyo-graphy (EMG), motor evoked potentials (MEPs), and electroencephalography

Essential Neuromodulation. DOI: 10.1016/B978-0-12-381409-8.00012-7

(EEG). But it also may simply manifest in the knowledge that the physiology itself is the main concern. For example, one might know the general anatomy of the occipital nerve, the medial and lateral branches, as they course and branch in the suboccipital region, up and through the parieto-occipital area, but it is the knowledge of how the physiological interface of a transverse electrode placement will potentially capture and modify the nerve's activity that helps more in placing one or two horizontally oriented 1×8 leads. And because of this, and because of the discomfort and difficulty in trying to use local anesthesia in the region, thereby altering the ability of the stimulation to obtain patient feedback if awake, one can typically place the leads under general anesthesia without patient feedback, or with local anesthetic (for a trial) and not worry about patient feedback.

This example is purposely chosen to illustrate this point because it is so removed from the typical emphasis placed on the physiological bases of using microelectrode recordings (MER) and stimulation in deep brain stimulation (DBS) for movement disorder targets, or cortical mapping techniques in motor cortex stimulation (MCS), where the reliance on physiology is obvious. In a different context, we have almost completely abandoned awakening patients to place dorsal column stimulators. Using a technique relying on stimulation through the lead and examining the evoked EMG is faster, more comfortable, and correlates specifically with the physiological target of the appropriate dermatomal levels in the patient – stimulating the *same* fiber tracts that are stimulated when the patient is awakened and asked about parasthesias. This technique avoids the sometimes confounding aspects of waking patients up from sedation, prone, sometimes confused, and occasionally misconstruing that we are asking them about where their pain typically is located rather than where they are feeling parasthesias, not to mention the risk of oversedation without a secured airway [1–3]. This stimulation-EMG technique is entirely dependent on the underlying physiological target – it does not rely on the fluoroscopic imaging, or the patient's compliance, or the ability to see the anatomy directly, the surgeon often falsely 'assured' the electrode is oriented as desired. None of these approaches alone is reliable, as the cord may be altered in its rotation within the canal, the fluoroscopy image notoriously can have parallax problems [4] or, as mentioned, the patient may not be relating accurate enough details to place the lead because of medication, confusion, discomfort, or any combination thereof.

In emphasizing the physiological target, rather than imaging, or anatomy per se, the surgeon places him/herself within the center of the goal, rather than just outside of it, hoping they capture it. Some have recently done excellent work in exploring the potential of placing DBS leads, for example, using high resolution, high-strength magnetic resonance imaging (MRI) better to visualize the subthalamic nucleus (STN) [5]. Although there is a strong correlation between the anatomical STN as a target and the benefit derived with DBS in Parkinson's disease, it is still quite unclear whether or not the actual target of the stimulation that provides the benefit is the STN itself, fibers in the H2 field of Forel that course over the dorsal aspect of the STN, the zona incerta, or whether the benefit derives from the *nature* of the stimulation itself (frequency, pattern, amplitude, pulsewidth) in any of these locations, or in stimulating a combination of them. Because these underlying mechanisms are still not worked out definitively, it is premature to rely exclusively on anatomical targeting – although one can often obtain a very good result using this technique because of the high degree of overlap and leeway in both programming and field spread with present DBS systems. In a related example, even if one can see the globus pallidus pars interna (GPi) extremely well on a preoperative or intraoperative MRI scan, testing with a microelectrode within or near the optic tract is still necessary to determine distance of the final electrode tip from the tract to prevent current spread and visual disturbances postoperatively.

The same principle supports this in the placement of an electrode to perform motor cortex stimulation. We rely on a combination of locating the N20 reversal potential with SSEPs and motor-evoked EMG responses using a ball probe and a train of five pulses (see [6]), mapping the M1 region without opening the dura, and without looking at the cortex directly or its anatomic orientation. But some groups rely on functional MRI exclusively to place the lead. Again, without direct *physiological* confirmation, accurate placement of the lead to effect the desired result becomes compromised in a higher percentage of cases [7,8].

BEING ATTUNED TO NUANCES (TISSUE, PHYSIOLOGY, AND THE PATIENT)

There are always multiple concerns during surgery that require vigilance on the part of the surgeon, who is continuously filtering his or her environment to determine relevance in the case at hand. As the saying 'the devil is in the details' goes, so goes this level of attention to nuances in neuromodulation surgery. Focus on concerns that are widely encountered, no matter what the neuromodulation application, would include details of the following: anchor positioning and suturing, locating IPG placement, how much tissue is dissected or removed for lead placement, the attention to detail of the neurophysiology staff and technicians (do they notice when anesthesia has changed? do they understand what effects it will have on the physiology?), company representatives supplying redundant and alternative devices and accessories, and the degree of reliance on the industry representative's knowledge of how to assess systems and test stimulation. Some of these are beyond the scope of this text – but some are essential elements of surgical techniques, and will be covered.

Anchors are an extremely important part of preventing lead migration, catheter migration, and lead breakage – all of which potentially require revision surgeries. Interestingly, The Neuromodulation Foundation (www.neuromodfound. org), incorporated only in 2008, discusses a review of spinal cord stimulator lead migration as a complication, including issues of encapsulation, dural suturing and anchor issues within the discussion. The following were conclusions and recommendations from their review:

- Incidence: surgical plate/paddle electrodes resist migration after encapsulation
- Time to appearance of symptoms: immediate, i.e. before encapsulation
- Treatment: non-invasively reassign contact combination if possible; if ineffective, revise electrode
- Usual resolution and impact on therapy: minor displacement usually can be addressed non-invasively; major displacement requires revision
- Risk reduction: some surgeons suture surgical plate/paddle electrodes directly to the dura, but this requires exposing a larger area, which is problematic, and might add mechanical stress. Some use an anchoring/strain relief sleeve to secure the emerging lead wire to the spine. Using absorbable sutures eliminates the continued focal stress that can be caused by non-absorbable sutures after the electrode becomes encapsulated. During system implantation, avoid increasing mechanical stress by avoiding unnecessary bends of small radius and superfluous connectors. Subject to patient preference and surgical judgment, avoid crossing a mobile joint or segment with subcutaneous lead wire or extension cable; e.g. a thoracic electrode encounters more stress and strain if connected to an upper buttock pulse generator than if connected to a lateral abdominal generator.

This review classifies evidence for these recommendations as level B – using their unique reworking of traditional levels of evidence grading, incorporating practical aspects of care like 'only option' and data supporting an advantage of risk-benefit analysis if it exists. Their level B includes well-designed clinical studies (prospective, non-randomized cohort studies, case-control studies, etc.), randomized control trials (RCTs) with design problems, and/or weighing risk versus potential benefit with expert consensus revealing a good likelihood of a favorable outcome. Level B review becomes a 'recommendation' from them.

I would add several comments on their assessment. On their first point (and this is addressed somewhat by their second point), it is important to remember that once a paddle lead has become encapsulated (within the first 3–6 weeks), it is very unlikely that it can migrate at all. This is not true for percutaneous leads, which may migrate cephalad to caudad even years after implantation. But further, it is important to realize this fact when discussing problems in pain coverage with patients or other care-givers after surgery. If coverage wanes, often the patient (and others) will ascribe it to a fall or other mishap (e.g. motor vehicle accident) and cannot be assuaged from this conclusion until an x-ray is obtained. This x-ray will almost perforce necessitate the ability to compare it to an original postoperative film, or at least a film of the lead when stimulation was working well. Therefore, it behooves one to have such a comparison film available on every patient. Additionally, though, it is unlikely that there will be any migration of the lead noted on such a film if the lead is a paddle and has been in place long enough – under such circumstances, there is more chance that scar has thickened, a wire has broken, or there is fluid in a connection, than a lead migration.

On their third point, while it is no doubt best to try to reprogram the lead initially, it is often the case that coverage still will not be obtained (though there is still stimulation felt), and the lead may still appear to be perfectly well-positioned. In such cases, it can be difficult in determining how to revise such a lead. One should consider moving it further cephalad or caudad, so that scarring thickness changes on the dura under the lead play less of a role in shunting current differentially, but advising the patient that it can recur and create the same problem again.

On the last point in their recommendation, I would suggest that suturing the lead to the dura is a viable option, but not recommended specifically for preventing migration per se, but rather to make sure the lead stays in direct proximity to the dural surface over its entire length. In some cases where the distal end of the paddle lead continues to divert right or left, perhaps because of a midline keel of sorts under the lamina, suturing the tip of the lead can be helpful in preventing this diversion after closing, until it encapsulates. Some paddle leads are easier to suture than others – Medtronic tripole leads have a nice intrinsic mesh within the silastic insulation and a slight margin within which a suture will hold well (4-0 Neurolon, for example). Finally, it is vital to make note in the chart of the fact that a lead was sutured to the dura, in case the same or another surgeon has to remove it, so it is not pulled out in a way that tears the dura.

Anchors themselves can obviate much of the need to worry about the strain on the lead directly from bending or where the lead wire crosses a flexion position of the body (e.g. neck, waist). Titanium cinching anchors work well and seem to prevent kinking and focal stress on the wire itself. However, traditional silastic anchors can work well to avoid such suture stress points if sutured correctly, and the newer titanium anchors can add a small additional cost. Recently, I had a patient with a medication pump who lost significant weight after the original surgery and could manipulate her pump in her abdomen over and over – a pump 'twiddler'. Eventually, the catheter had either dislodged or was kinked; somehow the medication was not reaching the intrathecal space. Upon surgical exploration, it was appreciated that the anchor was still sutured to the fascia where I had left it and, surprisingly, the catheter did not move within the anchor when tugged – what happened instead was that the twisting of the catheter had caused the catheter material itself to fail and it sheared off just on the pump side of the anchor.

Anchors for vagal nerve stimulation (VNS) specifically can often lead to anterior neck pain if not sequestered below the platysmal tissue layer and are sutured so that they point into the neck rather than out toward the skin. This principle also applies to spinal cord stimulation (SCS) anchors, either thoracic or cervical, wherein pain or discomfort may occur if the anchors are secured too close to the subdermal layer, outside of the fascia. MCS 'anchors' typically can be reliably made by simply tacking the lead wire in a short loop (without kinking it) on the dura before the wire exits a burr hole from under the bone flap. The lead itself is also sutured down to the dura in several locations. DBS has available a device at the burr hole itself to secure the lead wire – these seem to work fine for the most part, although in thin scalped individuals, they can place increased pressure on the underside of the galea and create a higher propensity for scalp erosion.

IPG location is also important for patient comfort, convenience, and for mitigating problems with lead dislodgement or migration. Surprisingly, surgeons place IPGs in many locations of the body, often without confirming with the patient beforehand where the patient might want it, or need it, as patients with limited shoulder range of motion, for example, cannot even reach certain locations to use their controllers, or rechargers. Typically, there are only two appropriate locations to place IPGs thoracic leads, two locations to place IPGs cervical or occipital leads, and one general location to place IPGs DBS, VNS, or MCS leads. For thoracic or lower leads, primary preference should be to place the IPG in the upper buttock region, above the area of the ischial tuberosity where the patient will sit and below the region of the 'belt line', always avoiding two important features – the side a male patient may carry a wallet routinely, and the side on which the patient may prefer sleeping which may or may not be a pressure concern. In a patient with some extra subcutaneous tissue, the nearby posterior flank may be very acceptable for leads in this location as well, taking care not to make the position too lateral or imposing on the costal processes or scapula. For cervical leads, DBS, MCS, occipital nerve stimulation (ONS), VNS, and a potential variety of head peripheral nerve stimulation (PNS) leads, the subclavicular region (2 finger-breadths below the clavicle with the incision in line with the clavicle, and the pocket always kept superficial to the pectoralis fascia) is preferred, although the flank region as described just previously is also a possibility. I have avoided bringing these leads, with necessary extension wires, all the way to the buttock area. However, the subclavicular region often forces awkward lateral positioning in the operating room (OR) for lead placement, or the need for closing, undraping, repositioning and redraping laterally to place the IPG. For DBS and MCS this is not an issue, and may not be for certain locations of head PNS (e.g. supraorbital nerve stimulation), but for cervical SCS and ONS it should be thought out carefully.

As a final note on IPG implantation, certain IPGs currently require a particular tissue depth, often of 1–2 cm, no more, and no less, for optimal communications and recharging ability. Consideration should be made for contraction of the tissue during the healing and encapsulation process. Adipose tissue may resorb quickly, the capsule tighten, and within 6 weeks or so the IPG is only the thickness of the dermis away – causing pain and tethering and resulting in impossible recharging. The appropriate depth can be less than simple to achieve in many patients, especially if the patient is obese. Care must be taken as well in hemostasis of the pocket before closure, and in the configuration of the wires or catheter deep to the IPG or pump before closure – preventing kinking and migration of wires over the top of the IPG.

On the issue of tissue dissection and removal in the process of placing electrodes, typically in thoracic laminotomies for SCS, an incision can be made which is often as small or smaller than many made for the anchoring portion of percutaneous-to-permanent incisions – in other words, they can be just as minimally invasive as a percutaneous lead placement, aside from the muscle dissection itself. The removal of bone for the laminotomy, likewise, should be minimized for lead placement such that the lead just has enough room to be aligned appropriately and not have a propensity to migrate from pressure against an edge of bone or ligament. But this amount of bone removal can typically be achieved by only dissecting the muscle free from one laminar side (usually chosen to be the side that the patient may need more coverage on – but this should not be a major factor in deciding the side – for instance, surgeon preference can be just as important in this regard). The laminotomy will in no way lead to instability, even if one entire facet joint is removed, or even if a full laminectomy is needed to place the lead adequately. Because of the muscle dissection, and reclosure of the muscle fascia, these patients usually have more postoperative discomfort than a permanent conversion of a percutaneous lead. However, the entire process should take an hour or so of OR time and most patients should still be able to leave the same day.

Performing cervical laminotomies for paddle lead placements often encounters difficulty placing the lead unless small but adequate openings are made over each of several of the lamina in order to manage alignment of the paddle lead, help the lead progress in the epidural space, and even to pull the lead into position if necessary. In both the thoracic region and cervical region, it should be remembered that retrograde placement is always reasonable if coverage or simply placement of the lead at all requires it. Little has been published regarding cervical lead placement [9,10], but both percutaneous and paddle leads may be placed either anterograde or retrograde throughout the cervical spine from the occiput above C1 and below, within the limitations of prior surgeries and scar tissues. Leads do not need to be placed only at the most superior levels possible. In most cases, a paddle lead extending from C4–7 may be more appropriate for coverage of pain over C6–T1 dermatomes than a C1–2 lead.

Some have wondered whether MRI of the spine region to be accessed should always be obtained prior to surgical lead placement. Certainly, in any patient in whom myelopathy is suspected, adequate imaging and evaluation for safety of lead placement in the epidural space should be performed. However, in patients who are asymptomatic, it is very unlikely, even after prior fusion or decompressive surgeries, that there will be a concern. Additionally, one should consider the idea that a decompression may be performed in order to place the lead (in fact it may *need* to be performed due to prior surgery there) and this can eliminate the canal narrowness and the concern for creating excess stenosis in many cases. Anecdotally, in over 250 paddle lead new placements or revisions, including

over 50 cervical leads, we have had no removals or injures related to cord compression, in a 10-year period. Some of these patients already coincidentally had
prior MRIs of related regions of the spine, but new scans were obtained for this
reason only rarely alone.

ASSURING THAT INTRAOPERATIVE RESOURCES ARE ADEQUATE FOR DECISION MAKING

This final category extends across several broad but converging areas involved
in performing neuromodulation surgeries. Surgical decision making in neuromodulation ranges from the self-evident to the radically uncertain, but can
be bolstered by planning. Once a patient has been identified for surgery, the
assumption being that an appropriate indication has been found and work-up
performed, circumstances can arise wherein appropriate therapy will be compromised unless adequate resources are available.

On the surface, one might consider that 'resources' in this case means
'implantable devices', but only sometimes is that also the case. Consider, for
example, the situation of a placement of an MCS, and the craniotomy is performed after making measurements on the scalp, or integrating a functional
MRI (fMRI) into a navigation system, or both, and the time has come for physiological mapping. The strip electrode, typically the one to be used for implantation as well, is moved around on the dura, looking for a reversal of the phase
in the waveform consistent with the N20 SSEP. A small grid may be utilized
for this as well, although the jackbox and setup will need to account for such a
method. In any case, factor in that the pain for the patient extends up into the
lower face area, where it is quite severe, but also into the hand and forearm,
following a stroke 3 years prior. Initially, no N20 phase reversal is obtained, at
least any that is consistent and unequivocal. The surgeon begins already to question the location of the craniotomy – is it too anterior, or posterior? Is it large
enough to move the electrode to the central sulcus under the edge of the bone,
or should it be extended – but which way? A decision is made to try the motor
mapping first. The appropriateness of the anesthetic technique is checked and
confirmed, after the surgeon asks the technician looking at the SSEPs to discuss
this with the nurse anesthetist who has replaced the attending anesthesiologist in
the room for this critical juncture of the case. The ball probe is moved around as
amplitudes are adjusted and communication between the physiology team and
the surgeon eventually confirms where the face region seems to be, inferiorly,
and where the lowest thresholds for the hand region might be located, all on
the surface of the dura. While apparently satisfactory signals are rechecked and
confirmed, and no seizure has occurred, the strip electrode is not large enough
to cover both areas, especially with any leeway in electrode configuration.

What are reasonable solutions? Adequate planning would not only account
for placing two leads side by side or juxtaposed in tandem, but the discussion
with the patient about two IPGs or a single dual channel IPG with two 2×4 leads,

instead of the planned single 2×8 lead that has a similar footprint, also needs to have occurred before the surgery. Does the industry representative have these alternative devices with them? Is the patient prepped and draped for this possibility? Can this therapy use a rechargeable IPG, or only a non-rechargeable system, presently? These are all important aspects of the case that need to have been worked out ahead of time to render appropriate decision making in the OR.

A further example may suffice. While working with an orthopedic joint surgeon to place a peripheral nerve stimulator along the common peroneal nerve, above a postoperative neuromatous region that followed several prior orthopedic procedures, the plan is to bring the lead wire, or extension wire, up to the ipsilateral lower buttock region in the traditional location. However, once exposure is made, the appropriate region of the nerve dissected free and adequate securing of the lead and a strain relief loop with anchors is accomplished, it appears that there is now a problem – the realization that the patient has another device already implanted in an awkward lower buttock area. This was not appreciated in positioning and prepping the patient because the scar was well healed and a tattoo partially covered the area. A chart review confirmed that the patient had had a bladder stimulator (Interstim) placed in the past but it was unclear whether she still used it or whether it was still connected. Moreover, it was in the way of where the new IPG for the PNS needed to go.

While fluoroscopy was not needed in the original surgery, an x-ray was called for. But the patient was not positioned on the OR table in a manner conducive to obtaining an appropriate view on a cross-table lateral, and an anteroposterior (AP) was impossible because it was not an x-ray compatible table. The patient had to be moved by staff from under the drapes just enough to allow for the x-ray to be performed. The x-ray confirmed that the IPG was no longer connected to any lead. Alternative confirmation could have been made with a simple set of sterile needle electrodes or surface electrodes preoperatively to see if the device was turned on. Also, a decision could have been made preoperatively with the patient fully involved in terms of replacing this old IPG and moving the new IPG into a more appropriate location on the same side, or placing it on the other side if needed. Postoperatively, the patient admitted she had forgotten about the Interstim device. In any case, decisions need to be made, and proper planning ahead of time can eliminate major problems with device selection, location of incisions, appropriate removal or revisions, intraoperative testing and analysis of the physiology, and appropriate coverage for the patient, whether SCS, PNS, MCS, or DBS.

CONCLUSIONS

Several groups have addressed details and nuances of DBS, MCS, ONS, VNS, and SCS surgeries. This chapter emphasizes three principles in this regard: emphasizing the physiological target, being attuned to anatomical and patient subtleties, and assuring the availability of intraoperative resources necessary to decision making. The surgeon should have more than passing interest in neuro-modulation itself. This interest facilitates the attention to detail that is required in these cases. Managing the integration of MER in DBS surgery is a study in itself of sustained focus and attention to detail, supervising multiple disciplines, understanding the probabilistic facets of the neurophysiology, the patient condi-tion, the mechanics of the stereotactic equipment, the skill sets for using tar-geting software and appropriate positioning, the finer aspects of the burr hole, cannula, and electrode placements, and the role that medication and anesthesia may play in ultimately making the final decision that the electrode should go here or there in the end. DBS using MER, MCS with the use of SSEPs and motor mapping, and to some degree all other neuromodulation surgical tech-niques, requires this focus, these principles, and a deep belief in the therapy, along with its nuances and inadequacies, in order to perform the surgery.

REFERENCES

1. Bhananker SM, Posner KL, Cheney FW, et al. Injury and liability associated with monitored anesthesia care: a closed claims analysis. *Anesthesiology*. 2006;104:228–234.
2. Skipsey IG, Colvin JR, Mackenzie N, et al. Sedation with propofol during surgery under local blockade. *Anesthesia*. 1993;48:210–213.
3. Remifentanil 3010 Study Group. Remifentanil versus propofol as adjuncts to regional anesthe-sia. *J Clin Anesth*. 1998;10:46–53.
4. Petilon J, Hardenbrook M, Sukovich W. The effect of parallax on intraoperative positioning of the charite artificial disc. *J Spinal Disord Tech*. 2008;21:422–429.
5. Forstmann BU, Anwander A, Schafer A, et al. Cortico-striatal connections predict control over speed and accuracy in perceptualdecision making. *Proc Natl Acad Sci*. 2010;107:15916–15920.
6. Arle JE, Shils JL. Motor cortex stimulation for pain and movement disorders. *Neurotherapeu-tics*. 2008;5:37–49.
7. Picht T, Wachter D, Mularski S, et al. Functional magnetic resonance imaging and cortical map-ping in motor cortex tumor surgery: complimentary methods. *Zentralbl Neurochir*. 2008;69:1–6.
8. Bartos R, Jech R, Vymazal J, et al. Validity of primary motor area localization with fMRI ver-sus electric cortical stimulation: a comparative study. *Acta Neurochir (Wien)*. 2009;151:1071–1080.
9. Vallejo R, Kramer J, Benyamin R. Neuromodulation of the cervical spinal cord in the treatment of chronic intractable neck and upper extremity pain: a case series and review of the literature. *Pain Physician*. 2007;10:515–516.
10. Arle JE. *Surgical techniques in placement of the penta paddle lead in the cervical spine*. Las Vegas: Abstract-Poster presentation at North American Neuromodulation Society meeting; 2010.

Commentary on Surgical Techniques

Phillip Starr

UCSF, Department of Neurological Surgery, San Francisco, CA

Dr Arle provides a perspective on technical nuances, as well as underling principles and attitudes, relevant to the implantation of devices for neuromodulation. Many procedures in functional neurosurgery are new, and many new indications are under exploration. Thus, technical approaches are not standardized. Physiological localization is important for many functional neurosurgical procedures. However, for subcortical structures, anatomic targeting alone can be considered if several criteria are met: there exists a highly stereotyped relationship of structure to function for the target structure; the target structure's boundaries can be imaged with precision; and the chosen stereotactic guidance method can place the device at the image defined target with a very high degree of accuracy. For deep brain targets, at this time, the accuracy of standard stereotactic frames or 'frameless' neuronagivation-guided systems is such that a real time intraoperative method of correct electrode placement is needed.

As this chapter indicates, hardware-related complications such as infection, device migration, and device breakage are the bane of neuromodulation procedures. It is very important for practitioners to know the details of the hardware system implanted, anchor devices appropriately, understand how to place strain relief loops, and make sure there is sufficient tissue for device coverage. A recent multicenter trial of deep brain stimulation for Parkinson's disease, in which all surgeries were performed by experienced implanters, showed a hardware infection rate of 9.9% [1]. Development of smaller hardware components may decrease this risk in the future.

REFERENCE

1. Weaver FM, Follett K, Stern M, et al. Bilateral deep brain stimulation vs best medical therapy for patients with advanced Parkinson disease: a randomized controlled trial. *J Am Med Assoc.* 2009;301:63–73.

STUDY QUESTIONS

1. What training, residency and/or fellowship level, would best serve the 'field' of neuromodulation and in what medical disciplines would this be best accomplished? Why would improvements in this training improve access and quality of care in deployment of the therapy?

2. Make a list of the most common targets of stimulation in the subdisciplines of neuromodulation currently (e.g. Vim thalamus, dorsal columns, the vagus nerve, etc.) and consider what may be common to all of them. What about such commonality limits our ability to consider other targets of neuromodulation?

Chapter 13

Trials and Their Applicability

Louis J. Raso, MD
CEO, Jupiter Intervention Pain Management Corp. Jupiter, FL

INTRODUCTION

We as neuromodulators have a unique advantage. We are able to apply a technology that is both safe and testable. There are not many procedures that medicine offers that lets the patient test the therapy for a short period of time before sitting down with the practitioner and discussing whether this is the best option for the patient. This gives the patient and physician a tremendous advantage and opportunity. The patient has already failed conservative therapy, does not have a surgically correctable lesion, is psychologically stable, and has the cognitive ability to participate in the trial and implantation process.

The patient should have undergone appropriate diagnostic studies including computed tomography (CT) scans, magnetic resonance imaging (MRI) studies, and electromyography/nerve conduction velocity (EMG/NCV) etc. Treatable or correctable pathology should have been corrected. Once the above criteria have been met, the patient is ready for a trial of neuromodulation. I will discuss the two most common techniques, spinal cord stimulation (SCS) and intrathecal drug therapy.

INTRATHECAL INFUSION TRIALS

One of the biggest advances in the treatment of pain of malignant origin is the use of intraspinal narcotics. Due to this change, less neurodestructive surgery is performed as a palliative procedure on this patient population. With the success of intrathecal therapy for malignant pain, it was a smooth transition to utilizing this therapy for chronic non-malignant pain.

Over the past decade, intrathecal infusion pumps have become technologically sophisticated, and both Food and Drug Administration (FDA) approved drugs and other medications are being commonly used to treat a variety of chronic painful conditions. Basically, there are six types of implantable drug delivery systems ranging from a simple percutaneous catheter system to a totally implanted programmable pump with the ability to administer boluses of medications. Each system has advantages and disadvantages. It is the responsibility of the practitioner to determine the optimal and simplest system that will adequately treat the pain patient. With each system, cost increases as the

Essential Neuromodulation. DOI: 10.1016/B978-0-12-381409-8.00013-9

complexity of the system increases and the cost is likely to be scrutinized by insurers in the future. Once the system is implanted, monthly expenditures continue as the need for medication refills and refill supplies (i.e. needles, tubing, syringes) continue. The practitioner and patient are linked together and will likely remain in a long-term relationship.

The availability of implanted pumps for intrathecal infusion has provided an important alternative therapy for patients unable to tolerate side effects of systemically administered medications. A concentration gradient forms when a substance is infused into the subarachnoid space of the spinal cord, with the concentration being highest at the tip of the catheter and decreasing rapidly as the distance from the catheter tip increases. Drug concentrations should be lower in the CSF surrounding the brain and extremely low in the peripheral tissues. This is the reason for decreased side effects and sedation with intrathecally administered medications.

Prior to proceeding with the trial, the patient needs to have a thorough evaluation and correct diagnosis. The patient's current therapeutic regimen needs to be analyzed and deemed appropriate. The extent of the pain problem and disease and the likelihood of progression need to be taken into consideration. Any oncologic therapy in the case of malignant pain needs to be optimized prior to proceeding with the trial. A pre-implantation trial of spinal opioids is necessary to determine whether an implantable system will adequately relieve the patient's pain. Not all pain is relieved by spinal opioids. It is required that the opioids relieve the patient's pain on two separate occasions during the trial period and that the degree of relief be greater than 50%. In addition, the duration of relief should last at least twice the duration of the half-life of the therapeutic agent. In the case of morphine, the duration of relief should last 8–12 hours.

Trials of opioids may fail for a number of reasons. The medication may have been deposited in the wrong location; there may be psychological barriers to the testing process, i.e. major depression or a severe anxiety disorder. In addition, the incorrect dose of medication may have been chosen or the patient may have developed an extreme tolerance to opioids prior to the trial period. The pain may not be responsive to spinal narcotics, i.e. central pain syndromes and, if any questions remain, a trial of a placebo injection, may clarify the situation. The response to acute drug administration is predictive of long-term outcome and success in chronic intraspinal therapy.

Pump selection

Two major classes of implanted pumps have been designed for intrathecal infusion therapy. Constant flow pumps rely on a fairly constant pressure exerted by a gas on a diaphragm, forcing a constant stream of medication from the reservoir through a small orifice. Programmable pumps are extremely accurate and reliable. Any infusion rate within the volume capabilities of the pump may be programmed. In addition, complex continuous infusions with a daily bolus and/or varying infusion rates throughout the day may be programmed.

Indications

Patient selection for continuous intrathecal morphine therapy is not as well defined as for spinal cord stimulation. Most candidates are patients with intractable pain conditions that have failed to respond to conservative therapies and have failed chronic opioid therapy due to either excessive side effects and/or inadequate analgesia. Candidates should have a life expectancy of at least 3 months before an implantable pump is considered. There are no other simple rules for choosing appropriate candidates. Nearly all patients will receive substantial relief from a test dose or infusion trial. Many will develop significant tolerance to intrathecal opioids after 1–2 years of infusion therapy such that pain relief is only mildly improved from pre-implant levels.

Choosing patients for intrathecal baclofen therapy is much more straightforward than selection for intrathecal morphine therapy. Patients with spasticity of central origin who have failed oral baclofen therapy and at least one other appropriate antispasmodic agent are suitable candidates. The types of patients vary from those who will be able to ambulate more easily if the therapy is successful to those who will remain bedridden but will be easier to care for.

Patients are generally referred by a neurologist or neurosurgeon for an intrathecal baclofen trial. Single shot intrathecal test doses are given in a monitored setting where the patient's response to the injection can be observed on an hourly basis for 8 hours. Response may be rated with the Ashworth scale and by observing functional ability. Some patients rely on the spasticity of certain muscle groups for ambulation and arm movement. Function in these cases may worsen as spasticity improves. Test dosing begins at 50 μg. Doses of 75 μg, and 100 μg may be given on subsequent days if the initial dose produces no results. If the patient does not respond to a 100 μg test dose, implantation is not a viable option.

Patient Selection

Patient selection is the most important part of the pre-implant process. Experience indicates that a structured approach to the pre-implant phase facilitates the best possible experience for patient and practitioner. The clinician should establish a checklist algorithm to ensure that each patient has experienced a complete pre-implantation evaluation. Careful selection of patients improves outcome results and builds trust in the patient–clinician relationship.

Important components of the pre-implantation checklist include determining if the patient meets appropriate criteria. Appropriate criteria include:

- Ineffective oral analgesia with multiple oral or transcutaneous trials including dose titrations
- Intolerable side effects despite adequate rotation of opioids
- Intractable spasticity unrelieved by oral antispasmodics with improved Ashworth scores at baclofen test dose
- Access to care
- Functional analgesia during temporary trial infusion

- Psychological stability and realistic post-implant goals
- Patient acceptance.

An appropriate pre-implant checklist includes:

- Does the patient meet appropriate patient selection criteria?
- Has the patient been cleared by a knowledgeable psychologist?
- Did the patient achieve a 50% or greater reduction in pain during the trial infusion?
- Did the patient have appropriate occupational and physical therapy evaluations during the trial that showed acceptable functional gains?
- Did the patient have preoperative teaching?
- Was the infection risk assessment completed and discussed with the patient and family?

The patient must be physically able to have the pump and catheter implanted. In some cases, the patient may have had very extensive spinal or abdominal surgery which can increase the level of surgical complexity. The patient must be evaluated for access to the intrathecal or epidural space, and access to a site suitable for pump implantation. Patient positioning may be an issue secondary to anatomical factors. The patient will need to be placed in the lateral decubitus position with the pump side up during the implant procedure.

The risk of intraspinal catheters and pump implantation should not be ignored. The risks are manageable and, in experienced hands, are limited. General risks include infection, post-dural puncture headache, catheter-related epidural infections, granuloma formation, dose escalation and tolerance, and serious withdrawal symptoms due to pump or catheter failure (clonidine and baclofen).

NEEDLE PLACEMENT FOR INTRATHECAL PUMP PLACEMENT

Introduction

An intrathecal pump system is implanted via a sterile surgical procedure performed under local, regional, or general anesthesia. The implant procedure typically lasts from 2 to 4 hours. Prior to the procedure, a complete preoperative physical examination should be performed and the patient should be educated as to the procedure and the associated risks.

Preoperative Preparation

Before the implantation procedure, the physician and patient need to spend some time deciding on the side and location of the pump. It is usually placed in the left or right lower quadrant of the abdomen. It is placed so that it does not contact the iliac crest, pubic symphysis, ilioinguinal ligament, or the costal margin.

Preoperative antibiotics are given in the holding area. Usually, a cephalosporin is adequate and should be completely infused prior to the patient's transportation to the operating suite.

Anesthesia

Implantation of the catheter and subsequent pump placement may take place with the patient under general or local anesthesia with monitoring. Local anesthesia, in conjunction with sedation, is often preferred in the outpatient setting. When general anesthesia is chosen, the use of muscle relaxants is usually avoided until after the catheter is passed into the intrathecal space.

Procedure

Fluoroscopic Guidance

Fluoroscopy is utilized throughout the placement of the needle and passage of the catheter into the intrathecal space. It should initially be used to identify correctly the proposed lumbar level of entry.

Positioning

The patient should be positioned in the lateral decubitus position on the operating table with the side of implantation upward. At this time, C-arm fluoroscopy is brought into view and placed to obtain easy access to multiple views in different planes. Position the C-arm to permit an anteroposterior view allowing easy access to identification of the lumbar levels and the intrathecal space. Because this operation necessitates a middle lower-back incision for placement of the intraspinal catheter and a low abdominal incision for placement of the totally implanted subcutaneous drug administration system, both areas must be draped for surgical access. Split drapes are placed above and below the prepped back, flank, and lower quadrant of the abdomen.

Percutaneous Placement and Cut-Down Technique

The first task of this operation is intrathecal placement of the spinal catheter via a 15-gauge Tuohy epidural needle. There are two different techniques that the operating surgeon may perform. The ideal level for entry in CSF is below the conus medullaris, most commonly at the L3–L4 or L4–L5 interspace. In my opinion, the needle should never be introduced into the thecal sac above L2, for fear of it damaging the spinal cord. There are very few circumstances requiring entry above these levels. In a purely percutaneous approach, the needle is placed in the CSF prior to any surgical incision while, in a cut-down technique, an incision is made from L2 to L5 down through the subcutaneous tissues to the lumbar supraspinous fascia.

Needle Angle

The needle angle for entry into CSF is similar to placement of an SCS lead. Passage of the intrathecal catheter is facilitated if the angle is less than 30 degrees. If too steep an angle is obtained, the catheter is more difficult to pass and may be sheered at the needle tip.

Paramedian Approach

When performing most single shot CSF procedures, we classically utilized a midline approach. Unfortunately, this technique lends itself to a steep needle angle making passage of the catheter more difficult. A paramedian approach starting inside the pedicle line of the lumbar level below the desired entry level allows for a shallow entry with less of a chance of neural injury. Using sterile technique, mark the needle entry location parallel to the vertebral pedicle approximately 1–2 cm off of the midline and 1–1{1/2} vertebral levels below the interlaminar space through which the needle will pass. For example, using the pedicle of L4 as an entry point, aim the needle towards the midline at the L2–3 interlaminar space. Orient the bevel of the 15 T-gauge spinal parallel to the dural fibers and insert the needle under fluoroscopy. As the needle passes through the epidural space, a loss of resistance may be noted. The next loss will be accompanied by free flow of CSF following removal of the stylet. Following confirmation of placement with fluoroscopy, orient the needle bevel cephalad to permit passage of the catheter. The catheter is then passed to the level corresponding to the level of the pain generator.

Adverse Events

Adverse events of intrathecal therapy are classified as either system related or catheter related. System-related complications include cessation of therapy due to end of service life or component failure of the pump, change in flow performance or characteristics due to component failure, inability to program the device due to programmer failure or loss of telemetry, and catheter access failure due to component failure.

Catheter-related complications include, but are not limited to, changes in catheter performance due to catheter kinking, catheter breakage, complete or partial occlusion, catheter dislodgement or migration, or catheter fibrosis or hygroma. Hygroma can be difficult to diagnose and is sometimes confused with seroma or hematoma. Careful fluid analysis may give insight to the origin of the problem but, in some situations, the mixture of CSF with blood or serum makes the diagnosis difficult and a surgical exploration is needed to confirm the diagnosis.

Testing for therapy failure includes careful attention to end volume to computer predicted volume at pump refill, plain films of the catheter and pump, contrast studies of the catheter using the side access port, and rotor testing by x-ray analysis. Nuclear medicine studies of the pump using labeled xenon and other agents have been described but are not practical for clinical use.

Conclusions

The placement of the spinal needle for implantation of an intrathecal pump is the most critical step of the procedure and is also the step with the highest risk of neuronal injury. It is a skill that needs to be mastered if the implanting surgeon is to become comfortable with the therapy. Proper positioning, fluoroscopic guidance, and a paramedian approach with a shallow needle angle, all add up to a successful procedure and outcome.

The decision to use a spinal drug delivery system should build on the previous aggressive and optimized use of more conservative modalities. The patient's pain-related diagnosis, other medical diagnoses or general health, previous treatment, and future potential treatment options are all considered in the process of evaluating the patient for spinal drug delivery. There are many advantages to spinal drug delivery. Intrathecal opioid doses of 300 times the dose of oral morphine can be achieved. Clinicians have the choice of programmable or cost-effective, non-programmable pumps. The systems are completely implantable and are capable of combinations of simple infusions to a complex infusion pattern. The clinician also has the ability to add adjuvant drugs, resulting in decreased opioid requirements. The amount of drug delivered is precise and there is the future ability to deliver patient controlled analgesia (PCA) doses through the intrathecal space.

SPINAL CORD STIMULATION TRIALS

Introduction

Like intrathecal infusion trials, patient selection for trialing of spinal cord stimulation is essential to achieving a good outcome with permanent implantation. Trials can be informative to both the physician and patient and provide a significant amount of information about potential success of stimulation, location of electrodes, array design and type of generator required. The ultimate goal of the procedure will be to relieve pain by applying electrical stimulation to cause paresthesias covering and overlapping the areas of pain. The stimulation should not be painful and there should be no motor effects. Stimulation will not affect acute pain. Spinal cord stimulation is a reversible mode of neuromodulation that impairs vibratory sensation. It is necessary to perform a trial of sufficient length to forecast long-term success and identify a failure. A trial of 5–7 days is generally sufficient to provide the needed information while reducing the infection risk.

Preoperative Considerations

It is important to conduct a subjective review of an individual patient's functional, cognitive, and behavioral status. The patient needs to be able to tolerate the prone position. A complete discussion of common complications and informed consent needs to take place. Patients need to be aware of the potential for post-dural puncture headaches (1% risk of dural puncture), dural insult with the needle or lead, potential epidural blood patch, infection, bleeding, neurologic injury due to nerve or spinal cord trauma, inability to access the epidural space, and intolerance to the paresthesia. Patients need to be instructed on the trial process and be comfortable with the hardware and use of the equipment and the family needs to be present for the teaching. Patients need to sign a list of postoperative instructions.

Patients should be evaluated for the presence of coexisting diseases. Careful consideration should be given to morbid obesity, diabetes mellitus, coagulopathy, single lead pacemaker or pacemaker/defibrillator, and smoking.

SCS Patient Selection

Patient selection will be essential to developing a successful neuromodulation practice and increased patient satisfaction. Patients need to be evaluated for various factors. They need to have a chronic painful condition that has affected their daily life and their ability to maintain employment. They should have failed conservative (non-operative) management and, if mechanical pain exists, it should not be surgically correctable. The patient should be readily acceptable of the procedure and be fully cooperative and have the physical ability to manage the SCS system. The patient also needs to have a psychological evaluation by a trained therapist. Appropriate diagnostic studies prior to considering a SCS trial may include MRI or CT imaging, EMG, etc. If at all possible, an MRI is preferable. It is not generally necessary that a thoracic MRI is needed prior to a percutaneous trial, but if a laminectomy trial paddle lead is planned, a preoperative thoracic MRI is essential. The MRI and imaging studies will allow a correlation of the primary complaint to pathology seen on the imaging studies. The physician should be aware of multifocal pain complaints and all treatable/correctable pathology ruled out. The patient should have failed all conservative and other interventional therapies, and verification that this chronic condition has had a significant impact of pain on the patient's quality of life and activities of daily living.

SCS psychological Clearance

Every patient will need a complete psychological evaluation prior to proceeding with a trial of stimulation. This will necessitate a referral to a therapist and at least one complete therapy session. The psychologist will need to ensure that there are no acute contraindications to therapy, such as an acute psychosis, personality disorder, or untreated depression. There can be no active history of drug or alcohol abuse or illicit drug use. It should be part of the physician evaluation that the patient's pain complaints are felt to be real and related to diagnosis and that the patient understands, in general terms, the procedure (risks/benefits), and that they have reasonable expectations/motivation. It may be appropriate that in-office testing can be performed if the patient is well known to the office and a full evaluation is not felt to be necessary.

Percutaneous Versus Tunneled Trial Leads

A patient may undergo two types of trials. It is usually the implanting physician who decides whether the trial leads will be purely percutaneous or tunneled laterally to a stab wound and this decision is made prior to the trial. A percutaneous trial is a temporary trial as opposed to a tunneled or permanent trial. A permanent lead requires that it be performed in an operating room, whereas a temporary lead may be performed in an office-based fluoroscopy suite.

With a percutaneous trial, the leads are placed thru an epidural needle and advanced to the appropriate level. They are then connected to an external battery source and intraoperative testing is performed and the patient is brought to the recovery room for final testing. The trial typically lasts 5–7 days and, at the end of the trial, the temporary leads are removed and discarded. At that time, a detailed discussion takes place between patient and physician and results of the trial are evaluated and, if successful, a permanent implantation is scheduled.

With tunneled trial leads, implantation must be performed in an operating room. After successful placement of the leads and adequate intraoperative testing, the leads are sutured to the paraspinal or supraspinous ligament and are coiled into a midline incision with adequate undermining to allow a restraining loop to be coiled without any undue tension. They are connected to an extension and the extension is tunneled to a lateral stab incision at the opposite side of the planned generator pocket. After a trial lasting 3–7 days, the patient returns to the operating room and will either have the leads removed surgically or the extension will be removed and the leads will be tunneled to the newly created generator pocket.

When comparing the two techniques, it becomes obvious that percutaneous trials are less invasive requiring only needle insertion, while tunneled trials involve a surgical incision with the associated discomfort. The discomfort associated with a tunneled trial may require a longer trial or may cloud the evaluation of the patient's level or percentage of pain relief. This less invasive technique seems to be easier to obtain consent and patients tend to be more accepting of this procedure. They have already undergone many epidural injections and they have an understanding that the stimulator lead placement is similar to an injection with the exception that a 'wire' will be left in place for 5–7 days. If, during the trial, it is determined that a different location is required for permanent implant, it tends to be easier to alter the planned implant with a percutaneous trial.

If difficult anatomy is anticipated or encountered, a tunneled trial should always be considered. It may be necessary to include a tunneled trial in the operative consent with potentially difficult patients. There is a perceived increased risk of infection with a tunneled trial due to reopening the midline incision and allowing bacteria to enter thru the incision and also thru the extension site. In the USA, percutaneous trials are the more commonly performed trial.

Trial Evaluation

There are generally accepted criteria for what constitutes a successful trial. The patient should have obtained greater than 50% pain reduction from preoperative levels. This can be measured on a visual analog scale or by the perceived percentage reduction. Studies show that if a patient obtains greater than 70% relief, there is a better chance of long-term efficacy and success.

It is important to include functional improvement in the evaluation of the success of trial stimulation. It is important that specific milestones are discussed in the preoperative visit and that they are documented prior to the trial. Then, during removal of temporary leads, it is important to evaluate whether these functional milestones were met. These milestones can be something as basic as walking in the mall, sitting thru a meal at a restaurant, playing golf, or going to the movies. It is important to include family members, spouses or siblings in the evaluation. Once the evaluation is completed, the decision is made whether or not to proceed to implantation.

Operative Technique

Patient Positioning

The positioning of the patient is essential to being able easily to place the electrode in the epidural space. The trial is usually performed in the outpatient setting, should reproduce the work and home environment, and reproduce the effect of the permanent system. The level of approach dictates the positioning. For cervical placement, it is recommended that the patient be prone with pillows under the chest and the neck in the neutral position. The arms are in the neutral position at the patient's side. A pillow is placed under the patient's legs for comfort. Alternatively, a foam wedge can be placed under the patient's chest. In this case, the arms are placed on padded arm boards at a 90-degree angle. The patient can also be placed in the lateral position. For thoracic placement, it is recommended that the patient be prone with pillows under the abdomen. The patients can also be placed in the lateral position.

Patient Prep

The patient should be prepped and draped in an area wider than the proposed surgical site. The prepping solution is whatever the facility has chosen as beneficial. Occlusion drapes can be helpful and are often impregnated with prepping solutions such as iodine. Draping should be wide enough to include the planned surgical field. Fluoroscopy should be used in imaging the placement of both the needle and the lead. Anteroposterior (AP) and lateral views are beneficial in lead positioning, and no implant should be performed without fluoroscopic control. The C-arm should be draped prior to being positioned over the patient. Once aligned over the patient, anatomic landmarks should be identified. Obtaining a true AP image will enable the placement of both the needle and the lead. Align the C-arm in order to correct the parallax of the image. This is done by squaring off the end plates of the vertebral bodies and bisecting the pedicles with spinous processes. Verifying the anatomical level is done by identifying the last rib normally located at T12 and/or counting up from the iliac crest. A typical entry site for percutaneous leads is the L1–2 interspace when the electrodes are positioned at T10 or above.

Anesthesia

Anesthesia for SCS procedures varies from local anesthetics to general anesthesia. The ideal situation is that in which the patient can provide coherent feedback during the intraoperative testing. Often a combination of sedation and local anesthesia is used. Once the entry site is selected, local anesthetic is administered around the site in part to control bleeding and also to provide preemptive anesthesia.

Needle Placement

Using a paramedian approach, the epidural needle is inserted at an angle no greater than 30 degrees. Using an angle steeper than 30 degrees will hinder the passage of the leads and increase the risk of lead damage during insertion and manipulation. Confirm entry into the epidural space with the loss of resistance technique. Be careful for minimal loss of resistance, especially in the cervical spine or the elderly. Even the most skilled implanters will occasionally get a 'wet tap'. If that happens, entry at a different level is appropriate or, depending on the size of the 'wet tap', it may be prudent to return on a different day. Using fluoroscopy, the lead is passed through the epidural needle. The epidural needle can be rotated to change the direction of the beveled tip and control the direction of the lead. Once the first lead is inserted through the needle and advanced to the desired spinal level, the second lead is placed in the same fashion as the first, starting with insertion of the epidural needle. To facilitate directional control, the stylet handle is rotated while the lead is advanced. Lead location is verified by fluoroscopy to determine placement at the desired level. The lead contact typically is several levels above the desired area for concordant paresthesia. Intraoperative testing confirms that the electrodes are positioned correctly and the parameters are adjusted to cover the patient's pain area with paresthesia. This testing is done with the patient awake and alert so that he or she can provide feedback as to the location of the paresthesia coverage.

In preparation for testing, the trial cables are introduced into the field and connected to the leads. During testing, different electrode combinations and electrical settings are tried until the patient's painful areas are adequately covered. If the desired coverage cannot be obtained, it may be necessary to reposition the leads and test again in order to achieve the necessary coverage. Once the desired lead position has been confirmed, the trial stimulator is turned off and the trial cables disconnected from the leads.

Conclusions

The criterion by which the trial period is deemed a 'success' or a 'failure' has historically been a reduction of pain by 50%. In addition to the overwhelming painful sensations, chronic pain affects many aspects of a patient's life – psychological function, physical function, self-care, social interactions, and work status. Many physicians now consider improvements in activity, quality of life, and analgesic consumption during the trial just as predictive of favorable long-term outcomes as pain relief. Fifty percent pain reduction alone may inadequately describe the impact of the therapy. The pain should be neuropathic in nature. It is also important to ensure that the patient is not forced to fit a pain syndrome. Do not pressure the patient to accept the implant if the patient says that 'maybe I felt good' or if the patient had problems with the trial controller. One size does not fit every patient. It is important to design the correct system for the patient. Following a successful trial, the patient is brought back to surgery for the implant procedure. As a safety check – check stimulation, if only briefly, in the recovery room, especially with percutaneous leads. You will want to verify that the electrode did not move and that paresthesia is the same as it was in the operating room/procedure room. If the coverage is inadequate, the patient can be returned to the operating/ procedure room and the lead can be repositioned. Transferring the patient from the operating room/procedure table to the stretcher is a common cause of lead migration, especially if the patient has been positioned on a surgical frame. The highest risk for bleeding in the epidural space is in the first 24 hours. Patients should be monitored for changes in neurological function. A patient who complains of continuous paresthesia when the device is not activated should be evaluated immediately. Patients with epidural hematomas will complain of numbness and severe back or leg pain followed by weakness. This requires immediate neurological evaluation.

SELECTED READING

1. Waldman S. *Interventional pain management*. 2nd edn. Philadelphia: W.B. Saunders; 2001.
2. Synchromed II Clinical Reference Guide, Surgical Procedures, April 2004. Medtronic Inc.
3. Waldman S. *Atlas of interventional pain management*. 2nd edn. Philadelphia: W.B. Saunders; 2004.

Commentary on Trials and Their Applicability

Y. Eugene Mironer, MD

Managing Partner, Carolinas Center for Advanced Management of Pain, Greenville, SC

Neuroaxial opioids and neurostimulation are both very popular techniques of neuromodulation for chronic pain. The outcomes of the use of these modalities are quite good and gradually improving with greater experience and enhanced technologies. However, the results are still not where we would like to see them. We have especially stringent demands for these treatments. The reason for this is the unique ability to perform a trial of the modality prior to an offer of definitive therapy.

The success of neuromodulation, just like with the majority of other invasive modalities of treatment, largely depends on the proper selection of candidates. It involves, not only the nature of the pain condition, but also a psychological profile of the patient, presence of concomitant conditions possibly even with an overriding priority, anticipated progression of the disease, socioeconomic factors, etc. After the process of selection is completed, the patient is exposed to the trial of the selected modality. An adequate choice of the technique and duration of the trial, as well as proper assessment of the results, is crucial in the overall final results of the treatment. In the next step, the proper selection of technology, type of surgery, position of the hardware, etc., will help to accomplish desirable outcomes.

Before proceeding to a more detailed discussion on some aspects of the trial process, it is very important to stress a key understanding about the trial in general. It does not help to select good candidates for the definitive treatment, but rather helps to eliminate poor candidates. No trial, even the most successful, can be an absolute predictor of the great future outcomes. With this in mind, let us look closer on certain important choices in the trial of neuromodulation modalities.

NEUROAXIAL OPIOIDS

There are three different methods of performing a trial for neuroaxial opioid delivery: single injection, epidural infusion and intrathecal infusion.

The single injection option is widely used, despite being less reliable than continuous infusion [1]. It has a higher possibility of being affected by a placebo response, and does not permit access to the degree of pain relief during different levels of daily activity. Besides, the single injection option does not give an opportunity to titrate the dose to a desirable therapeutic level.

There are some advantages and disadvantages in using an epidural versus intrathecal infusion during the trial. While the epidural route reduces the risk of post-dural puncture headache and is easier to titrate, the intrathecal route is technically easier and closely reproduces the future treatment. In some patients with previous back surgeries, the intrathecal route is the only possible option. While some literature suggests a higher incidence of adverse effects during epidural infusion [2], another source shows no difference between the two options, except for cost-effectiveness [3].

The setting of the trial is also a subject for consideration. A single injection trial can easily be performed in ambulatory settings, which is not always the case for an infusion method. While an ambulatory setting will help to reduce the cost, it will also increase the risk due to the lack of continuous observation of the patient. This will also interfere with timely titration of the drug.

Duration of the trial can last from one day to months. In the latter case, there is a need to tunnel the catheter to reduce the possibility of infection. A longer trial leaves the possibility for more than one drug infusion, or for the use of a control/placebo infusion [4,5]. Nevertheless, neither an increased duration of the trial, nor the use of a placebo has proven more selective or efficacious [4,6].

The choice of the drug for the neuroaxial infusion trial is, at the same time, an easy choice and a difficult dilemma. On one hand, there is no variety of drugs approved by the FDA for use in implantable pumps. On the other hand, in current practice, many patients require a mixture of different agents to help achieve adequate pain relief. Due to this, the variety of choices for drug selection can stretch from using only morphine for all patients who are not allergic to it, to combining multiple agents from the beginning of the trial.

Finally, a decision has to be made regarding the starting dose of the neuro-axial drug and the appropriate reduction of oral opioids during the trial. This should be based on the degree of opioid tolerance in the patient, overall health, concomitant diseases, age, etc. Obviously, it is always preferable to have the initial dose too low rather than too high. When opioids other than morphine are used, it is imperative to know the relative potency conversion, which is different in neuroaxial delivery from the relative opioid potency in parenteral delivery. The author finds most of the existing sources on the subject confusing and, quite often, misleading, and would like to offer his version of relative intrathecal/parenteral opioid potency (Table 13.1).

NEUROSTIMULATION

There are a few different techniques that can be used to perform spinal cord stimulation trials. The lead can be inserted percutaneously or through a surgical laminotomy. The latter approach should be used only when it is technically impossible to perform the former one. The percutaneous trial may consist of simple placement of the lead or may involve a surgical incision to anchor it and to add an extension. The 'permanent' trial eliminates the need to reinsert a new

TABLE 13.1 Relative potency and starting intrathecal doses of commonly used opioids

Opioid	Parenteral relative strength (morphine = 1)	Neuroaxial relative strength (morphine = 1) (References)	Suggested starting intrathecal dose
Morphine	1	1	1 mg
Hydromorphone	6	2–3 [7,8,9]	0.5 mg
Meperidine	0.125	0.03 [10,12]	25 mg
Methadone	1	0.5 [11,13]	2.5 mg
Fentanyl	100	10–15 [10,14]	50 μg
Sufentanil	1000	82–100 [10,15]	10 μg

lead and to search again for the 'sweet spot'. However, based on reports in the literature, trial success rates, even in the best series, indicate that one out of five patients will require two surgical interventions – placement and removal of the lead – which will not help the pain condition.

The duration of the trial is another subject for consideration. A very short trial will reduce specificity, while a very long one will unnecessarily increase the risk of infection. At least one study suggested no reduction in specificity with a 3-day trial [16].

Current practice in neuromodulation includes different arrays of leads used during the trial – from a single lead to three leads. No reliable science shows the advantage of one method over the other. Nevertheless, it is absolutely clear that, if the same coverage can be achieved with a lower number of leads, then it will not only reduce the cost of the trial, but also decrease the surgical time and the rate of complications. It is also important to make a selection of the trial technique based on the plans for future implantation. For example, if the permanent implantation will be done with a surgical laminotomy placement, then a time consuming and labor intensive search for the 'sweet spot' may be unnecessary. The location of the laminotomy paddle lead may be completely different than a complex percutaneous trial array.

Maximum possible coverage of the pain area is very important in achieving success during the trial. A recent development in neurostimulation is the increased use of subcutaneously positioned leads for peripheral nerve field stimulation [17]. This allows for improved coverage in a variety of conditions, but mainly in axial low back pain. An even newer discovery is spinal–peripheral neurostimulation – an interaction between spinal cord stimulation and peripheral nerve field stimulation [18]. With these in mind, the use of these new modalities during the trial is a consideration, especially in patients with predominantly back pain.

The final challenge in the neurostimulation trial is the evaluation of its result. The commonly accepted standard of success is 50% or greater reduction of pain, as reported by the patient [19]. The standard method of measuring this is the use of the visual analog scale. However, one study showed a loss of sensitivity with the use of the visual analog scale versus scaling pain relief [20]. Changes in activities of daily living and a reduction in the consumption of analgesics are very important, but not compulsory components of a successful trial.

The unique ability to perform a trial that will replicate the planned modality of neuromodulation places an additional responsibility on the physician. The trial should be well thought out, planned for each individual patient and the method of neuromodulation, and provide the maximum amount of objective information. This will assure an improvement in the outcomes of neuromodulation therapy.

REFERENCES

1. Paice JA, Penn RD, Shott S. Intraspinal morphine for chronic pain: a retrospective, multicenter study. *J Pain Symptom Manage*. 1996;11:71–80.
2. Deer T, Winkelmuller W, Erdine S, et al. Intrathecal therapy for cancer and nonmalignant pain: patient selection and patient management. *Neuromodulation*. 1999;2:55–66.
3. Anderson, V.C., Burchiel, K.J. & Cooke, B. (2000). Randomized screening trial of intraspinal morphine for selection of patients for chronic opioid therapy. Presented at the Worldwide Pain Conference, San Francisco, CA, July 2000.
4. Krames ES, Olson K. Clinical realities and economic considerations: patient selection in intrathecal therapy. *J Pain Symptom Manage*. 1997;14:S3–S13.
5. Maeyaert J, Kupers RC. Long-term intrathecal drug administration in the treatment of persistent noncancer pain: a 3-year experience. In: Waldman SD, Winnie AP, eds. *Interventional pain management*. Philadelphia: W.B. Saunders; 1996:447–456.
6. Doleys DM. Psychological assessment for implantable therapies. *Pain Digest*. 2000;10:16–23.
7. Mironer YE, Grumman S. Experience with alternative solutions in intrathecal treatment of chronic nonmalignant pain. *Pain Digest*. 1999;9:299–302.
8. Parker RK, White PF. Epidural patient-controlled analgesia: an alternative to intravenous patient-controlled analgesia for pain relief after cesarean delivery. *Anesth Analg*. 1992;75:245–251.
9. Liu S, Carpenter RL, Mulroy MF, et al. Intravenous versus epidural administration of hydromorphone. *Anesthesiology*. 1995;82:682–688.
10. Van den Hoogen RHWM, Colpaert FC. Epidural and subcutaneous morphine, meperidine (pethidine), fentanyl and sufentanil in the rat: analgesia and other in vivo pharmacologic effects. *Anesthesiology*. 1987;66:186–194.
11. Mironer YE, Tollison CD. Methadone in the treatment of chronic nonmalignant pain resistant to other neuroaxial agents: the first experience. *Neuromodulation*. 2000;4:25–31.
12. Tamsen A, Hartvig P, Fagerlund C, et al. Patient-controlled analgesic therapy: clinical experience. *Acta Anaesthesiol Scan (Suppl.)*. 1982;74:157–160.
13. Eimerl D, Magora F, Shir Y, et al. Patient-controlled analgesia with epidural methadone by means of an external infusion pump. *Schmerz/Pain/Douleur*. 1986;7:156–160.

14. Chrubasik J, Wust H, Schulte-Monting J, et al. Relative analgesic potency of epidural fentanyl, alfentanil and morphine in treatment of postoperative pain. *Anesthesiology.* 1988;68:929–933.

15. Van der Auwern A, Verborgh C, Camu F. Analgesic and cardiorespiratory effects of epidural sufentanil and morphine. *Anesth Analg.* 1987;66:999.

16. Mironer, Y.E. Efficacy and selectivity of short-term percutaneous spinal cord stimulator trial. Presented on APM 17th Annual Scientific Meeting, San Diego, CA, November 1998.

17. Barolat G. Subcutaneous peripheral nerve field stimulation for intractable pain. In: Krames E, Peckham PH, Rezai A, eds. *Neuromodulation.* Elsevier; 2009.

18. Mironer, YE, Hutcheson, KJ, LaTourette, PC, et al. Prospective study of the interaction between epidural and subcutaneous leads: a new breakthrough method in providing axial pain coverage. In materials of: Neuromodulation: 2010 and Beyond. NANS, Las Vegas; 2009: 172.

19. Turner JA, Loeser JD, Bell KG. Spinal cord stimulation for chronic low back pain: a systematic literature synthesis. *Neurosurgery.* 1995;37:1088–1096.

20. Mironer, YE, Skoloff, E & Hutcheson, JK. Scaling pain relief vs. visual analog scale in assessment of spinal cord stimulation treatment success: a prospective study. In Materials of NANS Annual Meeting, Las Vegas, NV, December 2008: 154.

STUDY QUESTIONS

1. Consider how different reimbursement scenarios might alter the frequency and method of performing trials in neuromodulation. Is there an optimal way trials should then be used?
2. Does it seem feasible to develop a universal trial paradigm – accurate across all patients? Why or why not?

Chapter 14

Limiting Morbidity

Konstantin V. Slavin, MD

Department of Neurosurgery, University of Illinois at Chicago, Chicago, IL

Surgery of any kind carries risk of complications and, to date, there is no surgical intervention that is completely safe. As ever, the goal of the surgeon in this regard has been twofold – to prepare the patient for the possibility of complications, as a part of preoperative preparation and informed consent and, perhaps more importantly, to develop strategies to minimize or avoid complications to begin with. Since even the best orchestrated process may result in adverse events, the surgeon's job is to know how to deal with these events once they occur so that negative end-results, measured by morbidity and mortality, are minimized.

When it comes to neuromodulation, an attractive feature of the field, along with reversibility, adjustability and testability of action, is its non-destructive nature. Most devices that are implanted to modulate activity of the nervous system, whether electrical or chemical neuromodulation, require relatively straightforward interventions – and yet the most difficult and, possibly most important part of the process has often been considered to be patient selection, as the proper selection of appropriate candidates greatly affects the short- and long-term outcome of neuromodulation procedures.

So how does one limit morbidity associated with placing neuromodulation device(s) in the human body? Mainly knowing of possible complications and procedural traps beforehand and taking steps to avoid them, goes a long way. This principle applies to all stages of the implantation procedure, from choice of best anesthesia and most appropriate neuromodulation equipment, to postoperative care of neuromodulation patients and troubleshooting during the follow-up period.

Listed below are the main groups of complications that will be discussed along with the recommended steps for their avoidance:

1. Infection
2. Hemorrhage
3. Injury of nervous tissue
4. Cerebrospinal fluid leak
5. Placing the device into the wrong compartment
6. Hardware migration
7. Hardware erosion

Essential Neuromodulation. DOI: 10.1016/B978-0-12-381409-8.00014-0

8. Hardware malfunction/fracture/disconnection
9. Granuloma formation
10. Other issues (radiographic safety, drug overdose, etc.).

The list of issues presented here is not all-inclusive. Other complications and side effects may occur at every stage of neuromodulation – and readers are encouraged to approach each patient and each intervention individually so special patient- and procedure-related peculiarities might be taken into account.

INFECTION

Although it would be expected for neuromodulation procedures to have a higher infection rate due to the presence of implanted devices and frequent need in keeping some part of the system externalized (like electrodes or extensions during neurostimulation trials and temporary epidural or intrathecal (IT) catheters used in screening for subsequent pump insertion), the incidence of infection may be equal or slightly higher than that observed in most surgical interventions. This may be, at least in part, explained by the elective nature of neuromodulation procedures. Since insertion of the neuromodulation device is usually done in a planned fashion, there is always a time interval that would allow the implanter to check the patient for ongoing infection or bacterial colonization, and to treat this infection to decrease the chances of the new device becoming colonized.

This would include examining a prospective patient for signs of infection and getting routine laboratory tests (peripheral white blood cell count, urinalysis) for active infection – and postponing surgery, including trial procedures, until any infection is treated adequately. In addition, in patients with known exposure or infection with methicillin-resistant *Staphylococcus aureus* (MRSA), it is generally recommended to obtain nasal swabs and treat those who are positive with chlorhexidine showers until the colonization and/or infection are eliminated. It is usually recommended to follow one's institutional protocol for addressing MRSA infections if one exists.

In addition to the above measures, a recent comprehensive review of the literature and post-marketing surveillance data suggested controlling blood glucose levels in the perioperative period and stopping tobacco use for 1 month prior to surgery would further mitigate against infection [1]. In general, the presence of healing surgical incisions or open wounds serves as a reason for postponing the elective neuromodulation surgery – with the possible exception of non-infected pressure sores in patients with severe spasticity who are scheduled for IT baclofen pump insertion, as these sores are unlikely to heal until spasticity is aggressively treated.

In terms of antibiotic prophylaxis, there is no consensus among neuromodulation implanters regarding the duration, route of administration and choice of antibiotic regimen. An almost universally accepted practice of administering preoperative antibiotics, usually second generation cephalosporins, within 2 hours of making a surgical incision represents more of a general standard that applies to most surgeries done in the operating room. We do not change such routines in our neuromodulation procedures and the choice of antibiotics is usually determined by the patient's allergies and local hospital guidelines.

In order to sterilize the surgical field, we clean the surgical area with two sets of preparations containing iodine povacrylex (0.7% available iodine) and 74% isopropyl alcohol (DuraPrep, 3M, St Paul, MN) or, if no contact with nervous tissue is anticipated, 2% chlorhexidine gluconate/70% isopropyl alcohol formulation (ChloraPrep, CareFusion, San Diego, CA). We also routinely cover the entire surgical field with iodophor impregnated adhesive (Ioban, 3M) unless the patient is known to have an allergy to any of the mentioned components.

Postoperative antibiotic administration is probably the most controversial aspect of care as implanters do not agree with each other on the most rational, safe and effective way to approach infection prophylaxis. Specialists from the field of infectious diseases generally recommend keeping postoperative prophylactic antibiotic administration to a minimum, but most implanters give it either over 24 or 72 hours after device insertion, or for the duration of the trial. We are not aware of any scientific evidence that such practice reduces the risk of infection but, in our own practice, we continue prescribing oral antibiotics to all neuromodulation implant patients for the duration of the trial, when the part of implanted device is externalized, and several days after the permanent implantation. This way, patients with IT pumps and vagal nerve stimulators (VNS) receive a total of 5 days of postoperative antibiotics, those with deep brain stimulation (DBS) where trial or inter-stage interval lasts for 3–5 days receive 8–10 days of oral antibiotics, and those with spinal cord stimulators (SCS) and peripheral nerve stimulators (PNS) where trials are usually 7-days long end up getting a total of 12–14 days of oral antibiotics. With clear understanding of the lack of scientific support for this particular approach, we have been reluctant to change it since our incidence of infection after all neuromodulation procedures is well below 1%.

A recently suggested use of local antibiotic administration around the site of implantable pulse generators may be an attractive option for reduction of surgical infection [2] but, so far, this technique has not gained expected popularity.

Handling of implantable hardware deserves separate mention – and here one may extrapolate recommendations for handling ventriculoperitoneal shunts published almost 20 years ago [3]. These, among other things, include suggestions to do implant surgery as a first case of the day or as early during the day as possible, minimizing the number of people in the operating room, obtaining meticulous hemostasis and quality skin closure and, perhaps equally important, opening sterile packaging of the implanted hardware at the last moment [3].

One of the avenues that has not been explored so far would be to coat implantable components with antibiotic, similar to currently used shunt catheters, ventricular drains and venous access devices – and I am sure the device manufacturers have already considered (or are considering now) this as an additional approach further to reduce infection rates.

Once a device is implanted, infections do happen unpredictably – sometimes due to hematogenous spread of microorganisms from some other identifiable source, sometimes from local skin dehiscence or erosion over the implanted components, and sometimes from an unknown source in the absence of bacteremia. The recommended approach in either case would be to remove the implanted hardware and then re-implant it a few months later. Although multiple reports exist in the literature documenting successful salvage of infected neuromodulation hardware, our preference is to remove the device(s) altogether [4] as systemic antibiotics do not penetrate all crevices and material interfaces of implants, and therefore eradication of infection is typically very challenging.

HEMORRHAGE

Hemorrhage may occur either during or after an operation – and since bleeding involving areas in or near the central nervous system may produce severe and sometimes irreversible neurological deficits, prevention of hemorrhage is extremely important in all neuromodulation procedures. There are several steps that have to be undertaken in order to limit hemorrhage-related morbidity; they include preoperative patient preparation, surgical planning, and procedural details aimed at obtaining adequate hemostasis.

In terms of patient preparation, it is extremely important to check each patient for the presence of coagulopathy or thrombocytopenia. This is usually accomplished by obtaining blood coagulation tests (prothrombin and partial thromboplastin times) and a complete blood count that includes platelets. Anticoagulants, such as heparin, enoxaparin and warfarin, should be stopped ahead of time – and the coagulation tests should be repeated and confirmed to be normal so there are no intraoperative surprises. We prefer stopping heparin 6 hours prior to elective surgery, low-molecular weight heparin (such as enoxaparin) 24 hours prior to surgery, and warfarin 5 days prior to surgery, sometimes bridging the patient with intravenous heparin for 2 days before surgery. In terms of antiplatelet agents, the existent guidelines for neuraxial procedures by the American Society of Regional Anesthesia and Pain Medicine postulate that these medications, including aspirin and other non-steroidal anti-inflammatory drugs (NSAIDs), thienopyridine derivatives (ticlopidine and clopidogrel) and platelet glycoprotein IIb/IIIa antagonists (abciximab, eptifibatide, tirofiban) exert diverse effects on platelet function and that pharmacologic differences make it impossible to extrapolate between the groups of drugs regarding the practice of neuraxial techniques [5]. Based on analyzed current literature, these guidelines recommend stopping ticlopidine 2 weeks and clopidogrel 1 week before intervention. Platelet aggregation normalizes 48 hours after abciximab and 8 hours after eptifibatide and tirofiban discontinuation. They also state that aspirin and other NSAIDs do not create a level of risk that will interfere with neuraxial procedures [5]. These guidelines, however, differ from usual surgical practice and we routinely ask our patients to stop aspirin 10 days prior to surgery, and ticlopidine and clopidogrel 2 weeks prior to surgery, including insertion of an IT catheter or pump, DBS, SCS, VNS or PNS electrodes or generators.

In order to reduce the risk of hemorrhage during the procedure, common-sense dictates a need to minimize manipulation with sharp instruments and needles during device insertion. However, it is very important to maintain adequate visualization of the surgical field and, sometimes, keeping incisions too small may be counterproductive as they would limit adequate access for hemostatic instruments. Hemostasis at open incisions is best accomplished with electrical coagulation and, in the beginning of surgery, before any metal implants are in place, monopolar coagulation with a Bovie device is perfectly appropriate (with the known need to avoid coagulation of skin and excessive tissue charring). Once the electrode, pumps or generators are inserted into the body, the use of monopolar coagulation is not recommended any more because of potential distant thermal and electrical effects. At this stage, bipolar coagulation should be used to stop bleeding.

In intraspinal procedures, it is not recommended to try to overcome resistance if it is encountered during electrode or catheter insertion forcefully. It is also not recommended to insert percutaneous electrodes at or through the levels where epidural space is filled with scar tissue as it makes surgery both difficult and dangerous. In these situations, where there was a previous laminectomy or laminotomy, there is a risk of not only hemorrhage but also possible dural penetration, particularly if the dura was violated during the original intervention.

For patients with previous surgery at the site of planned electrode location, it would be more prudent to consider placing paddle electrodes rather than trying to insert percutaneous ones. A procedure that seems to be particularly prone to cause epidural hemorrhage however, is a revision or removal of paddle-type SCS electrodes. For these patients, a somewhat longer postoperative observation may be needed as epidural hematomas may develop and/or become symptomatic as late as several days after the surgery.

For intracranial DBS electrode insertion, it appears that one of the ways to reduce the incidence of intracranial hemorrhage is to place electrodes through a gyrus – sulcal insertion has increased the incidence of cortical and subcortical hemorrhages [6]. The incidence of ventricular hemorrhages also increases with transventricular trajectories [6] and, if possible, the trajectory for electrode insertion should pass next to the ventricle rather than through it. Although male gender, age and diagnosis of Parkinson's disease were found to be significant risk factors for intracranial hemorrhage, hypertension seems to be the most consistent risk factor [7]. Additionally, the use of microelectrode recording (MER) in DBS targeting resulted in a higher incidence of intracranial bleeding [8] and the highest incidence of hemorrhage was observed in those who had hypertension and MER. We use MER routinely for DBS surgery and, in order to reduce the incidence of intracranial hemorrhage, we try to minimize the number of electrode passes and pay special attention to blood pressure control during surgery with a dedicated anesthesiologist or nurse anesthetist monitoring patients undergoing DBS surgery and treating elevated blood pressure while avoiding sedation or agents that would adversely affect MER findings.

Needless to say, a decision to remove a neuromodulation device and/or evacuate a hematoma depends primarily on the degree of symptomatic impairment. Asymptomatic hemorrhages do not require surgical intervention but have to be monitored clinically and radiographically. Those who present with symptoms of cord compression have to be decompressed emergently. Management of intracranial hemorrhages depends on the size and location of the bleeding as well as on the risks and benefits of possible surgical intervention.

INJURY OF NERVOUS TISSUE

The principle of neuromodulation involves non-destructive reversible modification of neural function through electrical, chemical or other means. The main advantages of surgical neuromodulation as opposed to non-surgical approaches, such as systemic medications or transcutaneous stimulation, are the high selectivity of action and significantly lower amount of electrical energy or chemical needed to achieve a similar response. This comes from direct proximity of surgical neuromodulation devices to their respective targets within the nervous system – the brain, spinal cord, peripheral and cranial nerves, and cerebrospinal fluid (CSF). Unfortunately, such proximity comes with a price – namely, the risk of inadvertent damage of the nervous system during implantation of neuromodulation components.

Of all instances of nervous tissue injury related to neuromodulation, most occur during the insertion procedure, and each particular location carries its own set of precautions and concerns. Only some of these concerns are listed here. In general, they include direct injury of the nerves, nerve roots, spinal cord, and brain, compression of the neural structures by the device or the scar around it, and long-term neural injury from the presence of the implanted device.

During insertion of SCS electrodes, care must be taken to avoid injury of the spinal cord and the nerve roots. Keeping the patient awake during these procedures solves only part of this concern since the mechanical irritation of nerve roots is usually perceived by the patient as sharp pain or acute discomfort – but spinal cord compression or even penetration are usually painless. Therefore, use of intraoperative fluoroscopy is recommended to keep the electrode in the middle third of the spinal canal away from the nerve roots. It is also recommended that one avoid using excessive force during electrode advancement as this may cause significant focal injury to the underlying spinal cord. Entering the epidural space with percutaneous electrodes is usually safer over the midline under the cephalad lamina or the base of the spinous process as the epidural space is usually largest in that location.

Dural penetration during implantation of percutaneous SCS electrodes is associated with multiple problems. CSF leak, spinal headaches and electrode migrations are discussed in the next sections, but additional risk involves potential direct injury of the spinal cord which may be unrecognized until the patient's neurological function is tested or until the stimulator is turned on.

Compression of the spinal cord by a SCS electrode is less likely to occur with percutaneous electrodes and more common with paddle leads. Here, the important issue is to assess the size of the spinal canal and presence of any kind of stenosis, as well as clinically symptomatic or asymptomatic compression prior to the electrode implant. Interestingly enough, in those cases where a patient suffers neurological deficit after having a paddle electrode placed over the spinal cord in the presence of a compressing lesion, it is possible, in retrospect, to conceive that some or all of these symptoms that led to electrode insertion may have been indeed caused by that compressive phenomenon. Therefore, it should be mandatory to perform adequate preoperative imaging with MRI or CT myelography prior to placing paddle-type electrodes in order to avoid these situations.

Finding symptomatic or asymptomatic compression, or even subclinical stenosis, at the level of paddle insertion should not prevent SCS placement, but it may be prudent to combine electrode insertion with decompression through a laminectomy if the concern is established.

Another possible mechanism of spinal cord injury during paddle electrode insertion is related to inadequacy of dissecting tools used in the insertion process. Existing dissectors, sometimes called dural separators, carry a certain risk of spinal cord injury themselves as the material they are made of is either too hard (metal or hard plastic covered with thin layer of softer plastic or silicone) producing undue pressure either under the instrument tip deep under the lamina, or under the curvature of the tool directly in a laminectomy/laminotomy opening, or too soft (made of soft plastic attached to the rigid handle) to the point that this plastic tip buckles and folds over itself when resistance is encountered. With this in mind, new insertion tools are now being developed – putting most of the advancing force to the lateral recesses of the spinal canal rather than its middle portion, and shifting possibly injurious manipulation away from the spinal cord itself.

When an IT catheter is implanted, the injury to the nervous tissue may be caused by the needle through which the catheter is introduced or by the catheter itself. In order to reduce the possibility of unrecognized injury, use of fluoroscopy is strongly recommended as the midline of the lower lumbar spinal thecal sac has the most room between nerve roots. Also, the needle insertion site should be chosen in a way that the needle tip enters the spinal canal below the level of L1 (below the spinal cord conus). In addition, some implanters prefer keeping the patient awake during needle insertion so the patients' complaints of discomfort, pain and their neurological function may guide the operator in the direction and depth of needle placement. We routinely perform catheter and pump implantation under general anesthesia however, as we feel the concern of cord or root injury is outweighed by a chance of a patient's voluntary or involuntary movement during needle insertion causing the needle tip to injure the underlying neural structures.

In terms of catheter direction, we prefer advancing it in a cephalad direction as the pattern of CSF circulation inevitably creates a drug concentration gradient thereby supporting the old postulate that the catheter should end at the level where medication is needed. The argument of those who aim the catheter in a caudal direction is that the risk of cord injury with that approach is immensely lower and that the catheter tip granuloma (see below), should it form, will not be compressing the spinal cord. This approach has rationale, particularly when one considers the possibility of intramedullary catheter insertion [9]. With this in mind, we suggest checking free CSF flow from the catheter – without active aspiration – at every stage of catheter insertion, including its advancement through the needle, anchoring, tunneling, etc. This approach ensures catheter patency and the subarachnoid tip, position of the catheter tip and eliminates the possibility of an intramedullary or 'extrathecal' catheter location. It must be mentioned, however, that a patient whose catheter was placed into the spinal cord by a less skilled implanter did not develop any symptoms from it immediately. Leg weakness came several days later, apparently as a result of intramedullary morphine infusion, and the symptoms resolved once the catheter was removed [9].

For DBS cases, brain injury other than from hemorrhage described earlier is quite rare, although non-hemorrhagic infarctions have been described in the past. Cortical and subcortical non-hemorrhagic lesions usually develop from superficial coagulation of the pial surface. In addition to this risk, coagulated pia also becomes harder and requires more force to advance cannulas and electrodes into the brain, therefore, we recommend using focal application of Gelfoam (Pfizer, New York, NY) with thrombin rather than electrocoagulaiton if minor cortical bleeding is encountered.

Postoperative imaging studies in DBS patients frequently reveal changes in T2 images [10] and this may or may not be an asymptomatic finding. In our experience, development of changes on T2 and FLAIR MRI images often correlates with the number of passes during DBS implantation and, therefore, one should strive to minimize MER and DBS electrode passes as the brain may not be as 'forgiving' as was previously thought. Moreover, penetration of the brain with microelectrodes and DBS electrodes produces some transient neurological effects – sometimes temporarily improving their condition due to a 'microlesioning' effect [11], but making initial programming more difficult and possibly necessitating additional follow-up visits for reprogramming once this improvement wears off.

An injury to the peripheral nerve during PNS procedures may be more likely if the nerve is tethered to surrounding tissues due to scar from previous interventions. This was noted many years ago, when PNS devices were implanted through the open exposure procedure [12], but the concern remains even now when most PNS electrodes are inserted via the percutaneous approach. It is possible, at least theoretically, to reduce the chance of nerve injury if the nerve course is identified with ultrasound guidance. Live ultrasound may help track the insertion needle during PNS electrode placement in extremities [13], inguinal area [14], and occiput [15] and this approach seems to be gaining popularity among implanters according to some recent unpublished surveys.

In addition to surgical injury, neural tissue damage may occur during long-term follow up of neuromodulation patients. Sometimes this is related to ongoing progression of their underlying degenerative or neoplastic disease, sometimes it is caused by a cord compression due to catheter-tip granuloma but, in rare cases, the cause of neurological dysfunction may be the scar tissue that formed around the electrode over the many years. A recently published description of two such cases [16] illustrates such a possibility and underscores the relative youth of the entire neuromodulation field and our lack of knowledge regarding its long-term effects due to the fact that SCS and PNS have been around for less than 45 years and IT pumps, DBS and VNS were introduced even more recently.

Delayed neurological deficit was one of the main reasons for the old open approach for PNS to be abandoned or modified as the nerves targeted by PNS would exhibit perineural fibrosis [17] or nerve ischemia [18], particularly if the electrode were wrapped around the nerve to avoid migration and assure better contact with tissue. In many cases, however, an open approach is taken because a neurolysis is first performed.

CEREBROSPINAL FLUID LEAK

As the majority of neuromodulation procedures are done outside CSF-containing spaces, CSF leaks occur almost exclusively with IT pumps. However, even though we are not aware of any CSF leaks in PNS and VNS procedures, both SCS and DBS patients have been described having CSF leaks in the postoperative period.

When one implants a catheter into the subarachnoid space, CSF is expected to leak around the catheter, simply because the needle that we use for catheter insertion is larger that the catheter itself. However, CSF leaks out even more if the hole created by the needle is not partially occupied by the catheter. Therefore, it is imperative to minimize the number of dural punctures during IT catheter insertion and the use of fluoroscopy along with adequate positioning of patients on the operating table serve as important adjuncts during the procedure.

The previously suggested purse-string suture around the site where the catheter penetrates the fascia is intended to decrease external CSF leak [19], but it has to be placed before removing the needle through which the catheter is inserted. More important, however, is a multilayer tight closure of surgical incisions that promotes fast and complete healing without over-tightening of the sutures that may result in tissue necrosis and inhibit the healing process.

One has to keep in mind that CSF may leak not only directly from the sub-arachnoid space through a true fistula but rather from disconnected or punctured IT catheters. Since catheters do not 'heal', it is strongly recommended to avoid handling them with any sharp instruments. It is also important to be careful with needles and sutures around implanted hardware as even a minor violation of catheter integrity may result in persistent CSF leakage.

The skin closure may be complicated in cachectic or elderly patients, although reported patients who developed CSF leaks after DBS surgery were relatively young – a 62-year-old patient with Parkinson's disease [20] and a 51-year-old patient with tremor [21].

In addition to external CSF leaks, CSF may accumulate in subcutaneous tissues resulting in 'seroma' and this may occur under the lumbar catheter insertion site or, more commonly, around the pump or generator, or become absorbed in the epidural space or soft tissues. In the latter case, the only symptom of CSF leak may be post-dural puncture headaches ('spinal headaches') and, in some cases, these headaches may become quite disabling. A standard approach to management of spinal headaches includes administration of caffeine, orally or intravenously, analgesics, aggressive hydration and placement of abdominal binder. If these interventions fail, epidural injection of autologous blood is considered the next treatment step [22].

There are many issues related to CSF leaks (both internal and external). First, these headaches may be very severe, limiting the patient's mobility and producing a great deal of discomfort. Sometimes these headaches are associated with nausea and vomiting necessitating patient admission to a hospital for intravenous hydration and pain control. Second, the CSF leaks significantly increase risk of infections. CSF leakage is a well-known risk factor for surgical site infections [1,20] and an untreated CSF fistula may indeed result in the development of meningitis. Finally, external CSF leaks and seroma formation usually mean that medication intended for IT delivery leaks out instead of staying in the IT space and this translates into development of 'underdosing' symptoms that may even reach a degree of withdrawal. This phenomenon is best treated with an immediate restart of oral or parenteral analgesics or antispasmodics that have to be continued until the CSF leak is controlled.

In terms of prevention of pseudomeningocele or lumbar seroma, it is important to overcome the temptation to insert the catheter through the level of a previous laminectomy. Obviously, lack of bone over the thecal sac makes catheter insertion much easier, but the same lack of bone means lack of epidural fat that could tamponade the hole in the dura, with or without the catheter traveling through it.

If the CSF leak does develop, and there is no reason to suspect a hole or disconnect in the catheter, one may consider surgical repair of the leak with either oversewing the fascia or even placing a suture on the dura itself. Prior to that, one may try an epidural blood patch as it has a high success rate – staying aware of the catheter or electrode location so that the needle does not damage hardware in place. Since the epidural space may be filled with CSF, it may be difficult to discern it from intrathecal space, and therefore radiographic guidance with contrast injection prior to introduction of autologous blood has been recommended [23].

PLACING THE DEVICE INTO THE WRONG COMPARTMENT

In using the term 'wrong compartment' we do not mean missing the intended target for stimulation or drug infusion but rather placing the device into a wrong anatomical location altogether. For example, SCS electrodes are designed to be placed in the dorsal epidural space, overlying the dorsal columns of the spinal cord. The choice of this location is not serendipitous: in the past, these electrodes have been placed into the intrathecal space but were difficult to maintain and keep in place and CSF leakage along the electrode track was rarely salvageable. In the epidural space, both anterior and lateral locations have been tried but eventually abandoned in favor of the current dorsal location, primarily due to the predictability of effect and wide therapeutic window.

During electrode implantation, direct visualization of the dura with paddle electrodes serves as confirmation of correct localization for electrode placement but, with percutaneous electrodes, the situation is less straightforward. Over the years, we have seen perfectly appearing radiographs associated with lack of any benefit from stimulation. In each case, however, the electrode(s) were outside of the dorsal epidural space – they were either anterior to the thecal sac or extraspinal.

The only way to appreciate the wrong location of the electrode before the procedure ended would be to get a lateral radiograph of the spine and, at the beginning of one's career, it is strongly recommended that a confirmatory fluoroscopy image for each percutaneous electrode be made to verify that it projects immediately anterior to the inner surface of the laminae and distant enough from the posterior edge of the vertebral bodies.

The intrathecal position of the electrode is usually deduced from CSF dripping from the needle hub. Two other clues for intrathecal electrode location are the pattern of electrode advancement – when the electrode array moves from side to side during longitudinal manipulation rather than 'snaking' forward when the electrode follows its curved tip – and the very low threshold of stimulation, usually below 1 V or 1 mA, suggesting that there is no resistance to stimulation usually provided by intact dura. In situations like this, it is recommended to withdraw the catheter and the needle and then re-enter the epidural space one level higher or at the same level, but from the opposite side. Re-entering the epidural space one level lower may not work as the electrode may 'find' the hole in the dura as it is advanced towards its target.

Similarly, low thresholds but no CSF leak would be observed if the electrode ends up in the subdural space – a potential space between the dura and arachnoid membrane. This is a very rare occurrence, but most experienced implanters would recall a case or two when this peculiar situation occurred.

Just as one would not want the SCS electrode in the subarachnoid space, the IT catheter is definitely preferred to be in it. An intramedullary catheter location was described in the previous section – it is usually discovered due to deleterious effects of local toxicity that infused medications exert on the spinal cord tissue. In contrast to this, 'extrathecal' location of the IT catheter (epidural or extraspinal) is usually encountered in situations when the patient requires significantly higher than expected doses of medication for achieving the desired effect. This is not surprising as the equianalgesic dose of epidural opioids is about 10 times higher than with IT delivery. For baclofen, epidural administration of medication results in minimal, if any, improvement in spasticity – greatly differing from the patient's response observed during the IT trial. As mentioned earlier, checking free CSF flow from the distal catheter end at every stage of implantation would provide sufficient assurance of IT location of the catheter tip.

When it comes to brain stimulation, a lack of desired stimulation effects or absence of expected MER patterns should make one recheck the coordinates and compare them with those chosen in the stereotactic frame. Because the cannula and electrode may bend, it may be worthwhile in some instances to take an anteroposterior radiographic view of the electrodes in addition to the usual lateral view of the head.

HARDWARE MIGRATION

Neuromodulation devices are often implanted into or across highly mobile parts of the human body (neck, thoracolumbar segment of spine, etc.), and it is not surprising when these devices move over time.

Usually, migration of SCS, PNS or DBS electrodes would mean loss of stimulation benefits. It is generally recommended to interrogate the device and check the impedance of each contact before submitting the patient to radiographic testing. Normal functioning of the system suggests viability of the battery and adequacy of the recharge procedure, and impedances within the expected range indicate lack of disconnection or short circuit. A simple reprogramming may be tried as the next step and, since most migration occurs at a relatively short distance, it may be possible to recover the benefits of stimulation by simple changing polarity of the electrode contacts, moving active contacts in the direction opposite the migration. This is why we routinely try to position our electrodes in a way that active contacts are close to the middle of the electrode array and minor migration in either direction may be dealt with by simple reprogramming.

We also recommend keeping a hard copy of radiographs with the initial electrode location in different projections. These images should include identifying markers for the determination of spinal level, such as first or last rib, etc. In cases of suspected migration, baseline images help to determine if the electrode(s) has moved from its original position. The migration may occur in both an outward and inward direction and one should be ready to find electrodes at a long distance from their original site.

The best way to prevent migration is to anchor the electrode in place using appropriate devices. The usual trade-off for anchors is the risk of migration versus risk of fracture and, in the past, most anchors would be either too loose (such as standard silicone anchors) or too strong, such as Twist-Lock anchor (Medtronic, Minneapolis, MN). The newer generation of anchors seems to be better suited for long-term use. They now combine relative softness with strong grip onto the electrode outer surface. These new anchors are named Titan (Medtronic), Cinch and Swift-Lock (St Jude Medical Neuromodulation, Plano, TX). Alternatively, a medical adhesive may be used inside the silicone anchor to keep the electrode in place.

Needless to say, if the anchors are not sutured appropriately, or if they are sutured to loose tissue, or if the sutures absorb over time, the migration risk increases. Therefore, it is recommended to use 2-0 non-absorbable sutures and attach anchors to the underlying fascia. It is also recommended for implanters to practice their suturing technique prior to their implant procedures.

There is a myth that paddle electrodes do not migrate. Although extremely rare, migration of paddle electrodes can occur and, in some series, the rate of paddle electrode migration is higher than that for percutaneous electrodes [24]. For this, one should be prepared to replace such an electrode with a similar or slightly larger model to prevent recurrent migration. In some cases, it is recommended to perform a full laminectomy and then suture the electrode directly to the dura but, fortunately, this is required very rarely. Finally, one should avoid keeping a large segment of electrode between the anchor and the fascial entry as this segment is usually responsible for electrode migration if the anchors remain in place.

For DBS, using a Stim-Lock device (IGN – Medtronic) is usually sufficient to prevent migration. Before this device became available, there were problems with a locking cap provided with the electrode. However, some surgeons have had few, if any, problems with the original cap. Some implanters have been using cement to fill the bur hole around the DBS electrode or modified mini-plates to secure the electrode to the skull. In one of the DBS series, electrode migration occurs with a frequency of 5.1% per patient and 3.2% per electrode [20] while another series did not mention any migrations at all [21].

With IT catheters, migration is best prevented by placing an anchor close to the fascial penetration site, attaching it to the lumbosacral fascia and using heavy non-absorbable suture material [19]. Having a second anchoring point and using a connector between two catheter pieces as an anchor may be another way to prevent catheter migration.

Migration of generators and pumps occurs rarely but, in some situations, there is a higher risk for this to happen. An example of such a situation would be a subfascial placement of an intrathecal pump in children or adults where the intrathecal pump may migrate into the peritoneal cavity [25].

PNS electrodes, on the other hand, migrate very frequently. This most likely is related to the fact that none of the routinely used electrodes or anchors on the market today are designed or approved for this application. Therefore, in order to avoid placing bulky anchors into the subcutaneous plane, implanters frequently use simple 'drain-like' ('Roman sandal' style) sutures holding the electrodes to subcutaneous fascia. In many cases, this approach works, but unsurprisingly, the migration rate in some PNS series reaches 50%.

HARDWARE EROSION

Another well-known complication of neuromodulation is erosion, and usually occurs when device components are located too superficially under the skin. Such erosion frequently results in device infection. Cases of sterile erosion that can be fixed with simple suturing of tissues over the exposed hardware are extremely rare. Every component of the neuromodulation system, however small or soft, may erode. This complication has been observed with electrodes and catheters, generators and pumps, anchors and connectors.

General precautions to prevent erosions are often straightforward: avoid superficial placement of any device components; close soft tissues over the devices in multiple layers; keep the profile of every implant as low as possible; make the depth of device implantation at the allowable maximum; and try to avoid placing hardware over hard surfaces (such as rib cage or iliac crest). The depth of device implantation may be limited due to telemetry or recharging limitations (1.0–1.5 cm for most rechargeable neurostimulation generators) or due to difficulty with pump refill if it is placed too deep in the fat or under abdominal fascia. However, one has to avoid stretching the skin over the implanted device as this seems to be the most serious predisposing factor for skin hardware erosion. This means that pockets for each device should be large enough to prevent tightness of tissues next to the implanted pump or generator. It is also recommended to avoid placing devices directly under incision lines [1] as this may not only increase the risk of wound dehiscence but also create additional areas of discomfort and irritation during the postoperative period.

It is also prudent to consider smaller devices in thinner or smaller patients – a smaller size pump or smaller generator is likely to be associated with less risk of erosion. One has to keep in mind, however, that in currently used Synchromed-II pumps (Medtronic), the 40 mL pump is 26 mm thick and has a overall volume of 121 mL, and the 20 mL pump is 19.5 mm thick and has a volume of 91 mL (so doubling of the pump reservoir volume comes with only a 33% increase in volume and height).

Development of lower profile connectors seems to have lowered the incidence of erosion in DBS systems and, for PNS and SCS systems, a general trend now is to avoid connectors altogether and consider plugging electrodes directly into the respective generator.

HARDWARE MALFUNCTION/FRACTURE/DISCONNECTION

In the absence of hardware migration, the main reasons for device malfunction are disconnections and fractures of device components. These problems can occur in all types of electrodes, catheters, extension cables and connectors.

In most cases, malfunction of the electrical neuromodulation system may be assessed by checking its impedance. Very low impedance would indicate a 'short circuit' within the system, usually indicating internal fracture of the wiring inside the electrode or the extension cable so that the wires are touching each other and electrical impulses do not reach the nervous tissue. Very high impedance, on the other hand, indicates an 'open circuit' suggesting that components have become disconnected or that the internal wiring of the electrode or extension cable has broken apart. In some cases, such violation of electrical circuit integrity or continuity may be confirmed with radiographic imaging but, in most cases, the problem is solved only by revision and individual testing of each device component.

To avoid fractures and disconnections in neuromodulation devices, one has to be very accurate with each connection or stress point along the path of the device components. Over-tightening of the holding screws may result in breaking the contacts (these days, with torque wrenches supplied with each device component, it is very difficult, but not impossible, to over-tighten the connections), while under-tightening may result in loosening the connection and its eventual pullout. Use of strain-relief devices and sleeves allows one to reduce the chance of electrode kinks that would eventually produce metal fatigue and fracture. Similarly, in placing extra loops of electrode or extension cable, one should try to create smooth loops rather than sharp turns and bends in each hardware component. While older anchors carried some risk of crushing the electrode if the holding suture was too strong, newer anchors (described earlier in the section on migrations) are designed in a way that even very tight closure will not damage the external or internal electrode structure.

Breakage (fracture) rates were much higher for paddle-type electrodes than for percutaneous electrodes (12.4% versus 6.8%) and, in both categories, the incidence of fractures was higher for cervical than thoracic electrodes [24]. One of the suggestions in this regard is to avoid anchoring paddle type electrodes, but that may result in a higher migration rate and, so far, the benefit of non-anchoring for fracture avoidance is rather unclear. Another suggestion is to remove more of the caudal spinous process during paddle electrode insertion as this would reduce the angle that wire takes exiting the epidural space.

In DBS practice, most fractures occurred due to migration of the connector between the electrode and the extension cable [20]. Interestingly, in our personal experience of more than 200 consecutively implanted DBS electrodes, we have not had any electrode fractures, probably related to high vertex placement of the electrode–extension connectors.

Another issue that may be contributing to device fracture (and possibly migration) is the location of the generator. A frequently used gluteal location of the generator is associated with a short distance between the electrode anchor and the generator pocket. The main benefit of this location, other than a short path between two incisions, is that it may be prepared with the patient in the prone position – the main reason why many implanter prefer it. The lumbar area, however, is quite mobile, and with every bend forward, the distance between mid-lumbar midline and buttock may even double.

We routinely place all pumps and generators into the abdominal wall, 15 cm away from the midline, when thoracic and lumbar areas are addressed (and use infraclavicular location for all DBS, VNS, cervical SCS and upper body PNS generators). The segment between the mid-thoracic anchoring point for the electrode and lateral abdominal generator location is significantly longer, but with the patient's every movement, the entire mid-section of the body moves as one and the stretching is rather minimal. A disadvantage of our approach is the need for the patient to be in the lateral position for device internalization, but since all our trial electrodes are implanted as if they were permanent, the internalization procedure does not require patient collaboration or fluoroscopy and may be safely done with the patient laying on their side.

Interestingly enough, a recent study on volunteers confirmed our suspicion about difference in distance changes associated with different generator locations for occipital PNS. The gluteal generator position fared worse than the abdominal when the electrode was tunneled from retromastoid or occipital midline areas [26].

GRANULOMA FORMATION

First mention of catheter tip granulomas took place in 1991 when a single case report described an inflammatory mass forming around the tip of an intrathecal catheter and producing cord compression with resultant paraplegia [27]. Eleven years later, a series of 41 patients was analyzed showing that all patients who had this problem received very high doses and high concentrations of opioids, morphine or hydromorphone [28]. Later that year, a consensus statement was published summarizing recommendations on detection and management of these inflammatory masses [29]. Most of these recommendations deal with treatment of granulomas – a subject outside the scope of this chapter.

The pertinent issues related to the cause of granuloma formation may be summarized as follows: the most common etiology is drug dosage or concentration, particularly that of opioids; the granulomas are unlikely to be caused by an infection; the granulomas are not caused by trauma of catheter insertion or allergic reaction to catheter material; and no masses were reported in patients who received baclofen as the only intrathecal medication [30]. There were also recommendations dealing with prevention of granuloma formation: for example, it was stated that while catheter placement in the lumbar versus the thoracic spinal canal cannot be relied upon to prevent the development of a granuloma, lumbar placement theoretically might mitigate the neurological consequences if it occurs, primarily because the spinal cord, more susceptible than the cauda equina to permanent injury from extrinsic compression, ends in the upper lumbar region [29]. It was also recommended to keep the drug dose and concentration as low as possible for as long as possible while still achieving adequate analgesia [29].

Our general approach for prevention of granulomas is exactly this – restraint in raising the dose of IT medication and keeping the daily dose between 1 and 8 mg of morphine sulfate. Among other things, this allows us to use commercially available – and approved for use with Synchromed pumps – morphine sulfate with concentrations of 10 mg/mL or 25 mg/mL. These concentrations and doses seem to be associated with extremely low incidences of granuloma formation, whereas the majority of the reported granuloma cases received compounded morphine solution with concentrations higher than 25 mg/mL or hydromorphone solution. So far, there has not been a single granuloma in more than 100 patients with IT pumps implanted and managed in our institution. All patients with granulomas that were treated by us were referred to us from other facilities.

With baclofen IT administration, it appears that, despite initial reports of several patients that were found to have catheter tip masses while receiving baclofen only, these masses turned out to be precipitated baclofen rather than granuloma. These patients were receiving IT infusion of baclofen with concentrations higher than baclofen solubility in CSF with physiological osmolality values [30].

An anecdotal observation recently described the presence of granuloma and intramedullary abscess in the same patient receiving IT infusion of morphine at a rate of 32 mg/day [31]. Both problems were treated at once – with granuloma resection and abscess drainage and, over time, the patient regained some strength, improving from paraplegia to the level of ambulating with an assistive device. The IT granuloma did not harbor any infection whereas the intramedullary abscess and the pump reservoir were both positive for *Streptococcus*. It was therefore hypothesized that development of the granuloma resulted in increased requirement for opioids that in turn resulted in more frequent refills of the IT pump with compounded, rather than factory-made, morphine preparation. These frequent refills may have introduced bacteria into the pump and from there it spread into the subarachnoid space and the spinal cord [31].

OTHER ISSUES

One issue rarely mentioned in the context of neuromodulation is x-ray exposure for the patient during neuromodulation procedures, with more attention usually (and understandably) paid to protection of the operator and surgical team. The higher sensitivity of pregnant patients and fetuses to radiation exposure usually serves as a reason to postpone device implantation, including a trial, until the child is born. We also routinely screen all women of childbearing age with pregnancy tests prior to surgical interventions, and the concern includes not only radiation risks but also the issues related to anesthesia and surgery.

To decrease exposure to x-rays, it is recommended to use lower settings for fluoroscopy, consider built-in diaphragms and collimators, avoid prolonged periods of 'live' fluoroscopy, using spot checks instead, and to move the cathode tube farther from the patient, placing the image intensifier closer to the working field. I also suggest that the surgeon themselves control their fluoroscopy devices in order to reduce exposure time and also to decrease surgical time by eliminating inevitable delays related to communication with radiology technicians that are being asked to start and stop fluoroscopy during device insertion, positioning and subsequent location checks after anchoring, tunneling, etc.

The possibility of drug overdose with the IT systems is real but rarely observed – most likely due to a serious attitude toward high dose analgesics in the health-care community. It has been said that all or most issues with medication over- and underdose are based on human errors – the IT pumps themselves are very unlikely to malfunction. But, in our institution, we routinely recheck the concentration of drug to be injected into the pump and, by all means possible, prefer using factory-made (and approved for pump use) drug preparations.

All other complications, such as venous thromboembolism, pulmonary and urinary infections, allergic reactions to devices and medications, side effects of anesthesia, etc. have to be dealt with as in non-neuromodulation cases. Routine surgical practice should not be changed simply because neuromodulation is being used.

In general, following the same step-by-step protocol in all neuromodulation procedures, whether stimulation trial, IT pump implantation, device troubleshooting, battery replacement, pump refill, or anything else, seems to be the best way to decrease the number of mistakes and thereby lower morbidity associated with neuromodulation. There are no minor complications as there are no minor neurosurgical interventions – every complication, no matter how small, has to be dealt with very seriously. Not only do we have to resolve all new issues, but we have to take all steps possible so that complications do not recur.

REFERENCES

1. Follett KA, Boortz-Marx RL, Drake JM, et al. Prevention and management of intrathecal drug delivery and spinal cord stimulation system infections. *Anesthesiology*. 2004;100:1582–1594.
2. Miller JP, Acar F, Burchiel KJ. Significant reduction in stereotactic and functional neurosurgical hardware infection after local neomycin/polymyxin application. *J Neurosurg*. 2009;110:247–250.

3. Choux M, Genitori L, Lang D, Lena G. Shunt implantations: Reducing the incidence of shunt infection. *J Neurosurg*. 1992;77:875–880.

4. Slavin KV. Commentary to P.G. Liechty et al. The use of a sump antibiotic irrigation system to save infected hardware in a patient with a vagal nerve stimulator. *Surg Neurol*. 2006;65:49–50.

5. Horlocker TT, Wedel DJ, Rowlingson JC, et al. Executive summary: Regional anesthesia in the patient receiving antithrombotic or thrombolytic therapy. American Society of Regional Anesthesia and Pain Medicine evidence-based guidelines, 3rd edn. *Reg Anesth Pain Med*. 2010;35:102–105.

6. Elias WJ, Sansur CA, Frysinger RC. Sulcal and ventricular trajectories in stereotactic surgery. *J Neurosurg*. 2009;110:201–207.

7. Sansur CA, Frysinger RC, Pouratian N, et al. Incidence of symptomatic hemorrhage after stereotactic electrode placement. *J Neurosurg*. 2007;107:998–1003.

8. Gorgulho A, De Salles AA, Frighetto L, Behnke E. Incidence of hemorrhage associated with electrophysiological studies performed using macroelectrodes and microelectrodes in functional neurosurgery. *J Neurosurg*. 2005;102:888–896.

9. Slavin KV. Intramedullary placement of intrathecal catheter. Report of a rare complication of intrathecal therapy. *Neuromodulation*. 2006;9:94–99.

10. Ryu SI, Romanelli P, Heit G. Asymptomatic transient MRI signal changes after unilateral deep brain stimulation electrode implantation for movement disorder. *Stereotact Funct Neurosurg*. 2004;82:65–69.

11. Mann JM, Foote KD, Garvan CW, et al. Brain penetration effects of microelectrodes and DBS leads in STN or Gpi. *J Neurol Neurosurg Psychiatr*. 2009;80:794–797.

12. Nielson KD, Watts C, Clark WK. Peripheral nerve injury from implantation of chronic stimulating electrodes for pain control. *Surg Neurol*. 1976;5:51–53.

13. Huntoon MA, Burgher AH. Ultrasound-guided permanent implantation of peripheral nerve stimulation (PNS) system for neuropathic pain of the extremities: original cases and outcomes. *Pain Med*. 2009;10:1369–1377.

14. Carayannopoulos A, Beasley R, Sites B. Facilitation of percutaneous trial lead placement with ultrasound guidance for peripheral nerve stimulation trial of ilioinguinal neuralgia: A technical note. *Neuromodulation*. 2009;12:296–301.

15. Skaribas I, Aló K. Ultrasound imaging and occipital nerve stimulation. *Neuromodulation*. 2010;13:126–130.

16. Dam-Hieu P, Magro E, Seizeur R, et al. Cervical cord compression due to delayed scarring around epidural electrodes used in spinal cord stimulation. *J Neurosurg Spine*. 2010;12:409–412.

17. Kirsch WM, Lewis JA, Simon RH. Experiences with electrical stimulation devices for the control of chronic pain. *Med Instrum*. 1975;4:167–170.

18. Nashold Jr BS, Goldner JL, Mullen JB, Bright DS. Long-term pain control by direct peripheral-nerve stimulation. *J Bone Joint Surg Am*. 1982;64:1–10.

19. Follett KA, Burchiel K, Deer T, et al. Prevention of intrathecal drug delivery catheter-related complications. *Neuromodulation*. 2003;6:32–41.

20. Oh MY, Abosch A, Kim SH, et al. Long-term hardware-related complications of deep brain stimulation. *Neurosurgery*. 2002;50:1268–1276.

21. Umemura A, Jaggi JL, Hurtig HI, et al. Deep brain stimulation for movement disorders: morbidity and mortality in 109 patients. *J Neurosurg*. 2003;98:779–784.

22. Eldrige JS, Weingarten TN, Rho RH. Management of cerebral spinal fluid leak complicating spinal cord stimulator implantation. *Pain Pract*. 2006;6:285–288.

23. Huch K, Kunz U, Kluger P, Puhl W. Epidural blood patch under fluoroscopic control: non-surgical treatment of lumbar cerebrospinal fluid fistula following implantation of an intrathecal pump system. *Spinal Cord.* 1999;37:648–652.

24. Rosenow JM, Stanton-Hicks M, Rezai AR, Henderson JM. Failure modes of spinal cord stimulation hardware. *J Neurosurg Spine.* 2006;183–190.

25. Vanhauwaert DJ, Kalala JP, Baert E, et al. Migration of pump for intrathecal drug delivery into the peritoneal cavity. *Surg Neurol.* 2009;71:610–612.

26. Trentman TL, Mueller JT, Shah DM, et al. Occipital nerve stimulator lead pathway length changes with volunteer movement: An in vitro study. *Pain Pract.* 2010;10:42–48.

27. North R, Cutchis P, Epstein J, Long D. Spinal cord compression complicating subarachnoid infusion of morphine: Case report and laboratory experience. *Neurosurgery.* 1991;29:778–784.

28. Coffey RJ, Burchiel KJ. Inflammatory mass lesions associated with intrathecal drug infusion catheters: report and observations on 41 patients. *Neurosurgery.* 2002;50:78–87.

29. Hassenbusch S, Burchiel K, Coffey RJ, et al. Management of intrathecal catheter-tip inflammatory masses: a consensus statement. *Pain Med.* 2002;3:313–323.

30. Deer TR, Raso LJ, Coffey RJ, Allen JW. Intrathecal baclofen and catheter tip inflammatory mass lesions (granulomas): a reevaluation of case reports and imaging findings in light of experimental, clinicopathological, and radiological evidence. *Pain Med.* 2008;9:391–395.

31. Vadera S, Harrop JS, Sharan AD. Intrathecal granuloma and intramedullary abscess associated with an intrathecal morphine pump. *Neuromodulation.* 2007;10:6–11.

Commentary on Limiting Morbidity

Alon Y. Mogilner, MD, PhD

North Shore-LIJ School of Medicine, Great Neck, NY, USA

As Dr Slavin mentions in his chapter, one of the main features of neuro-modulation that has attracted both practitioners and patients is its 'revers-ibility, adjustability … and non-destructive nature'. Although few (myself included) will argue with the overall veracity of this statement, try explain-ing it to a patient (or family of a patient) who has suffered a devastating intracerebral hemorrhage after a deep brain stimulator implantation, paraly-sis or paraplegia following spinal cord stimulator placement, or meningitis following intrathecal pump implantation! Yes, neuromodulation is a gener-ally low-risk procedure and, because of that fact, is now being performed by many physicians from many specialties with a wide variance in their respective training and experience. As such, those practitioners specializ-ing in neuromodulation surgery have the opportunity to learn and hone their technique both from their own personal experience and from compli-cations referred from their colleagues in interventional pain management and (even) neurosurgery.

The author has done an excellent job of summarizing the pitfalls of these procedures, and has provided advice as how to best avoid these complications. Some thoughts from our own experience follow.

INFECTION/EROSION

Prior to the advent of peripheral neurostimulation procedures such as occip-ital, supraorbital, and subcutaneous field stimulation for lower back pain, it could safely be said that the vast majority of device infections involved the generator or pump, i.e. the largest implanted foreign body and thus the largest nidus for infection. With the increased use of peripheral stimulation, leads and anchors designed for deeper placement in the body are now being placed quite superficially and thus the risk of erosion and secondary wound infection in these cases is, in my experience, a much greater concern than IPG-site infection. For example, percutaneous SCS leads with a pointed tip were never designed to be placed in the supraorbital subcutaneous tissue. At the very least, patients with a thin scalp may complain about the protrud-ing tip from a cosmetic standpoint, while at the worst, the tip may erode from the skin, necessitating urgent surgical intervention to prevent infection.

On the other hand, when such infections occur, these situations represent secondary infections rather than true surgical-site infections, and the organisms involved are usually more benign skin flora such as coagulase-negative *Staphylococcus*, and prompt identification and treatment may in fact allow salvage of the hardware. I instruct all my patients, particularly those with trigeminal branch stimulators, to notify me immediately if any changes occur over the hardware, as immediate identification and surgical revision in these cases is indicated.

In the case of a true surgical-site infection unrelated to a skin erosion, I agree with Dr Slavin that complete removal of all hardware is the only way to guarantee complete eradication of the infection. I am sure that many, if not all, implanters have attempted partial hardware salvage, such as removing an extension lead and IPG while leaving the stimulating lead in place, only to find that initial success is met weeks to months later with recurrence of the infection, necessitating removal of all the hardware.

CSF LEAK FOLLOWING INTRATHECAL CATHETER PLACEMENT

Not infrequently, as mentioned, patients may complain of positional (spinal) headaches for days to weeks after intrathecal catheter placement. In the absence of any evidence of subcutaneous collection or external CSF leakage, our policy has been to follow the standard recommendations of bed rest, caffeine, oral analgesics and fluid intake mentioned in the chapter. Patients with persistent and incapacitating headaches lasting over one week are then offered a fluoroscopically-guided epidural blood patch, with the epidural needle placed at the same level, but from the contralateral side, as the catheter entry. Commercially-available fibrin glue (Tisseel, Baxter Pharmaceuticals) is an alternative to autologous blood, which expands quite rapidly and thus requires smaller volumes (in our experience, approximately 5 mL) than blood itself.

When a small subcutaneous CSF collection is present which does not threaten the wound, we have resisted the temptation to proceed to early surgical revision, and have found that use of an abdominal binder may facilitate resorption of the collection over a period of weeks to months. Failure of these collections to resolve should prompt an investigation into the possible causes, including, but not limited to, previously undiagnosed hydrocephalus, not uncommon in the traumatic brain injury population referred for intrathecal baclofen therapy. In such cases, it has been our procedure to cut and remove the intrathecal catheter portion, as it is highly unlikely that a reinforcing purse-string suture around the catheter will provide any improvement over the original purse-string suture placed by an experienced surgeon. A new intrathecal catheter is placed at a different spinal level and spliced into the distal catheter, with purse-string sutures placed both around the new catheter as well as the old catheter fascial entry point.

COMPLICATIONS FOLLOWING SCS PADDLE LEAD IMPLANTATION

As eloquently mentioned in the chapter, it is not uncommon in the community for a paddle electrode to be placed in the epidural space, particularly in the thoracic spine, with the only preoperative imaging being the fluoroscopic images from the trial! The reasons for this are, in my mind, quite obvious: patients with the most common pathologies/indications for SCS are those with lumbar post-laminectomy syndrome, who upon arrival to surgery may have had multiple lumbar spine MRI and CT scans, but have never had their thoracic spine imaged. Furthermore, many of these patients may arrive at the surgeon's office following a successful percutaneous SCS trial by another physician, literally 'aching' to get a permanent implant, but without a thoracic spine MRI. In many cases, obtaining the MRI may result in a delay of the implant due to insurance approval requirements, resulting in an unhappy patient (and perhaps referring physician).

Experience has taught me and others that the prudent course of action is to obtain appropriate spinal imaging (MRI or CT myelogram) prior to paddle placement. In another common scenario, a patient is referred with an indwelling permanent percutaneous SCS system for a revision to a paddle electrode. Since an MRI cannot be performed, the question arises as to the need of subjecting the patient to an invasive myelogram – to that I answer a resounding 'yes'.

Finally, whereas paraparesis/paralysis is the most dreaded possible outcome of cord compression, implanters should be aware of more benign, yet clear sequelae of neural compression. Both in my own experience and in that of others (Giancarlo Barolat, personal communication), patients with clinically undiagnosed stenosis have presented with severe, unremitting, band-like neuropathic pain following lead placement which does not resolve after appropriate medical management including steroids and oral and intravenous analgesic therapy. In such cases, the patient should be taken back to the operating room for a complete laminectomy and decompression over the electrode, followed by either suturing the electrode to the dura and/or placement of fibrin glue over the electrode to prevent lead migration. This should result in near immediate resolution of symptomatology. I completely agree that an aggressive laminectomy/ decompression should be considered *a priori* in any patient with evidence of spondylosis/stenosis on imaging.

LEAD FRACTURE

In my experience, the incidence of lead fracture, both of SCS paddle-type leads as well as DBS leads, has decreased dramatically after we routinely began to place the extension connector as proximal as possible to the stimulating lead itself. In the case of SCS paddle leads, this means placing the extension connector directly in the laminectomy incision itself rather than distally, i.e. between the laminectomy incision and the IPG incision. The recent advent of 'direct connect' SCS paddle leads, 60 cm or greater in length, in which the lead tail plugs directly into the generator, removes that connection as a possible 'stress fracture' site, and thus will likely lower the incidence of SCS lead fracture.

Unlike SCS, all DBS systems utilize extension leads, and will be likely to continue to do so. As such, and analogous to the SCS paddle lead situation, I routinely place the extension connector as close to the cranial incision as possible (the 'high vertex' placement mentioned by Slavin), thus allowing for the least amount of tension on the DBS lead itself. I thus avoid placing the connector lower, in the parietal or retromastoid region as others do.

Continued improvements in hardware will undoubtedly result in a further reduction in overall morbidity of these procedures. However, one must never forget the fact that limiting morbidity begins with appropriate patient selection, followed by appropriately performed surgery in experienced hands and continued clinical follow up.

STUDY QUESTIONS

1. Should more research monies be spent on eliminating complications in neuromodulation, or in new therapy delivery?
2. Consider that neuromodulation is often used in cases that have exhausted all other forms of treatment for a disorder – i.e. the disorder is 'refractory' to best medical or other treatment. Given the risks associated with neuromodulation, whether thought to be high or low, decide if it makes sense both clinically and ethically to consider using neuromodulation earlier in the treatment course.

Part V

Troubleshooting and Repair

Chapter 15

Intraoperative Evaluation

Jay L. Shils, PhD[1] and Jeffrey E. Arle, MD, PhD[2]

[1]*Director of Introperative Monitoring, Lahey Clinic, Burlington, MA*
[2]*Director, Functional Neurosurgery and Research, Lahey Clinic, Burlington, MA;*
Associate Professor of Neurosurgery, Tufts University School of Medicine Boston, MA

INTRODUCTION

As neuromodulation therapies become a treatment standard for a variety of disorders including medically refractory Parkinson's disease (PD), essential tremor, dystonia, pain syndromes, epilepsy, psychiatric disorders, and other future indications, more is being learned about the longevity and function of the implantable stimulator and its components. A literature review describes a 15–30% failure rate that includes both infection and device failure [1–9]. As neuromodulation becomes more prevalent due to increased disease penetration and as the number of medical conditions that are treatable with implantable neuromodulation devices increases, the total number of device failures will also rise. With this in mind, a systematic method for trouble-shooting these failures is necessary in order to minimize both neuromodulation 'downtime' and the number of invasive actions required to identify and replace failed components.

Surgery to isolate and fix device malfunctions takes time, is expensive and exposes the patient to additional risk. Therefore, it is important to evaluate completely the patient who is responding poorly to neuromodulation before manipulating his/her device surgically. The potential causes of a poor response to neuromodulation include badly placed leads, an incorrect initial diagnosis, poor stimulator programming, and a worsening disease state [10]. If specific symptoms or electrophysiological data derived through device interrogation do not suggest a device failure (see below), these clinical issues must be ruled out before assuming that a device malfunction exists. However, even when it is clear that a malfunction is present, it is essential to make every possible attempt to localize the fault non-invasively before embarking on surgical interventions.

Unless a failure mode, such as a lead fracture, is visible on x-ray, locating short or 'open' circuits in system components is very difficult with current manufacturer-supplied hardware and software. Intermittent system problems are especially difficult to locate, and the differentiation of an intermittent problem from a *pseudo* problem can be nearly impossible.

Essential Neuromodulation. DOI: 10.1016/B978-0-12-381409-8.00015-2

PRINCIPLES OF ASSESSMENT

The system

The neuromodulation system consists of:

1. a multicontact stimulating lead
2. a combination implantable pulse control system and self-contained power supply (IPG)
3. an extension cable that connects (1) to (2) (see Chapter 9).

At this time, there are two Food and Drug Adminstration (FDA) approved leads for deep brain stimulation (DBS): Medtronic models 3387 and 3389, with other manufacturers running or getting ready to run clinical trials. For spinal cord stimulation (SCS), there is a much wider choice of approved electrodes and implantable pulse generators (IPGs) from multiple manufacturers. The Appendix contains a list of all approved systems with images.

In all cases, the lead needs to be secured. For DBS, this occurs where it exits the skull, and for SCS this occurs with a friction suture cover. For DBS, the excess length of lead wire is coiled beneath the scalp and connected to an extension wire [11], which is thicker and more durable than the lead. For SCS, the lead may or may not use an extension wire. Conductors in the Medtronic DBS and SCS extensions are made from silver core MP35N. Each conductor is coiled and set in an individual cylindrical opening which reduces the chance of shorting.

The DBS circuit-Paradigm for all fault testing

The DBS extension is passed through a subcutaneous tract that traverses the retrosigmoid sinus region and neck to an ipsilateral subcutaneous pocket in the subclavicular area of the upper chest. There the extension is connected to the IPG. For a dual stimulation device, each lead is connected to a single extension wire via a 'Y'-adapter ('Y' in shape only – all contacts are still individual). A silastic cover (boot) is placed over the lead–extension connection and two suture ties are placed on each end creating a water-tight seal for the connection (Fig. 15.1). The connector screws are made from titanium and the connector blocks from stainless steel. The extension insulation is silicone rubber and polyurethane while the connector block is sealed in silicone rubber and siloxane-coated silicone rubber. The maximum resistance of the complete extension wire is $7\,\Omega$.

In order for current to flow through an electrical circuit, the circuit needs to be configured in a closed loop. The electrical circuit that contains the DBS system is depicted in Figure 15.2. The power source provides a constant voltage pulse of potential V that, when activated, sends a current around the circuit. The current (I) is determined by the potential (V) and the impedance (Z) that the potential needs to overcome.

$$I = \frac{V}{Z} \tag{15.1}$$

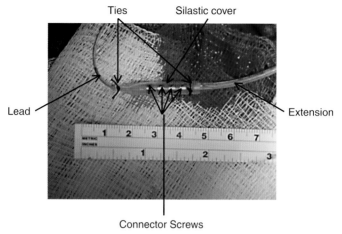

FIGURE 15.1   Example of the connection between the lead and the extension. The lead slides into a connector that has 1 to 4 screws for securing the lead in the connector. A silastic cover is placed over the connection to keep fluid from interfering with the electrical contacts. Finally ties are placed on either end of the silastic cover to make the connection watertight.

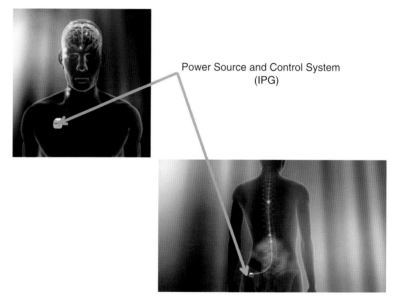

FIGURE 15.2   A graphical representation of the DBS and SCS systems. For the DBS system the IPG is usually placed in the chest, but can also be placed elsewhere if the patient desires or for other medical reasons. For the SCS system the IPG is usually placed in the upper buttock or abdomen.

For neuromodulation, impedance includes both the circuit resistance and the effects of capacitance and inductance at the biomechanical interface of the electrode and tissue. Therefore, the total circuit impedance is composed of three elements:

1. the connections between the system components
2. the impedances of the conductors (wires) used in both the extension cable ($<7\,\Omega$ [12]) and the lead ($<100\,\Omega$ [11])
3. the brain–body–electrode interface, which contributes the largest impedance.

It is also important to note that the impedance of tissue varies with the stimulation frequency. Thus, comparing impedances over time is only useful if the same test frequency and pulsewidth were used. In general, as the frequency increases, the measured impedance of the biologic material decreases [13–15]. For example, for an intact DBS system, normal impedance values for a test pulse of $210\,\mu s$ at $30\,Hz$, when referenced to the IPG case (i.e. monopolar configuration), should range between 600 and 2000 ohms with a current between 9 and $25\,\mu A$ using 2.0 Volts. This is true for electrodes located within the subthalamic nucleus (STN), globus pallidus pars interna (GPi), and ventral intermedius (VIM) when using the Medtronic model 8840 programmer in the electrode impedance test mode, not during therapy measurement testing. Future systems may use different test parameters and will yield different normal impedance values. Also, normal impedance values in other brain regions may differ from those observed in these three areas.

The literature describes very little anatomical change at the electrode–brain interface as a consequence of chronic DBS [16–18]. Therefore, one may conclude that a major change in the measured electrical properties of impedance and current over time will most likely occur at the other two primary circuit impedance points (i.e. the conductors and connection points). It should be noted that within 3 months of implantation there may be large changes in impedance, likely due to surgical healing. Three types of electrical failure modes have been identified in implanted neuromodulation systems:

1. foreign body accumulation at the connection points
2. an 'open' circuit (i.e. a break in the circuit path)
3. a 'short', which is a new unexpected and unwanted circuit pathway between what should be independent circuit elements.

An internal failure of the IPG is exceedingly rare, but possible. However, locating a problem in the IPG is more complex and is arrived at through a process of elimination, when all other testing, to be described below, fails to localize the failure.

Under the 'open' circuit condition, current is unable to flow due to a break in the pathway (Fig. 15.3). If the circuit is completely 'open', the measured current will be zero and the impedance will be 'infinite'. If the circuit contains an intermittent 'open' and 'closed' condition, current will flows some of the time (transient mode failure), during the time in which current flows through the circuit may appear normal. Intermittent 'open' circuits are very difficult to troubleshoot, and may only be found during the actual 'open' period. An intermittent 'open' circuit could be seen when a break in the conductor leaves the two ends in close

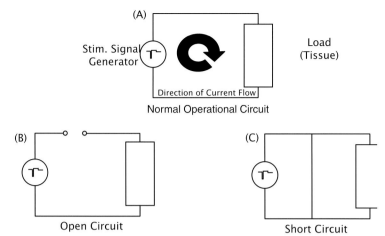

FIGURE 15.3 A schematic representation of current flow in an electrical circuit. (A) shows a normal circuit. (B) shows an open circuit where no current can flow into the load. (C) shows a short circuit where current will be shunted away from the load via the short.

proximity. When the ends are in contact, the circuit will function normally, however, if the extension or the lead are moved (e.g. while turning the head), the ends separate and the 'open' circuit condition occurs.

In patients with quickly reacting symptoms such as tremor, which varies quickly in relation to the state of stimulation, the ability to diagnose an intermittent 'open' circuit is easier than in patients whose symptoms change more slowly. If the intermittent condition is very brief, no abnormality may be detected. Patients with brief intermittent 'open' circuits may derive benefit from stimulation, but the results will be suboptimal. Therefore, if a patient presents with an unexplained reduction in therapeutic efficacy, but the system appears to be functioning properly, one must consider a transient mode failure.

In a 'short' circuit situation, current is shunted away from the electrode contacts in the brain. This is because the new circuit pathway, created by the 'short', is of lower impedance and 'draws' current away from the lead tip. For the internalized stimulation system, there may be multiple short-circuit types. The first type involves a break ('open'-circuit) in the extension or lead insulation. The wires on the IPG side of the break may touch each other causing the current to flow only in the electrical circuit and not in the body. Under this condition, one will measure very low impedance and very high current during the therapeutic test (see below). The terms 'high' and 'low' are used because the normal values depend on the therapeutic settings being employed, the target tissue, and the device model. Thus, it is critical to look at the therapeutic parameters at each visit so that a reference exists for each particular patient.

A second type of 'short' occurs if the insulation between the conductors in the extension breaks down and the conductors begin to short due to contact with biological tissue. Since there is no 'open' circuit, some of the current flows back to the IPG via the shorted wire, while the rest flows to the conductors in the brain. As a consequence, some 'inactive' contacts may transmit current, stimulating unintended areas of the brain. Short circuits can cause excessive current flow because, when the impedance trends toward zero, the current will exceed the maximum desired, rapidly draining the power source. One dangerous problem with 'short' circuits, and the high current that results, is that this high current may break down the insulation, causing additional unwanted current paths. Also, higher current can generate heat at the site of the short which will, in turn, heat adjacent tissue, generating potential burns. In Figure 15.4 burns can be seen on the extensions wires removed from this patient.

A third type of 'short' circuit condition can arise when fluid enters the connection between the extension and the lead or the extension and the IPG. The fluid can act as a conductor, shunting the current away from the stimulating electrode surfaces to other unintended contacts. In monopolar configurations, the shunted current may activate an alternate conductor, again sending current to an inactive electrode, stimulating an area of brain inadvertently.

Monopolar and bipolar stimulation behave, to some extent, differently in failure modes due to the differing return pathways for the current. During monopolar stimulation, the return is the casing of the IPG. If the lead and extension wire

FIGURE 15.4   Example of a burn inside the lead insulation.

insulations are intact (i.e. the break is inside an intact insulation), the insulation will create a very high resistance thus still allowing the current to flow to the lead and not from the break point to the IPG. However, if the insulation has a break, current will escape from the opening back to the IPG and the patient may feel a 'shock' at the break point. For the 'short' circuit situation in a monopolar configuration, the current will be split between the two shorting leads if only one lead is active. No changes will be seen if both of the leads are active, assuming the insulation is not broken. If the short is between the connector and the IPG, the patient may feel intense pain at the IPG site. A note of caution: under normal circumstances thin patients may feel a sensation at the location where the IPG is implanted during monopolar stimulation, which may be mistaken for a short circuit. However, interrogation of the system will reveal normal impedance. If the insulation is broken, a shock may be felt at the break point. If the impedance at the break point is lower than that of the electrode contacts in the brain, current will pass from the break point to the case, taking the path of least resistance.

During bipolar stimulation, multiple types of 'open' circuit situations can occur. If the insulation is intact, no current will flow in the circuit. If, however, the insulation surrounding one of the conductors is broken, current will flow along two pathways. The first pathway is from the insulation break to the reference electrode. The second is the intended pathway between the active and reference electrodes.[1] The amount of current flowing at the break will depend on the relative impedance in each pathway. In fact, no problem may be noticed by the patient in the case where the impedance at the break point is very large. If both conductors are broken, current will most likely flow at the break point.

'Short' circuits also present in multiple ways depending upon the state of the insulation and the state of the conductors at the location of the 'short'. If the insulation and the conductors are fully intact, minimal current will get to the brain because the impedance at the short is very low. If the insulation is intact and multiple electrodes are being employed for therapy, two conditions could arise. First, if the short is between the active contact and another (an electrode not used for the patient's particular therapy program), or the reference and another, the current will be split between the normal circuit and the new path, stimulating an unintended region of the brain. Second, if the short is between two active contacts or between two reference contacts, no difference will be seen. If the insulation is broken, the current has multiple pathways it can travel and the current to the electrodes will most likely be reduced due to the low power supply resistance.

Note that in contrast to constant voltage devices, the internal resistance of constant-current stimulators must be very high to ensure that the power supply and not the load controls the current delivered. In every day life, load typically controls current. For example, when we turn on a brighter light at home, the

---

1. At first one may think that, if there is a break in the wire and conductor, then no current will get to the target. Yet, if there is fluid in the conductor between the breaks, then a current pathway may exist for energy to get the target.

lamp draws more current from the power supply. In a patient, the electronics need to operate in reverse, so that when there is a short, the power supply will automatically 'limit' the current to a safe level. In the home, there are fuses and circuit breakers for protection.

NON-INVASIVE TESTING

When evaluating a patient with a reduction in stimulation efficacy, signs of a potential device failure include:

1. a sudden change in the therapeutic benefit of stimulation
2. strange electrical shocks along the circuit pathway
3. a sudden onset of muscle contractions
4. a sudden onset of continuous or intermittent paresthesias
5. a sudden change in vision
6. battery depletion long before expected.

The techniques and methods for troubleshooting a malfunctioning DBS device fall into two categories: non-invasive testing performed in the clinic, and invasive testing performed in the operating room (OR).

Initial testing is performed with the clinical patient programmer. Observe and record the following:

1. device state (on/off)
2. number of activations since the previous visit
3. percentage of on-time since the previous visit
4. battery voltage
5. therapeutic impedance
6. therapeutic current
7. monopolar impedances
8. monopolar currents
9. bipolar impedances
10. bipolar currents
11. battery charging.

All eleven of these details are discernable with the programmer. It is critical that at the end of each visit the internal counters in the implanted IPG are reset so items 2 and 3 above are accurate. If the device is 'off' the clinician must attempt to determine why and when it turned off. One way to estimate the length of time the device was shut off, if a singular event, is to determine the amount of time the device has been off. This can be estimated by observing the date of the last reset and subtracting the total hours used from the total hours since the last reset and calculating the intervening hours by:

$$\sim \text{Days Since Off} = \frac{(\text{Number Of Days Since Reset} \times 24) - \text{Hours Used}}{24} \qquad (15.2)$$

The patient should be questioned about specific events within a few days of this estimated time. Ask the patient about recent travel, shopping or other excursions. Ask if they were near large power lines or electrical substations or close to large neon signs. Ask whether they used power tools or welders. Finally, ask about any impacts they may have taken to the IPG area. Newer stimulation systems (i.e. post 2003) are less susceptible to external magnetic interference, but the patient's recent history to such exposure should be recorded. If the patient cannot recall any specific event (the most common case) turn the device back 'on' and run both the electrode specific impedance and current checks as well as the therapeutic parameter check. If all parameters are within the normal range, no other changes are necessary. Recommend that the patient keep a diary of potential causative external events (as cited above) in order to have a record if the same situation arises in the future.

If the battery voltages test at a reasonable level (see device specifications), the next step is to examine the circuit integrity. To accomplish this, both the electrode circuit test (each electrode and each electrode combination is checked) and the therapeutic test (current and impedance at the therapeutic settings) need to be performed. Both are needed since an acceptable therapeutic test will not always identify a short circuit involving an 'inactive' electrode, which could be the cause of paresthesia, contractions, visual problems, or poor therapeutic results due to the stimulus being shunted from desirable to undesirable brain areas. When performing the therapeutic parameters test, major impedance and current changes (i.e. >200 ohms and $>20\,\text{mA}$, respectively) since the last visit will indicate a problem.[2] However, remember that during the first 3–6 months after implantation, changes of this order may be observed as a result of the normal brain healing process.

The results of the bipolar component of the electrode testing program help one to make sense of abnormalities or changes found during the therapeutic test. During bipolar testing it is critical to look at all of the electrode combinations, not just those involving active contacts. A low impedance value and a high current value between *any* electrode pair that includes an active electrode indicates that a *non-planned* current is being delivered. For the case where the active electrode is paired with an inactive electrode, current is being delivered to an area of the brain that should not be receiving any current. For the case where the active electrode is paired with another active electrode of opposite polarity, an inappropriate amount of current is being delivered. Either of these cases will require replacement of some system component. If the short is between two inactive electrodes, no changes may be required. If no abnormal values are noted then more investigation is necessary.

---

2. Some manufacturers do not provide specific impedance and current values. Instead, they give a numeric value indicating the status of the lead and the battery life. Thus, it is critical to have a copy of the device specifications for reference.

Monopolar impedances and currents should also be investigated using the electrode test program as the problem may not involve a specific electrode pair but could reside within a single conductor or within the IPG switching matrix. An 'open' circuit (very large impedances and very low currents) in electrodes that are inactive could indicate a future problem in an active electrode or an intermittent fault. Intermittent faults are generally the most difficult to localize. This is because they may not show as faults during normal testing with the clinical programmer. A break in an inactive electrode is good evidence that there may be a transient fault in an active lead in the event of a sudden change in therapeutic benefit. Another way to identify intermittent faults is to manipulate the lead connector lightly under the skin while asking the patient if he/she feels any changes. If the intermittent fault is causing motor or sensory phenomena this technique has a good chance of exposing it. When manipulating near or at the break point, the patient may experience sharp paresthesias or contractions. If no paresthesias or contractions are found when manipulating the lead or extension wire, but a transient fault is still suspected, a lateral and anteroposterior (A-P) x-ray of the chest, head and neck may be useful in locating a troubled area in the lead or extension. A potential troubled area is one where there is a sharp bend in the lead or extension, or one where the wires appear to be broken. It has been the authors' experience that near the connector, the wires may appear broken even when they are intact. Utilizing information from the x-ray, push on the wire at the point where the x-ray indicates a problem and then re-test the system using the electrode test program. Manipulating the lead and extension at the break point may cause a change that can be detected. It is especially important to do this testing at the connections (i.e. the lead to extension connection and the IPG to extension connection). X-rays may also be used to determine if there are large breaks in the lead or extension wire (Fig. 15.5).

Non-invasive active testing

In order to identify more complex failures, such as transient failures, or better to localize a failure, we developed a technique that utilizes either an oscilloscope or an intraoperative neurophysiological monitoring system with electromyography (EMG) software to visualize the electrical pulse traveling through the circuit. By analyzing the shape and amplitude of the recorded wave one may determine:

1. whether or not a fault exists
2. the type of fault
3. potentially the exact location of the fault.

Figure 10.12 shows the theoretical shape of the IPG wave. The testing is performed in both monopolar and bipolar modes. This technique is based on the principles underlying surface EMG and far field evoked potentials. The electrical potential generated in the muscle or nerve synapse forms an electric field (Fig. 15.6) that can be recorded on the skin. The measured impedance (ignoring normal lead properties) is that from the electrode contact at the target and the

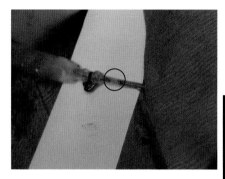

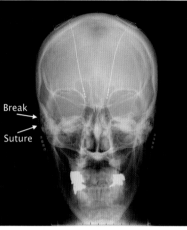

Break

Suture

FIGURE 15.5 Example of a break in the lead as seen on both A-P skull film and after explanting the system. It should also be noted that suture for the silastic cover can also show as a break.

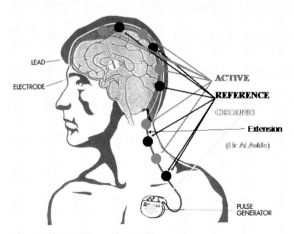

Medtronic Activa™ Tremor Control Therapy

LEAD

ELECTRODE

ACTIVE

REFERENCE

GROUND

Extension

(Use At Ankle)

PULSE
GENERATOR

FIGURE 15.6 DBS testing example. These are the locations of the test points on the surface of the skin. The ground is placed over the IPG while the active and reference leads are moved over the wires.

case in instances where monopolar stimulation is used. This impedance consists of all the energy that is dissipated in all the tissue along the pathway, including the skin. Thus, by placing recording leads on the skin, the voltage gradient between the two leads can be recorded. If all the wired connections are intact then signal that is recorded on the surface of the skin will be similar for all leads and skin locations measured.[3] If there is a break in one of the wired elements, then the potential that is generated on the skin during stimulation with that lead will be different as compared to the other leads. Depending upon the relationship of the 'open' circuit to the recording electrodes (i.e. distance from the 'open' circuit, space between the active and reference electrode, and whether or not the 'open' circuit is located between the recording electrodes or not), the signal will either be larger or smaller than when it is between the other leads. In the normal intact stimulation system, the primary signal is a far field signal that is generated at the electrode– tissue interface. If there is a break in the lead or a short between wires, then a new *synaptic* point is created. This new point will divert some, or all, of the signal from the electrode tip–tissue interface. If the signal diversion is kept completely inside of the insulation and the diversion is an 'open' circuit, the far field amplitude at the electrode-tissue interface will have a smaller amplitude than the normally operating system. If the signal diversion is kept completely inside of the insulation and the diversion is a short-circuit bringing an inactive electrode into play, the far field amplitude at the electrode–tissue interface may be larger in either the positive or negative direction during bipolar stimulation or in both directions during monopolar stimulation. This is due to the larger spread of energy in the brain target tissue. If the signal diversion is kept completely inside of the insulation and the diversion is a break with a short to another contact, the far field amplitude at the electrode–brain interface may show either an increase in amplitude or a decrease in amplitude, but a change will be noted. If the diversion is at a point where the insulation is broken, there will likely be an increase in the signal at the point of the break for an 'open' circuit, or the amplitude will decrease as the recording electrodes are moved away from the short in the case of a short circuit.

During monopolar stimulation the circuit pathway includes the IPG, extension, lead, and body back to the IPG (see Fig. 15.3A). Monopolar testing is performed in two different steps (note that by monopolar we mean the IPG stimulation configuration). The first step is to place the reference electrode on the skin over the closest point to the electrode. For example, in DBS, this is where the lead enters the brain. The active test electrode is then moved along the active recording electrode along skin overlying the system pathway. Large changes in the recorded voltage indicate that the active recording electrode is near the area of a failure. The second step is to separate the active and reference electrodes by about 3 cm and move the pair along the system. While moving

3. There will actually be a slight amplitude variation due to the impedance differences in the leads. This difference will be well below any change that is detected from a fault.

the electrodes, carefully palpate the wire between the leads (if possible). If the wires are intact, the recorded amplitude should change only 5–10% owing to geometry and normal body impedances. Larger changes indicate the site of a break. This procedure should be done while activating each of the stimulation system contacts in a monopolar configuration.

One problem with this technique occurs when the break is very close to the IPG or is within the IPG–extension connector. This is because the IPG generates the signal and therefore it is difficult to determine if an increase in the recorded amplitude is resulting from a break or simply reflects proximity to the stimulus source. Figure 15.7 shows an example of an 'open' circuit test when the active and reference electrodes are placed on either side of a transient break in the extension. When the wire was manipulated the 'open' circuit occurred. The patient in the case report section later in this chapter is an example of such a situation. A variation of this testing method, for a DBS system, is performed by keeping the reference electrode at one end of the circuit and moving the active lead along the extension wire and lead external to the skull. When the active lead and the reference point are on the same side as the break, the potential will be very small. When the active lead comes close to the break point, the potential will start to increase rapidly.

Testing for faults when the system is in a bipolar configuration allows for better localization of short circuits. When a short exists, two, three, or up to all wires may be implicated. If all four leads are shorted, the signal passing through the wire will encounter a nearly zero impedance and thus the field generated in the

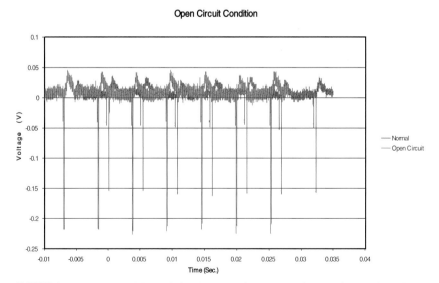

FIGURE 15.7 Examples of the DBS signal recorded from a system in a simulated body load and from a surface recoding where the impedance was 560 ohms. The initial negative peak is the cathodal stimulation amplitude that is represented by the voltage value that is set on the programmer. The positivity represents the charge balanced component of the wave form.

surrounding tissue will be too small to be detected if it exists at all. If the short exists at a point where there is a break in the insulation or at a connection point, due to the introduction of some biological tissue or fluid, the signal may be large enough to be detected up at the surface of the skin. If there is a question as to whether a short exists, all bipolar combinations need to be checked. If it is already known that a short exists, then the shorted wires should be checked relative to a good wire (shorted wires as cathode, good wire as anode). By placing the active and reference surface electrodes furthest from the IPG, recording a signal, and then manipulating the wire through the skin while moving along the wire, one may see a change in the signal at the break area. The purpose of the manipulation is to break and make the short, which will change the impedance and thus the signal picture.

INTRAOPERATIVE TESTING

Despite our best efforts, there are still instances where non-invasive testing cannot indicate the defective element to replace. For a DBS system replacing the brain lead obviously poses the greatest risk and inconvenience to the patient. Therefore, it is imperative that every effort be made to rule out faults in the other system components before proceeding with lead replacement. When necessary, we perform invasive testing under general anesthesia with endotracheal intubation. Unless the IPG is known to be defective, or the battery needs replacement, the first incision is made at the extension–lead connection so that the lead may be tested independent of the other system components. The incision is made directly over the connector and is of sufficient length (typically 3 cm), to provide enough room to remove and replace the silicone cover boot. The incision should avoid crossing or 'T'-ing into a prior incision made during the original placement. Care must be taken at every step not to damage the lead (even cleaning and drying it with a gauze pad can inadvertently catch and pull a contact free of its connector). It is best to avoid monopolar cautery, use of which may cause heating of the implanted electrodes and injury to the surrounding brain. It is essential to note every detail of the tissue and the state of the hardware when it is initially encountered. In particular, fluid type, amount, and location, as well as the exact configuration of the wires, boot, and sutures may be important and should be appreciated before disturbing the hardware for evaluation. Avoidance of local anesthetic injections is recommended as inadvertent needle damage to a wire or the introduction of fluid within the boot can confound the intraoperative evaluation.

The extension–lead connection is examined first. We have sometimes found that fluid becomes trapped within the plastic boot that is placed over the lead–extension connection, causing a short between the connections. The exterior of the boot can be gently patted dry. Do not press too firmly when drying the boot as fluid can be forced out thereby erasing the evidence. The two sutures on each end of the boot should be removed and the boot slid off of the connector. The connection should be opened using the small hexdriver supplied by the manufacturer. Both the lead and extension connection should be dried. Using a small suction tip, the inside of the female end of the connector can also be dried. The boot should be checked for cracks, holes or other defects, and replaced if any defect is found. In rare cases, each tiny set screw should be removed and cleaned and its connector threading cleaned and dried as well.

If the boot is intact and no fluid is observed in the connection, the extension should be the next component tested. It is tested by using the implanted IPG as the power generator and recording signal on the exposed end. We use small sterile alligator clips to attach to the small connections on the exposed end of the extension. The wires attached to the clips are passed from the sterile field and connected to an oscilloscope across a $1\,k\Omega$ resistor. The IPG programming head is placed in a sterile bag and positioned over the IPG. All electrode combinations are tested. For these tests, the IPG parameters are set to 2.0 Volts, 60 Hz, and $210\,\mu s$. During each test, the surgeon manipulates the extension through the skin. Both shorts and 'open' circuits will cause a reduction in the signal observed on the scope and will show a flat or much reduced trace. If this occurs while testing any electrode combination, the extension wire should be replaced. The next step is to test the brain lead itself. For DBS patients this is somewhat more difficult because we do not have access to the full lead, specifically the end that is in the brain. In the authors' experience, we have encountered only one case where the lead was defective *inside* the skull. Two areas of the lead that are accessible and that should be checked when investigating the lead are the area near the extension–lead connection and the area near the cranial locking mechanism. For SCS patients, more of the lead is accessible, yet, until all accessible lead areas are tested, care should be used to keep the electrodes from moving on the spinal dura.

If the lead and the extension are intact, and the boot had no fluid in it, the IPG must be exposed to test its integrity. The continuity of the IPG–extension connection is tested in a similar fashion to the lead–extension connection, using small alligator clips attached to the battery end connector of the extension wire. During the test, the extension connector should be manipulated to elicit any transient fault. The IPG should be inspected for fluid and cleaned. If the circuit is intact the sole remaining possibility is that the IPG has an intermittent fault and needs to be replaced.

TESTING METHODOLOGY

Initial interrogations of devices with suspected malfunctions are performed with either the particular manufacturer's external programmer/testing device. All contacts are tested in both monopolar and bipolar stimulation modes. For DBS, when testing in monopolar configurations, the presence of identical impedances and currents (to within $\pm 5\,\Omega$ and $\pm 1\,\mu A$) in any two leads suggests a short circuit, which should be further investigated with bipolar testing. For SCS and motor cortex stimulation (MCS) patients, this condition can be met without indicating a failure. During bipolar testing (same test parameters as described above) impedances $<200\,\Omega$ and currents $>200\,\mu A$ indicate a short between the tested leads.

When a fault is identified or if an intermittent fault is suspected, we proceed to our non-invasive detection technique. Nicolet skin recording electrodes (model 019-420800, Viasys Healtcare, Madison, WI) are preferred. The reference and ground electrodes (for the EMG/IOM machine (Nicolet Viking IV, Madison, WI)) are placed over the IPG and then over the lead–extension connector. Prior to placing the leads, the skin is prepped with alcohol and dried. Skin prep is not used in order to minimize the risk of skin breakdown over the length of the system. Thus far, all measured skin electrode impedances were below $1\,k\Omega$ when using the EMG/IOM machine. The oscilloscope (Tektronix model TDS 3032, Beaverton, OH) impedances measured with an impedance meter (Grass Model F-EZM5, Grass Telefactor, Warwick, RI) were similar to the values given above. The instrumentation is set up as follows: the time base of the oscilloscope is set to $100\,\mu s$/div and $2\,ms$/div. The time base of the Nicolet Viking IV is adjusted to $10\,ms$/screen. The output of the IPG is adjusted to 2.0 Volts. The pulsewidth and frequency are unchanged from the values of the clinically effective settings. For some EMG machines, it may be easier to see the wave when using a larger pulsewidth. It is important to note that, in this configuration, the anodal component of the charge balanced pulse will have a larger amplitude and shorter pulse length and will therefore look a little different than the waves depicted in Figure 15.7.

The contacts are tested in sequential order starting at the most ventral contact or most cranial contact. The scale of the recording device is adjusted so the full wave is visible on the screen. A short segment of each contact is recorded, noting both the amplitude and phase of the signal. After each contact is tested, the amplitudes are compared. If no major amplitude changes (greater or less than approximately 25% of most of the leads) are observed, each contact is tested again. During the second test, the extension and connector are all tapped with a finger. In later tests, both of these techniques were combined. The skin over each element is also gently manipulated. Any system amplitude and phase perturbations are investigated further. Figure 15.7 shows an example of such a signal change. In many cases, the location of the failure can be found by gentle manipulation of the system.

For an 'open' circuit with two faulty conductors, moving the active test electrode over the implanted extension and lead will record a large reduction in amplitude at the site of the break during manipulation. This is due to the two open leads shorting during the manipulation. For a single wire 'open' circuit, one may or may not observe a large amplitude decrease. For the single lead 'open' case (where the lead has penetrated through the insulation only), the test needs to be performed in a bipolar mode and the stimulation voltage may need to be increased to 3 or 4 volts. Care must be taken when doing this to avoid causing the patient any discomfort.

EXAMPLES

The patient is a 75-year-old male with PD. The patient underwent staged bilateral STN DBS implants in the fall of 2000. Two and a half years later, the patient presented with electrical shock sensations down the anterior and posterior aspects of his leg and arm. He stated that these sensations were some-times associated with movement of his neck, but not one specific movement. Manipulation of the battery and extension wires failed to reproduce the symptoms. No abnormalities were found with the programmer. We hypothesized that a transient short was present due to the paresthesias in the leg and arm. Since the sensations were on the left side it was felt that the problem was with the right IPG.

Figure 15.8 shows the results of the EMG signal tests. A reference lead was placed over the IPG and the active lead was placed over the lead–extension connector. When the IPG was set to monopolar mode (Fig. 15.8A–D), contacts 0 and 3 showed a full signal while contacts 1 and 2 exhibited an amplitude reduction of >50%. Bipolar testing was then performed. Figure 15.8E shows the results of the bipolar test with no pressure on the extension. Figure 15.8F shows the results of the test when pressure was applied over the right mastoid region, the only place where pressure elicited this change. Figures 15.8G and 15.8H show the same type of testing on the left IPG with no changes in the signal. Based on these results, the right extension was replaced with a resolution of the difficulties.

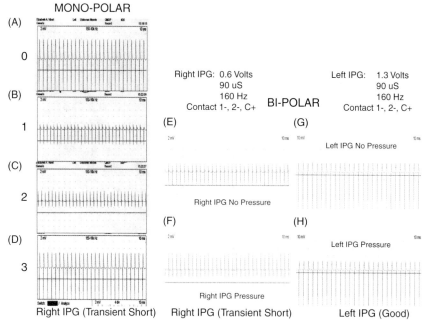

FIGURE 15.8 This series of tests shows how a transient short is seen on the output of an EMG machine as compared to the good side. It was thought that reduction in amplitude was due to a partial short where there was still some continuity between the IPG and the lead, but the wires were touching in the insulation also.

CONCLUSION

The testing paradigms described in this chapter have been developed through more than 15 years of experience with implantable stimulation systems. Our current testing algorithm is exhibited in Figure 15.9. Clinic testing times range from 30 minutes to 4 hours depending on how complex the fault mode is and if they are a function of the disease (i.e. does the patient need to have the disease symptoms to notice the failure?). Unfortunately, even after working through all of the testing described herein, there are still a small number of cases where the failure may not be localized non-invasively. Often, in such cases, simply opening the system and cleaning the connections has restored proper device function. We have assumed that, in these cases, failure was due to fluid in the connector, but were not able to prove this. Nevertheless, we are typically able to

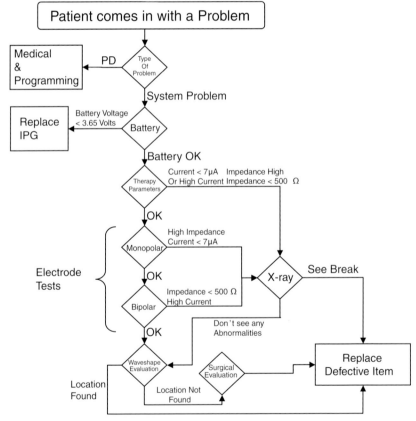

FIGURE 15.9 This flow chart is designed to offer a general pathway through trouble shooting starting from the simplest non-invasive tests to the most complex invasive tests. Even though the most complex tests may offer instantaneous results, we feel that the invasiveness to the patients is not warranted until all other options have been met.

localize system faults accurately while minimizing the number of surgeries and the number of surgical incisions needed to fix the problem.

REFERENCES

1. Beric A, Kelly PJ, Rezai A, et al. Complications of deep brain stimulation surgery. *Stereotact Funct Neurosurg.* 2001;77:73–78.
2. Oh MY, Abosch A, Kim SH, Lang AE, Lozano AM. Long-term hardware related complications of deep brain stimulation. *Neurosurgery.* 2002;50:1268–1276.
3. Umemura A, Jaggi JL, Hurtig HI, et al. Deep brain stimulation for movement disorders: morbidity and mortality in 109 patients. *J Neurosurg.* 2003;98:779–784.
4. Joint C, Nandi D, Parkin S, Gregory R, Aziz T. Hardware-related problems of deep brain stimulation. *Mov Disord.* 2002;17(Suppl. 3):S175–S180.
5. Lyons KE, Wilkinson SB, Overman J, Pahwa R. Surgical and hardware complications of subthalamic stimulation. A series of 160 procedures. *Neurology.* 2004;63:612–616.
6. Constantoyannis C, Berk C, Honey CR, Mendez I, Brownstone RM. Reducing hardware-related complications of deep brain stimulation. *Can J Neurol Sci.* 2005;32:194–200.
7. Goodman RR, Kim B, McClelland III S, et al. Operative techniques and morbidity with subthalamic nucleus deep brain stimulation in 100 consecutive patients with advanced Parkinson's disease. *J Neurol Neurosurg Psychiatry.* 2006;77:12–17.
8. Turner JA, Loeser JD, Bell KG. Spinal cord stimulation for chronic low back pain: a systematic literature synthesis. *Neurosurgery.* 1995;37:1088–1095.
9. Turner JA, Loeser JD, Deyo RA, Sanders SB. Spinal cord stimulation for patients with failed back surgery syndrome or complex regional pain syndrome: a systematic review of effectiveness and complications. *Pain.* 2004;108:137–147.
10. Okun MS, Tagliati M, Pourfar M, et al. Management of refereed deep brain stimulation failures: a retrospective analysis from 2 movement disorders centers. *Arch Neurol.* 2005;62:1250–1255.
11. Medtronic. Medtronic model 3387 and 3389 DBS Lead kit for Deep Brain Stimulation implant manual 2006.
12. Medtronic. Medtronic model 7482 DBS Extension kit for Deep Brain Stimulation implant manual 2006.
13. Webster JG. *Medical instrumentation: application and design.* New York: John Wiley and Sons; 1995.
14. Shils JL, Patterson T, Stecker MM. Electrical properties of met al microelectrodes. *Am J Electroneurodiagnost Technol.* 2000;40:143–153.
15. Geddes LA. *Electrodes and the measurement of bioelectric events.* New York: John Wiley and Sons; 1972.
16. Burbaud P, Vital A, Rougier A, et al. Minimal tissue damage after stimulation of the motor thalamus in a case of chorea-acanthocytosis. *Neurology.* 2002;59:1982–1984.
17. Haberler C, Alesch F, Mazal P, et al. No tissue damage by chronic deep brain stimulation in Parkinson's disease. *Ann Neurol.* 2000;48:372–376.
18. Caparros-Lefebvre D, Ruchoux MM, Blond S, Petit H, Percheron G. Long-term thalamic stimulation in Parkinson's disease: postmortem anatomoclinical study. *Neurology.* 1994;44:1856–1860.

Commentary on Intraoperative Evaluation

Jeffery E. Arle, MD, PhD

Director Stereotactic and Functional Neurosurgery section, Lahey Clinic, Burlington, MA;
Associate Professor of Neurosurgery, Tufts University School of Medicine, Boston, MA

Although I am co-author on this chapter with Dr Shils, my co-editor of the book overall, I wish to emphasize that it is Dr Shils who has so well documented and organized the details of troubleshooting DBS system failures, both in terms of methodology and in technique. His experience in this regard spans the resulting complications and puzzles following over 750 DBS lead placements with multiple different surgeons, both more and less experienced, over 15 years. It is my belief that there is no better analysis of this aspect of neuromodulation, that of sorting out and fixing system failure, than this. Despite this high level of praise, I feel the reader deserves to consider additional perspective on what has been described, because it will round out the discussion to include other aspects of neuromodulation beyond DBS, and because it will hopefully also lend the complementary views of a surgeon often involved in the same endeavor. Troubleshooting is an important and under-addressed topic in neuromodulation and, as such, should be read intently by anyone involved in placing or caring for patients with such devices.

First, I wish to emphasize that what has been described for DBS herein also applies almost in its entirety to SCS, vagal nerve stimulation (VNS), and peripheral nerve stimulation (PNS). The philosophy of the approach is the same. The methods of analysis are very much the same. The considerations in terms of fixing the problem, while different in detail at times, are basically otherwise the same. As pointed out in the early part of the chapter text, there are risks to the patient and to the system in question by re-operating on it, simply replacing components because one does not want to take time to gain information in other ways beforehand. Even more egregious are clinicians who may decide to replace entire systems in question with a device from a different company, only because they refuse to work with the currently implanted device company. Certainly, the patient may want to change devices, or a device may have a new or different feature that allows one to solve a clinical problem for a patient in a unique way and changing to that device might be advised on clinical grounds.

This kind of thinking has at times occurred in evaluating spinal cord stimulators that seem to be failing. Perhaps the patient has had difficulty with one company in terms of reprogramming service, or personality interaction. Now the IPG hurts them, only when 'on', and sticks out a little. Maybe the coverage is not complete and there is a possible break in the insulation of the lead or possible shorting within contacts on the lead. In any case, the system likely will be explored and replaced completely, repositioning it at the same time. In situations like this, the patient may request using the device of a different company without any conflict of interest on the part of the surgeon. But, simply replacing an entire system for another company's system when all the patient needs is a battery changed is indefensible.

Some aspects of troubleshooting a failed device must take into account the overall circumstances of the patient and their goals. A device may have been placed with the best of intentions, with reasonable forethought and testing, only to find that later it is not helping, or no longer helping, and the patient wants the device removed. It is important to determine the real reason the patient wants the device taken out. Perhaps they never really needed the device, or never needed *that* device. One should be reluctant to remove a device if it is not hurting the patient in and of itself, even if it is not being used by the patient anymore. There is some risk to the patient in removing a device – typically infection at a minimum, but in truth, there could be stroke, seizure, abscess, hemorrhage, or anesthetic complication removing DBS systems; carotid artery, jugular vein, recurrent laryngeal nerve, sternocleidomastoid muscle or vagus nerve injury removing VNS systems; spinal cord injury, dural tear with CSF leak, or hemorrhage with SCS removal; and nerve injury from PNS removal. So every effort should be made in discussing these aspects of the process with the patient beforehand. It is not unreasonable simply to leave a device in place and turn it off if it is no longer used but does not bother the patient. In contradistinction, it is important to consider, particularly in DBS systems, that if the patient is deriving significant benefit from a device, and it looks like it might need exploration, removing and/or replacing the device may be a greater risk, or loss of benefit, than leaving the device as is, and working with the defect, even in cases of infection at times, where debridement and antibiotics without removal is worth attempting.

Although different tissues have differing ranges of expected impedance values if tested in vivo, such knowledge should not dissuade the clinician from testing SCS, VNS, and PNS systems and deducing whether or not the values found are still clearly outside of any normal range. Most of the values discussed in the text here apply in these other systems, with only slight variation. As such, the clinician must take on the responsibility of making such determinations to the best of their ability in a stepwise application of logical deduction. Only in this way can trauma and risk to the patient be minimized.

Finally, and perhaps most importantly, the clinician involved in neuromodulation is best served by first *embracing* the entire concept of neuromodulation overall and, as the prime corollary to this, understanding that a part of neuromodulation involves being able to handle the complications and solving the mysteries of failure when they occur – and they *will* occur. The surgeon may not be the one who programs most of the devices. They may not be adjusting medications of the patient much. They may not see the patient in most of their visits back to the institution. This mode of thought, however, with patient safety and benefit foremost, will lead to the most responsible use of neuromodulation overall.

STUDY QUESTIONS

1. Define a decision algorithm for troubleshooting all devices regardless of type or application. Would it need to include extra equipment for testing? If so, what would be needed at a minimum?
2. Consider the various types of neuromodulation practice (solo private pain, hospital-based DBS, academic multiapplication) and the resources, in terms of personnel and equipment necessary to handle the troubleshooting aspects of care.

Part VI

Postoperative Management

Chapter 16

Programming – DBS Programming

Jay L. Shils PhD

Director of Intraoperative Monitoring, Lahey Clinic, Burlington, MA

INTRODUCTION

One of the greatest advantages of neuromodulation therapies, and in contrast to ablative procedures of the past is that the anatomical and physiological substrate of the therapy may continue to be manipulated long after the implant surgery is accomplished. The adjustment of stimulation parameters postoperatively is called *programming* and allows the clinician to modify the effect of the implanted device. It is the goal of programming to optimize the delivery of the electrical therapy by changing specific parameters of the stimulation signal. Even though each area of neurostimulation therapy affects different clinically relevant modalities, and the specific neural elements that are targeted with stimulation may be different, the underlying approach behind programming is similar. When approaching the neuromodulation patient, the clinician needs to know the goals of the programming session (e.g. why surgery was performed and the expected reduction in symptoms), the potential side effects (both acceptable and unacceptable), the specific neural elements involved, and the time course for the benefits of the therapy to take effect. This chapter will describe a common methodology for programming and then use examples encountered in common neuromodulation programming situations.

OVERVIEW

Programming is the term given to the spatial and temporal adjustment of the stimulation signal parameters used to treat specific neurologic disorders. In all approved electrical delivery systems, there is a standard set of parameters that can be adjusted which are:

1. amplitude
2. pulsewidth
3. frequency
4. polarity (Fig. 16.1).

Essential Neuromodulation. DOI: 10.1016/B978-0-12-381409-8.00016-4

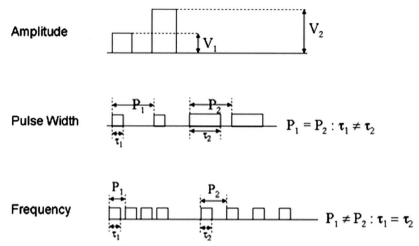

FIGURE 16.1   Stimulus waveform parameters.

The clinical relevance of each of these parameters is discussed in Chapters 6, 7, and 8 of this book. Most systems (Boston Scientific, Valencia, CA; Medtronic Inc., Minneapolis, MN; St Jude Medical, Plano, TX) allow for the activation of at least one of a multiple of electrical transfer surfaces (interfaces) that make contact with tissue. These interfaces are known as the electrodes. The polarity of the electrodes includes either the cathode or the anode and both are needed for a complete circuit. The cathode is defined as the electrode which is negative in the initial phase of the stimulus waveform (see Chapters 6, 7, and 8 for a more detailed description of these parameters). The polarity of the electrodes is the most commonly modified parameter, other than amplitude. Pulsewidth and frequency tend to be adjusted later either to reduce adverse effects or to focus the effect of neural stimulation on a specific neural element (e.g. cell body, axon, and axon size) [1–3].

When first approaching a programming session, irrespective of the therapy modality, the clinician and the patient need to discuss the desired benefit and also the acceptable adverse effects. Even though this has been discussed prior to the surgical intervention, it is important to keep these two opposing stimulation effects in the forefront during the application of the therapy. Also, the mindset during initial programming is different than during follow-up programming or complication and change assessment; during initial programming the process is evaluation while during follow up the process is fine-tuning, while during troubleshooting it is evaluation and hardware troubleshooting, described in Chapter 15, is searching. In all, however it is only four parameters that are accessible by the programmer.

Figure 16.2 shows a flow diagram representing the overall pathways for successful programming. Therapies but may differ on the internal details of each decision or action point, but the main points in the figure are similar for all therapies.

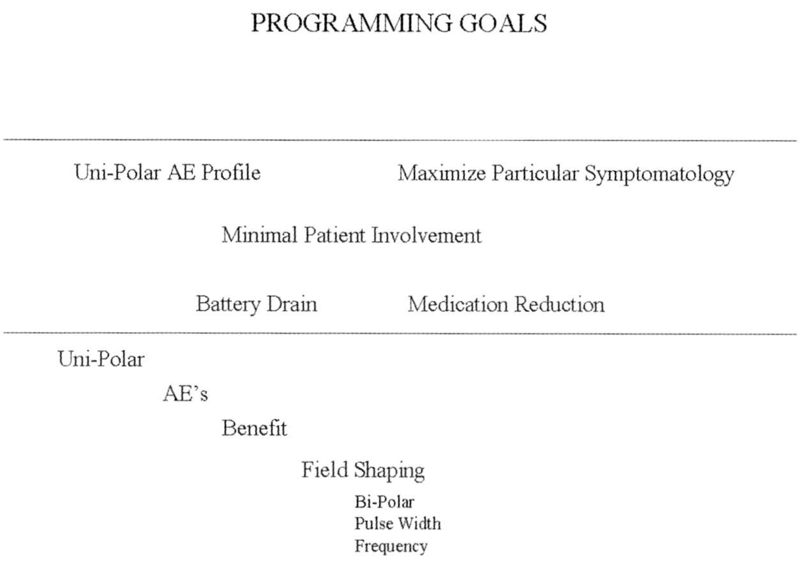

PROGRAMMING GOALS

Uni-Polar AE Profile   Maximize Particular Symptomatology

Minimal Patient Involvement

Battery Drain   Medication Reduction

Uni-Polar

AE's

Benefit

Field Shaping

Bi-Polar
Pulse Width
Frequency

FIGURE 16.2   Programming goals and cncepts.

PROCESS

Hardware Evaluation

All neuromodulation systems rely a continuity of stimulation current from the stimulation generator (more commonly called the pulse generator) to the tissue, so it is critical that implanted hardware be functioning. The implanted hardware, as described in Chapters 6, 9, 10, and 11, includes the pulse generator (PG), the electrode, or tissue–stimulation interface, and the connections between the two. Many new devices also include external devices that can recharge the implanted battery and also allow the patient to adjust various parameters on their own. At all visits, including the initial visit, testing of this hardware should be the first agenda item during a patient programming visit. The time required for this is minimal yet, as described in the troubleshooting chapter (Chapter 15), it is both helpful in determining basic issues such as the lack of proper charging, or creating a record of the patient specific normative electrical parameters of impedance and current which can be helpful at future visits. Hardware evaluation consists of three components:

1. system continuity
2. battery state and longevity
3. utilization history.

System Continuity

Continuity assessment involves passing a known quantity of current, or generating a known voltage difference across the active and reference leads, from the pulse generator and then recording the variation in potential or current across the system. The method of testing is somewhat different for each manufacturer and thus the normative values are different. Also, since the implanted devices are in different tissues for different therapies and thus may overlie areas of differing biological material (e.g. CSF, gray matter, white matter, dura, blood), the exact values of the impedance can be highly variable. Thus, it is not the exact values that are important but their consistency, after an initial period of adjustment. Immediately after implant, and for about 3 months, tissue impedance changes due to the damage from the implant and the healing process. Devices that are passed through tissue, such as deep brain stimulation (DBS) leads, cause more damage than devices that are placed directly on the surface of the tissue, such as for peripheral nerve stimulation, or on neural coverings, such as spinal cord stimulation devices. Even with these devices there are impedance changes due to scarring and normal biologic foreign body reactions, yet they are less damaging under normal circumstances [4–6]. By recording these values at each visit, the data can be used to evaluate potential continuity breaks, short circuits, or even device movements, which although unlikely, are possible [7]. The details of localizing the points of these failures are described in Chapter 15.

Battery State and Longevity

Since there are so many different programming configurations, the exact life of the stimulation power supply cannot be determined. In all present devices, the power supply is a battery. For primary cell batteries (non-rechargeable), the described life span is about 5 to 7 years yet, in practice, it is usually between 2 and 5 years for DBS devices, and between 2 and 7 years for spinal cord stimulator systems. Factors that affect battery life are described in Chapter 11 yet, in general, the higher the amplitude of stimulation, the more time the stimulator is on (larger pulsewidth and higher frequency), and the greater area to which the stimulator is delivering energy, the shorter battery life will be. Also, in order to get more energy out of the batteries, specialized circuitry, in some devices, is designed to activate at certain values and will increase energy consumption, thus depleting the battery sooner. For rechargeable batteries, the life span is on the order of 9 years yet, as the battery gets older, the time between charges may get shorter, thus the amount of time the device can deliver therapy with the same charge is reduced. Finally, as battery use continues, the amount of energy output is decreased. This decrease can be either smooth and slow with the shape of a long hill, or quick with the shape of a cliff. Each manufactuter allows for checking of the battery status. Some devices, such as the Medtronic Soletra™, give battery output in volts, while others, such as the St Jude Libra XP™, give codes relating to the device status. Early systems, such as the Soletra, have slow or ramp-shaped battery energy depletion curve so, even though the manufacturer recommended battery replacement, at a device indication of 'low', our center found that when the battery voltage was at 3.65 volts or less a battery replacement was performed in order to assure constant therapy. As new batteries are introduced, the value at which the battery should be replaced needs to be adjusted. The new Activa™ (Medtronic) primary cell implantable pulse generator (IPG) has a more stable output and thus does not need to be replaced as soon. One important note is that with devices having a sharp energy reduction curve versus time, the time between implants needs to be watched more closely since there will be a quick reduction in therapy if the battery is not replaced. Finally, in order to increase shelf-life, some batteries include a special oxidative layer to impede small transfers of electrons from one polarity to the other. This layer can interfere with impedance and device testing when the battery is at certain values (e.g. the St Jude Libra™).

Utilization

As technology advances, the information stored in the device and accessed from the clinician programmers increases. Original devices (such as the Medtronic Itrel II and III) described length of time since last data reset, percent of device 'on' time, and number of device activations, or on/off cycles. Even these limited data were helpful in determining if the device was inadvertently shut 'off' or was inappropriately cycling on and off. For example, if a patient came into the clinic complaining of reduced stimulator efficacy and during interrogation of the device it was noticed that device was in the 'off' state, one could, utilizing the total time since the last programming session (assuming the device was reset) and the percentage of time the device was 'on' in that period, calculate the approximate date the device went 'off' (assuming one activation).

$$\text{Days since device OFF} = \frac{(\% \text{ device on})(\text{total time since last reset})}{24} \quad (16.1)$$

With this information, discussion with the patient and or family members can, in most cases, determine the exact situation that caused the device to turn off and then avoid it in the future. Newer devices are able to keep more detailed information of the exact times systems are in either the 'on' or 'off' state, charging status, program utilization, multiple program percent usage, and patient adjustment times. One major problem that patients with rechargeable systems run into is improper or incomplete charging. A good example of a regular charging cycle is shown in Figure 16.3 where you can see that the patient charges their system every other day to 100% charge thus never allowing for a low stimulation output potential. Using this graph, the clinician can determine if the device is being charged and if not get an understanding of why. If, for example, the patient states they charge every day it is most likely that the patient is not placing the charger over the IPG properly and thus the clinician can re-educate the patient on proper placement of the recharger to be able to get more efficient system utilization.

4 Week Battery Charge Profile

Battery Charge Profile

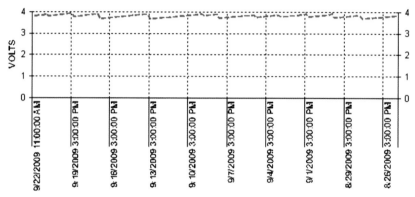

FIGURE 16.3   Patient utilization outputs.

INITIAL PROGRAMMING SESSION

The initial programming session should include a complete therapeutic evaluation. Even if an intraoperative evaluation has been performed (which is recommended when beneficial and adverse events can be evaluated), this initial session creates the roadmap for all future sessions. During this session, each contact is independently (if the system allows) evaluated for both the beneficial and adverse effects of stimulation. For example, during programming of the movement disorder patients, testing should be done in different medication states. Some of the effects of stimulation, both beneficial and adverse, may take hours to days to become evident. When starting a programming session, the patient needs to be relaxed and in a stable state. For movement disorders, having the patient in the medication 'off' state is recommended to maximize the beneficial effects of stimulation. For initial pain programming, the pain should be felt by the patient, but should not be so excessive (unless there is no other option) that it can interfere with the patient's ability to report results. For other stimulation therapies, such as vagal nerve stimulation for epileptic seizures, DBS for Tourette's syndrome or psychological disorders, adverse effects may be the only initial conditions that can be detected due to both the time course of the therapy and the duration of the symptoms or there may be only one state that is 'safe' for the patient.

All devices are delivered in a standard configuration from the manufacturer. Prior to starting the programming session, a complete hardware check should be performed. This includes checking the continuity of all electrodes and the battery voltage and current. Due to variations in manufacturers and also between devices, detailed procedures for these steps are defined in the device's user manual. It is important to note the initial amplitude, pulsewidth and frequency used for these tests since they are the ones that should be used at all times. For subthalamic nucleus (STN) and globus pallidus pars interna (GPi) procedures, we recommend 210 μs, 30 Hz and 2 volts. For ventral intermediate nucleus (VIM) procedures, we recommend 210 μs, 30 Hz, and 1 volt, for motor cortical stimulation, we recommend 410 μs, 30 Hz, and 4 volts, for spinal cord stimulation, we recommend 210 μs, 60 Hz, 2 volts. For other deep brain procedures, the 210 μs, 30 Hz, and 2 volts values should be sufficient. Vagal nerve stimulation systems have an automated protocol for evaluating the impedance and battery life. Older devices sent the energy out over a continuous period, while the newer devices send pulses out that are much less detectable by the patient. After, or prior to the hardware checks, a patient evaluation should be performed.

Initial Programming

For movement disorders therapy, it is recommended to start with a pulsewidth of 60 μs and a frequency of 185 or 135 Hz for all initial evaluations. These higher frequencies (>100 Hz) are chosen for movement disorders since they correspond with the best results [8–10] yet, for other conditions, lower frequencies appear to be better. For instance, weight control [11], pain [12], and dystonia [13] have shown a benefit when using lower frequencies. Studies investigating DBS treatment for psychiatric disorders have found that a wide range of frequencies offer benefit depending upon the disease and brain location stimulated [14]. This variation in beneficial ranges indicates that, as new treatments become available, optimization of programming may not be easily realized early on. We recommend initial programming be done in monopolar mode and at least 2 weeks after lead implant to give the brain and the patient time to recover from the procedure [15]. This programming session should be done in the patient's 'off' medication state if medication reduction or cessation is one of the end point goals. For some conditions, this 'off' state testing may not be easily achieved, such as in some dystonia patients and psychiatric patients and thus 'on' medication initial programming may be needed.

After selecting the initial electrodes, stimulation testing can begin. Starting at 0 volts or mA go up in 0.1 volt or mA increments until the first adverse effect is noted or until 2.0 volts or mA is reached. Note the adverse effects (AEs) if detected. Perform an evaluation of the relevant patient functions (this will depend on the DBS therapy modality). For all therapy modalities, adverse events need to be noted and the level at which they occur. Depending upon the therapy, the early ability to notice beneficial effects may be limited. For Parkinson's disease (PD) and essential tremor (ET) procedures, the beneficial effects may be noticed almost immediately, or at least within a few days. For example, the effects in PD occur so early that we can use a modified version of the Unified Parkinson's disease rating scale (UPDRS) part III. When testing a PD patient during either STN or GPi programming, we test rigidity, bradykinesia, and tremor at all stop points (0.5 volts or mA), but will test gait and balance at 1.0 V (or mA) increments. It is important to note transient adverse effects since they can indicate potential changes in the electrode over time (see next paragraph below). Importantly, these transient effects are common and the patient should be informed as such. After documenting positive and negative effects, increase the energy in 0.1 V (mA) steps until the next half volt or mA point is reached. For PD, repeat the UPDRS III scoring and also note the AEs. If AEs are noted before this point is reached, stop increasing the energy and note the type of effect. If the effect is permanent, stop increasing the energy. After testing the first contact (most distal), sequentially test all other contacts. It is recommended that a short rest period be given to the patient between each electrode to minimize the chances of fatigue. Figures 16.4 and 16.5 show the most common side effects encountered during STN and pallidal stimulation for PD.

One important feature of some side effects is that they can be transient in nature, and that the transient time course can be on the order of days. For example, PD patients may become hypophonic and dysarthric during the session, yet over a 7–14 day period these problems become very much reduced. Adverse effects almost never completely resolve, but they are manageable by the patient. Gait disturbances and even falling are a common adverse effect noted by STN DBS patients. During programming sessions it is very difficult to evaluate this effect and the patient and caregiver need to be made aware of this. It is critical to inform the patient to keep track of potential side effects and let them know they should be careful when doing standard activities such as walking early after both the initial programming session and follow-up programming sessions. Additionally, in patients who were having benefit for a long period of time and start to develop an AE, with no change in electrode position or change in diagnosis, then other reasons should be investigated such as spine problems or accommodation.

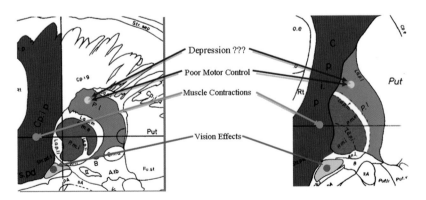

FIGURE 16.4   GPi stimulation adverse events.

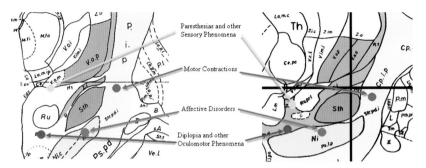

FIGURE 16.5   STN stimulation and adverse events.

For PD and ET deep brain stimulation patients there is no reason to stimulate at greater than 4.0 volts or 4.0 mA during this testing period. Initial programming, has been able to reach good therapeutic benefit in PD. After a few years of DBS in ET patients the stimulation may need to go greater than 4.0 volts due to accommodation[1] [16]. Since the tremorgenic zone is located very close to the VIM/ventral caudal (VC) border [17,18], even low levels of stimulation can adversely affect the VC nucleus. During initial programming in tremor patients, these tend to be more transient sensory affects, felt most in the hands and face. When increasing the stimulator energy, sensations in the area where the tremor is the worst are a very good sign for reduction of tremor. The most common adverse side effect of VIM stimulation is disarthria or sensory parasthesias in the hand and/or face. Transient or low level parasthesias are found to be acceptable to patients, while disarthria is typically problematic. Trying complex electrode configurations (Fig. 16.6) can help increase local stimulation while reducing effects on more distant structures, as of the writing of this text, there are still no ways directionally to focus the stimulation field. The size of the stimulation field also can be the cause for having to use stimulation amplitudes greater than 4.0 Volts or mA. Since some targets are large or the DBS electrode is not in the center of the functional target, more energy is needed to reach the area of therapy. Larger nuclei such as the GPi, and procedures involving cortical structures can require very large energies. Voltages for cortical structures can require up to 7.0 volts, although one needs to be cognizant about generating seizures.

Once all contacts on one laterality have been tested, the patient should be given a 15–30 minute rest period. At this time, the clinician can review the data and choose the most optimal setting. Once again, this setting is the one with the greatest improvement in disease-specific symptoms while inducing none or minimal adverse effects. In some cases, it may be impossible to get rid of all

1. For tremor, it is the goal intraoperatively to be able to reduce or even stop tremor with the microelectrode stimulation. When this occurs, we know that the electrode is in the center of the tremorgenic zone.

(a)   (b)

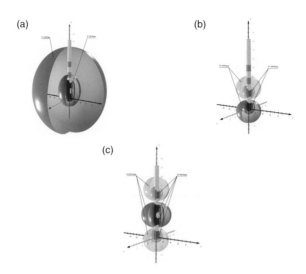

(c)

FIGURE 16.6 Representations of the voltage field for some common multielectrode stimulation configurations. Cooler colors (green and blue) represent the cathodal stimulation side of the circuit while warmer colors (red and yellow) represent the anodal side of the circuit.

disease-specific symptoms completely with or without some adverse effects. Other than for tremor cases this is an acceptable initial programming condition since some of the beneficial effects (as well as some of the adverse effects) may take time to manifest. For the tremor case, the contact with the greatest tremor reduction should be chosen even with minor parasthesias, or speech irregularities since it has been the author's experience that these adverse effects often dissipate over time. If stimulation is bilaterally applied, then once the rest period is over, the second side should be evaluated in the same manner as the first. It is recommended that this second side evaluation be performed with the first side stimulator in the 'off' condition. Even though there are some bilateral effects from a unilateral stimulation [19] independent testing should be performed.

Once second side testing is performed and documented, the patient should have both stimulators turned 'on' to the optimal and combined effects sent to the waiting area for at least 30 minutes. Since, as mentioned above, there are some bilateral effects from unilateral stimulation, it is important to make sure that the combination of both stimulators is not causing any adverse effects. If there are cases of multiple types of stimulators in patients for varying reasons, it is important to test all devices at the maximum possible patient-controlled settings. For movement disorders, one should perform the disease appropriate rating score in this 'on' stimulator, 'off' medication condition. For some diseases, it is not recommended to perform this initial test in the 'off' medication condition due to the slow nature of the effect (i.e. dystonia) or potential adverse effects of stopping medication abruptly (i.e. neuropsychiatric). For these conditions, it is also near impossible to determine beneficial effects at this session, thus adverse effects are all one can see. Yet, for the diseases where the change in the patient is quick, when going from the 'off' medication state to the 'on' medication state, the 'on' state should be checked at this setting. To do this, the patient should be given their normal dose of medication and asked to wait until the medication 'kicks in'.

Once in the 'on' medication state, both stimulators should be turned 'off' and a disease-specific score should be performed as a baseline. Each each contact should be evaluated, in a method similar to the 'off' state condition. When evaluating the best settings for the stimulator using both the 'on' and 'off', the 'optimal' setting can become skewed. This happens if the optimal setting for the 'off' and the 'on' condition are far apart, which is luckily a rare condition. Both 'off' and 'on' adverse effects (rigidity, parasthesias, dyskinesias, gait, speech, pain) tend to occur with the same medication state, thus it is usually only beneficial effects that need to be considered. Since medication, and stimulation are attempting to accomplish similar goals, in cases where there is a great difference between the two medication cases, the 'on' medication condition takes precedence. For the most part it is not recommended to stop quickly or reduce medications at this visit. In many cases, we may not even change medications until the next visit. By that time, the patient has some idea of how the stimulator will respond. Also, when sending the patient home we recommend that they do not engage in much for 24 hours. It has been our experience that patients tend to get very tired or even feel a little weaker the day of this session. The next visit is usually made 1 month later, with the caveat that the patient can contact us if they have any difficulties. We will make office time available to these patients as some negative effects can be both scary and a little debilitating. In many cases, a phone call or e-mail can determine that the problem is either not related to the stimulation, or transient in nature and thus avoid having the patient come to the office. Some situations however, such as excessive 'on' medication dyskineisias, new freezing, swallowing problems, new burning pains, or loss of facial sensation require a visit.

FOLLOW UP

There are multiple reasons follow up visits occur. During the early sessions, the patient's therapy may need to be optimized, especially if the results take time to manifest. Once the therapy is optimized, the patient should be seen routinely to assure that the device is functioning properly, that batteries do not need replacement, and for general follow up. Trouble shooting is another reason for follow up and is covered in Chapter 15. Finally, since most of these DBS procedures are not cures but therapies, the stimulation may need to be adjusted from time to time in order to maintain a beneficial result. It should be noted that the speed of follow up can have an impact on both the ease of determining what needs to be done and in minimizing the hardship on the patient. Battery changes are part of DBS therapy, even with rechargeable batteries, and thus there should be a protocol for battery evaluation and change-outs.

The time course from therapy problem onset to when they should be seen is somewhat complex. Even though the patient may perceive the problem as stimulator related, it may not be. Due to external factors potentially affecting the therapy, it is unwise to have the patient come in the day they notice a change in therapy (unless the therapy failure has a potential to initiate a life-threatening situation, such as therapies involving psychiatric conditions). Yet changes lasting more than a few days, unless correlated to a specific sickness or patient induced stimulator change, should be evaluated quickly. A good rule of thumb should be to investigate therapy reductions if they last longer than about 4 days. Bringing the patient in sooner may initiate a programming change that is less optimal than the programming that the patient already has. It is common for the patient to come to the clinician's office and be in a better state than at home, which is one result of the placebo effect, and can mask the true result of a programming change [20][2]. If one waits too long (greater than 7 days), recovery of the therapy may be harder. One important effect of increasing the time to program adjustment is in decreasing the ability of the patient and the caregivers to remember events that could have affected the therapy.

After determining that the reduction in therapy is related to the stimulator, there are four pathways to follow depending upon the therapeutic reduction. These four pathways are:

1. increase or decrease in amplitude
2. a. move from monopolar to bipolar configuration
   b. change the active electrodes
3. change the pulsewidth
4. change the frequency.

The easiest change relates to the amplitude of the stimulation. One common misconception that patients have is that more stimulation is better. Stimulation can either be too much or too little. If patients come in with a reduction of therapy right after initial programming, or for some conditions later on, an increase in stimulation amplitude may be able to re-optimize the therapy. On the other hand, excessive amplitude may cause delayed adverse events or, with the addition or reduction of medications, cause new adverse events that can easily be removed by lowering the stimulation. Early in programming, lower stimulation values may be necessary to reduce adverse effects and, initially, these lower amplitudes may offer beneficial results yet, over time, these beneficial effects may wane due to the small number of neurons being affected by the electric field. With time, the parts of the brain that are initially adversely affected by the

---

2. An exception to this rule is during the initial programming period where searching for adverse effects and beneficial effects can take some time to demonstrate. These effects usually occur within the first 48 hours after a change and thus should be reversed as soon as possible.

stimulation may accommodate to the stimulation, requiring amplitudes while having less affect on the parts of the brain that cause adverse effects. This affect is most easily seen in DBS in the VIM for tremor. Initially turning the stimulator 'on' causes many patients to get a parasthesia sensation due to stimulation effects on the sensory nucleus of the thalamus yet, over time (seconds to minutes), the bothersome sensations dissipate. Yet, if the stimulation is too high the sensations never dissipate.

Item 2 is more commonly defined as 'field shaping', yet it really consists of two parts – polarity and location[3]. At present all systems only allow for cathodal monopolar stimulation. Monopolar stimulation uses the case as the reference and thus the electric field is a uniform sphere (Fig. 16.7) around the electrode (assuming the case is far from the electrode). The advantage of such a stimulation pattern is that you can cover more tissue volume while the disadvantage is this extra volume can include tissue that should not be stimulated. Field shaping utilizes other combinations of electrodes and electrode polarities to reduce the extensiveness of the stimulation volume while still delivering therapeutic stimulation into an appropriate volume. Sometimes during a programming session, as the stimulation is increased transient beneficial effects can be noted (e.g. reduction in tremor, pain decrease, warm sensation, reduction in rigidity, 'rush of happiness') for varying periods of time, but eventually disappear. As the stimulator is turned up, these beneficial effects occur on a permanent basis, yet there are additional adverse effects at these new levels. These adverse effects are due to excessive spread of the electrical field. Figure 16.6 shows the field distribution of multiple electrode configurations. There is a 50–70% stimulation area volume decreases as stimulation changes from monopolar (see Fig. 16.6A) to bipolar (see Fig. 16.6B). Simply moving to a bipolar configuration (i.e. bringing the anode from the case to one of the electrodes) reduces the field spread while keeping the intensity the same at a smaller area. When changing from a monopolar to a bipolar configuration, the reduction in intensity volume may be enough to stop the adverse effects, but now the amount of necessary tissue may not be stimulated. By adding more cathodes or anodes, the overall shape of the electric field can be adjusted. One common technique used is to surround a cathode with two anodes. Since the cathodal electrode is the active electrode, it is possible to block adverse stimulation by surrounding the active stimulation field with two anodes (see Fig. 16.6C). In addition to these electrode combinations, it may be possible just to change the cathodal electrode position, thus moving the center, and maximum, of the stimulation field. Many of the manufacturers now offer a method of 'moving' the electric field via a joystick-like control. The user chooses a starting configuration, and then via this 'joystick'

3. It is interesting to note that future DBS devices and present spinal cord devices have combined this therapeutic control effect into a software controlled pulse *steering* function that automatically adjusts electrodes and polarities in response to the movement of a joystick system {Boston Scientific}.

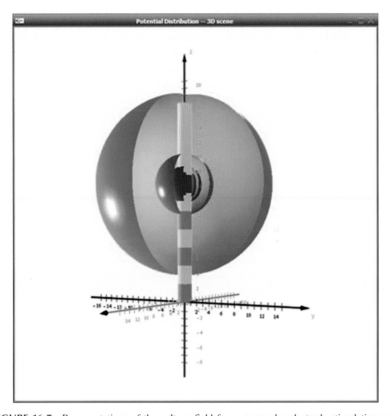

FIGURE 16.7 Representations of the voltage field for a monopolar electrode stimulation configurations. The anode is the case of the IPG. Cooler colors (green and blue) represent the cathodal stimulation side of the circuit while warmer colors (red and yellow) represent the anodal side of the circuit.

can *push* the field ventral, dorsal, left, or right and the programmer calculates the percent of stimulation going to each contact and when to turn a contact on. This has been shown to be very useful for cochlear implants [21] and is finding a role in spinal cord stimulation (SCS). There is still no consensus for deep brain stimulation, though the next generation DBS leads will likely allow for non-cylindrical symmetry in the stimulation field and thus the potential for more useful field shaping and steering. So, for example, if beneficial effects are reached at a specific stimulation level during DBS of the STN, but there are also some corticobulbar effects on speech, it will be possible to shut down the lateral side of the electrode which faces the corticobulbar fibers while still having the medial, anterior, and posterior areas reaching full stimulation.

Adjustments of frequency and pulsewidth are presently more nebulous when it comes to cortical and subcortical stimulation. Data published by Benebid et al [10] show that low frequency (<100 Hz) exacerbates tremor and PD symptoms while high frequency stimulation (>100 Hz) diminishes tremor and PD symptoms, yet work by Alterman and colleagues [13] demonstrates that for dystonia low frequency (60 Hz) stimulation is also beneficial and uses less energy. Frequency can affect the system in two specific ways. Krauthamer et al describe how high frequency stimulation may not allow a cell completely to recover to its resting membrane state, polarize the cellular membrane, cause ionic imbalances, cause modifications to the metabolic load, and effect nerve excitability [22]. Their findings also demonstrate that, in some unmyelinated neurons, this rate can be as low as 190 Hz. On the other hand, lower frequency stimulation can act as a pacer causing the cell to fire with each stimulus pulse. If the *abnormal* cell is firing too fast this sync pulse will not stop the cell from firing fast, it will just add a pulse of a specific frequency to its output and potentially knock out a single pulse due to antidromic collision.

The other way stimulation can affect the system is by causing action potentials on the axon (other than at the initial segment) between nuclear groups. Depending upon the strength of the stimulation[4], an action potential can be initiated in either all or some fraction of communicating axons. These 'externally' generated action potentials will travel in both directions, thus potentially annihilating action potentials from the cell body and also adding their own action potentials in the receiving cell. The amount of information lost from the presynaptic cell (whether aberrant noise of true information) depends on the location of the stimulator, the length of the axon, the frequency of the stimulator, and the geometry of the axonal system. For movement disorders, the frequency range of choice is 130 Hz and higher. For dystonia, the question is more open ended and more studies need to be performed, yet we recommend starting at 130 Hz and, if no benefit is really noticed in the first 3 months, then going to a lower frequency like 60 Hz. For cortical stimulation, there is a wide range of choices [23–26] from 40 Hz to 180 Hz. There is some unpublished work from the Meglio group in Rome, Italy, that found 80 Hz to be an optimal stimulation frequency for movement disorder treatment via motor cortex stimulation (MCS) due to conditioning the motor system at 80 Hz demonstrated the lowest threshold facilitation of First Dorsal Interosseus muscle activity.

Grill and Mortimer [27] theoretically determined that shorter pulsewidths should offer better spatial selectivity than larger pulsewidths. The results of the clinical studies on the effects of small changes in pulsewidth are less clear. Other than avoiding the *voltage doubling level* of some of the devices, pulsewidth has not played a major role in DBS treatment of movement disorders. A study by Andrade et al found no correlation between pulsewidth and reduction in UPDRS III scores in GPi DBS for PD [28] while another study by Kuncel et al described pulsewidths between 60 μs and 450 μs, as having the lowest probability of

---

4. Very high stimulation will completely shut the system down, but this when only very close to the neural element.

predicting tremor response or adverse effects in VIM DBS [8]. On the other hand, Woods et al described a correlation between larger pulsewidths ($>120\,\mu s$) and cognitive decline in ET patients who had VIM DBS [29]. Moro et al did the most detailed study of stimulation parameters in STN DBS and PD and found some correlation to increased pulsewidth and improvements in rigidity yet, for bradykineisia, there were similar results from $60\,\mu s$ and higher [9]. Wu et al studied the effect of pulsewidth size on the control of dyskineisias in GPi DBS and found that higher pulsewidths allowed for low stimulation amplitude for achieving control [30]. Grill and Mortimer [27] have nicely demonstrated that, for very low pulsewidths ($<200\,\mu s$), stimulation is more likely to act on the axon, while at pulse widths $>200\,\mu s$, the effect is seen more at the cell.

The theory behind adjusting the pulse and its effect on stimulation are related to the chronaxie of the system. There is a non-linear relationship between the length of the stimulation pulse (pulsewidth) and the amplitude of the stimulation and the ability to depolarize the cell enough to generate an action potential. This relation varies for different nervous tissue. By adjusting the pulsewidth, one can theoretically reduce the amplitude of the stimulation and get the same effect for a specific piece of neural tissue. This effect is very useful in the spinal cord where the axonal tracts and nervous tissue are aligned in very specific patterns, while in the brain, the geometry of all of the structures make focusing the energy to specific structures much harder with varying parameters, such as pulsewidth and frequency, while field steering may be a important option to help in this area.

MAINTENANCE

Once optimal settings are achieved, patients should follow a standard maintenance schedule. This schedule is based on the type of battery, either primary cell or rechargeable, the disease state being treated, and amount of patient control given outside of the office. When planning office visits for neuromodulation therapy, the battery type is easiest to plan for. Theoretically, present rechargeable systems should last on the order of 9 years or more, while primary cells last on the order 3–5 years when using standard values. We recommend a battery evaluation every 6 months for primary cells for the first 3 years and then a battery evaluation every 3 months after that. This evaluation should include a reading of the battery voltage and an evaluation of the output currents and impedances, if possible. Each device manufacturer displays battery usage in a different way. The point at which the battery should be replaced depends upon the load curve over the life of the battery as described in the battery longevity section. The criticality of a device failure is dependent on the disease being treated. Device failure or therapy failure in a major depression patient is more significant than failure in a tremor patient and thus these patients need to be followed more closely and a potential system failure needs to dealt with emergently. This is why pre-emptive device replacements should be considered and maintenance visits should be included in this discussion. If a patient using a rechargeable system is noticing a shorter time interval between charges, then replacement needs to be considered. If patients feel comfortable evaluating their system (given the capabilities of some of the newer devices) then maintenance visits can be extended out.

One critical element to consider during this maintenance time is the accommodation to stimulation, the continual progression of the disease, and the fact that, as the patient ages, other medical conditions can be confused with device failure. Consider the example of a patient who returned to the office after 2 years of excellent STN DBS control of his PD. At this time, he started to complain of feeling unsteady on his legs and a worsened gait. Over a 6-month period of continual stimulator adjustments thinking this was due to worsening of the patient's PD, no improvement was noted, in fact only worsening was noted. While talking at one visit he stated that he used to do helicopter skiing on all terrains. We performed an MRI of his back (this was before the MRI contraindication with DBS) and noticed that he had very significant lumbar stenosis and instability. After surgery on his back, we were able to go back to his original settings and he was back to doing well. This example demonstrates the importance of considering the complete patient at every programming visit. The other area to consider is accommodation as described above. For example, MS tremor can demonstrate an accommodation to stimulation. In the author's experience, adjustments in both pulsewidth and frequency can sometimes help in accommodation of MS patients, but there can come a point where a *stimulator vacation* is needed, and even that may not help. By turning the stimulator off for a few months, the brain may 'reset' itself and then stimulation can help again.

ALTERNATIVE PROGRAMMING METHODS

As the number of electrode contacts increases, the volume of stimulation increases, the focality of each electrode decreases, and programming becomes much more complex. Some examples of these new electrodes are the microelectrode arrays that have been described in treating spinal cord injury, blindness, and hearing loss [31–36]. In addition to these, present therapies are exploring electrodes that will split standard DBS cylindrical lead into halves, thirds, or quarters (US patent 7,212,867) or using electrodes with up to 32 contacts, such as the NeuroNexus Deep Brain Stimulating Array (NeuroNexus, Ann Arbor, MI), yet spanning almost the same length as the current Medtronic 3387 DBS lead. One advantage of becoming aware of the methods described above is the fact that these methods can be incorporated into expert programming systems or can utilize neural network-based systems to manage their programming. Utilizing compound nerve action potentials from cochlear implants to train a single layer perceptron, Charasse et al [37] were able to demonstrate that artificial neural networks (ANN) are capable of programming the cochlear implant threshold circuitry as well as physicians, critical as the number of detectors increases.

Using patient responses related to programming parameters, one can program more complex neural networks that can converge to a desired result, or at least present the clinician with a set of potential settings that can then be tested to look for the optimal one. This in turn can reduce the amount of programming time from hours to minutes. As time progresses and long-term changes in the patients are correlated to settings and then entered into the ANN system, these variables can also be accounted for.

Field steering is also an important consideration in programming complex electrode arrangements. This option has been implemented in SCS systems using the Precision Plus™ methodology (Boston Scientific, Valencia, CA) the Target Stim™ (Medtronic Inc., Minneapolis, MN), and Dynamic Multistim™ (St Jude Medical, Plano, TX) yet, at the writing of this chapter, only Boston Scientific has independently controlled power sources. By utilizing simplified user interfaces, such as a joystick, directional arrows or even touch screen interfaces, the clinician, programming professional, or even the patient can guide the electrical field to optimize therapy. For example, during the patient's initial visit, the clinician and patient can determine an initial setting, following the guidelines above, then the clinician can set limit parameters, similar to present devices, but with the added control of spatial limits. So, if during the initial programming session, certain electrodes caused adverse effects at low values, the clinican can lock these electrodes out or, in more complex situations, the clinician can lock out specific electrodes at specific values. Once these parameters are entered, the patient can then go home and modify the programming themselves as new effects are noted and as specific lifestyle-related events are encountered.

PATIENT CONTROL

Some form of patient control exists in most systems. For many pain syndromes, and more minimally in movement disorders syndromes, it is almost necessary to give the patient some form of dynamic control of their device. Accommodation to the stimulation, such as tremor therapies or in situations where patient movement can affect the 'location' of the stimulation such as SCS for pain, are examples of having some type of dynamic (or in one realization patient) control is helpful. For other therapies, such as epilepsy treatments (both DBS and vagal nerve stimulation (VNS)), and neuropsychiatric therapies, either not enough is known about the dangers of excessive stimulation, or the therapy window is very tight and patient parameter control would be detrimental to the therapy. Patient control falls into two specific categories: (1) parameter adjustment and (2) program selection. Stimulation amplitude is the primary parameter control for patients. When giving the patient the ability to increase stimulation amplitude, the limit values should be tested in the office before sending the patient home. This testing is more critical for some therapies, such as MCS, DBS, or VNS due to types of side effects such as a seizure. Yet, for therapies such as SCS, direct sensation due to an increase in stimulation amplitude gives the patient necessary feedback to avoid going too high. At present, SCS is really the only therapy where frequency and pulsewidth limits are made use of, but as more is learned about MCS, these should play a larger role.

As the complexity of IPGs increases, the number of user selectable programs is growing. Programs are set by the clinician and then, via a patient *programmer*, the patient can activate a specific program. The programs are complete stimulation paradigms that include contacts, amplitude, pulsewidth, and frequency. Inside each of these programs there are usually multiple subprograms that all act in concert to affect multiple spatial areas simultaneously. Patients can also, in some devices, turn each of these subprograms on and off. Once again, with these complexities and choices, the clinician needs to assess both patient needs, and desired level of involvement. It is recommended that one try to begin with as little complexity as necessary to treat the disease state and then slowly bring in other *user controllable* features as needed. Over the course of therapy, it is also important to investigate frequently the amount of patient-initiated changes occuring. Excessive patient involvement may be necessary, but it more often represents a non-optimal programming situation, or a failed therapy in our experience. A patient who states they can never find a good setting is typically a patient in whom the programs and therapy need to be evaluated.

CONCLUSION

Neuromodulation programming is a multifaceted and multiphased procedure. Each disease state looks for different therapeutic endpoints, yet the programming paradigm for all neuromodulation therapies can be separated into similar categories. Initial programming involves understanding the relationship of the device and the neural tissue. Follow-up programming involves *fine-tuning* of the delivery of the therapy. Maintenance involves tracking device life, and patient involvement with the therapy. Troubleshooting is another area of programming, but is more fully developed in Chapter 15. Present approved cranial neuromodulation therapies still utilize only a small window of the potential range of stimulation devices, accessible in relationship to the spatial range of the electric field and the resonant properties of the neural elements.

REFERENCES

1. Rank B. Which elements are excited in electrical stimulation of mammalian central nervous system: A review. *Brain Res*. 1975;98:417–440.
2. McIntyre CC, Grill WM. Excitation of central nervous system neurons by nonuniform electric fields. *Biophysl J*. 1999;76:878–888.
3. McIntyre CC, Grill WM. Selective microstimulation of central nervous system neurons. *Ann Biomed Eng*. 2000;28:219–233.
4. Lempka SF, Miocinovic S, Johnson MD, Vitek JL, McIntyre CC. In vivo impedance spectroscopy of deep brain stimulation electrodes. *J Neural Eng*. 2009;6:1–11.

5. Sun DA, Yu H, Spooner J, et al. Postmortem analysis following 71 months of deep brain stimulation and the subthalamic nucleus for Parkinsons disease. *J Neurosurg.* 2008;109:325–329.

6. Pilitsis JG, Chu Y, Kordower J, Bergen DC, Cochran EJ, Bakay RA. Postmortem study of deep brain stimulation of the anterior thalamus: case report. *Neurosurgery.* 2008;62:E530–E532.

7. Shils JL, Alterman RL, Arle JE. Deep brain stimulation fault testing. In: Tarsy D, Vitek JL, Starr PA, Okun MS, eds. *Deep brain stimulation in neurological and psychiatric disorders.* Humana Press; 2008.

8. Kuncel AM, Cooper SE, Wolgamuth BR, et al. Clinical reponse to varying the stimulus parameters in deep brain stimulation for essential tremor. *Mov Disord.* 2006;21:1920–1928.

9. Moro E, Esselinh RJA, Xie J, Hommel M, Benabid AL, Pollak P. The impact on Parkinson's disease of electrical parameter settings in STN stimulation. *Neurology.* 2002;59:706–713.

10. Benabid AL, Pollak P, Gao D, et al. Chronic electrical stimulation of the centralis intermedius nucleus of the thalamus as a treatment of movement disorders. *J Neurosurg.* 1996;84:203–214.

11. Halpern CH, Wolf JA, Bale TL, et al. Deep brain stimulation in the treatment of obesity: A review. *J Neurosurg.* 2008;109:625–634.

12. Rasch D, Rinaldi PC, Young RF, Tronnier VM. Deep brain stimulation for the treatment of various chronic pain syndromes. *Neurosurg Focus.* 2006;21:E8.

13. Alterman RL, Miravite J, Weisz D, Shils JL, Bressman SB, Tagliati M. Sixty hertz pallidal deep brain stimulation for primary torsion dystonia. *Neurology.* 2007;69:681–688.

14. Nuttin BJ, Gabriels LA, Cosyns PR, et al. Long-term electrical capsular stimulation in patients with obsessive-compulsive disorder. *Neurosurgery.* 2003;52:1263–1274.

15. Moro E, Poon Y-YW, Lozano AM, Saint-Cyr JA, Lang AE. Subthalamic nucleus stimulation: Improvements in outcome with reprogramming. *Arch Neurol.* 2006;63:E1–E7.

16. Hariz MI, Shamsgovara P, Johansson F, Hariz G-M, Fodstad H. Tolerance and tremor rebound following long-term chronic thalamic stimulation for parkinsonian and essential tremor. *Stereotact Funct Neurosurg.* 1999;72:208–218.

17. Hua SE, Lenz FA, Sirh TA, Reich SG, Dougherty PM. Thalamic neuronal activity correlated with essential tremor. *J Neurol Neurosurg Psychiatry.* 1998;64:273–276.

18. Lenz FA, Normand SL, Kwan HC, et al. Statistical prediction of the optimal site for thalamotomy in Parkinsonian tremor. *MovDisord.* 1995;10:318–328.

19. Walker HC, Watts RL, Guthrie S, Wang D, Guthrie BL. Bilateral effects of unilateral subthalamic deep brain stimulation on Parkinson's disease at 1 year. *Neurosurgery.* 2009;65:302–310.

20. Wickramasekera I. A conditioned model of the placebo effect: predictions from the model. *Biofeedback Self Regul.* 1980;5:5–18.

21. Berenstein CK, Mens LH, Mulder JJ, Vanpoucke FJ. Current steering and current focusing in cochlear implants: comparison of monopolar, tripolar, and virtual channel electrode configurations. *Ear Hearing.* 2008;29:250–260.

22. Krauthamer V, Crosheck T. Effects of high-rate electrical stimulation upon firing in modeled and real neurons. *Med Biol Eng Comput.* 2002;40:360–366.

23. Lefaucher J-P, Drouot X, Cunin P, et al. Motor cortex stimulation for the treatment of refractory peripheral neuropathic pain. *Brain.* 2009;132:1463–1471.

24. Pagni CA, Albanese A, Bentivoglio A, et al. Results by motor cortext stimulation in treatment of focal dystonia. Parkinson's disease and post-ictal spasticity. The experience of the Italian study group on the Italian Neurosurgical Society. *Acta Neurochiurg Suppl.* 2008;101:13–21.

25. Arle JE, Apetauerova D, Zani J, et al. Motor cortex stimulation in patients with Parkinson disease: 12-month follow-up in 4 patients. *J Neurosurg.* 2008;109:133–139.

26. Arle JE, Shils JL. Motor cortex stimulation for pain and movement disorders. *Neurotherapeutics*. 2008;5:37–49.

27. Grill WM, Mortimer JT. The effect of stimulus pulse duration on selectivity of neural stimulation. *IEEE Trans Biomed Eng*. 1996;43:161–166.

28. Andrade P, Carrillo-Ruiz JD, Jimenez F. A systematic review of the efficacy of globus pallidus stimulation in the treatment of Parkinson's disease. *J Clin Neurosci*. 2009;16:877–881.

29. Woods SP, Fields JA, Lyons KE, Pahwa R, Troster AI. Pulse width is associated with cognitive decline after thalamic stimulation for essential tremor. *Parkinsonism Relat Disord*. 2003;9:295–300.

30. Wu YR, Levy R, Ashby P, Tasker RR, Dostrovsky JO. Does stimulation of the GPi control dyskineisia by activating inhibitory axons?. *Mov Disord*. 2001;16:208–216.

31. Mushahwar VK, Jacobs PL, Normann RA, Triolo RJ, Kleitman N. New functional electrical stimulation approaches to standing and walking. *J Neural Eng*. 2007;4:S181–S197.

32. Bamford JA, Putman CT, Mushahwar VK. Intraspinal microstimulation preferentially recruits fatigue-resistant muscle fibers and generates gradual force in rat. *J Physiol*. 2005;569:873–884.

33. Normann RA, Greger BA, House P, Romero SF, Pelayo F, Fernadez E. Toward the development of a cortically based visual neuroprosthesis. *J Neural Eng*. 2009;6:1–8.

34. Sekirnjak C, Hottowy P, Sher A, Dabrowski W, Litle AM, Chichilnisky EJ. Electrical stimulation of mammalian retinal ganglion cells with multielectrode arrays. *J Neurophysiol*. 2006;95:3311–3327.

35. Tatagiba M, Gharabaghi A. Electrical evoked hearing perception by functional neurostimulation of the central auditory systems. *Acta Neurochir Suppl*. 2005;93:93–95.

36. Shepherd RK, McCreery DB. Basis of electrical stimulation of the cochlea and the cochlear nucleus. *Adv Oto-Rhino-Laryngol*. 2006;64:186–205.

37. Charasse B, Thai-Van H, Chanal JM, Berger-Vachon C, Collet L. Automatic analysis of auditory nerve electrically evoked compound action potential with an artificial neural network. *Artific Intelligence Med*. 2004;31:221–229.

Commentary on Programming – DBS Programming

Jeffrey E. Arle MD, PhD

Director Functional Neursurgery and Research, Lahey Clinic, Burlington, MA,
Associate Professor of Neurosurgery, Tufts University School of Medicine, Boston, MA

The rise of DBS systems and the careful expansion of indications and targets within the brain have left in their wake the problem of programming these patients. Dr Shils has developed a deep understanding of DBS programming nuances over many years of broad experience. But he is also a practitioner with a background in electrical engineering and clinical neurophysiology. He has worked closely with high-level centers of excellence in DBS surgery as the technology and indications were being pioneered. As such, we learn a great deal from him in this chapter.

I struggle to see, though, how the future will adapt to allow a wide array of clinicians to program optimally and adjust medications for their patients without completely disrupting the already strained time commitments both practitioners and patients already have.

We will eventually need programming interfaces that automate or 'intelligently' solve some of the programming challenges with information from real-time exam changes, or other data from the patient in adjunctive testing, to make decisions – probably on the fly. This will have to come if not for the simple reason that many newer indications have no reliable immediate alteration in the patient (e.g. depression, obesity, anorexia, addiction, and so on). There may be secondary messenger systems related to transmitter up- and downregulation that require days to weeks to be modulated by DBS – how will the ever more limited time clinicians have be able to allow them to program such patients adequately? Moreover, how will they be programmed optimally from one institution to another? We will likely have a small but active community of research centers doing a small number of the potential cases for these indications who each develop their own manner of solving these problems with little ability to cross-pollinate the information because of time and funding constraints, as well as professional propriety concerns which may also play a role.

The future of DBS is bright, but it is becoming more apparent that the future of research in *using* DBS, in *programming* these systems, will also be bright. Design modifications that allow clinicians to make more creative changes in the parameter space, or allow data-driven algorithms to do it for them, will keep research busy for many years.

STUDY QUESTIONS

1. How do time and resource constraints play roles in managing patient visits, their frequency, urgency, and reasons?
2. It is not too difficult to obtain improvement from many neuromodulation therapies, including DBS. Consider how one would best know, however, whether a patient has been *optimally* programmed.

Chapter 17

Programming – SCS

Gabi Molnar, MS, Lisa Johanek, PhD, Steve Goetz, MS and John Heitman, BS
Medtronic Neuromodulation, Minneapolis, MN

INTRODUCTION

Spinal cord stimulation (SCS) is indicated for use in patients with chronic, intractable pain of the trunk and/or limbs, including failed back syndrome and complex regional pain syndrome. Leads are placed in the epidural space at the segment of the spinal cord that, when electrically stimulated, produce a perceived paresthesia that is comfortable and that covers the patient's area(s) of pain.

Targeting of SCS therapy is guided by patient feedback based on their descriptions (magnitude and quality) and locations of the paresthesia sensations. Stimulation-induced paresthesia covering the patient's area of pain is a statistically significant predictor of the success of SCS [1]. Therefore, the initial goal when programming for SCS is to maximize overlap of the perceived paresthesia with the areas of patient-reported pain. Fortunately, as stimulation parameters change, the associated paresthesia perception also changes rapidly. Thus, the patient's perception of paresthesia serves as a quick way to assess the effects of stimulation. While paresthesias are an immediate effect of the stimulation, pain relief may take several days to weeks to assess. The principles guiding the programming of SCS devices are:

1. to maximize clinical benefit
2. to minimize side effects
3. to maximize device battery life.

ANATOMICAL CONSIDERATIONS IN SPINAL CORD STIMULATION

A main contributor to the successful treatment with SCS is accurate placement of electrodes to deliver current to the correct neural targets. Several anatomical factors are important to consider when programming electrodes for spinal cord stimulation. These include the organization of the dermatomes in the dorsal columns, the variations in cerebrospinal fluid (CSF) that determine the distance between the electrodes and the dorsal columns, and variations in the geometry of the spinal cord.

Essential Neuromodulation. DOI: 10.1016/B978-0-12-381409-8.00017-6

Dermatome organization

Sensory information from light touch and vibration stimuli is detected by specialized receptors in the skin and signaled through primary afferent axons. These axons form the dorsal roots as they enter the spinal cord and ascend through the dorsal columns toward sensory-receiving areas of the brain. Axons from more rostral dermatomes enter the dorsal columns more laterally, resulting in a topographic organization of the dorsal columns (Fig. 17.1). In the traditional representation (Fig. 17.1A), a dermatome is represented by a corresponding discrete 'band' of axons within the dorsal columns. However, a large amount of dermatome mixing has been found within the dorsal columns such that the 'bands' are not as discretely organized as previously thought (Fig. 17.1B).

Because SCS electrically excites these axons in the dorsal columns, a sensation of stimulation-induced paresthesia is detected by the patient. Furthermore, because the axons are arranged topographically, it is reasonable to target the axons associated with the painful area. However, specificity in stimulation can be difficult to achieve because of the dermatome 'mixing' shown in Figure 17.1B. This mixing may lead to differences in paresthesia mapping of consecutive dermatomes among patients as stimulation is moved laterally along the dorsal columns. Although stimulation can be focused to stimulate a subgroup of axons, multiple or overlapping dermatomes may be stimulated because of the dorsal columns' structure. Stimulation of the dorsal columns is necessary when large, widespread areas of pain must be targeted.

Deeper penetration of the electric field within the dorsal columns may increase the number of recruited fibers per dermatome. Since there are fewer afferent fibers from the low back compared to the leg, it is thought that deeper

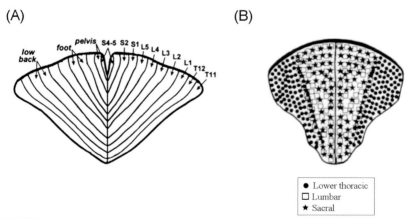

FIGURE 17.1 Topographic representation of dermatomes in dorsal columns. (A) The traditional dermatome representation within the dorsal columns is shown on a slice at the T11 spinal level (from [2]). (B) A large amount of dermatome mixing has been found within the dorsal columns as shown on a slice at the T9 spinal level (from [3]).

(A)

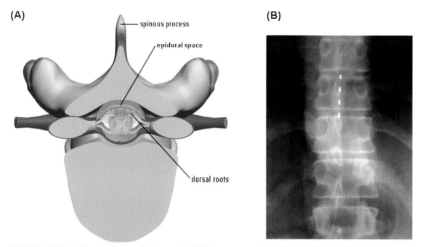

(B)

spinous process

epidural space

dorsal roots

FIGURE 17.2 Views of the spinal cord. (A) Transverse view of the spinal cord showing nerve roots entering the dorsal columns laterally. This creates a 'window' in the dorsal center of the spinal cord for a lead to be positioned within to capture deep fibers of the back without activating the lateral dorsal roots. Note that this 'window' is typically the width of the spinous process. Lateral placement of the lead will lead to activation of the dorsal roots. (B) Anteroposterior (A-P) view of the low thoracic spinal cord (from North et al, 2005). A lead on the midline, within the width of the spinous process, may be used for targeting low back fibers.

activation within the dorsal columns may help generate paresthesias in the low back [4]. However, stimulation of the back often accompanies stimulation of the legs because of the mixing of fibers in the dorsal columns. Placement of a lead on the midline will most likely activate fibers of the back (Fig. 17.2) [5].

Dorsal roots may also be stimulated during SCS. In contrast to stimulation of dorsal columns, which may elicit paresthesia sensations in multiple dermatomes, stimulation of a single dorsal root would result in single dermatome stimulation. In some instances, high-intensity stimulation of the dorsal root can cause unwanted motor activation through a spinal cord reflex pathway. In other instances, lateral targeting of an SCS lead may provide appropriate paresthesia over a focal area of pain. Stimulation of the dorsal roots may be preferentially achieved if the lead is located laterally (see Fig. 17.2).

Variations in the dorsal CSF layer thickness

One challenge of SCS is that the electrodes and the neural targets of stimulation are separated by a distance primarily determined by the thickness of the dorsal cerebrospinal fluid layer, or dCSF. This distance varies by vertebral level, and the variation in distance between the leads and spinal cord results in differences in stimulation thresholds for the perception of paresthesia (Fig. 17.3) [6]. This finding indicates that the farther the lead is from the dorsal columns (due to dCSF), the more energy will be needed to activate neurons in the dorsal columns.

Thickness of CSF Layers and Perception Thresholds

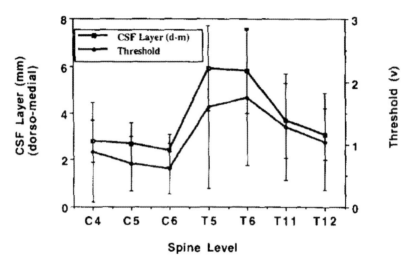

FIGURE 17.3   Dorsal CSF layer thickness (dCSF) correlates with perception threshold. Mean and standard deviation of dCSF and perception thresholds are shown at various vertebral levels (with permission from [6]).

Data also suggest that dCSF varies with patient position [7]. At the low thoracic level, the magnitude of the spinal cord movement between supine and prone positions was 2.2 mm at T11 and 3.4 mm at T12, likely due to the effects of gravity [7]. Several studies have shown that therapy amplitudes are significantly higher in the standing compared with supine positions [8–10]. These variations in dCSF create challenges to maintaining consistent electric field strength and location, and therefore consistent paresthesia sensations when patients assume different positions.

While dCSF is the main determinant of the distance between the electrodes and dorsal columns, lead placement variability within the epidural space may also account for threshold differences between patients. A lead placed against the dura will have a lower threshold for stimulation, compared with a lead placed more dorsally within the epidural space.

Variations in cord geometry

Based on a magnetic resonance imaging (MRI) study of 26 healthy male volunteers, several observations have been made regarding the geometry of the spinal cord and vertebrae [7]:

- 40% of the subjects have an asymmetrical position (right or left) of the spinal cord in the spinal canal (up to 1.5–2 mm offset of the spinal midline and vertebral midline)

- the mediolateral position of the spinal cord can vary by 0.5–1.0 mm within a few centimeters of its length
- some subjects have a rotated spinal cord
- some subjects have asymmetrical vertebrae.

These anatomical findings may explain why only 27% of patients with the lead on the radiological midline have bilateral symmetrical paresthesia [11]. Further, even with contacts placed 0.5–3 mm from the midline, stimulation-induced paresthesias were felt on the contralateral side of the body in 11% of patients. These observations highlight the importance of distinguishing between the radiological midline and the physiological or functional midline for SCS to obtain the desired paresthesia responses.

PROGRAMMABLE ASPECTS OF STIMULATION

In addition to proper anatomical targeting, successful SCS depends on properly set stimulation parameters. Modern neurostimulation devices provide a high degree of configurability of stimulation parameters. This configuration process, typically called 'programming', allows clinicians to vary the location of paresthesia within a patient's body by selectively activating different contacts on an electrode array and to manage the perceptual quality of stimulation within a region of paresthesia. Stimulation is delivered by trains of electrical pulses between two or more contacts on an electrode array. The amplitude, pulsewidth, and the rate of the pulse train (or the period between pulses) may be controlled and adjusted by the physician and/or patient (see Fig. 17.5A).

Stimulation parameters

Electrode polarity

Electrodes on a lead can be configured as anodes (positive potential relative to a reference) or cathodes (negative potential relative to a reference). At a minimum, at least one electrode must be configured as a cathode with a second electrode configured as an anode to create a closed electrical circuit. Current flows from the anode to the cathode, and stimulation is initiated in fibers near the cathode as this is the location where the largest depolarization typically occurs [12]. Thus, in general, the cathode is placed at the spinal level maximally to stimulate the corresponding dorsal columns. The probabilities of targeting different body areas as a function of vertebral level of the stimulating cathode are shown in Figure 17.4. The variation in probabilities is likely due to anatomical and lead placement differences between patients.

In contrast to effects near the cathode, neural elements placed under anodes are hyperpolarized. Even so, actual suppression of action potentials is unlikely. The selection of anodes and cathodes is one way to shape the electric field.

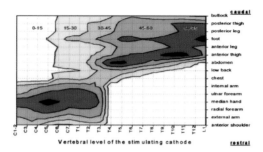

FIGURE 17.4 Probability-of-paresthesia contours as a function of body area and vertebral level of stimulating cathode using a single epidural lead (with permission from [29]).

Some typical configurations used in programming SCS leads include an adjacent bipole and a guarded cathode.

- Adjacent bipole: this is a cathode with an adjacent anode. Preferential stimulation of the dorsal columns is achieved with a narrow spacing between electrodes [13,14]. The electric field also tends to penetrate deeper and be broader within the dorsal columns compared to wider spaced electrodes because the extent of the field is more limited laterally and more confined with narrow spaced contacts, thereby increasing the threshold for dorsal root fiber activation [15]. In a patient with low back and leg pain, successful activation of the deeper fibers of the back will very likely activate the leg fibers as well. If too many cathodes are activated simultaneously, the electric field will be more superficial, which may limit the ability to activate the deeper fibers of the back. Leads with 'compact' or 'subcompact' electrode spacing (<4 mm spacing between electrodes) are designed to facilitate deeper activation of the dorsal columns.
- Simple bipole: this is a cathode with a non-adjacent anode. Using electrodes widely spaced from each other creates a field that is longer and shallower within the dorsal columns. This elongated vertebral coverage is useful for covering legs or arms, where length of coverage is desired but depth is not needed. For example, instead of generating paresthesia in one area of the leg, it may be possible to expand it to the entire leg. Leads with 'compact' or 'standard' electrode spacing (>4 mm spacing between electrodes) are designed to enhance this property and facilitate more coverage.
- Guarded cathode: this is a cathode bordered by anodes just above and below. Computer modeling has shown that a guarded cathode on the midline provides maximal recruitment of the dorsal columns [14] and clinical studies have observed a patient preference for this configuration [1]. This configuration is often useful for covering pain in the back and primarily in one leg, or in both legs if using dual leads.

Percutaneous and surgical lead types may be involved when programming a patient. Percutaneous leads have varying numbers of electrodes. Typically, leads with four or eight electrodes are used. Compared to a four-electrode lead, the eight-electrode lead provides a larger vertebral spread and greater programming options, as well as more reprogramming options if lead migration or changes in painful areas occur. Multiple leads are frequently used to cover larger body areas with paresthesia. Dual leads covering left and right sides are most commonly used.

In some patients, three parallel leads may be used, usually with one cathode in the center flanked or even encircled by anodes to attempt tighter penetration into the dorsal columns and capture the deep fibers of the back. Surgical paddle leads are found in single, dual, and three or more column configurations, with up to 16 programmable electrodes.

Amplitude

Depending on the technology used by a given stimulator, pulse amplitudes may be specified in either milliamps (mA) or Volts (V). The range of programmable amplitudes is typically 0 to 25.5 mA or 0 to 10.5 V, although some devices are restricted to a subset of this range. Increasing amplitude will increase the number of fibers recruited by the electrical stimulation (though in a non-linear fashion). Amplitude is generally perceived by patients as the intensity of the paresthesia, but may also correlate with the total area of paresthesia generated. Stimulation below a given level for a patient, known as the 'perception threshold', cannot be felt and is thought to be non-therapeutic. Increasing amplitude above that threshold will intensify the stimulation felt and generally expand its extent until a point is reached at which either the intensity is no longer comfortable, the extent of the stimulation has reached areas where the sensation is not tolerated, or motor responses have been triggered [11]. This state is called the 'discomfort threshold'. The range between perception threshold and the discomfort threshold is called the therapeutic or usage range. A typical ratio between discomfort and perception is 1.4 [16].

Pulsewidth

The duration of the pulse, or pulsewidth, may be controlled within a range starting below 100 μs to as high as 1000 μs in some systems. Pulsewidth is a secondary factor in controlling the energy delivered, and an increase in this parameter will result in a larger number of activated fibers. Theoretical studies have found that short pulsewidths increase the threshold difference between activation of different diameter nerve fibers [17]. In addition, computer models of SCS have shown that increasing pulsewidth may result in activation of more medially located, higher-threshold small diameter fibers that may contribute to an increase in paresthesia coverage in a caudal direction [18]. A recent study has shown that an increase in pulsewidth results in an increase in potentially therapeutic dorsal column paresthesia in patients with chronic pain [19]. In addition to increasing the area of stimulation-induced paresthesia, an increase in pulsewidth may result in an increase in the perceived intensity of the stimulation. Programmable ranges for pulsewidths are between 60 and 1000 μs. Typical values vary from patient to patient but often fall within the range of 400–500 μs [20].

Rate

The stimulation rate, or frequency, of stimulation is typically controllable within a range from less than 10 Hz (pulses per second) to as fast as 1200 Hz in some systems. Low stimulation rates, those below 30–50 Hz, can often be perceived as discrete impulses by a patient. This may or may not be well tolerated. Higher rates (>50 Hz), those at which pulses are no longer perceived as discrete events, are more commonly used for spinal cord stimulation [21,22]. In a study involving 171 patients (mean 7.1 years follow up), the average rate selected by patients was 62.7 ± 54.2 Hz (range 8–200 Hz) [22]. Increases in rate are sometimes perceived as increasing stimulation intensity, although this effect is typically secondary to amplitude and pulsewidth effects. One retrospective study involving 101 patients has suggested that frequencies >250 Hz may be used to regain pain control in some patients with break-through pain [23]. However, few data exist on the utility of high rate stimulation (>100 Hz) for spinal cord stimulation. While there may be exceptions, patients with complex regional pain syndrome (CRPS)-type pain usually prefer the smoother faster rates, whereas patients with deep, focal low-back pain tend to prefer the massaging effect of slower rates.

Stimulus mode

Different control technologies respond differently to changes in electrode–tissue impedance over time. Current controlled systems (CC) attempt to adjust their drive voltage to maintain a constant current output. Voltage controlled systems (CV) maintain a fixed voltage and allow the current delivered to comply with changes in impedance. Neither control mechanism has demonstrated clinical evidence of superiority for long-term pain relief.

Electrode control

Systems control individual contacts either by multiplexing a single stimulation source to multiple contacts simultaneously (i.e. single-source systems), thereby sharing a given set of stimulation parameters among them, or by providing individual control over each contact in an array (i.e. multiple-source systems). Systems that support independent control of simultaneous pulses require a specific amplitude to be set for each active contact on an array.

Additional control options

Modern stimulation systems provide a number of other controllable aspects of stimulation in addition to those characterizing a basic pulse train. These allow targeting of multiple pain sites, control of transitions into and out of stimulation, energy conservation mechanisms, and options for patient control.

Interleaved pulse stimulation

For patients with bilateral or complex pain patterns, modern neurostimulation systems allow configuration and control over separate stimulation regimes for each pain region. Such independence is typically achieved by providing pulse trains with different characteristics (electrodes, pulsewidths, amplitudes, and sometimes rates) on a time interleaved basis (Fig. 17.5B). Thus, one pulse train may be optimized to target paresthesia in one area using one set of stimulation parameters, and another pulse train may be optimized to target paresthesia in a second area with another set of stimulation parameters. For example, a patient with bilateral pain may need one pulse train to be programmed to generate paresthesia on the right side of the body, while a second pulse train may be programmed to generate paresthesia on the left side.

Gradual initiation/cessation of stimulation

Some patients find abrupt transitions while initiating or ending stimulation to be uncomfortable. This discomfort is most common on initiation of stimulation. Most stimulators address this problem by providing a means gradually to ramp pulse amplitude toward a target over a period of time (see Fig. 17.5C). Additionally, some stimulators provide a similar control over cessation of stimulation, ramping amplitudes towards zero when shutting off. The time period for such controlled transitions is configurable in some systems, but typically occurs over a period of a few seconds (2 to 8 seconds).

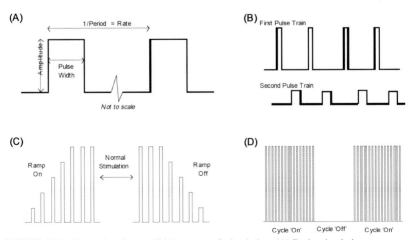

FIGURE 17.5 Examples of controllable aspects of stimulation. (A) Basic stimulation parameters of amplitude, pulsewidth, and rate. (B) Interleaved pulse stimulation. (C) Gradual initiation and cessation of stimulation. (D) Cycling of stimulation.

Duty cycle

Many stimulators provide the capability to cycle stimulation on and off, often with controllable on and off periods to allow management of the duty cycle (the time stimulation is on or off). Times may range from fractions of a second to minutes (see Fig. 17.5D). Cycling of stimulation is valuable for managing total energy use (either for longevity of primary cell batteries or management of time between recharges). Cycling can, in some cases, be poorly tolerated, especially in patients for whom transitions are uncomfortable. However, in patients who have a persistence of stimulation effect or carry-over, it can be a useful feature.

Patient control

Multiple stimulation regimes may be used to improve targeting of paresthesia to desired areas [24]. Pain patterns and intensities, as well as stimulation effects, can vary outside of the clinic on a daily or hourly basis. As an example, stimulation intensity is well known to be dependent on posture [8–10].

Modern stimulation systems address these variations by providing the patient a degree of control over their therapy. Most systems allow patients to turn stimulation on or off, control the parameters (amplitude, pulsewidth, and rate) of a given interleaved pulse train, and to switch between different regimes of stimulation (i.e. from one group of interleaved pulse trains to another) to match activities or to adjust sensation. Thus a patient may have one or more stimulation regimes, each consisting of one or more interleaved pulse trains that they may select to optimize their stimulation.

Systems support varying numbers of stimulation groups (regimes of interleaved pulse trains) and the number of simultaneously available pulse trains. Modern systems typically allow patient selection between at least four stimulation regimes, while some allow up to 26, with support of between four and eight simultaneously available pulse trains.

Many systems allow clinicians selectively to enable or disable the degree of control a patient has better to match the patient's needs and capabilities. Further, some systems allow specific limits to be set on specific parameters, such that a patient may only adjust a stimulation parameter (such as amplitude, pulsewidth, or rate) within limits set by the programming clinician.

Impedance

One additional tool is the ability to check the impedance of a stimulation circuit. The wires have an inherent resistance (impedance) against the flow of electrical current, measured in ohms. If there are connection problems, an impedance test will show the continuity status. If there is a short circuit, a very low number (ohms) will be displayed. This may suggest a breach in the insulation and the patient may sense a loss of paresthesia. If there is an open circuit, a very high number will be displayed. If possible, it is important to check the connections when an open circuit is found to make sure that each connector (device to extension, and extension to lead) is secured. A patient may sense a change in paresthesia or, if only one anode or cathode is used, a loss of paresthesia. If the connection has good continuity, the numbers will be in a broad middle range. The variation in impedance within this middle range may be caused by several factors, primarily the thickness and electrical conductivity of scar tissue around the lead, the electrical conductivity of the tissue immediately surrounding the lead, and the dorsoventral location of the lead within the epidural space [25,26]. Electrodes suspected to be involved in a short- or open-circuit should not be used for stimulation. Specific values that determine a short- or open-circuit vary by the system involved.

Energy conservation strategies

Stimulation parameters that reduce energy consumption are desirable to preserve battery life in primary cell devices and to increase the interval between battery recharge in rechargeable devices. In general, therapy needs dictate the choice of stimulation parameters. If multiple combinations of stimulation parameters provide the patient with adequate pain relief, the combination with the lowest energy consumption may be selected for use. The following may be considered if a reduction in energy consumption is desired [27,28]:

- decrease amplitude, pulsewidth, rate
- decrease duty cycle
- decrease number of cathodes
- decrease number of programs.

INTRAOPERATIVE PROGRAMMING

In this scenario, the lead itself can be moved into a better position if it is off target, followed by reprogramming until paresthesia coverage is optimized. An impedance test should first be conducted in order to ensure system integrity. The following algorithm is proposed for this setting.

1. Using an anteroposterior (AP) radiographic view as a guide, identify the likely electrode(s) to be activated by correlating their location with the dermatomes they are theoretically covering.
   - If the targeted painful areas are the *low back and legs*, generally the lead should be placed as precisely on the midline as possible around T8 to T9. Due to patient-to-patient variations, the active electrodes may rest anywhere from T8 to T10 to obtain this type of coverage [5,29,30].
   - If the targeted painful area is *legs only with no back component*, there is slightly more latitude for being off-midline with the electrodes. The more lateral the electrode placement, the more lateral is the fiber recruitment. In general, to obtain sciatic coverage, the electrode needs to be within T9 to T11 inside the shadow of the spinous process. To obtain outer leg/hip coverage, the electrode needs to be just on or outside the lateral edge of the spinous process. If the electrode is placed more laterally than that,

stimulation may result in unwanted radicular, rib or chest/abdominal stimulation or, if near the dorsal roots, may result in motor responses instead of the sensory 'tingling' sensation.

- If the targeted painful area is the *upper limbs*, side of the neck coverage usually occurs at C1, shoulder around C2, arm around C3 to C5, hand coverage C5 to C6, and much lower than C6 to C7 tends to recruit the axilla. Cervical stimulation is more forgiving of leads being placed lateral to the midline than are low thoracic lead placements. When cervical electrodes are too close to the midline, it is not unusual for a patient to complain of feeling paresthesia in their feet; those fibers travel more medially compared to the arm fibers. Placing the electrodes slightly off midline helps to target the upper limbs.
- If the patient is to be under general anesthesia for the final lead placement (not recommended for percutaneous leads), they would ordinarily have had a prior stimulation trial using percutaneous lead(s) that identified the patient's specific optimal electrode position for the subsequent implantation.

2. Identify the location of the cathode. The programmed amplitude should be set at 0 mA or V before making changes to the electrode configuration or other stimulation parameter settings. The rate of stimulation may be initially set at a value between 30 and 60 Hz. Sometimes, sedation may blur a patient's perception of the paresthesia boundaries if faster rates are used. Begin with programming an adjacent bipole or guarded cathode.

- When targeting the *low back and legs*, a high pulsewidth (i.e. 450 μs or higher) is generally used to provide a large recruitment of neural elements and activate the spinal cord's deeper back fibers and produce paresthesias in the low back. If using a three-column lead configuration, ideally, the cathode will be activated on the middle column, surrounded by anodes on all sides to permit fine-tuning of the depth capability. If using dual leads, two aligned bipoles or guarded cathodes may be used in order to create a field that is more symmetrical between the leads; however, the penetration of the electric field in the dorsal columns is higher with one lead compared to two leads [31]. Some patients have a small 'window' of dorsal column fibers between the left and right dorsal roots (i.e. innervating the legs) such that that the roots become overstimulated. In this situation, the patient may not tolerate an increase in amplitude due to uncomfortable sensations in the legs and, at the same time, still feel little or nothing in the back. In these patients, if the electrodes are known to be centered on the midline, not much more can be done to capture the back, even with a guarded cathode surrounded by anodes and variations in pulsewidth.
- If targeting the *legs with no back pain component*, lower pulsewidths (120–330 μs) and simple bipole configurations may be used.
- If targeting the *upper limbs*, even lower pulsewidths may be used (90–180 μs) and active electrodes may be further apart.

3. Adjust the amplitude of stimulation in 0.1 V or mA increments slowly from zero initially until the patient's perception and discomfort thresholds are established. After these thresholds are known, increase amplitude to approximately 1 V or mA below their perception threshold to allow for a quicker pace, especially for patients that may require higher amplitudes.
   - Clinically relevant stimulation involves a choice of amplitude that falls within the range between perception threshold and discomfort threshold. The range between perception and discomfort amplitudes should not be too low (lower than 0.5 V or mA) as there will be limited ability to vary the intensity of stimulation for threshold adjustments [11,32,33].
4. Assess programs
   - If the lead is close to the target, concordant paresthesias will be obtained, but these may not completely cover the desired area, which may require some reprogramming. Increasing the pulsewidth may widen the area and/or intensity of stimulation-induced paresthesia. Changing the configuration from an adjacent bipolar to a guarded cathode may also increase the area of paresthesia.
   - If the lead is not near the target, concordant paresthesias cannot be obtained, and thus the lead will need to be repositioned appropriately. Repositioning cues are logical and depend on the distance and direction of the patient's paresthesia from the target: the lead moves up if higher coverage is needed; likewise, the lead moves lower if coverage is needed lower. The lead is moved left or right depending on which direction the lead needs to move laterally.
5. Adjust rate. The rate is titrated based on patient preference.
6. Add other programs if needed. More programs may be used if additional paresthesia coverage is needed or if different stimulation parameters are needed during different postures or activities of daily living.

Modern neurostimulators have the capability of shifting the electrical field longitudinally along a lead and laterally to an adjacent lead toward an available electrode. This electrode will 'pull' the stimulation incrementally between the two electrodes to center it. This may be helpful in situations where an electrode cannot be placed exactly on the target.

POSTOPERATIVE PROGRAMMING

In this scenario, the leads cannot be moved as the lead and device have been permanently implanted. Programming should aim to deliver paresthesias as close as possible to the patient's pain target(s). Use cues from patient descriptions of their pain target(s) and where they currently feel paresthesia to try initial logical parameters for stimulation. The guidelines for stimulation parameter titration discussed in the intraoperative section are also useful for postoperative and follow-up programming sessions.

- Cathode location: note that the location of the cathode primarily correlates with the patient's description of the paresthesia location. Depending on the laterality of the lead placement, it is possible to generate paresthesia in dermatomes located at or below the cathode level. An adjacent bipole or guarded cathode configuration may be used as a starting point.
- Stimulation parameters: the rate of stimulation may be initially set at a value between 30 and 100 Hz. The pulsewidth may be set at 450 μs for a low back and legs target, or to 240 μs for a legs with no back component or upper limb target.
- As a starting point, it is also possible to map the 'envelope of coverage' by test stimulating the outermost electrodes on a patient's leads (most distal and most proximal electrodes). If the anatomical target is within those boundaries, correct coverage is likely to be obtained. If the lead(s) has migrated too far out of position for coverage to be appropriate or if there is a system disconnect (as confirmed by an impedance test and radiograph), a revision surgery will be required to correct lead placement.

Some of the newer neurostimulators have the capability to allow patients to make a range of vertical electrode adjustments of their configuration using their own patient programmer. Thus stimulation may be shifted longitudinally along a lead and laterally between two (or more) leads.

The next section describes case scenarios that may be helpful during postoperative reprogramming.

Case scenarios – troubleshooting programming

Single lead: coverage is only slightly off optimal coverage

The controllable range is up and down the lead, which will be good if the patient's existing stimulation is vertically aligned with the pain. Following the patient's cues, activate the electrodes adjacent to the ones currently in use, shifting the configuration up or down the lead in the direction desired. Keeping a log may help keep track of electrode combinations and what areas they are covering as different combinations are tried. Fine-tuning the pulsewidth by increasing its value usually helps to expand coverage.

Single lead: coverage has moved vertically

If the lead has migrated rostrally, activate electrodes (as cathodes) that lie below the currently activated ones. Alternatively, if the lead has migrated caudally, activate electrodes above the currently activated ones, until stimulation has been restored to the correct target. If the existing program is already using the very top or bottom electrode on the outer margin of the painful area, surgical reposition of the existing lead or replacement lead will be required. Some practitioners believe in trying to cover the ideal location or 'sweet spot' of the spinal cord target with the sweet spot of the lead. The sweet spot of the lead refers to deploying the electrodes in the middle of the lead, without using any of the most

proximal or distal electrodes on the lead. This approach has not been systematically studied.

Single lead: coverage has moved laterally

If a patient's existing program is laterally off target, there are fewer options.
In this instance, increasing only the pulsewidth may help re-establish good
coverage. However, increasing the pulsewidth could activate fibers that may
increase the area of coverage, but also may intensify sensations in currently
covered areas. Thus, while satisfactory coverage may be obtained by increasing
pulsewidth, it could intensify the off-target spot too much thereby offsetting any
improvement in widening the coverage.

If the target is not the low back, one option is to try stretching out a simple bipole over a longer span of electrodes, with inactive electrodes between
the anode and cathode. This makes the field more elliptical and shallower. A
guarded cathode configuration has reasonable depth for back fibers, and is also
an option to try. Frequently, surgical repositioning or the addition of a new second lead in the desired spot is the solution for lateral displacement.

Dual or triple leads: coverage is too diffuse or too narrow

In this situation, more programming options exist compared with a single lead,
but the principles remain the same: direct the stimulation to the targeted areas.

- If the coverage is *too diffuse*, the problem is probably that too many lateral
  cathodes have been activated. Methodically eliminate the outer active electrodes to eliminate the spurious coverage and try reducing pulsewidth.
- If the coverage is *too narrow*, start by widening the pulsewidth; quite often
  this is all that is required. If more coverage is needed, electrodes may be
  activated in the direction that the patient requests. If an anode is added in
  an outer position of the lead, the patient will likely perceive the stimulation
  increase as milder than if the outer position is a cathode – a cathode intensi-
  fies sensations more than adding an anode.
    - Quite often *multiple programs* are needed if these solutions prove insuf-
      ficient. If back coverage and a wider area are needed, multiple programs
      with specific targets may be used. A typical example is a focused low
      back program that delivers high pulsewidth via a single cathode and
      simple anodes for the back (adjacent bipole or guarded cathode), with
      another program that will spread either symmetrically or asymmetrically
      to the legs, hips, etc. Sometimes, even a third program that targets hip/
      flank areas may be used. The patient can run one to four programs in iso-
      lation or concurrently. Three column configurations (three percutaneous
      leads in parallel or a three column surgical lead) that are well centered
      at the correct vertebral elevation provide more programming options for
      specific paresthesia targeting to a patient's painful areas.

Dual or triple leads: coverage has moved vertically

If programming a surgical paddle lead, the electrodes all move as a unit, so if paresthesia coverage has moved to more rostral dermatomes, activate lower electrodes at the appropriate level. If coverage has moved down to more caudal dermatomes, activate higher electrodes. It would be unusual not to be able to re-establish coverage as paddle leads tend not to migrate too far. Otherwise, surgical revision is needed.

If the leads are percutaneous, determine how many leads have moved. If one is still in a good position, its active electrodes would not change, but the other lead would need the active electrodes changed to cover the original anatomical location. If a recent radiograph is available, the electrode locations may be determined. Otherwise, methodical shifting of paresthesia up or down the lead is necessary to obtain the original coverage. If the percutaneous leads have moved beyond the range to re-establish coverage, surgical revision is required.

Dual or triple leads: coverage has moved laterally

When programming a surgical paddle lead, the electrodes move as a unit; lateral electrodes and pulsewidth are the main parameters to control. Higher pulsewidths may be used to recruit sufficient fibers in the desired direction without unduly intensifying paresthesia in the unwanted direction. Activating electrodes on the second column, if paresthesia is needed in that direction, may also help re-establish coverage. Other options with two columns frequently work well: side-by-side cathodes and an anode on only one column (this directs stimulation in the direction of the cathode with no anode), or its inverse: side-by-side anodes with the cathode on one side (this directs stimulation in the direction of the cathode).

If the surgical paddle lead is a three-column lead, there is better chance for achieving good coverage because of the wider lateral span of electrodes and less likelihood of significant lateral migration of the paddle. With this type of lead, activation of electrodes laterally in the desired direction may be sufficient. Assigning a cathode to the lateral electrode will result in more pronounced paresthesias in that direction compared to an anode.

With percutaneous leads, mapping by test stimulation may be needed to determine how many leads have moved. Similar programming principles will apply using cues from the patient to re-establish good paresthesia coverage of the pain areas.

Single or multiple leads: intermittent, inconsistent, loss, or overstimulation

If the patient's concern is that the stimulation is intermittent, inconsistent, or suddenly starts with a jolt or abruptly stops for no recognizable reason, an impedance test should be performed to confirm if there is a fracture or disconnect in the system. If the impedance test shows an open circuit, and adjacent electrodes are not useful in obtaining coverage, a lead conductor may have fractured and surgical revision will be needed. Occasionally, there can be a bad connection which could create a current path through body fluids, such that when the patient has the system on, he or she feels the paresthesia subcutaneously at the site of the disconnect (either the location of the neurostimulator or the extension–lead connection) and not in their usual painful area. If the power is off, they feel nothing. These instances will need surgical revision to re-establish a good connection.

If there does not seem to be a connection problem, it is possible that the 'shocks' or 'jolts' are due to changes in position. It is easy for a patient, especially if they are more sensitive than average, to mistake 'shocks' or 'jolts' that can occur as the result of normal positional changes for the more serious problem of a bad connection. This type of uncomfortable stimulation occurs because of the change in the spinal cord's proximity to the scarred-in lead when the patient changes position, mostly between lying and standing. Patients may demonstrate to themselves that it is the spinal cord that moves with position changes by very slowly changing position from lying down to standing while modulating the intensity gradually from weak to strong with no abrupt stops or starts.

Consideration of the actual lead position if one or more leads have moved

Sometimes percutaneous leads migrate. This problem can often be solved by reprogramming different electrodes [34]. Large lead migrations may be suspected if the patient detects paresthesias in locations remote from the original targeted area. A radiograph of the lead locations is useful to determine whether reprogramming may be useful or a surgical revision is needed.

1. If a radiographic view is not available, knowledge of the actual lead position(s) is more challenging. A starting point is testing the outermost electrodes to identify the 'envelope of coverage'.
   - Single lead example: for one lead, activate the top two electrodes and the bottom two electrodes (adjacent bipole). If the lead is askew, it can be assumed that the bottom electrodes recruit the upper left leg and the top two electrodes recruit the right rib cage area. Thus, this lead is resting with the bottom electrode over the left dorsal column at about T8 to T9 inside the left spinous process shadow and the top of the lead is tilted off to about T7 past the right dorsal column about midway between the outside of the right spinous process shadow and the nerve root entry area. Since this is a partially sideways span of almost two vertebral segments, this is most likely a lead with eight electrodes with a large spacing between electrodes. Based on the patient's physiological responses to the stimulation mapping and the clinician's understanding of the anatomic dermatomes, a 'mental x-ray' of the lead position can be created. In this

patient, good back/leg coverage could be obtained using the second or third electrodes above the bottom one, as the lead's trajectory passes those two electrodes over the midline at about T9. Additional information regarding the magnitude of the lead movement relative to its original postoperative position may help decide whether a lead revision is necessary.

- Multiple lead example: if two or three percutaneous leads have been placed, the same process may be repeated: test stimulate the four or six outermost electrodes to identify the 'envelope of coverage'. If the target is inside the envelope, the likelihood is good for getting appropriate coverage by following the cues from the patient. If the target is not within the envelope, the leads are out-of-range and a revision will be necessary. A radiograph of the leads will be needed to confirm their location.

2. If a radiographic view is available, it can make this process much quicker because it will be easy to identify the electrode location and thus directly program the most likely electrodes to activate.

Multiple lead options

Using more than three leads is less common but may be useful when targeting more complex pain patterns. Cervical and thoracolumbar epidural SCS leads can together lengthen the coverage through the back. One or two leads could be placed in the cervical region, and another one or two leads placed at T9 or lower to cover a long painful area because the leads may be programmed to create a very long stimulation field. Each lead may have multiple cathodes and anodes activated. Multiple leads may be used for whole back coverage or coverage of pain in both upper and lower extremities.

CONCLUSIONS

The initial programming of SCS systems involves adjusting stimulation parameters to maximize overlap of paresthesia with the painful areas. While paresthesia concordant to the painful area is necessary, it is insufficient to assure pain relief [1]. Understanding the anatomy and the effects of electrical stimulation provides a foundation for creating a rational selection of stimulation parameters. Nonetheless, achieving appropriate selection can be complicated by a number of factors, such as variable lead placement, anatomical differences between patients, and occasional patient inability to provide accurate feedback on their paresthesia and pain. Despite these factors, a wide range of stimulation parameters may be controlled to optimize SCS therapy in a patient-specific manner. A better understanding of the effects of SCS stimulation parameter changes will lead to more effective and efficient programming strategies.

REFERENCES

1. North RB, Ewend MG, Lawton MT, et al. Spinal cord stimulation for chronic, intractable pain: superiority of 'multi-channel' devices. *Pain*. 1991;44:119–130.
2. Feirabend HKP, Choufoer H, Ploeger S, et al. Morphometry of human superficial dorsal and dorsolateral column fibres: significance to spinal cord stimulation. *Brain*. 2002;125:1137–1149.
3. Smith MC, Deacon P. Topographical anatomy of the posterior columns of the spinal cord in man. *Brain*. 1984;107:671–698.
4. Oakley JC. Spinal cord stimulation in axial low back pain: solving the dilemma. *Pain Med*. 2006;7(S1):S58–S63.
5. Barolat G, Massaro F, He J, et al. Mapping of sensory responses to epidural stimulation of the intraspinal neural structures in man. *J Neurosurg*. 1993;78:233–239.
6. He J, Barolat G, Holsheimer J, et al. Perception threshold and electrode position for spinal cord stimulation. *Pain*. 1994;59:55–63.
7. Holsheimer J, den Boer JA, Struijk JJ, et al. MR assessment of the normal position of the spinal cord in the spinal canal. *Am J Neuroradiol*. 1994;15:951–959.
8. Abejon D, Feler CA. Is impedance a parameter to be taken into account in spinal cord stimulation? *Pain Physician*. 2007;10:533–540.
9. Cameron T, Alo KM. Effects of posture on stimulation parameters in spinal cord stimulation. *Neuromodulation*. 1998;1:177–183.
10. Olin JC, Kidd DH, North RB. Postural changes in spinal cord stimulation perceptual thresholds. *Neuromodulation*. 1998;1:171–175.
11. Barolat G, Zeme S, Ketcik B. Multifactorial analysis of epidural spinal cord stimulation. *Stereotact Funct Neurosurg*. 1991;56:77–103.
12. Ranck JB. Which elements are excited in electrical stimulation of mammalian central nervous system: a review. *Brain Res*. 1975;98:417–440.
13. Holsheimer J, Wesselink WA. Effect of anode-cathode configuration on paresthesia coverage in spinal cord stimulation. *Neurosurgery*. 1997;41:654–660.
14. Manola L, Holsheimer J, Veltink P. Technical performance of percutaneous leads for spinal cord stimulation: a modeling study. *Neuromodulation*. 2005;8:88–99.
15. Alo KM, Holsheimer J. New trends in neuromodulation for the management of neuropathic pain. *Neurosurgery*. 2002;50:690–704.
16. Holsheimer J. Effectiveness of spinal cord stimulation in the management of chronic pain: analysis of technical drawbacks and solutions. *Neurosurgery*. 1997;40:990–999.
17. Gorman PH, Mortimer JT. The effect of stimulus parameters on the recruitment characteristics of direct nerve stimulation. *IEEE Trans BME*. 1983;30:407–414.
18. Lee DC, Hershey K, Bradley K, et al. *Dorsal column selectivity in pulse width programming of spinal cord stimulation: computational model for the 'sacral shift'*. North American Neuromodulation Society Acapulco; 2007.
19. Yearwood T, Hershey B, Bradley K, et al. *Pulse width and dorsal root stimulation in spinal cord stimulation*. North American Neuromodulation Society Las Vegas, NV; 2009.
20. Sharan A, Cameron T, Barolat G, Evolving patterns of spinal cord stimulation in patients implanted for intractable low back and leg pain. *Neuromodulation* Date reg 5:167–179.
21. Barolat G, Ketcik B, He J. Long-term outcome of spinal cord stimulation for chronic pain management. *Neuromodulation*. 1998;1:19–29.
22. North RB, Kidd DH, Zahurak M, et al. Spinal cord stimulation for chronic, intractable pain: experience over two decades. *Neurosurgery*. 1993;32:384–395.

23. Bennett DS, Alo KM, Oakley J, et al. Spinal cord stimulation for complex regional pain syndrome I [RSD]: a restrospective multicenter experience from 1995 to 1998 of 101 patients. *Neuromodulation*. 1999;2:202–210.

24. Alo KM, Yland MJ, Charnov JH, et al. Multiple program spinal cord stimulation in the treatment of chronic pain: follow-up of multiple program SCS. *Neuromodulation*. 1999;2:266–272.

25. Alo KM, Varga C, Krames E, et al. Factors affecting impedance of percutaneous leads in spinal cord stimulation. *Neuromodulation*. 2006;9:128–135.

26. Butson CR, Maks CB, McIntyre CC. Sources and effects of electrode impedance during deep brain stimulation. *Clin Neurophysiol*. 2006;117:447–454.

27. De Vos CC, Hilgerink MP, Buschman HP, et al. Electrode configuration and energy consumption in spinal cord stimulation. *Neurosurgery*. 2009;65(6 Suppl.):210–216.

28. Kuncel AM, Grill WM. Selection of stimulus parameters for deep brain stimulation. *Clin Neurophysiol*. 2004;115:2431–2441.

29. Holsheimer J, Barolat G. Spinal geometry and paresthesia coverage in spinal cord stimulation. *Neuromodulation*. 1998;1:129–136.

30. North RB, Kidd DH, Olin J, et al. Spinal cord stimulation for axial low back pain. *Spine*. 2005;30:1412–1418.

31. Holsheimer J, Wesselink WA. Optimum electrode geometry for spinal cord stimulation: the narrow bipole and tripole. *Med Biol Eng Comput*. 1997;35:493–497.

32. Tulgar M, Barolat G, Ketcik B. Analysis of parameters for epidural spinal cord stimulation 2. Usage ranges resulting from 3,000 combinations. *Stereotact Funct Neurosurg*. 1993;61:140–145.

33. Tulgar M, He J, Barolat G, et al. Analysis of parameters for epidural spinal cord stimulation 3. Topographical distribution of paresthesiae – a preliminary analysis of 266 combinations with contacts implanted in the midcervical and midthoracic vertebral levels. *Stereotact Funct Neurosurg*. 1993;61:146–155.

34. Cameron T. Safety and efficacy of spinal cord stimulation for the treatment of chronic pain; a 20-year literature review. *J Neurosurg (Spine)*. 2004;100:254–267.

Commentary on Programming – SCS

Richard B. North, MD

The Sandra and Malcolm Berman Brain & Spine Institute, Baltimore, MD Professor of Neurosurgery, Anesthesiology and Critical Care Medicine (ret.), Johns Hopkins University School of Medicine, Baltimore, MD

Heitman and colleagues present an industry perspective of programming contemporary spinal cord stimulation (SCS) systems. The authors represent a single company, and implicit in their presentation is the assumption that the products of different companies are substantially similar. This is a significant, unstated limitation. Furthermore, it should be noted that the range of possible programs is limited by the capabilities of the devices presently made available to practitioners by industry in general, and that of these capabilities only a fraction has been studied by generally accepted scientific methods, viz., randomized controlled trials (RCTs).

Overlooked here is a body of published work on automated, patient-interactive methods that make every patient's programming session a blinded RCT, avoiding the bias inherent in face-to-face programming by an expert [1]. Applied to a series of patients, these methods have yielded high quality evidence of efficacy [2] (superior to the manual methods described by Heitman and colleagues) and of cost-effectiveness (prolonging the life of implanted primary cell batteries) [3,4], fulfilling the primary goals of programming, as articulated by the authors. This technology, however, remains to be promulgated by industry.

The task of SCS programming in contemporary practice, as broadly described in this chapter, is focused on the devices and electrode configurations practitioners implant. Thus a good deal of effort is devoted to programming systems with two columns of contacts ('dual leads'), which remain popular despite evidence that, for the common indication of low back pain, they are technically inferior to a single column placed accurately in the physiologic midline. This evidence includes the modeling work cited by the authors as well as RCTs using the automated, computerized methodology cited above, applied to percutaneous as well as to surgical 'paddle' electrodes [5,6].

The capability of producing interleaved pulse trains, a useful programming tool, is now available in the products of all three SCS manufacturers. Well before this was available in commercial products, we studied interleaved pulse trains using the above methodology and found that they provided increased pain/paresthesia coverage and that the effects were due, at least in part, to an increase in the pulse repetition rate [7]. Interleaving pulse trains are just one example of novel capabilities, exceeding those currently available in commercial devices, which require further research and development.

It is difficult to achieve paresthesia coverage of pain involving the posterior neck and mid-low back with central somatosensory stimulation in general and SCS in particular [8]. Heitman and colleagues believe that 'cervical and thoracolumbar epidural SCS leads can together lengthen the coverage through the back'. Indeed, 'the leads may be programmed to create a very long stimulation field', but it remains to be demonstrated that the physiologic response and patient perception of paresthesia are thereby lengthened or enhanced or that axial pain syndromes are better treated in this fashion. This useful summary of current practices reminds us not only how far we have come, but also that we still have a long way to go in designing and programming SCS systems.

REFERENCES

1. North RB, Nigrin DA, Fowler KR, Szymanski RE, Piantadosi S. Automated 'pain drawing' analysis by computer-controlled, patient-interactive neurological stimulation system. *Pain*. 1992;50:51–57.
2. North RB, Sieracki JN, Fowler KR, Alvarez B, Cutchis PN. Patient-interactive, microprocessor-controlled neurological stimulation system. *Neuromodulation*. 1998;1:185–193.
3. North RB, Calkins SK, Campbell DS, et al. Automated, patient-interactive spinal cord stimulator adjustment: A randomized, controlled trial. *Neurosurgery*. 2003;52:572–579.
4. Khaless,i AA, Tayor RS, Brigham D, North RB. Automated vs. manual spinal cord stimulator adjustment: A sensitivity analysis of lifetime cost data from a randomized controlled trial. *Neuromodulation*. 2008;11:182–186.
5. North RB, Kidd DH, Olin J, et al. Spinal cord stimulation for axial low back pain: A prospective, controlled trial comparing dual with single percutaneous electrodes. *Spine*. 2005;30:1412–1418.
6. North RB, Kidd DH, Olin J, Sieracki JN, Petrucci L. Spinal cord stimulation for axial low back pain: a prospective, controlled trial comparing 16-contact insulated electrode arrays with 4-contact percutaneous electrodes. *Neuromodulation*. 2008;9:56–67.
7. North RB, Kidd DH, Olin J, Sieracki JM, Boulay M. Spinal cord stimulation with interleaved pulse trains: A randomized, controlled trial. *Neuromodulation*. 2007;10:349–357.
8. Holsheimer J, Barolat G. Spinal geometry and paresthesia coverage in spinal cord stimulation. *Neuromodulation*. 1998;1:129–136.

STUDY QUESTIONS

1. Describe the underlying physiology that is changing as a result of changes in pulsewidth. How might these changes be helpful in obtaining therapeutic benefit?
2. How should one proceed in programming a patient who has significant 'positionality' with stimulation? Are there technological advances that could mitigate positionality in spinal cord stimulators?

Chapter 18

Safety Concerns and Limitations

Paul S. Larson, MD[1,2] and Alastair J. Martin, PhD[3]

[1]*Associate Professor, Department of Neurological Surgery, University of California, San Francisco, CA, USA*

[2]*Chief, Neurosurgery, San Francisco VA Medical Center, San Francisco, CA, USA*

[3]*Adjunct Professor, Department of Radiology and Biomedical Imaging, University of California, San Francisco, CA, USA*

Neuromodulation is an essential part of functional neurosurgery and an important therapeutic modality for many patients suffering from medically refractory disorders. Deep brain stimulation (DBS), motor cortex stimulation (MCS), vagal nerve stimulation (VNS), spinal cord stimulation (SCS), and peripheral nerve stimulation (PNS) have become commonplace, as well as other devices such as phrenic nerve stimulators that are used in smaller numbers. As a result, the safety implications of implanting these devices must be well known to the neurosurgeon and communicated clearly to the patient and their caregivers. Every device has its own specific design characteristics and internal pulse generator (IPG) components and, given the variety of systems and manufacturers that exist, this chapter will focus on general safety concerns and limitations for neuromodulation devices as a whole. Where appropriate, circumstances specific to a particular device will be discussed. Finally, the field of neuromodulation is a rapidly evolving landscape with new innovations occurring at an ever-increasing pace. New devices as well as modifications to existing devices are constantly under development, so one must keep this in mind when considering the safety issues discussed here.

PATIENT ACTIVITY AND ENVIRONMENTAL CONCERNS

Postoperative activity

Patient movement can cause device-related complications anytime after implantation. Lead migration due to patient movement is a well-described complication in SCS and PNS such as occipital nerve stimulators [1–3]. These systems are particularly susceptible to lead migration for several reasons. First, they are placed in relatively mobile parts of the body, which exposes the leads and lead extensions to frequent and varied degrees of strain. Second, the ability to anchor

Essential Neuromodulation. DOI: 10.1016/B978-0-12-381409-8.00018-8

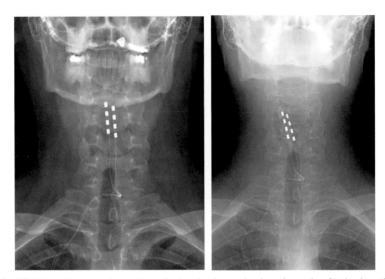

FIGURE 18.1 Migration of a paddle-style dorsal column stimulator 3 months after implantation. (A) Immediate postoperative films were obtained as a baseline study, showing the device in good position in the midline. (B) Films taken after the patient turned his head suddenly to look at something behind him. He reported a sudden onset of new paresthesias and loss of efficacy.

the leads in a secure fashion to fixed, bony structures is limited. Finally, the move toward percutaneous placement has resulted in cylindrical electrodes as opposed to paddle electrodes, which have a wider profile and may resist 'pull-out' more than cylindrical devices [4]. However, even wider paddle electrodes are not immune to migration (Fig. 18.1).

Authors and device manufacturers almost universally recommend having SCS patients limit bending, twisting, stretching, lifting objects greater than 5–10 pounds (2.3–4.5 kg) and other strenuous activities for upwards of 8 weeks after surgery. The goal is to give the system components time to scar into place as much as possible. Some recommend occipital nerve stimulator patients wear a cervical collar for 10 days after surgery to remind them to limit their neck movements [5]. Sudden loss of efficacy and/or sudden appearance of a stimulation-induced adverse event such as paresthesias are clues that lead migration may have occurred. Plain films, particularly when compared to immediate post-implant baseline studies, are the best way to make the diagnosis.

Patients are not immune to movement-related complications beyond the immediate postoperative period. The same 'scarring down' that occurs in the first few weeks and months after surgery may make device migration less likely, but it also creates more strain on the components when sudden or significant movements are made. A DBS patient in our practice had been implanted for years with no device-related issues. He was swimming in a lake, and reached his arms up to be pulled out of the water into a boat by his friends. Although he noticed nothing unusual at the time, he gradually began to notice a loss of efficacy on one side

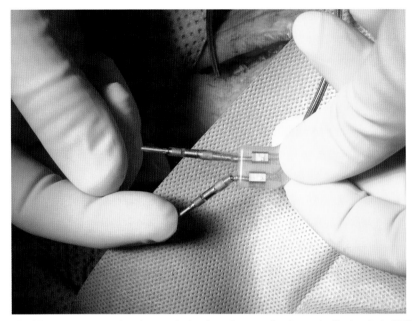

FIGURE 18.2 Intraoperative photograph showing a fracture of a plug at the distal end of a DBS lead extension. The patient was pulled out of the water into a boat by friends while swimming in a lake.

of his body. Device interrogation suggested an open circuit, although plain rays were unrevealing. At surgical exploration he was found to have a fracture of one of the lead extension plugs where it entered the IPG (Fig. 18.2).

'Twiddler's syndrome'

Twiddler's syndrome occurs when the IPG rotates or twists within its subcutaneous pocket. There are two recognized forms of this phenomenon; spontaneous and external [6, 7]. In spontaneous twiddler's syndrome, the IPG moves without external manipulation by the patient. Factors that predispose a patient to spontaneous twiddler's syndrome include subcutaneous pockets that are too large, become filled with fluid such as seromas, or IPGs that are not anchored. In these circumstances, even normal patient activity can cause twiddling. External twiddler's syndrome is caused by digital manipulation of the IPG by the patient, usually soon after implantation before the device has had time to scar down.

Either form of twiddler's syndrome can lead to a number of device complications. The IPGs of many devices must face a certain way for external programmers to be able to interrogate them. If the IPG flips such that it is facing the 'wrong' way, the device cannot be accessed. More worrisome are fractures in the lead extensions or plugs that enter the IPG. This is particularly true in devices such as DBS systems with a dual channel IPG, where two lead extensions are

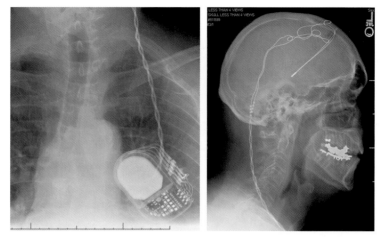

FIGURE 18.3   AP and lateral radiographs demonstrating an example of spontaneous twiddler's syndrome in a DBS patient. In this case, the patient had a fluid collection around his IPG in the immediate postoperative period. He suffered a fractured lead extension that required surgical replacement.

running in close proximity. The wires can twist around each other and result in a tremendous amount of strain on the lead extensions (Fig. 18.3). Fractures of the lead extension or plug can result in localized paresthesias, loss of efficacy, and/ or current drain from an open circuit with rapid depletion of the IPG. Strategies to prevent twiddler's syndrome include taking care when creating subcutaneous pockets, tacking the IPG to the wall of the pocket with a non-absorbable suture, and patient counseling. In the past few years, we have placed many of our DBS pulse generators under the pectoralis fascia. These subfascial pockets have much less potential space than subcutaneous pockets, and the incidence of IPG movement and seroma formation seems to have been reduced significantly.

Security and anti-theft systems

Security systems are common in airports, department stores, libraries and other public places, and are therefore a frequently encountered source of electromagnetic interference, (EMI). EMI is the disruption of normal function in an electronic device by the electromagnetic field created by a second electronic device. The likelihood of EMI is related to the amplitude and/or frequency of the external electromagnetic field, as well as its proximity to the patient. The EMI from security devices may be strong enough to turn the device on or off, and can even reset an IPG or change its settings in some instances. Unfortunately, it may be hard for patients to avoid these systems entirely, and they are frequently constructed such that it is impossible for the patient to maintain a safe distance from the system when passing through them. Recommendations are mixed on how to handle encounters with these devices. For VNS, security systems 'should not' affect the IPG, but it is recommended that patients keep the IPG at least 16 inches (40 cm) away [8]. Other manufacturers advise patients to avoid security systems when possible by asking to be hand screened by security personnel, with a specific request not to use the hand-held metal detector directly over the IPG. If patients do walk through a security gate, they should temporarily turn the IPG off and pass through as close to the center point of the system as possible (i.e. equidistant from each side of the detector).

Other sources of EMI

Other sources of EMI exist in the environment but are less likely to cause problems for patients. High voltage power lines, microwave transmitters, citizen band or ham radio antennae and other communication equipment are potent sources of EMI; in almost all cases, these devices are fenced-off or have limited physical access to the general public in accordance with Occupational Safety and Health Administration (OSHA) and other governmental agency guidelines. As long as patients respect physical barriers and warning signs, these devices should pose no danger. Other unusual but strong sources of EMI include arc welders, resistance welders, and induction furnaces or burners; patients should avoid getting close to these devices. One case of a DBS patient experiencing symptomatic EMI when driving a hybrid car was recently reported [9]. The symptoms resolved when the patient vacated the front seat of the vehicle or when the IPG was turned off. Such examples underscore the importance of teaching patients to use common sense when they suspect that EMI or some other source of interference is altering the function of their device. In these situations, patients should move away from the suspected source of interference and/or turn their IPG off temporarily.

Some sources of EMI are not necessarily strong but come in close proximity to an IPG during regular use. Examples include dental drills, ultrasonic probes or cleaners (also common in dental offices) and electrolysis wands. Because these sources of EMI are frequently encountered in a dental or medical office, patients are often lulled into a false sense of security that these devices are safe. One manufacturer specifically states that the IPG should be turned off when near these devices, and that a distance of at least 6 inches (15 cm) should be maintained between the device and the IPG [10]. Finally, strong magnets can interfere with the function of some IPGs if they get close enough. Many older IPGs were designed with magnet switches in them, while others like the VNS device have a separate, magnet-induced mode of operation. Common household items that contain strong magnets include stereo speakers and the doors of some refrigerators or freezers. Putting an IPG up against such items is not recommended. Because magnets can change the stimulation mode in VNS, these patients should be specifically warned to stay at least 8 inches (20 cm) away from strong magnets, hair clippers, vibrating devices, anti-theft tag deactivators and loudspeakers [8].

MEDICAL IMAGING

One of the most important safety considerations from a medical standpoint is imaging of the patient after the device is implanted. The need for imaging may arise from a desire to confirm the position of an electrode immediately after implantation, or to diagnose a migration or damage to the device after a patient has lost efficacy. The patient may also develop another medical condition that necessitates diagnostic and/or interventional studies. Because the need for imaging may occur emergently as the result of an accident or other acute event, it is important to educate not only the patient but also their caregivers about what type of imaging is safe and what could cause potential problems. If nothing else, they should be provided with emergency phone numbers for either the implanting team, the device manufacturer's support line, or both. This can avoid device-related complications or withholding of imaging that is safe and would benefit the patient by diagnosing a serious or treatable condition.

Plain radiographs, standard fluoroscopy and computerized tomography (CT) are safe and do not interfere with the normal functioning of neuromodulation systems. Radiographs produce very clear, artifact-free images of the device itself, and are the study of choice for detecting a broken wire, possible disconnection or gross migration of a device component (Fig. 18.4). One manufacturer's labeling for SCS states that CT scanning can cause a momentary increase in the level of stimulation, which has been reported by patients as a transient

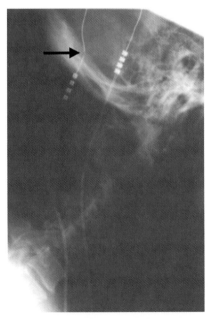

FIGURE 18.4 A plain radiograph demonstrating a complete DBS lead fracture, visible as a discontinuity in the lead just above the connector (black arrow).

electric shock or jolt-like sensation. They recommend turning the SCS off and programming the output to zero when performing CT [10].

Ultrasound is generally believed to be safe, although some manufacturers say that ultrasound done directly over system components may cause mechanical damage. We have used small, portable ultrasound units used by anesthesia in the operating room to locate the end of a DBS brain lead underneath particularly thick or edematous scalps with no damage to the device and no adverse events. Nuclear medicine studies are also quite common and do not appear to have risk in this patient population. Mammography in patients with IPGs in the chest is safe, but can pose difficulties with positioning the breast tissue in the imaging apparatus. The IPG can also obscure visualization of some of the breast tissue. Patients should warn the mammography technician about the presence of the device to avoid pain and potential damage to the hardware.

Magnetic resonance imaging

Magnetic resonance imaging (MRI) is more problematic for implanted electrical devices. MRI is generally considered the gold standard for intracranial imaging and most spinal imaging. For placement of DBS electrodes, postoperative MRI is extremely helpful in determining electrode location relative to the intended brain target; this facilitates programming of the device, as well as providing the implanting team with important feedback about their surgical technique. MRI is unique in that it uses static and pulsed gradient magnetic fields with pulsed radiofrequency (RF) energy to produce images. While this makes MRI biologically safer to tissues than the ionizing radiation used in plain radiography and CT, potential issues can arise when implanted hardware is present. Ferromagnetic attraction, the phenomenon that can cause sudden and strong pull on ferrous objects as they approach the magnet bore, tends not to be problematic; most modern neuromodulation systems are made of non-ferrous components and therefore are not subject to these forces. However, device screening to assure that it does not contain ferromagnetic material is still mandatory.

Device heating and interference with the normal functioning of the IPG are the greatest concern when performing MRI in these patients. Neuromodulation systems generally have an IPG that is located at some distance from the actual site of stimulation. The two are connected by an insulated wire or wires which, by design, are longer than they need to be to accommodate different anatomical considerations or pulse generator locations, provide some degree of strain relief, allow technical ease in placement and subcutaneous tunneling, etc. As a result, the redundant portion of the wire usually contains random turns and loops that can act as an antenna, locally focusing the energy of the RF pulses and producing heat along the metallic components of the wire and in tissue adjacent to the wires. Since the stimulating surfaces are non-insulated metal directly in contact with neural tissue, any heating that does occur is realized at the metal–tissue interface [11–14]. Heating can also occur anywhere along the system, and electrical discontinuities such as a break in the lead or lead extension can be particularly problematic. In the brain parenchyma, temperature increases in the 5–7° Celsius range cause reversible tissue dysfunction; increases in excess of 8 °C may cause irreversible thermal injury depending on exposure duration [13,15,16]. The most common method of measuring energy absorbed by the body during MRI is to calculate the specific absorption rate, or SAR. SAR is directly related to the RF energy used to excite spins during a given MRI pulse sequence. Some pulse sequences, such as gradient echoes, generally have a low SAR (approximately 0.1–0.5 Watts/kg). Other sequences, such as spin echoes or steady state acquisitions, have an inherently higher SAR (upwards of 3.0 Watts/kg). In general, the higher the SAR for a particular MR sequence, the more potential for heating.

In 2005, Medtronic issued new safety guidelines for performing MRI in patients with implanted DBS systems based on two case reports of adverse events during MR imaging [17–20]. Both of these adverse events occurred under very specific and unusual circumstances. The incidents occurred in 1.0 T scanners, which are uncommon and use a different radiofrequency wavelength than standard 1.5 T magnets. One case involved the use of a whole body RF coil for spine imaging in a patient with an abdominally placed IPG. Use of a whole body RF coil exposes more of the implanted hardware to RF energy, which can result in higher energy deposition into the device. Moreover, this patient had an abdominally placed IPG, which means that a longer lead extension was used; this may have provided a more robust 'antenna' for focal absorption of RF energy. Interestingly, many groups in the past have used whole body RF coils to perform spinal and other studies in DBS implanted patients at 1.5 T without incident [21,22]. The other adverse event involved scanning a patient with externalized leads, also in a 1.0 T scanner. Again, other groups have scanned patients with externalized leads at 1.5 T without incident, and our group has been implanting DBS leads using interventional MRI since 2004, a technique that involves scanning with externalized leads [23]. The exact causative factors that led to these two patient injuries are still not clear.

As of 2010, the Medtronic guidelines for MRI scanning of implanted DBS systems include using only horizontal bore 1.5 T MRI systems, using only head transmit/receive coils that do not extend over the pulse generator site in the chest, entering the correct patient weight into the MR console (so the SAR for a given sequence is calculated correctly), limiting the gradient field to 20 Tesla/s or less, and using exam parameters that limit the displayed average head SAR (or applied SAR if known) to 0.1 W/kg or less for all RF pulse sequences. MRI should not be performed in the presence of a known broken lead or lead extension. It is also recommended that the electrode array be changed to a bipolar configuration and that the IPG not only be turned off but the amplitude also turned to zero prior to scanning; it is well known that the IPG can be switched on and off during scanning, which is why it is not considered sufficient merely to turn the device off. Avoiding sedation during scanning is also advised so patients can report any untoward events. Finally, if scanning is done with externalized leads, it is recommended that they be kept straight, down the center of the head coil, out of contact with the patient and wrapped in an insulating material [20].

The labeling for VNS also warns against the use of body coils and spine MRI, although there is a published case report of cervical MRI done emergently for spinal cord compression in a VNS patient without incident [24]. Cranial MRI with VNS is considered safe, provided that the following steps are followed as outlined in the device labeling:

1. use only a head T/R coil
2. set the outputs of the device (both regular output and magnet mode output) to zero
3. use only MR scanners that are 2.0T or less
4. keep SAR less than 1.3W/kg for a 70kg patient
5. keep the time-varying intensity less than 10Tesla/s [25].

There is a recent phantom study that shows imaging VNS at 3T may be safe as well [26]. All manufacturers of SCS systems warn against the use of MRI with their devices, although Medtronic does state that cranial MRI with a head T/R coil is okay [10]. One group has carefully studied patients with SCS undergoing spine MRI at 1.5T and, although some patients reported sensations of stimulation in the same distribution of their usual SCS and two reported mild heating around the IPG, no permanent or serious adverse events occurred [27].

MEDICAL AND SURGICAL INTERVENTIONS

Diathermy

Among the many medical therapies that patients may be exposed to, the most dangerous may be diathermy. Diathermy is the therapeutic, localized heating of soft tissues by electric current, ultrasound or high frequency electromagnetic radiation. It can be used to treat musculoskelet al pain, joint pain, tendonitis, and other inflammatory conditions. Two patients with DBS have suffered complications related to electrode heating after receiving diathermy, one with catastrophic results [28,29]. All forms of diathermy (shortwave, microwave or therapeutic ultrasound) are considered to be contraindicated in the presence of neuromodulation devices, regardless of the body site being treated. In addition to issues of heating, device damage may also occur in this setting.

Cardioversion, defibrillation and sensing pacemakers

In patients requiring cardiac defibrillation or cardioversion, the primary concern should be the patient's overall well-being and risk of death. In a life-threatening situation, these therapies should not be withheld. Guidelines have been suggested and include positioning the paddles as far from the implanted system as possible, trying to position the paddles perpendicular to the wires and using the lowest clinically appropriate energy [10]. In addition to acute situations where external cardioversion and defibrillation is necessary, the use of chronic implanted defibrillators has risen dramatically in recent years. Although some device labeling warns against the presence of both neuromodulation devices and implanted cardiac defibrillators and pacemakers that have a sensing function, there are numerous published reports of such devices coexisting without incident [30–32]. For DBS, it is recommended that the pulse generators be placed on opposite sides of the chest and that the DBS be programmed in bipolar mode to minimize the potential for interference between the two devices.

Surgery and electrocautery

Specific precautions should be observed in patients undergoing surgical procedures involving bipolar or monopolar electrocautery. Bipolar cautery with forceps involves current passing only between the tips of the instrument, therefore, it is considered safe for use in patients with implanted devices. Monopolar cautery involves current flow from the tip of the instrument through the patient to the grounding pad, which is placed somewhere on the body (typically the thigh). As a result, much higher electrical fields are generated in the body with monopolar cautery. Many surgeons routinely use monopolar cautery when operating on patients with implanted devices, but some guidelines should be observed. The stimulator should be turned off and, if the electrode is externalized, it should be disconnected from any external devices such as pulse generators. Some recommend placing the grounding pad as far from the stimulator components as possible, and avoiding the use of full-length grounding pads on the operating room table surface [10].

Lithotripsy

Extracorporeal shock wave lithotripsy (ESWL) is the use of strong sonic pulses to non-invasively treat renal or biliary calculi. Intermittent fluoroscopy or high output ultrasound are used to localize the stones and monitor treatment. The shock waves associated with this therapy are strong enough to disrupt or physically damage device components. Most manufacturers caution against the use of lithotripsy, particularly in patients with IPGs located abdominally or in the buttock region.

External beam radiotherapy

Patients may develop malignancies that require external beam radiotherapy (XRT) to a body region that contains some component of a neuromodulation device. Unfortunately, the potential effect of radiotherapy on these devices is not well known. The metal oxide semiconductor circuitry used in cardiac pacemakers is sensitive to the effects of radiation, and therapeutic stimulation thresholds can be altered due to thermal damage at the stimulating surfaces of pacemaker electrodes [33]. Various studies have described pacemaker malfunctions after cumulative doses as low as 10 Gy, prompting some to recommend this as the maximal dose that the system should be exposed to; others have recommended a maximum exposure of only 2 Gy for cardiac devices [33,34].

Even less is known about the effects of radiation on neurostimulation devices. One paper describes XRT to a squamous cell carcinoma of the neck in a Parkinson's patient that the authors actually felt benefited from the presence of DBS, as the patient was able to lie still during the treatment sessions. It was estimated that the total dose to the IPG in this patient was 7.5 Gy, most of it due to scatter. The device was shielded with external lead during some of the treatments [35]. While it does not seem prudent to withhold XRT in patients with neuromodulation devices in place, it does seem wise to minimize the dose to the IPG as much as possible, with either external lead shielding or other maneuvers. A DBS patient in our practice with an IPG implanted in the chest developed a breast mass on the same side. She underwent excisional biopsy and was diagnosed with carcinoma. When it was advised that she undergo XRT to the breast, we elected to reposition her IPG both to remove it from the radiation field and also to prevent it from interfering with future surveillance mammography. She completed her treatments with no adverse events.

Pain therapies

Transcutaneous electrical nerve stimulation (TENS) units pass electrical current into body tissues locally via an external pulse generator and pads that are applied to the skin. Since these devices are commonly used by patients with low back pain, and patients who have implanted with SCS for chronic back pain have IPGs and electrodes often occupying the same body regions, some SCS manufactures specifically state that the TENS pads should not be placed over any component of the SCS system [10]. A more invasive treatment for low back pain due to facet disease is radiofrequency medial branch neurotomy. This procedure has been performed in a DBS patient with an abdominally-placed IPG without complication [36].

Electroconvulsive therapy

Electroconvulsive therapy (ECT) is used in patients with severe, treatment refractory depression, catatonia, acute suicidality or psychosis, or in bipolar patients with severe mania. During ECT, electrodes are placed bilaterally in the frontal or temporal region, and electric current is passed between them to induce a seizure. Its use in our patient population may become increasingly important as DBS is now being used in clinical trials to treat patients with treatment resistant depression and other psychiatric disorders. The concern with ECT is current induction into the DBS system and heating of the electrodes, as well as potential electrode movement in the setting of repeated seizure induction.

Several case reports now describe ECT being successfully used in DBS patients, including two separate courses of treatment in the same patient, with no loss of efficacy and no electrode movement on post-treatment imaging [36–38]. The lack of heating-related complications may not be totally unexpected, as the current used to induce seizures in ECT is actually significantly less than the current used in cardioversion [36]. Authors recommend placing the ECT electrodes as far away from the DBS leads as possible.

CONCLUSIONS

A number of safety concerns and limitations must be considered after implanting neuromodulation devices. The good news is that, given the number of patients that have been implanted worldwide, the number of severe adverse events with these devices is actually quite low. Ongoing patient and physician education, good communication from device manufacturers and information dissemination via the medical literature should ensure that the highest possible safety standards are maintained.

REFERENCES

1. North RB, Kidd DH, et al. Spinal cord stimulation for chronic, intractable pain: experience over two decades. *Neurosurgery.* 1993;32:384–394 discussion 394-395.
2. Spincemaille GH, Klomp HM, et al. Technical data and complications of spinal cord stimulation: data from a randomized trial on critical limb ischemia. *Stereotact Funct Neurosurg.* 2000;74:63–72.
3. Trentman TL, Zimmerman RS. Occipital nerve stimulation: technical and surgical aspects of implantation. *Headache.* 2008;48:319–327.
4. Kapural L, Mekhail N, et al. Occipital nerve electrical stimulation via the midline approach and subcutaneous surgical leads for treatment of severe occipital neuralgia: a pilot study. *Anesth Analg.* 2005;101:171–174 table of contents.
5. Trentman TL, Zimmerman RS, et al. Occipital nerve stimulator placement under general anesthesia: initial experience with 5 cases and review of the literature. *J Neurosurg Anesthesiol.* 2010;22:158–162.
6. Geissinger G, Neal JH. Spontaneous twiddler´s syndrome in a patient with a deep brain stimulator. *Surg Neurol.* 2007;68:454–456 discussion 456.
7. Israel Z, Spivak A. A tremulous twiddler. *Stereotact Funct Neurosurg.* 2008;86:297–299.
8. Cyberonics. Epilepsy patient's manual for vagus nerve stimulation with the vns therapy system. from http://us.cyberonics.com/en/vns-therapy/healthcare-professionals/manuals/. 2008.
9. Chen C, Cole W, et al. Hybrid cars may interfere with implanted deep brain stimulators. *Mov Disord.* 2009;24:2290–2291.
10. Medtronic. Medtronic pain therapy: Using neurostimulation for chronic pain. from http://manuals.medtronic.com/wcm/groups/mdtcom_sg/@emanuals/@era/@neuro/documents/documents/wcm_prod042259.pdf. 2007.
11. Schaefer DJ. Safety aspects of radiofrequency power deposition in magnetic resonance. *Magn Reson Imaging Clin N Am.* 1998;6:775–789.
12. Rezai AR, Finelli D, et al. Neurostimulators: potential for excessive heating of deep brain stimulation electrodes during magnetic resonance imaging. *J Magn Reson Imag.* 2001;14:488–489.

13. Finelli DA, Rezai AR, et al. MR imaging-related heating of deep brain stimulation electrodes: in vitro study. *Am J Neuroradiol.* 2002;23:1795–1802.
14. Shellock FG, Cosendai G, et al. Implantable microstimulator: magnetic resonance safety at 1.5 Tesla. *Invest Radiol.* 2004;39:591–599.
15. Schueler BA, Parrish TB, et al. MRI compatibility and visibility assessment of implantable medical devices. *J Magn Reson Imag.* 1999;9:596–603.
16. Sawyer-Glover AM, Shellock FG. Pre-MRI procedure screening: recommendations and safety considerations for biomedical implants and devices. *J Magn Reson Imag.* 2000;12:92–106.
17. Spiegel J, Fuss G, et al. Transient dystonia following magnetic resonance imaging in a patient with deep brain stimulation electrodes for the treatment of Parkinson disease. Case report. *J Neurosurg.* 2003;99:772–774.
18. Henderson JM, Tkach J, et al. Permanent neurological deficit related to magnetic resonance imaging in a patient with implanted deep brain stimulation electrodes for Parkinson´s disease: case report. *Neurosurgery.* 2005;57:E1063 discussion E1063.
19. Medtronic. Urgent device correction - change of safe limits for MRI procedures used with medtronic activa deep brain stimulation systems. *J. Tremmel.* 2005.
20. Medtronic. MRI guidelines for DBS. 2006.
21. Larson PS, Richardson RM, et al. Magnetic resonance imaging of implanted deep brain stimulators: experience in a large series. *Stereotact Funct Neurosurg.* 2008;86:92–100.
22. Tagliati M, Jankovic J, et al. Safety of MRI in patients with implanted deep brain stimulation devices. *Neuroimage.* 2009;47(Suppl. 2):T53–T57.
23. Starr PA, Martin AJ, et al. Subthalamic nucleus deep brain stimulator placement using high-field interventional magnetic resonance imaging and a skull-mounted aiming device: technique and application accuracy. *J Neurosurg.* 2010;112:479–490.
24. Roebling R, Huch K, et al. Cervical spinal MRI in a patient with a vagus nerve stimulator (VNS). *Epilepsy Res.* 2009;84:273–275.
25. Cyberonics. MRI with the VNS therapy system. from http://us.cyberonics.com/en/vns-therapy/healthcare-professionals/manuals/. 2008.
26. Gorny KR, Bernstein MA, et al. 3 Tesla MRI of patients with a vagus nerve stimulator: initial experience using a T/R head coil under controlled conditions. *J Magn Reson Imag.* 2010;31:475–481.
27. De Andres J, Valia JC, et al. Magnetic resonance imaging in patients with spinal neurostimulation systems. *Anesthesiology.* 2007;106:779–786.
28. Nutt JG, Anderson VC, et al. DBS and diathermy interaction induces severe CNS damage. *Neurology.* 2001;56:1384–1386.
29. Roark C, Whicher S, et al. Reversible neurological symptoms caused by diathermy in a patient with deep brain stimulators: case report. *Neurosurgery.* 2008;62:E256; discussion E256.
30. Obwegeser AA, Uitti RJ, et al. Simultaneous thalamic deep brain stimulation and implantable cardioverter-defibrillator. *Mayo Clin Proc.* 2001;76:87–89.
31. Rosenow JM, Tarkin H, et al. Simultaneous use of bilateral subthalamic nucleus stimulators and an implantable cardiac defibrillator. Case report. *J Neurosurg.* 2003;99:167–169.
32. Schimpf R, Wolpert C, et al. Potential device interaction of a dual chamber implantable cardioverter defibrillator in a patient with continuous spinal cord stimulation. *Europace.* 2003;5:397–402.
33. Sundar S, Symonds RP, et al. Radiotherapy to patients with artificial cardiac pacemakers. *Cancer Treat Rev.* 2005;31:474–486.
34. Last A. Radiotherapy in patients with cardiac pacemakers. *Br J Radiol.* 1998;71:4–10.
35. Mazdai G, Stewart DP, et al. Radical radiation therapy in a patient with head and neck cancer and severe Parkinson's disease. *Clin Oncol (R Coll Radiol).* 2006;18:82–84.

36. Osborne MD. Radiofrequency neurotomy for a patient with deep brain stimulators: proposed safety guidelines. *Pain Med.* 2009;10:1046–1049.
37. Chou KL, Hurtig HI, et al. Electroconvulsive therapy for depression in a Parkinson´s disease patient with bilateral subthalamic nucleus deep brain stimulators. *Parkinsonism Relat Disord.* 2005;11:403–406.
38. Nasr S, Murillo A, et al. Case report of electroconvulsive therapy in a patient with Parkinson disease concomitant with deep brain stimulation. *J ECT.* 2011, 27(1):89–90.

Commentary on Safety Concerns and Limitations

Ali Rezai[1], Jeffrey E. Arle[2], MD, PhD

[1]*Ohio State Univeristy, Department of Neurosurgery, Columbus, OH, USA*
[2]*Director Functional Neurosurgery and Research, Lahey Clinic, Burlington, MA;*
Associate Professor of Neurosurgery, Tufts University School of Medicine, Boston, MA

Neuromodulation is among the most promising areas in medicine that can significantly improve the quality of life and functioning of patients with various chronic neurologic conditions. Worldwide, there are over 80 000 deep brain stimulation (DBS) implants and over 500 000 neuromodulation implants. Given the growing number and application of these implants, it becomes important for all practitioners to be educated and familiar with the various environmental-related interactions and the safety of these devices.

Drs Larson and Martin have provided an outstanding systematic review pertaining to mechanical stresses, electromagnetic interference, medical and surgical device interactions, medical procedures, as well as MRI-related safety issues with neuromodulation implants. More work will undoubtedly be needed to advance infection-resistant materials and streamlining the size and tissue interface necessary still to obtain the highest levels of efficacy, both of which address needed improvements in the local environment for implanted devices in this field. External environmental concerns have primarily revolved around MR compatibility, addressed herein. However, future concerns may lie in the interference and signal-to-noise issues involved in using multiple local external devices, media devices, and other medical therapies that pose risk to the efficacy and safety of neuromodulation implants. Some of these devices may interfere with communication parameters in wireless interfaces, airport security, and general local RF environments imposed by the growing consortium of devices being developed in the consumer and medical device areas.

This chapter, however, provides an essential foundation for safety and ongoing evaluation and management of neuromodulation implants for all related clinical personnel, engineers and neuromodulation device manufactures.

STUDY QUESTIONS

1. Determine the ways that patients may truly need an MRI scan after a neuro-modulation therapy has been implemented?

2. Consider all factors in defining what top three issues contribute most to poor or morbid outcome in neuromodulation and what steps can eliminate or mitigate against them?

Part VII

The End User

Chapter 19

Expectations and Outcomes

Guillermo A. Monsalve, MD, Chad W. Farley, MD, and George T. Mandybur, MD

Department of Neurosurgery, University of Cincinnati (UC) Neuroscience Institute and UC College of Medicine, and Mayfield Clinic, Cincinnati, OH, USA

EXPECTATIONS AND OUTCOMES

Since the resurgence of neurostimulation technologies in the 1990s, promising advances have been made in this field by altering nervous system function for relief of pain and other symptoms in select patients. Combined with a better understanding of the disease process, the use of electrical stimulation and lesioning of specific targets in the brain or spinal cord has provided many patients with the amelioration of symptoms and medication reduction, thus improving their overall quality of life. Deep brain stimulation (DBS) for movement disorders, spinal cord stimulation (SCS) for pain, motor cortex stimulation (MCS) for neuropathic pain, and vagal nerve stimulation (VNS) for epilepsy are described with respect to the end-user experience that includes patient clinical outcomes and perceptions.

DEEP BRAIN STIMULATION FOR MOVEMENT DISORDERS

Parkinson's Disease

Some of the initial attempts at long-term stimulation for chronic disease states were attributed to Shealy in the USA [1], and Bechtereva in Russia [2]. However, before stimulation efforts, other surgical measures had been used to effect modulation of the nervous system including the ligation of the choroidal artery for movement disorders [3] and targeted lesioning. Although DBS is currently used most often for movement disorders, there are many other indications. In this setting, implantation of stimulation devices generally targets the subthalamic nuclei (STN), globus pallidus internus (GPi), or ventralis intermediate nucleus (Vim) of the thalamus.

The exact placement of the stimulator has significant impact on the patient, resulting in symptom relief that is unique for each target. In terms of motor improvements, the targets and symptom relief can vary. For example, the target could include the GPi in cases of dystonia [4,5] or the Vim in cases of

Essential Neuromodulation. DOI: 10.1016/B978-0-12-381409-8.00019-X non-parkinsonian tremor. In patients with Parkinson's disease, targeting this subnucleus (Vim) does relieve the parkinsonian tremor but fails to modify the other chief symptoms of the disease [6]. Although debate exists what is the ideal target for various movement disorders, the target is largely based on the type of expected symptom relief. Target selection is then matched to the nature of symptoms experience of the patient. To ensure the optimal surgical outcomes for patients with movement disorders, the medical team should carefully select the patient, exhaust the available medical treatments, choose the best target and surgical technique [7], optimize the DBS settings [8,9], and properly manage the post-surgery medication regimens.

Deep brain stimulation for Parkinson's patients is complicated in part because of the spectrum of symptoms that are not just movement related. The stimulation effect reaches other aspects of the disease and does so to varying degrees depending on the target. Both the GPi and STN nuclei not only improve parkinsonian symptoms but also reduce drug-induced dyskinesias [10] but, in addition, the STN is thought to also reduce medication burden. In the 2009 COMPARE trial, the authors demonstrated that motor score improvements were similar whether targeting the GPi or STN [11]. Zahodne et al also noted that neither motor nor mood scores differed by these two targets, but the GPi target did demonstrate greater improvements in the subscale ratings related to mobility, activities of daily living (ADL), stigma, and social support. This relatively new claim warrants further research because the impact on patient quality of life is significant. With the negative impact of the disease on patient's cognition, Heo et al believed that bilateral STN stimulation might lead to slightly more detrimental effects on frontal lobe function and memory [12]. The authors inferred that improved outcomes with cognition might be obtained via GPi or unilateral targeting. The full impact of GPi versus STN targeting though is still an area of controversy, especially with respect to overall patient outcomes.

Results in quality-of-life studies have varied regarding the effects of surgical treatment in patients with movement disorders. And, the numbers of studies are sparse. Outcomes for movement disorders, particularly Parkinson's disease, have used the unified Parkinson's disease rating scale (UPDRS) part III, a standardized scale that primarily demonstrates the motor benefits subsequent to surgical intervention. Such measures can in part reflect subjective improvements from the point of view of the patient: for example, regaining motor dexterity and control (e.g. holding a cup of coffee without fear of spilling it). Admittedly, the patient's relief in these circumstances is subjective, difficult to measure and appreciate. In a long-term follow-up study of STN DBS, Krack et al found that there were significant improvements in the postoperative UPDRS scale, specifically ratings were 59% lower at 3 months and 54% lower at 5 years [13]. In this same study, the authors reported improvement in ADL functions by 49% compared with baseline functioning; they furthermore noted that, before surgery, most patients had depended on others to some degree but after surgery nearly all enjoyed independence throughout the entire follow-up period.

In a similar study assessing the psychological impact of deep brain stimulation, Schupbach et al demonstrated comparable results, that is, the UPDRS ratings showed 54% improvement at 5 years [14]. The authors noted that 10 of 20 patients were able to withdraw completely from all parkinsonian medications. Although the measures of neuropsychological and mood assessments were unchanged by the surgical procedure, cognitive decline was marked. However, this decline was attributed to the natural progression of the parkinsonian disease rather than the intervention performed. Another important aspect of this study demonstrated that daily dosing of levodopa was reduced by 58% and very likely significantly impacted the patient's perception of the disease. When medication regimens are simplified [15], patients can not only better control their disease symptoms but also notice significant impact on quality of life and cost effectiveness.

Essential Tremor

Neurostimulation for tremor can significantly benefit a patient's quality of life. Even while undergoing intraoperative testing, patients may feel tearfully happy about the prospect of better motor function. As discussed in other chapters, Vim or STN targeting is most often used in neurostimulation for essential tremor. Many of the research findings regarding tremor are included with information for Parkinson's disease. Sydow et al documented long-term relief of tremor 6 years after surgery in the 19 of 37 patients available for follow up; significant reduction in tremor score and improvement in activities of daily living were found compared with baseline or in the stimulation-off mode [16]. Zhang et al cited an 80.4% reduction in tremor and 69.7% improvement in handwriting in 34 patients with an average 56-month follow up [17]. Interestingly, between 57- and 90-months follow up, no statistical difference was found in functional ability when evaluating tremor and handwriting. Subtle adjustments in programming, primarily increases in voltage, were needed by many patients during the 5-year follow up. At 7 years postoperatively, Hariz et al noted decreases in the efficacy of DBS for tremor; however, a notable positive impact on quality of life and ADL functioning remained [18], especially for the patient's ability to eat and concerns with social life. The authors stated the more significant declines in the effects of DBS in most other areas, which began 6–8 years postoperatively, were likely due to aging, aging co-morbidities, and disease progression. At this endpoint, tremors significantly worsened when stimulation was turned off. Patients with tremors who had been unable to write their names or perform other common tasks found significant rewards with the renewed ability to function in ways often taken for granted by others.

Dystonia

Stimulation of the GPi is the most studied target for dystonia, both primary and secondary. In primary generalized dystonia (PGD), the mean improvement at

3- to 12-month follow-up ranged from 46% at 3 months and 80% at 12 months follow up in the Burke–Fahn–Marsden dystonia rating scale (BFMDRS) severity score, and 37% at 3 months, and 69% at 12 months follow up in the BFMDRS disability score [19–22]. At 2-year follow up, mean improvement ranged from 34 to 82% in the BFMDRS severity scores and 32 to 75% in BFMDRS disability scores [20,22–24]. At 3-year follow up, the motor improvement and quality of life (SF-36 questionnaire) observed at 1 year had been maintained [4]. Significant benefits of this therapy were evidenced by the improvements in general health and physical functioning at 12-month follow up [19] and 15% improvement in the unified dystonia rating scale (UDRS) which is statistically significant [25].

In children, the improvement reported at 6-month follow up was as high as 56% in the BFMDRS motor scores and 42% in the BFMDRS disability scores [26]. Compared with adults, the better outcomes in children were associated with DYT1-positive genetic status and with less motor impairment before surgery [27].

Regarding the neuropsychological outcomes, there have been no reported significant changes in measures of mood and cognition after pallidal stimulation in dystonia patients in short- and long-term follow up [4,19,28,29]. Some authors reported no significant reduction in the number of errors in the Wisconsin card sorting test (WCST) at 1-year follow up [28]. Others showed that bilateral GPi DBS clearly improved functional abilities and quality of life [29], and noted some improvements in concept formation, reasoning, and executive functions [4].

Contact location greatly impacted outcomes: overall clinical improvements were as high as 89% with posteroventral contacts versus only 67% with anterodorsal contacts in the pallidum [30,31]. There is also a chance of poor outcome because of lead misplacement. In addressing this topic, Ellis et al reported 12.8% improvement above the already-obtained improvement in the UDRS score after lead relocation [32].

Factors that predict poor outcome in generalized dystonia seem relate to a greater disability from symptoms (a high preoperative BFMDRS score) and long disease duration [33,34]. There was greater improvement in children with the genetic form DYT1-positive than in children with non-DYT1 forms. The volume of the GPi stimulated also influences the outcomes, the greater the GPi volume, the greater the degree of improvement [33].

For primary focal and segmental dystonia, improvements in the BFMDRS score are in the order of 64% at 3 months and 75% at 1 year 75% [35]. At 2-year follow up, ratings on the Toronto western spasmodic torticollis rating scale (TWSTRS) had improved nearly 60% for both disability and pain scores [23,36]. In patients with cervical dystonia, there was a 43% improvement in the TWSTRS severity score and a 59% improvement when both the disability and pain scores were combined [37]. A pilot study of bilateral pallidal stimulation in idiopathic cranial–cervical dystonia, BFMDRS motor and disability scores improved 72% and 38% at 6 months, respectively. Total TWSTRS scores improved 54% at 6-month follow up. Although the combined severity and disability subscores (BFMDRS) showed statistical improvement, the pain subscore only showed a trend toward improvement and was not statistically significant [38]. General health and physical functioning and depression scores improved significantly. Some negative changes in neuropsychological tests (memory and verbal skills) were observed, but did not impact daily life or employment [37].

For secondary dystonia (i.e. dystonias due to brain injuries), the reported outcomes of the GPi DBS were less promising than for other dystonias [38]. In a recent study that seems to refute some of these findings, Loher et al reported almost the same outcomes for both primary and secondary dystonias [39]. In tardive dystonia at 3–6 months after surgery, there were improvements of 74% in BFMDRS-M score, 89% in BFMDRS-D score, and 70% in abnormal involuntary movement scale (AIMS). In another study, quality-of-life improvements were significant in physical components and affective states [40].

In a mixed group of secondary dystonia patients (i.e. myoclonic dystonia, tardive dystonia, post-traumatic hemidystonia) who underwent treatment with bilateral GPi stimulation, improvements in the AIMS score ranged widely from 0 to 73.9%; the patient with no improvement had post-traumatic hemidystonia that temporarily improved after surgery but returned to the baseline findings some days after the DBS was turned on [21]. Some authors suggest that among the secondary dystonias, the drug-induced forms have potentially better outcomes compared with the secondary dystonias. That is, the BFMDRS severity scores improved 47.2% for the drug-induced group and 37% for the other mixed dystonias, and the BFMDRS disability scores were 54.6% and 34.4 %, respectively [9]. In patients with secondary dystonia, it is important to note that anatomical preservation of the basal ganglia is related to surgical outcome [41].

Previous reports of the usefulness of thalamic DBS to control dystonia have shown questionable results. Since then, GPi stimulation has gained greater acceptance in the treatment of this syndrome [42,43]. In a study of bilateral anterior dorsolateral STN stimulation in patients with predominantly cervical dystonia, significant improvement occurred in the motor, disability, and total TWSTRS scores. Outcomes were better in those who did not have fixed deformities. The mental component score of the SF-36 markedly improved, and neuropsychological function was not negatively affected as a result of surgery. However, there were no differences in the TWSTRS scores between stimulation-on and -off for the group as a whole [44]. In another study of patients with writer's cramp who underwent unilateral ventral oralis anterior/ventral intermediate (Voa/Vim) stimulation (one patient underwent both GPi and Voa/Vim DBS), Fukaya et al showed that BFMDRS scores improved 87.5% when the stimulator was turned on; this improvement was maintained at 2-year follow up. In the patient with dual DBS targets, the thalamic stimulation was superior to the pallidal stimulation [45]; however, superior results were obtained with pallidal stimulation in a previously Vim-DBS implanted patient with paroxysmal non-kinesiogenic dystonia [39]. These last two studies showed some interesting results, but need further study.

In our review of the literature, we found no consensus on programming settings for either primary or secondary dystonia. Usually GPi DBS required more amplitude than the GPi or STN DBS for Parkinson's disease, with wide ranges of pulsewidth and frequency settings for all groups [7,20,21,46]. Some authors suggested that the pulsewidth should exceed 180 μs (in dystonias), and the rate should be between 130 and 185 Hz (high-frequency stimulation) [9]. Other studies have reported significant improvement with low-frequency stimulation (50–60 Hz) [47]. In a study focused on the frequency of pallidal stimulation in primary dystonia, optimized stimulation at 130 Hz resulted in a 43% improvement in the BFMDRS score 6–12 months post-surgery. Quality of life measured through PDQ-39, EuroQoL1, and EuroQoL had significantly improved after surgery when measured in all of the scales. However, in this same study, a significant deterioration was observed at lower frequencies (0, 5, 50 Hz) in all patients [48].

After DBS, mobile, phasic dystonic movements respond rapidly and are predictors of good outcome, whereas fixed postures are less likely to improve and are predictors of poor outcome at 12-month follow up, mostly due to muscle contractures [34]. Although tonic components tend gradually to improve, some patients experience rapid improvement shortly after the DBS. Long-lasting benefits were not observed until 6–12 months later in most patients. The presence of microlesion effect immediately after pallidal DBS for dystonia also appears to be a good predictor of optimal clinical outcome, though this remains controversial [49].

SPINAL CORD STIMULATION FOR PAIN

As technology developed for electrical cardiac stimulation, these same principles were applied to stimulation of the nervous system, namely the spinal cord. This new technology found support from the gate theory of pain, which helps explain how the nervous system is affected in pain syndromes [50]. Although still somewhat controversial, the gate theory premise is the inhibition of small, unmyelinated pain fibers by the activation of large sensory nerve fibers. The spinal cord stimulator, placed over the dorsal columns of the spinal cord at approximately the mid-thoracic level, activates these large fiber neurons thereby to inhibit or diminish pain sensation. Good outcomes were reported with at least 50% reduction in pain; satisfaction was achieved in 47% of study participants 5 years after surgery [51]; 25% of patients returned to work after the implant. In the same study, many patients reduced or eliminated analgesics for pain and noted improvements in activities of daily living.

In the 2008 PROCESS study, Manca et al noted marked improvements in quality of life for patients with spinal cord stimulators [52]. While overall costs to the health-care system were higher, improvements in quality of life were seen by using the short-form (SF-36) and EuroQol-5D (EQ-5D) [53]. The EQ-5D consists of five questions, each relating to a different dimension that included mobility, self-care, and ability to undertake usual activity, pain/ discomfort and anxiety/depression. Each dimension has three possible levels of severity described as none or moderate or severe problems. Based on their combined answers to the EQ-5D questionnaire, patients can be classified into one of 243 health states. Each health state has an associated utility score on a 0 (death) to 1 (good health) scale. At mean baseline of EQ-5D, the PROCESS study participant scores were 0.15, which was considerably worse than the 0.31 for patients admitted to the hospital with ischemic strokes [52,54]. Considering the significant toll on the quality of life for patients with failed back syndrome, the potential for improvement on quality of life of these patients should not be underestimated.

Multiple studies have demonstrated superiority of SCS in comparison with other types of management. In a comparison with optimal medical management, Kumar et al demonstrated superior outcomes after SCS for failed back syndrome [55]. They reported that many patients not only found greater relief of neuropathic pain by implantation of SCS than conventional therapy, but also showed a significant increase in the quality of life. Kumar et al reported that 24 months after implantation, patients with SCS had greater satisfaction with treatment, and improved functional capacity and health-related quality of life [56]. Remarkably, 30% of patients returned to work after SCS therapy, including 4 of 37 patients who had been out of work for more than 2.5 years. In a 2007 comparison of reoperation versus SCS for failed back syndrome, North et al showed that SCS insertion was more effective and less expensive than re-operation for patients with failed back syndrome who had a previous surgery [57]. SCS was found to be most effective when patients avoided repeat surgery. Additionally, costs of SCS for patients with failed back syndrome were significantly less than for repeat surgery. Both factors play major roles in a patient´s quality of life.

MOTOR CORTEX STIMULATION FOR NEUROPATHIC PAIN

Motor cortex stimulation for neuropathic pain is a well-accepted practice that was introduced in the early 1990s by Tsubokawa et al [58]. However, as seen with SCS for pain, lack of uniformity in the surgical techniques used, pain pathologies enrolled, evaluation scales used, and even pain nomenclature make comparison of studies difficult [59–61]. Only a few statistically well-designed studies have assessed outcomes of MCS therapy with long-term follow up. In a randomized double-blind trial that included population of patients with mixed types of pain, central and peripheral pain, Velasco et al reported significant 40–80% pain improvement at 1-year follow up in all patients during the on-stimulation period [62]. The Bourhis scale and MPQ scores for the group decreased from 8.5 to 4.5 and from 133 to 40, respectively ($P < 0.01$) [62]. In another randomized controlled trial of patients with central and peripheral neuropathic pain, Smith et al reported >50% pain relief [63].

In a series of 32 patients who underwent MCS, Nguyen et al reported 77% pain relief in the patients with central pain, and 83.3% relief in the neuropathic facial pain group [64]. In a 4-year follow-up study including central and peripheral neuropathic pain, Nuti et al reported pain relief was excellent in 10% and good in 42% of the patients. Intake of analgesic medications decreased in 52% of patients and was completely withdrawn in 36% of patients. In the same study, 70% of the patients noted their satisfaction with the procedure, by saying that they would have undergone the same surgical intervention for relief of their pain [65].

In a literature review, Henderson et al reported rates of 40 to 100% relief of neuropathic facial pain relief from 3 to 28 months after MCS and 50% reduction in medications after the procedure [59]. In a series of 11 patients with thalamic pain, Tsubokawa et al reported that 73% of patients had excellent pain control during the trial period; the positive effects of the MCS were unchanged in 45% of the patients after more than 2-year follow up [58]. In a randomized controlled cross-over trial of MCS, Lefaucheur et al reported questionable results because the patient population was heterogeneous (i.e. different types and locations of peripheral neuropathic pain); however, during the open trial, there were significant improvements in visual analogue scale (VAS) and McGill pain questionnaire scores [66].

In a meta-analysis of the relevant studies on cortical stimulation and chronic pain, Lima et al reported a weighted-responder rate of 72.6% (95% CI, 67.7-77.4) in favor of the MCS [67]. Additionally, in another literature review, a good response was achieved of pain relief $\geq$40–50% in $\approx$ 50% of patients who underwent surgery and in 45% of 152 patients with a postoperative follow up $\geq$1 year [68].

Programming the stimulator is a very important factor in achieving good outcomes. However, with time, the initial benefits can be lost. Therefore, an intensive reprogramming can optimize the outcomes. Henderson et al showed that in patients with MCS, VAS scores declined from 7.44 to 2.28 with intensive reprogramming in chronically implanted MCS patients (mean 7.16 months (range 2–18)) [69]. In conclusion, MCS for chronic neuropathic pain is a reasonable and feasible option for select patients. Further studies are warranted better to define optimal patient populations and programming parameters.

VAGAL NERVE STIMULATION FOR EPILEPSY

Vagal nerve stimulation is one of the most recent developments in neurostimulation in selected patients with generalized epilepsy. VNS stimulation occurs via a specialized electrode wrapped around, typically, the left vagus nerve, and connected to an implanted pulse generator. The mode of action of this type of stimulation is thought to occur via retrograde activation of the vagus nerve into the brainstem and there causing the suppressing effects. This type of stimulation exerts a modulatory effect on cerebral neuronal activity. Since its approval in 1997, it quickly emerged as a well-accepted modality for the treatment of intractable epilepsy. In a study of 454 patients with generalized epilepsy, 43% of patients achieved a 50% reduction in baseline seizure frequency with the VNS [70] and achieved a long- lasting, even progressive benefit from long-term stimulation. Although VNS treatment may not allow a patient with seizures to resume normal employment or drive a car, the device may afford other significant quality-of-life improvements. Day-to-day providers who directly care for patients with VNS devices implanted have noted recovery is shortened in the immediate post-ictal period [71]. However, overwhelming support for this device among practitioners is somewhat lacking. Some caregivers helping with the patients with a VNS device may question its utility because debilitating seizures often continue and the sole prevention of generalization is less than ideal or insufficient to improve the quality of life for many patients and their families.

EXPECTATIONS OF THE PATIENT

Before Surgery

First, familiarity of the patient with the surgical and non-surgical team is advisable. Second, patients should understand the goals of the therapy, which symptoms will be targeted, what type of therapy will be pursued, and what technological options may best treat their disease. The surgeon also explains the steps of the surgery, ideally showing the samples of the leads, batteries, and other equipment. The patients will then have a better understanding of the potential complications and cosmetic implications associated with the surgery. In addition, a thorough review of the patient condition through a multidisciplinary approach is advisable. Most important, before surgery, patients should understand that further adjustments may be needed after implantation of the stimulator. Optimization of the therapy may take days, weeks, even months in the future. Programming the stimulator and drug optimization are as important as the surgery itself. Therefore, with education, the patient can better understand that therapeutic adjustments will likely be made after the initial postoperative period.

Other valid concerns expressed by patients and their families/caregivers are the costs and insurance coverage associated with the procedure(s). Although these procedures are routinely covered by most insurances (Medicare and Medicaid), coverage should be checked for each patient's insurance plan.

The Day of Surgery

Each step of the procedure should be explained during the operation so that patients feel confident with the team of nurses, surgeons, anesthesiologists, and neurologists involved in their care. Any patient discomfort (e.g. pain in the pin sites, micturition sensation during surgery, etc.) should be avoided. Patients and their families should feel both open and able to discuss if issues or concerns arise, and assured that these will be addressed.

Post-Surgery

In stimulators for pain and movement disorders, both rechargeable and non-rechargeable batteries are available, whereas vagal nerve stimulators use non-rechargeable batteries. Implications of these types of technologies need to be thoroughly discussed with the patient.

CONSIDERATIONS FROM A PATIENT´S PERSPECTIVE: PSYCHOLOGY OF PILL BURDEN

Neurostimulation techniques have the ability to improve patients' lives by decreasing the disability. The level of this disability is measured, in part, by how ill the patients feel. Some of these feelings are directly related to the disease, either limiting movements by decreased motor function or by inducing unpleasant sensations. By some standards, the amount of medication can be indirectly translated in 'how sick a patient feels'. For all of these therapies, the goal is to decrease the medications, whether a dopamine agonist, opiate, or anticonvulsant. With decreasing amounts of medication, patients can equate the success of the neurostimulation procedure with how much 'better' they feel. Compared with patients who adopt a passive role in their therapy, we have observed that patients who decrease more medications by pushing the system (i.e. constantly wanting better results) actually do better in the long run. However, patients with pain sometimes increase their medication usage, actually creating the impression that their condition is progressively worsening even when no changes are documented neurologically or radiographically.

Patients often ask to reduce or get off medications because of the perception that increasing the quantity of pills corresponds to increasing severity of disease. Accurate or not, this perception is important to the patient's overall outcome for any of the above-mentioned treatments. Is this perception due to the patient, therapy, or treating physician? It very well is all of the above. Without the will of the patient, nothing will happen no matter how hard the physician tries. The same goes for the procedure. In some patients, no matter how 'perfectly' placed the stimulator is, it still may not have the expected results.

Typically, patients go to the surgeon because of failure of other non-surgical therapies, such as medication or conservative measures. However, when surgical interventions fail to meet expectations or complications arise, the patient then reverts back to medical care. The failure of one leads to the other in a potentially continuous cycle until the patient achieves satisfaction with their outcome or becomes absolved in the disease state.

Education of the public and physicians is also important to the delivery of this contemporary care. Within the medical community, many physicians still regard these technologically advanced therapies as 'experimental' because of the lack of proper knowledge and education about neurostimulation. Yet, this attitude may prevent patients from coming to the forefront of medical care and gaining potential benefit.

CONCLUSIONS

The end user of neurostimulation therapies represents both the patient and physician, who both need to be involved with the care and maintenance of these devices. Physicians regulate the devices best to suit the patients' needs. Patients use the devices to ameliorate their disease symptoms. The better the communication between these end users, the better the therapy will be.

REFERENCES

1. Shealy CN, Taslitz N, Mortimer JT, Becker DP. Electrical inhibition of pain: experimental evaluation. *Anesth Analg*. 1967;46:299–305.
2. Bechtereva NP, Bondartchuk AN, Smirnov VM, Meliutcheva LA, Shandurina AN. Method of electrostimulation of the deep brain structures in treatment of some chronic diseases. *Confin Neurol*. 1975;37:136–140.
3. Cooper IS. Effect of anterior choroidal artery ligation on involuntary movements and rigidity. *Trans Am Neurol Assoc*. 1953;3:6–9.
4. Vidailhet M, Vercueil L, Houeto JL, et al. Bilateral, pallidal, deep-brain stimulation in primary generalized dystonia: a prospective 3 year follow-up study. *Lancet Neurol*. 2007;6: 233-229.
5. Hung SW, Hamani C, Lozano AM, et al. Long-term outcome of bilateral pallidal deep brain stimulation for primary cervical dystonia. *Neurology*. 2007;68:457–459.
6. Tasker RR. Tremor Parkinsonism and stereotactic thalamotomy. *Mayo Clin Proc*. 1987;62: 736–739.
7. Pinsker MO, Volkmann J, Falk D, et al. Deep brain stimulation of the internal globus pallidus in dystonia: target localisation under general anaesthesia. *Acta Neurochir*. 2009;151:751–758.
8. Starr PA, Turner RS, Rau G, et al. Microelectrode-guided implantation of deep brain stimulators, into the globus pallidus internus for dystonia: techniques, electrode locations, and outcomes. *J Neurosurg*. 2006;104:488–501.
9. Egidi M, Franzini A, Marras C, et al. A survey of Italian cases of dystonia treated by deep brain stimulation. *J Neurosurg Sci*. 2007;51:153–158.
10. Rezai AR, Machado AG, Deogaonkar M, Azmi H, Kubu C, Boulis NM. Surgery for movement disorders. *Neurosurgery*. 2008;62(Suppl. 2):809–839.
11. Zahodne LB, Okun MS, Foote KD, et al. Greater improvement in quality of life following unilateral deep brain stimulation surgery in the globus pallidus as compared to the subthalamic nucleus. *J Neurol*. 2009;256:1321–1329.
12. Heo JH, Lee KM, Paek SH, et al. The effects of bilateral subthalamic nucleus deep brain stimulation (STN DBS) on cognition in Parkinson disease. *J Neurol Sci*. 2008;273:19–24.
13. Krack P, Batir A, Van Blercom N, et al. Five-year follow-up of bilateral stimulation of the subthalamic nucleus in advanced Parkinson's disease. *N Engl J Med*. 2003;349:1925–1934.
14. Schupbach WM, Chastan N, Welter ML, et al. Stimulation of the subthalamic nucleus in Parkinson's disease: a 5 year follow up. *J Neurol Neurosurg Psychiatry*. 2005;76:1640–1644.
15. Ingersoll K, Cohen J. The impact of medication regimen factors on adherence to chronic treatment: a review of literature. *J Behav Med*. 2008;31:213–224.
16. Sydow O, Thobois S, Alesch F, Speelman JD. Multicentre European study of thalamic stimulation in essential tremor: a six year follow up. *J Neurol Neurosurg Psychiatry*. 2003;74: 1387–1391.
17. Zhang K, Bhatia S, Oh MY, Cohen D, Angle C, Whiting D. Long-term results of thalamic deep brain stimulation for essential tremor. *J Neurosurg*. 2010;112:1271–1276.

18. Hariz GM, Bomstedt P, Koskinen LO. Long-term effect of deep brain stimulation for essential tremor on activities of daily living and health-related quality of life. *Acta Neurol Scand.* 2008;118:387–394.

19. Vidailhet M, Vercueil L, Houeto JL, et al. Bilateral deep-brain stimulation of the globus pallidus in primary generalized dystonia. *N Engl J Med.* 2005;352:459–467.

20. Isaias IU, Alterman RL, Tagliati M. Deep brain stimulation for primary generalized dystonia: long-term outcomes. *Arch Neurol.* 2009;66:465–470.

21. Yianni J, Bain PG, Gregory RP, et al. Post-operative progress of dystonia patients following globus pallidus internus deep brain stimulation. *Eur J Neurol.* 2003;10:239–247.

22. Krauss JK, Loher TJ, Weigel R, Capelle HH, Weber S, Burgunder JM. Chronic stimulation of the globus pallidus internus for treatment of non-DYT1 generalized dystonia and choreoathetosis: 2-year follow-up. *J Neurosurg.* 2003;98:785–792.

23. Bittar RG, Yianni J, Wang S, et al. Deep brain stimulation for generalized dystonia and spasmodic torticollis. *J Clin Neurosci.* 2005;12:12–16.

24. Cersosimo MG, Raina GB, Piedimonte F, Antico J, Graff P, Micheli FE. Pallidal surgery for the treatment of primary generalized dystonia: long-term follow-up. *Clin Neurol Neurosurg.* 2008;110:145–150.

25. Diamond A, Shahed J, Azher S, Dat-Vuong K, Jankovic J. Globus pallidus deep brain stimulation in dystonia. *Mov Dis.* 2006;21:692–695.

26. Parr JR, Green AL, Joint C, et al. Deep brain stimulation in childhood: an effective treatment for early onset idiopathic generalized dystonia. *Arch Dis Child.* 2007;92:708–711.

27. Borggraefe I, Mehrkens JH, Telegravciska M, Berweck S, Bötzel K, Heinen F. Bilateral pallidal stimulation in children and adolescents with primary generalized dystonia – report of six patients and literature-based analysis of predictive outcomes variables. *Brain Dev.* 2010;32: 223–228.

28. Pillon B, Ardouin C, Dujardin K, et al. Preservation of cognitive function in dystonia treated by pallidal stimulation. *Neurology.* 2006;66:1556–1558.

29. Hälbig TD, Gruber D, Kopp UA, Schneider GH, Trottenberg T, Kupsch A. Pallidal stimulation in dystonia: effects on cognition, mood, and quality of life. *J Neurol Neurosurg Psychiatry.* 2005;76:1713–1716.

30. Houeto JL, Yelnik J, Bardinet E, et al. Acute deep-brain stimulation of the internal and external globus pallidus in primary dystonia: functional mapping of the pallidus. *Arch Neurol.* 2007;64:1281–1286.

31. Tisch S, Zrinzo L, Limousin P, et al. Effect of electrode contact location on clinical efficacy of pallidal deep brain stimulation in primary generalized dystonia. *J Neurol Neurosurg Psychiatry.* 2007;78:1314–1319.

32. Ellis TM, Foote KD, Fernandez HH, et al. Reoperation for suboptimal outcomes after deep brain stimulation surgery. *Neurosurgery.* 2008;63:754–761.

33. Vasques X, Cif L, Gonzalez V, Nicholson C, Coubes P. Factors predicting improvement in primary generalized dystonia treated by pallidal deep brain stimulation. *Mov Dis.* 2009;24: 846–853.

34. Isaias IU, Alterman RL, Tagliati M. Outcome predictors of pallidal stimulation in patients with primary dystonia: the role of disease duration. *Brain.* 2008;131:1895–1902.

35. Jeong SG, Lee MK, Kang JY, Jun SM, Lee WH, Ghang CG. Pallidal deep brain stimulation in primary cervical dystonia with phasic type: clinical outcome and postoperative course. *J Korean Neurosurg Soc.* 2009;46:346–350.

36. Yianni J, Bain P, Giladi N, et al. Globus pallidus internus deep brain stimulation for dystonic conditions: a prospective audit. *Mov Dis.* 2003;18:436–442.

37. Kiss ZH, Doig-Beyaert K, Eliasziw M, Tsui J, Haffenden A, Suchowersky O. The Canadian multicentre study of deep brain stimulation for cervical dystonia. *Brain.* 2007;130: 2879–2886.

38. Ostrem JL, Marks WJ, Volz MM, Heath SL, Starr PA. Pallidal deep brain stimulation in patients with cranial-cervical dystonia (Meige syndrome). *Mov Dis.* 2007;22:1885–1891.

39. Lohner TJ, Capelle HH, Kaelin-Lang A, et al. Deep brain stimulation for dystonia: outcome at long-term follow-up. *J Neurol.* 2008;255:881–884.

40. Gruber D, Trottenberg T, Kivi A. Long-term effects of pallidal deep brain stimulation in tardive dystonia. *Neurology.* 2009;73:53–58.

41. Eltahawy HA, Saint-Cy J, Giladi N, Lang AE, Lozano AM. Primary dystonia is more responsive than secondary dystonia to pallidal interventions: outcome after pallidotomy or pallidal deep brain stimulation. *Neurosurgery.* 2004;54:613–621.

42. Vercueil L, Pollak P, Fraix V, et al. Deep brain stimulation in the treatment of severe dystonia. *J Neurol.* 2001;248:695–700.

43. Krack P, Vercueil L. Review of the functional surgical treatment of dystonia. *Eur J Neurol.* 2001;8:389–399.

44. Kleiner-Fisman G, Liang GS, Moberg PJ, et al. Subthalamic nucleus deep brain stimulation for severe idiopathic dystonia: impact on severity, neuropsychological status, and quality of life. *J Neurosurg.* 2007;107:29–36.

45. Fukaya C, Katayama Y, Kano T, et al. Thalamic deep brain stimulation for writer´s cramp. *J Neurosurg.* 2007;107:977–982.

46. Toda H, Hamani C, Lozano A. Deep brain stimulation in the treatment of dyskinesia and dystonia. *Neurosurg Focus.* 2004;17:E2.

47. Ostrem JL, Starr PA. Treatment of dystonia with deep brain stimulation. *Neurotherapeutics.* 2008;5:320–330.

48. Kupsch A, Klaffke S, Kühn AA, et al. The effects of frequency in pallidal deep brain stimulation for primary dystonia. *J Neurol.* 2003;250:1201–1205.

49. Cersosimo MG, Raina GB, Benarroch EE, Piedimonte F, Alemán GG, Micheli FE. Micro lesion effect of the globus pallidus internus and outcome with deep brain stimulation in patients with Parkinson disease and dystonia. *Mov Dis.* 2009;24:1488–1493.

50. Melzack R, Wall PD. Pain mechansims: a new theory. *Science.* 1965;150:971–979.

51. North RB, Ewend MG, Lawton MT, Kidd DH, Piantadosi S. Failed back surgery syndrome: 5-year follow-up after spinal cord stimulator implantation. *Neurosurgery.* 1991;28:692–699.

52. Manca A, Kumar K, Taylor RS, et al. Quality of life, resource consumption and costs of spinal cord stimulation versus conventional medical management in neuropathic pain patients with failed back surgery syndrome (PROCESS trial). *Eur J Pain.* 2008;12:1047–1058.

53. Kind P. The EuroQoL instrument: an index of health-related quality of life. In: Spilker B, ed. *Quality of life and pharmacoeconomics in clinical trials.* Philadelphia: Lippincott-Raven; 1996:191–201.

54. Calvert M, Freemantle N, Cleland J. The impact of chronic heart failure on health-related quality of life data acquired in the baseline phase of the CARE-HF study. *Eur J Heart Fail.* 2005;7:243–251.

55. Kumar K, Taylor RS, Jacques L, et al. Spinal cord stimulation versus conventional medical management for neuropathic pain: a multicentre randomized controlled trial in patients with failed back surgery syndrome. *Pain.* 2007;132:179–188.

56. Kumar K, Taylor RS, Jacques L, et al. The effects of spinal cord stimulation in neuropathic pain are sustained: a 24-month follow-up of the prospective randomized controlled multicenter trial of the effectiveness of spinal cord stimulation. *Neurosurgery.* 2008;63:762–770.

57. North RB, Kidd D, Shipley J, Taylor RS. Spinal cord stimulation versus reoperation for failed back surgery syndrome: a cost effectiveness and cost utility analysis based on a randomized, controlled trial. *Neurosurgery.* 2007;61:361–368.

58. Tsubokawa T, Katayama Y, Yamamoto T, Hirayama T, Koyama S. Chronic motor cortex stimulation in patients with thalamic pain. *J Neurosurg.* 1993;78:393–401.

59. Henderson JM, Lad SP. Motor cortex stimulation and neuropathic facial pain. *Neurosurg Focus.* 2006;21:E6.

60. Gharabaghi A, Hellwig D, Rosahl SK, et al. Volumetric image guidance for motor cortex stimulation: integration of three-dimensional cortical anatomy and functional imaging. *Neurosurgery.* 2005;57(Suppl. 1):114–120.

61. Pirotte B, Voordecker P, Neugroschl C, et al. Combination of functional magnetic resonance imaging-guided neuronavigation and intraoperative cortical brain mapping improves targeting of motor cortex stimulation in neuropathic pain. *Neurosurgery.* 2005;56(Suppl. 2):344–359.

62. Velasco F, Argüelles C, Carrillo-Ruiz JD, et al. Efficacy of motor cortex stimulation in the treatment of neuropathic pain: a randomized double-blind trial. *J Neurosurg.* 2008;108:698–706.

63. Smith H, Joint C, Schlugman D, Nandi D, Stein JF, Aziz TZ. Motor cortex stimulation for neuropathic pain. *Neurosurg Focus.* 2001;11:E2.

64. Nguyen JP, Lefaucheur JP, Le Guerinel C, et al. Motor cortex stimulation in the treatment of central and neuropathic pain. *Arch Med Res.* 2000;31:263–265.

65. Nuti C, Peyron R, Garcia-Larrea L, et al. Motor cortex stimulation for refractory neuropathic pain: four year outcome and predictors of efficacy. *Pain.* 2005;118:43–52.

66. Lefaucheur JP, Drouot X, Cunin P, et al. Motor cortex stimulation for the treatment of refractory peripheral neuropathic pain. *Brain.* 2009;132:1463–1471.

67. Lima MC, Fregni F. Motor cortex stimulation for chronic pain: systematic review and meta-analysis of the literature. *Neurology.* 2008;70:2329–2337.

68. Fontaine D, Hamani C, Lozano A. Efficacy and safety of motor cortex stimulation for chronic neuropathic pain: critical review of the literature. *J Neurosurg.* 2009;110:251–256.

69. Henderson JM, Boongird A, Rosenow JM, LaPresto E, Rezai AR. Recovery of pain control by intensive reprogramming after loss of benefit from motor cortex stimulation for neuropathic pain. *Stereotact Funct Neurosurg.* 2004;82:207–213.

70. Morris GL 3rd, Mueller WM. Long-term treatment with vagus nerve stimulation in patients with refractory epilepsy: the vagus nerve stimulation study group E01-E05. *Neurology.* 1999;53:1731–1735.

71. Ardesch JJ, Buschman HP, Wagener-Schimmel LJ, van der Aa HE, Hageman G. Vagus nerve stimulation for medically refractory epilepsy: a long term follow-up study. *Seizure.* 2007;16:579–585.

Commentary on Expectations and Outcomes

Chris Hart, BA

Director of Urban and Transit Projects, Institute for Human Centered Design, Boston, MA

Editors' note: This is a special commentary chapter written by Chris Hart, who had bilateral GPi DBS to treat severe 'status dystonicus' after several months of hospitalization, during much of which he was in a heavily medicated coma and intubated without a tracheometry. He has cerebral palsy and developed this dystonic condition after surgery to perform an urgent abdominal exploration. Chris serves as the Director of Urban and Transit Projects at the Institute for Human Centered Design in Boston and serves on several Boards of Directors including as Treasurer of the Massachusetts Disability Law Center. It is worthwhile considering his observations with reference as well to patients who may have chronic pain, any variety of psychiatric disorder, or other movement disorders. We reproduce his perspective on the experience essentially unchanged, as it is a unique complement to the primary chapter.

If you asked me on April 27, 2006 what DBS was, I probably would have had to stop to think and eventually might have mustered some vague answer about hearing something about it being an experimental treatment for certain types of movement disorders like Parkinson's and a few of odd gimp conditions such as dystonia. If you asked me on June 20, 2006, in a whisper, I'd have asked how you liked my haircut and explained that my brain was now wired and had been reset thanks to DBS. Ask me what happened between leaving a meeting with my state's highway commissioner and June 18.... I really can't tell you. The pain I felt in the meeting, where even the commissioner noticed my intense discomfort, led to an emergency surgery immediately following a torturous day of diagnostic tests ordered by a cute resident. But I digress.

The surgery for an omental infarct, or rather the anesthesia level, as the anesthesiologist suspects, set off something in my brain, possibly related to the cerebral palsy I had acquired at birth. In the recovery room I began to experience intense uncontrollable movements and soon had every available nurse trying to hold me down. What followed was an induced coma that at least stopped many of the movements but didn't solve the problem. I also became a Petri dish for just about any infection found in a Boston hospital, spent 25+ days on a vent without a tracheostomy (thank god), and won the grand prize of double pneumonia, twice. While on some days I had up to 45 doctors and residents reviewing my case, most of the medical community, including those who are involved in the disability community and personally knew me, said to my parents, friends, and coworkers, 'Chris' run is done…He's had a good life, changed Boston, led the MBTA suit behind the scenes and wrote the settlement agreement that will change public transportation, and etc. but it is time to let go'. Somewhere in the midst of this adventure, my parents and friends learned of DBS and a video of me writhing around was sent up to the Lahey Clinic team in Burlington. Against much medical wisdom and before more errors could be made so I could croak, I was transferred to Lahey, still unconscious and on a vent. From there, I gather that the Lahey team mapped my screwed up brain enough to be pretty confident that DBS would work for me and the rest is history.

I woke up to a lot of exceedingly happy people and though I was still pretty drugged, I couldn't help but notice that the Red Sox were tanking (they had been in first place in April). As I gradually detoxed and my head began to clear, I began to absorb all that had happened (or in some cases, hadn't happened, but sure seemed real thanks to vivid hallucinations). While my memory is foggy of those early days, I recall waking one morning to someone carrying an electronic device and told me they needed to adjust the stimulators in my brain. Whether it was the detox process or the sheer emotional stress of all I had been through, I do remember strongly being absolutely terrified as I felt tingles, saw flashes of light, etc. even though, at some level, I knew the person was part of the team that saved my life. Today, just recalling that experience makes my hair stand up on end. My best explanation for what I felt that day is the following: having used about 13 of my nine lives, and realizing I was still alive, the last thing I wanted was someone messing with the one thing still seemingly working well – my brain.

Ignore for a moment that it was my brain that caused the whole mess to begin with and step into the world of disability where most of us with significant disabilities don't think in terms of being cured or having "disorders". It is a world where, at some level, most of us with significant disabilities acknowledge that we are more or less disabled by the environments with which we interact and with the attitudes of the people with whom we interact on a daily basis. It is a world where problem solving is a constant and where what works today might not work tomorrow or where (just as likely) what isn't working well today, might be fine tomorrow. Who knows why and, honestly, does it matter? After all, as long as your brain is working, everything else is solvable, right? Can't pee, insert a catheter. Can't swallow, insert a G tube. Hands/feet don't work, drive a chair with your head or puff 'n suck. Can't type or talk, there's technology for that too. Brain doesn't work, 'Houston, we have a problem!' Hence, my gut-level fear reaction.

Beyond these terrifying feelings, however, and with hindsight, I was just beginning to wrestle with the fact that the medical establishment seemingly failed me, to put it generously, and the implications of the failure to my life. The only reason I was still alive, aside from pure stubbornness, was a core group of friends (some of whom I hadn't seen in years but now had medical degrees), my boss, and my parents who kept catching medical errors and oversights, made phone calls and spent weeks at my bedside even as I steadily had slid downhill prior to the DBS surgery. For the condescending nurses and arrogant doctors who were so quick to write me off, my ICU room was plastered with photos of me speaking, press clippings where I was quoted, correspondence from government officials, etc. all to remind them that I indeed had a brain that worked just fine even if my body was a bit odd. Arriving at Lahey was my last shot and it worked for me. But any of us that have significant disabilities know that the medical system often fails us or can be all too quick to write us off, especially when friends/family aren't close and, consequently, we become much less confident in our caregivers.

As the days went by at Lahey and I came off my drug-induced high, I gradually got my mind around having wires in my brain, the fact that a little hand-held device could mess with my brain, and that I was a guinea pig for DBS. Sometimes the tuning improved my control, sometimes not, but on the whole things were looking up, I was finally eating solids, gaining weight, starting rehab for my emaciated body. I was growing accustomed to the buzz in my ear (something that I still hear from time to time) and my neck was loosening up. As my voice became stronger, people began to make the odd comment that my speech was actually clearer, more understandable. Additionally, I drooled less. Simply amazing, but alas within 6–8 months those improvements had returned back to previous levels. When my neurology team asked about rehab, I said that it was already the end of June and I was supposed to be speaking in Japan at the end of October, so we should all plan accordingly.

At rehab, I was working my butt off, reminding my therapists of my October trip and, against orders, I was even sneaking out to a couple of work meetings at City Hall. There, I was about to drive through the City Hall metal detector when something in my head screamed 'stop'…ah, yes, a reminder that I now have objects in my brain that don't respond kindly to magnets. It was then that I began to wonder what would happen if my DBS shut off. Truth to be told, none of us knew, and if you've suddenly had DBS, reading all of the caution notes from Medtronic might scare one into never going out. Certainly, staying home was not an option for me and after a few more brain tune-ups, as I affectionately call my programming sessions, I headed off to Japan for a couple of weeks.

Upon my return from Japan, I continued to wonder a lot about what would happen if the device turned off. After all, I had nearly been wanded by zealous security guards and just about everywhere I looked, I could see one or more devices that Medtronic warns all of us to avoid. In 2007, after a few more brain tune-ups, the Lahey crew let me turn my DBS 'off' on a Monday. Nothing happened. I was sent downstairs to get lunch – told to return in an hour. Ninety minutes later, I still felt normal, or what I describe as new normal, because the old normal ended in April 2006. Since I had my trusty restart kit and nothing was happening, I was sent home with clear directions: come back *any* time before or on Friday. Well, by Friday I was ready to turn the DBS back 'on'…I was clearly less coordinated and more jumpy. It was very gradual but noticeable to me within 48 hours. I was also fairly confident that the DBS is real and not a placebo…but then again my battery levels weren't going down either!

Turning the system back on again was a whole other experience. It wasn't painless and seemingly the way my system was programmed, it was a jolt. Aside from the typical flashes of light, seeing double, etc., there was a jolt and steadily intensifying pain from my neck up to my head. There and then I think everyone realized a need to slowly 'ramp up' when turning DBS on. Why on earth that isn't the default setting I don't know, but it's such an obvious thing to do.

Nearly five years later, my brain tune-ups are rare and, as far as I know, my DBS is working fine. I do wrestle with a couple of questions related to settings and battery life. With over 90,000 possible combinations of settings, has my complacency with the settings precluded experimenting to see what after improvements might be attainable? Why I've not had battery depletion or battery failures that others have had, I don't know but it seems odd. But I do know that in the last three months as the voltage has dropped precipitously, I've begun to feel different. Coordination is more problematic, my brain feels less clear, my cognition and memory seem to be impacted and perhaps more alarming, I feel depressed. As the date for my battery replacement surgery approaches, I find my energy level fluctuating, generally feeling not well and certainly much more emotional, especially feeling depressed. Can these symptoms be linked to the DBS or is it psychological? I don't know but I rather dread going into the OR even if I'm in good hands!.

While I generally do not worry about the DBS except when faced with a zealous TSA guy or when a doctor who doesn't know about DBS wants me to get an MRI. At a day-to-day level, the only time I really think about the DBS is if I hit my chest or the rare cases of random pain in my head near the prongs. The pain is random and shooting for 10–20 seconds then gone for days or weeks. Other times, I seemingly feel the wire shift on my neck but this is usually during strong sneezing (another activity Medtronic warns of, though I'm unclear how not to sneeze strongly). The wire shift is at times annoying and once in while can be painful but a little rubbing can usually get it back.

Clearly, life is not the same as it once was before DBS. Mentally, I've never fully regained my old self; writing and certain thinking skills do not come nearly as easily as they once did. Most people probably don't notice, but I do, and I know that my boss does, though she would probably never admit it to me. Are those skills impacted by DBS? Or, more likely, are they the result of the medical ordeal and the extraordinarily high doses of drug cocktails my body absorbed? We'll never know, but the old TV commercial, 'This is your brain on drugs' comes to mind.

Mentally, I also know that my time since 2006 has been gravy, albeit with a few lumps. I'm lucky that my 'good run' did not end and I can still leave the world a better place. I can't help but be thankful for each day. I went back to full-time work, probably a little too quickly, but there is so much to do. I sail, bike and can still ski, albeit tethered and not independently as I had previously – an unconfirmed rumor has it that I've had a few skiing spills since DBS and it still works. I'm working 50+ hours a week, though I lament my weak writing skills and lost eloquence. Finally, I also managed to meet a woman who will both put up with me and keep me in line. We have an agreement to keep each other around for 30 or 40 years, depending on which of us you ask...now if only my DBS battery would last that long!

STUDY QUESTIONS

1. What are the drawbacks in considering the patient perspective in design, application, and outcome with neuromodulation devices?
2. Consider ways patient perspectives and concerns, and clinician perspectives and concerns using neuromodulation can best be conveyed to third-party payors, administrators of health care, and legislators of health-care policy. How are these methods different from each other? How might they be improved upon?

Chapter 20

Neuromodulation Perspectives

Alim Louis Benabid, MD, PhD

CEA Clinatec, Clinatec, CEA Grenoble, Grenoble, France

Editors' note: The final chapter and its commentary section are meant to lend perspective, historical interest, and inspiration. Written by Professor Benabid from Grenoble, France, widely considered to be the 'father of DBS', it gives a flavor of the passion and poignancy he has brought to the field of neuromodulation. Dr Montgomery offers extensive counterpoint and guides the reader toward ever more sophisticated insight into mechanisms of brain function and how we might consider understanding neural function in general through the window of DBS. Nonetheless, these authors focus on DBS and we wish to encourage the reader to consider that, as useful as their content is in this regard, we believe it nevertheless leaves a framework for interpreting the relevant future of all neuromodulation.

INTRODUCTION

Nearly a quarter of a century ago the effects of deep brain stimulation at high frequency (DBS-HF) were discovered in a serendipitous manner. Its application to the treatment of movement disorders quickly expanded. It is validated for several indications (Parkinson's disease, essential tremor, dystonia, etc.). In some countries, DBS-HF is used as the first choice of surgical methods to treat severe cases of these diseases. The method itself has been extended to other indications and, most recently, has participated in the revival of psychosurgery. Industrial companies have quickly produced materials specifically designed for these new indications, mostly derived from the technology of cardiac pacemakers, and from the initial specific application to spinal cord stimulation and deep brain stimulation for pain. The number of publications has quickly increased, either for the development of the technique, the application to various indications and, more recently, for evaluation of the results in the scope of evidence-based medicine. In recent years, several multicenter double-blind studies have been produced. Alternative methods are still awaited but are yet not comparable in terms of results, particularly in the case of neural grafts. Gene therapy has been recently introduced which might be a challenge to DBS at high frequency, however, it is not yet ready for introduction to therapeutic use. This chapter does not intend to review DBS-HF over these 23 years, but expresses from the author some comments which he feels necessary to do at this point in time. The practice of this method since 1987, as well as the observation of the activity in this domain reflected in the published literature, has lead him to some thoughts, second thoughts, free opinions on several items and, of course, lays the path to perspectives.

SHOULD WE SAY DBS-HF OR SIMPLY DBS?

There is a persistent confusion in the literature about denominations. More and more frequently, DBS only is used in reports, and frequency is not even mentioned in the Method chapter. This does not acknowledge the dual effects of frequency, which are the most relevant characteristics of this method. I did not invent DBS, which has been used since the 1960s for pain and had been reported on several occasions for several targets in several indications decades ago, not to mention the use of electrical stimulation of superficial or deep structures during surgery for intraoperative localization purposes. These dual effects also were observed and reported on several occasions, but the respective definitions of low and high frequency were never clearly stated, and this did not come to the level of an application of frequencies higher than 100 Hz as a well-defined surgical method, mimicking lesions and then aiming at replacing them. What I claim is to have for the first time pointed out the striking different effects which were observed at low (below 100 Hz) and at high (above 100 Hz) frequencies, that high frequency DBS was closely replicating (though in a reversible and adaptable manner) the effects of lesions. It was our group who established DBS-HF (and not DBS *largo sensu*) as a well codified therapeutical method. We also claim that it is the demonstration in movement disorders that the method was reliable and safe which allowed its extension to other indications, particularly to psychosurgery, even if previous reports had suggested it.

Even if the reason of this frequency dependence of opposite effects remains to be understood, one should clearly mention at which frequency a study has been done or an effect has been observed. The term DBS is too large by itself and should not be used, except when the method of putting stimulating electrodes in the brain is concerned regardless of the frequency used.

TARGETING

Besides the selection of patients, targeting aims at the most prevalent criterion: the location of the electrode within the target. Ventriculography is, in our opinion, the gold standard. It shows the landmarks on which are based most of the coordinates and which allow comparison with the atlases. It is not a dangerous procedure once one knows how to perform it according to precise criteria, each of them aiming at improving the quality of the imaging and decreasing the risks of complications. Making a simple twist drill at 2.5 centimeters from the midline and 9 centimeters from the nasion most of the time allows entry to an avascular area with a minimum risk of bleeding. The perforation of the dura made by the tip of the twist, with some experience, is easy to perform. Rinsing this hole by soft injection of Ringer serum allows removal of the pieces of bone and to verify that there is no bleeding. Making a burr hole and opening the dura for coagulation of the pial vessels of the cortex does not really make sense, as only the superficial vessels are visible but not the deeper vessels in the sulci, which might also be hit and bleed. In addition, it is impossible to prevent the loss of CSF, which induces a brain shift and filling the hole with artificial CSF does not change it. Intraoperative x-rays as well as postoperative CT scans show the pneumocephalus. Tapping the ventricle on the right side must be done using a cannula with a stopper at 9.5 cm (with this length, there is no risk of missing the ventricle which is a question of orientation and also in the adult there is no risk of going below the level of the foramen of Monro and therefore no risk of hitting the basal ganglia structures). The good inclination is to aim at the imaginary center of the head and it is often easy to perceive the entry in the ventricle through the ependyma. Injection of 2 cm of air is mandatory to make sure that the CSF coming out from the cannula is not the CSF of the interhemispheric fissure (which is often dilated in parkinsonian patients) but the CSF from the ventricle, (which is clearly visible on the lateral x-ray showing the frontal horn as a round negative image with a horizontal limit between the air bubble and the CSF). Injection of 6.5 mL of Iopamiron® within two to four seconds allows good visualization of the lateral ventricle, the foramen of Monro and the third ventricle as well as the aqueduct of Sylvius. Sometimes, the anterior commissure is not easily visible and it is necessary to turn the patient upside down, which is not easy, or to add two to three more milliliters of air: this allows vision, in negative contrast, of the anterior commissure as well as the anterior ventricular features, such as the supraoptic recessus and the infundibulum as well as the pre-mammillary recessus.

The pictures from the anteroposterior view provide the location of the midline and particularly of the midline of the third ventricle, which might not coincide, but not always, with the midline of the cranium. This is important to delineate the laterality of the targets. The targets mentioned in functional neurosurgery are derived from previous anatomo-neurophysiological knowledge. But their coordinates are obtained from the observation of the clinical, therapeutical outcomes, which are the only valid criteria to be taken in account. Then, the coordinates of what we call the 'theoretical targets' are the mean and standard deviation of postoperative coordinates of the best clinically efficient electrode contact in the patients with the best outcomes. To avoid the important inter-individual variability, it is important to express these coordinates relative to some internal landmarks of the patients. Classically, since Guiot and Talairach, the internal landmarks available on ventriculography are the anterior–posterior commissure (AC–PC) distance and the height of the thalamus, the top of which corresponds to the floor of the lateral ventricle. Table 20.1 shows the coordinates values

TABLE 20.1 Coordinates and dimensions of the third ventricle

Target coordinates	Vim	STN	GPi
Units			
Ant-Post/PC (AC–PC length)/12	2.87 ± 0.37	5.16 ± 0.73	8.99 ± 1.04
Laterality mm (corrected)	13.77 ± 1.65	11.52 ± 2.11	16.65 ± 5.53
Height/AC–PC (height of thalamus)/8	0.65 ± 0.71	−1.30 ± 0.84	−0.85 ± 1.13
V3 dimensions mm (N = 197)	**Width of V3**	**AC–PC length**	**Height of thalamus**
Mean ± standard deviation	6.25 ± 2.31	24.82 ± 1.44	16.53 ± 1.57
Minimum/maximum	2.21/14.63	20.97/30.48	10.48/20.82

of the targets VIM (thalamic intermedius nucleus), STN (subthalamic nucleus) and GPi (globus pallidus pars interna). Other similar sets of coordinates can be derived for other targets such as the PPN (pedunculopontine nucleus) for instance and other structures such as the PHN (posterior hypothalamic nucleus), although these have been established on much smaller numbers and the AC–PC system might not be the best one, for instance for the PPN. The important message here is that aiming at a structure should be as independent as possible from the operator's choice. These statistical values represent a very good theoretical target, validated by a large experience. They are to be used for pretargeting, but the final determination of the individual target must be intraoperatively determined by other means. If one does not want to practice a ventriculography, these coordinates could be applied to the MRI midplane on T1-weighted images, as well as on T2-weighted images, which provide similar features, such as AC and PC for instance. The T2-weighted images are less credible because of the important variation of the image depending on the acquisition parameters.

It is surprising to see how often the authors, even using MRI and not ventriculography, do not take advantage of all the work that has been done to establish the statistical values referred to internal patient's landmarks. They produce results about their electrode location in millimeters, from the posterior commissure for instance, which does not take into account the variability of the AC–PC. Giving the position in millimeters with respect to the mid AC–PC is better than the anterior posterior value of the coordinate when the target is close to mid AC–PC, making the error smaller. However, determining the mid AC–PC requires knowledge of the positions of the AC and PC, therefore there is no excuse left for not expressing the AP coordinate to the ratio to the AC–PC distance. The variability of the height of the thalamus (HT) is also extremely important and expressing the distance from the AC–PC line in terms of millimeters and not in ratio of the height of the thalamus is subject to strong errors because of its variability. When the target is on the AC–PC level, this does not matter, but when it is about 5 mm below the AC–PC, this might make a significant difference.

MRI, particularly with the recent possibilities to go to 3 Tesla, and probably soon to 7 T, provides images which are extremely close to the anatomical definition and, for some targets, it is clearly the gold standard, particularly for the subthalamic nucleus which appears as a hyposignal, more or less surrounded by a little rim of hypersignal and clearly defined from the surrounding structures. Two remarks should be made about MRI use. The first one is related to the deformation of the image which depends on a large number of factors and particularly when the strength of the magnetic field is increased. The deformation of the image is a problem which has not been fully solved and there is no system capable of providing images with the certification that there is no deformation. This deformation varies from one patient to another, in part because of the influence of the changes in magnetic susceptibility of the patient, which introduces a fundamental imprecision on the location of the target particularly when this target is of small size such as the STN. There are several software programs that allow fusion of the MRI and the CT images (which themselves are supposed to present no distortion). However, these software programs producing this fusion are very different from each other and they are usually based on least square methods, which tend to accommodate the two sets of data with a minimal error, which does not mean any error and does not give certitude about the location of the target. The other point is that not all targets are so clearly visible on MRI. For instance, the PPN can be only located by exclusion of the surrounding structures, themselves not being always very well delineated. The STN has given us bad habits because of its visibility, which is reassuring. But for the targets that are not visible and, given the fact that, in the future, DBS-HF will be addressing many other brain structures which might not be clearly visible on MRI, this impedes slightly the value of MRI as a gold standard method.

ANESTHESIA

Under the pretext that surgery might not be comfortable or to make it more comfortable for the patients, some teams purport performing implantation under general anesthesia with the same quality of results. This seems difficult to believe as during this procedure one cannot at all check during the stimulation the side effects induced by the proximity of the electrode to some structures such as the lemniscus, or the pyramidal tract, or the fibers of the third nerve, when the STN is being targeted for instance. It is not possible either to observe the beneficial effects (which are the most important parameter to achieve the precise location of the electrode) or to deduce the proximity of functional structures from the side effects produced by the intraoperative stimulation (such as tingling in parts of the body suggesting the proximity of the lemniscus medialis, as well as flashlights when targeting GPi and trying to check the limit of the optic tract). Doing so means that one relies only on the prelocation with the imaging methods. This provides at best what is called the electrophysiological signature of the target, but does not permit obtaining the most important information which is the clinical benefit of the stimulation.

HOW TO CHOOSE BEST PRACTICE?

This is the ultimate surgical question. How do we do it? How should we do it? What are the criteria that make us do it this way and not this one? There are several, not of equal importance.

The most stringent one should be the benefit of the patient: a combination of beneficial effects on the symptoms, of risks taken during surgery, of discomfort throughout the procedure. The benefit for the patient will be what will persist for years: saving a few hours in the operating room (OR) might mean several years of insufficient improvement. Then speed in the OR must be carefully justified. So far, there is no miracle recipe, and the fastest methods are quite rarely the best and most efficient. Imaging is one clear exemplary issue: CT? CT+MRI? MRI+ventriculography? Ventriculography alone? MRI alone? All have drawbacks used as excuses not to perform them: lack of availability in the center, cost, deformation, risks or lack of practice (for ventriculography most often), etc. – patients do not want them!

If I had to do it in the simplest way, I would set a facility with x-ray ventriculography and intraoperative micro-macro stimulation along five tracks in an awake patient. More complex, I would add microrecording. More would be MRI. But why should I accept to do so?

Should we accept to perform procedures in a substandard manner because of the lack of equipment, the lack of time availability in the OR, the lack of training of the surgeon, the lack of multidisciplinary skills in the team, etc.? Certainly not: functional neurosurgery is not an emergency discipline where, like on the battlefield, the aim is to save life first. Even if the patients are on waiting lists, this does not warrant doing 'fast and cheap', which quite often means 'suboptimal'. Functional neurosurgery is not one of the many skills a neurosurgeon might have to fit within a global practice, it is a dedicated specialty, which should be handled by a dedicated multidisciplinary, well-trained and well-equipped, team, even within a larger wide spectrum neurosurgical department. Even if DBS-HF is not an experimental procedure anymore for several indications, it still needs careful evaluation within the team to guide the evolution of the method and the collection of high quality data to this purpose.

Electrophysiology is another burning issue: microelectrode recordings (MER) or not MER? Stimulation or no stimulation? Single pass or multiple tracks? Sequentially or all at the same time?

All have drawbacks used again as excuses (the same) not to perform them: not available in the center, cost, time consuming, risks or lack of practice (for MER most often) – patients do not want them!

The procedure which is going to be described has been progressively developed in our department since the discovery of the therapeutic effects of deep brain stimulation at high frequency in 1987. We have implemented the procedure to the current level in order to solve the problems that were encountered, the global aim being to improve precision, efficiency, and safety. When needed and when available, new methods and techniques were introduced. In 1987, we had at our disposal a Talairach stereotactic frame, a bi-orthogonal teleradiological setup, and a Radionics system combining stimulation and lesioning modules. We subsequently added an x-ray digitizing table, and an experimental robotized stereotactic arm from a local company AID. The next step was to include angio-localizers and a microrecording setup derived from our laboratory microrecording devices, micromanipulators and tungsten microelectrodes from FHC, adapted to our needs. The next version of the robot, called Neuromate, integrated IVS neuronavigation software, angio-localizers and a homemade designed stereotactic frame made by DIXI and combining the features of the Talairach frame and MRI compatible localizers. The final design of the dedicated stereotactic room included flat digital detectors for x-rays, directly connected, together with the digital MRI images, to the neuronavigation software.

Our routine procedure is split into several subsequent steps:

● Step 1: the patient is installed under general anesthesia on the frame, using bone screws allowing repositioning of the patients during subsequent steps on the following days. Ventriculography with iodine contrast medium provides the ventricular landmarks (see ventriculography) used for pretargeting. The same day or the day after, the patient has a stereotactic MRI (T1-, T2-weighted images, 3D volumic acquisition with gadolinium contrast injection). During the next 2 days, these data are fused to refine the definition of the target, by matching the theoretical coordinates derived from the ventriculographic landmarks with the actual visualization of the target structure on the MRI, when available and visible. These final coordinates are then fed into the neuronavigation software, allowing the choice of the target point and of the entry point, to avoid the cortical vessels as well as the structures within the brain between the entry point and the target (for instance vessels in the ventricular wall)

● Step 2: the patient is reinstalled under local anesthesia using the bone screws and the numbers taken on the frame pins during step one. The robotic arm or the goniometer of the frame is set using the data provided by the neuronavigation program. A burr hole is made at the entry point. The dura is not opened to prevent CSF loss. A five-channel electrode holder (BenGun) is introduced into the burr hole against the dura and five parallel guide tubes are introduced by simple punctures of the dura. The electrodes are manually promoted towards the target, and recording and stimulation are started about

10–15 mm before the center of the target, and stopped 10–15 mm beyond this point depending on the structure. It is extremely important to perform simultaneous recording of neural activity and stimulation induction of the beneficial effects and of the side effects, over the five electrodes, every 5 mm along the track between these two points. In addition, the introduction of the final electrode to replace the microelectrode corresponding to the best track is helped by the presence of the four remaining microelectrodes which keep the brain tissue still and prevent the electrode traveling into the wrong track. On the contrary, when the exploration is done by performing one track after another, five points situated at the same depth would be explored at rather large intervals during which the patient's condition might evolve significantly. This would not allow making a pertinent decision concerning which track should be used to introduce the final chronic stimulating electrode. The length along which the typical neuronal spiking activity is observed is usually used as a criterion of correct trajectory. The observation of beneficial effects as well as the absence of significant side effects is another criterion of adequate placement. This is not sufficient to make a decision as, if only one such track is performed, there is no evidence that it is the best one and that a better one might be among the four other tracks if they were not performed. The classical argument in favor of only one track is the risk of bleeding increasing with the number of tracks. Actually, this is not demonstrated at all. This does not prevail on the necessity that we have to provide our patients with the most efficient placement within the functional target. Similarly, not only using one track, but also performing the exploration using directly the chronic electrode is highly criticable, as the rigidity of the DBS electrode does not allow ensuring a strictly linear progression, which increases the risk of missing the target, and also does not allow recording of the typical neuronal activities which are the signature of the target.

• The benefit of precisely positioning the electrode must not be lost by an insufficient fixation on the skull. Various systems, and more or less expensive, are provided, some of them inducing a displacement of the electrode through the process of fixation, which cannot be controlled if intraoperative x-rays are not performed. From the beginning, we have been using a simple method by a ligature anchored into the bone and embedded in dental cement to obliterate the burr hole. This provides a highly secure fixation ensuring the absence of electrode displacement during the whole follow up of the patient and efficiently preventing the intracranial propagation of extracerebral infection.

• This second step is terminated by the folding of the distal part of the electrodes under the skin in a subgaleal pouch, dissected directly against the skull.

• Step 3 consists of the connection of the distal part of the electrodes to the subcutaneous extension which would be passed under the cervical skin down to the subclavicular area where the IPG (implantable programmable generator or pulse generator) is placed in the subcutaneous pouch dissected against the aponeurosis of the pectoral muscle. This apparently simple part of the procedure is of paramount importance, as this is the source of most of the postoperative complications: fracture of the cables, skin erosion, and infection. The extracranial part of the electrode must be placed under the galea which would protect it and prevent it moving towards the surface. The connector between the electrodes and the extension must be also placed under the galea and at the level of the convexity of the skull, and not at the level of the neck where the head movements will rapidly break the electrode, but not the extension which is much more resistant. The passage of the extension under the skin of the neck must be deep enough to prevent adherence to the superficial hypodermic region, which induces strong adherences, is unaesthetic, reproachful, rarely painful, and often disturbing for the patient. We perform step 3 under general anesthesia, 2 or 3 days after step 2, mainly because of the lengthy duration of step 2.

SUICIDE AND COGNITIVE EFFECTS

Although no systematic study has been performed, there is no statistical difference in the frequency of occurrence of suicide in operated and non-operated parkinsonian patients who already had a high rate as compared to the general population. Similarly, depression has been reported in operative patients while, at the same time, it is said that DBS-HF of the STN improves the depressive tendency of parkinsonian patients. This has to be related to the time after surgery where there is a clear increase in depressive mood up to constituting a depressive state for a few months. This is usually well responsive to mild pharmacological treatments with, for instance, clozapine. When one pays attention to the social and familial context, it is quite striking to find that most of the time this is related to a brutal return to a 'normal' type of life of the patient, who was, up to the operation, severely disabled and depending on the family and caregivers. This has been also observed in patients after major cosmetic surgery (for breast surgery for instance in particular), cardiac transplants, or even prisoners recovering freedom after 10–15 years of jail. In all these situations, different as they may be, one common denominator is the strong 'benefit' inducing a huge challenge in their lifestyle to which they have not been really prepared. They have to face a new situation and, particularly, in family life, most of the time they cannot recover their professional activities, and for those who are coming out of jail, society is not necessarily well receptive and many items of life have changed (habits, cultural events, even changes in social concepts). This does not happen when the effect of surgery has been poor and the clinical status of the patient is not significantly changed keeping them in their usual dependent situation. For the same reasons, changes in personal marital status are quite often observed, particularly patients divorcing their spouse to start a new life, of being abandoned by their spouse who would like also to enjoy a new life, the duty to the patient not so necessary.

MECHANISM OF ACTION

The mechanism, despite a very large number of teams devoted to unraveling the mechanisms, is not yet fully understood. There is a tendency for consensus about functional inhibition which could be using various phenomena such as inhibition of firing, activation of inhibitory structures, jamming the neuronal activity, interfering with abnormal oscillations induced in the systems. This might explain why the effect of DBS-HF is comparable in several targets and is efficient on very different symptoms. However, the action is clearly not to create new functions but rather to disturb the abnormal neural network behaviors, mostly being cycling oscillatory behavior of neural loops and networks, which may induce most of the time a bursting neuronal firing, and an increase in oscillation in the beta band frequency. One must stress the importance of the dual frequency response of these systems, low-frequency tending to excite or increase the activity of neurons while frequencies above 100 Hz have mostly an inhibitory-like effect in groups of neurons. This dual effect is not observed in fiber bundles where both low and high frequencies are excitatory. A recent example of this is found in the pedunculopontine nucleus (PPN) which has to be stimulated at low frequency, around 20 Hz, which means it has to be excited and its activity enhanced, as opposed to most of the other targets which need to be stimulated at high frequency, above 100 Hz. The most frequently used frequency, 130 Hz, corresponds to the minimal value at which these inhibitory-like effects are observed, and are not significantly better at higher frequencies up to several thousand hertz, which shows that it is not necessary to waste electrical energy at frequencies higher than 200 Hz. The dissemination of the method to a large neurosurgical community tends to present the method as a routine recipe, forgetting the crucial, pivotal, importance of the frequency which, most of the time, is not even mentioned in the method chapters of clinical papers. The example of PPN should stress the necessity of precisely stating this parameter and its value as, in the future, other targets might also necessitate being stimulated at low frequency.

It is always surprising that DBS-HF inactivation is so well tolerated, and that the symptoms of the disease are erased while the normal function of the stimulated networks is preserved. The side effects are, in general, not a consequence of the inactivation of the normal functioning of the network but, on the contrary, the creation in neighboring structures particularly in passing fibers of unwanted symptoms or activations which are then considered as side effects. As this observation can be replicated, in a more brutal manner, by lesions instead of high-frequency stimulation in the same target, this has led us to the concept of futile systems. Futile systems mean that these systems, brought into anatomy by evolution, or not erased during evolution, which in general does not know how to erase a previously created structure, are not absolutely indispensable to the correct execution of the normal activity of the individual, as opposed to primary systems, the lesion of which induces deficits (such as primary cortices, major fiber bundles, structures in the spinal cord, etc.).

Coming from the hypothesis that high-frequency stimulation may regulate or downregulate the abnormal hyperactivity of structures such as STN, we have raised the hypothesis that downsizing the hyperproduction of glutamate by STN (which might be participating in a vicious loop acting in the process of neural degeneration of dopaminergic cells in the substantia nigra compacta and reticulata) might be a favorable method to slow down the neurodegenerative process of the disease or even to reverse some of the previous effects. Several experimental works, including ours in rats and in monkeys, have provided data supporting this hypothesis. The clinical observation is more difficult to obtain. In our own data, we have seen over a series of 89 patients consecutively operated in STN, that 25% continued to impair; 36% of them had a UPDRS score in the off–off situation being maintained within the ±15% range of imprecision of this scale, and 38% had improved their off–off UPDRS scale, 50% of them continuing over 5 years. Besides the fact that this was not a prospective study, the data might be biased by the well-known fact of the persistent benefits of treatments which could account for this apparent improvement. Other clinical trials have not shown this improvement, but the clinical stages of those patients were too advanced as this was probably the case in the only study where the effects were observed using positron emission tomography (PET) studies. It might be possible that this neuroprotective effect could be obtained only in very early cases where, however, the destruction of the dopaminergic cell pool is already important. This might be why so far the experimental data have not been transferred to human patients.

SHOULD WE GO BACK TO LESIONING AND UNILATERAL IMPLANTATION?

It has been commonly observed in methods which have been established over a long enough period of time to deviate towards degraded versions which are presented as logical and even modern evolutions. This is a case for the temptation to return to lesioning methods as well as a tendency to prove that unilateral methods could be sufficient.

Both are based on the desire to reduce costs, to reduce duration of the surgeries and maybe even the invasiveness and, but never mentioned, to avoid the learning curve of a method not currently acquired by the practitioners.

Returning to lesioning would mean that one has forgotten or one ignores the reason why the stimulation at high frequency method, once being discovered, was developed, refined, documented to replace the previous technique used in functional neurosurgery which was to destroy, using various methods (electrical burning, heating, cooling, destroying by alcohol, or oil, etc.). The lesioning methods had one main attractive characteristic, which is to be apparently faster to execute (once the target is reached, coagulation is done usually for a few minutes, the electrode is withdrawn, the skin is sutured and the surgical involvement of the practitioner is terminated) and need no subsequent tuning, programming, changing of IPGs, etc. One totally forgets the other side of the coin which is that the size of the lesion is difficult to control (when it is too small one has to re-operate the patients, when it is too large one may have side effects or even complications) and, above all, this is irreversible, even if along time some deficits tend to decrease but also the benefits tend to decrease. One may argue that the difference between the precedent era of lesioning and the present time is the advent of modern imaging. I do not consider this argument as valid as, during the lesioning era, the good teams were using precise, even if not so sophisticated as MRI, radiological determination of the target by particularly ventriculography (see the specific comment on this method in this chapter) and also were highly skilled in electrophysiological exploration and intraoperative clinical evaluation of the effect of the treatment. In addition, if these risks were more or less acceptable in large targets, such as the thalamus or the internal pallidum (although they had close relationships to important bundles such as the lemniscus medialis, the internal capsule, the optic tract), this becomes more acrobatic in the STN nucleus. Moreover, it has been rapidly seen since the beginning of the method that bilateral implantation was much more tolerated than the bilateral lesions. Also, lesioning cannot in any case benefit from the major advantage of stimulation which is its reversibility but also its adaptability by simple manipulation of the amplitude of the current. To finish, we are currently witnessing a fast development of the method to new indications but also to new targets, some of them being situated in very sensitive areas such as in the brainstem as is the PPN. There is currently a tendency to implant several targets at the same time, either to explore their comparative effects or their cumulative effects. This practice is based on the fact that, in the case of implantation in a too dangerous or an inefficient target, the stimulation can be stopped and even the electrode can be removed. This approach is absolutely unthinkable; one cannot make for testing several lesions in different targets which might significantly increase the risks.

The proposal by some teams to operate only one side, at least for the targets involved in movement disorders, is not supported by the large experience one may have with bilaterally implanted patients in STN, GPi, Vim, where it easy to stop one side or the other and observe comparatively the benefits versus the bilateral stimulation. One knows that, for movement disorders in severe patients, stimulation on one side may induce about 10% of contralateral improvement which, most of the time, is insufficient. Implanting only one side of STN if the symptomatology is too asymmetrical does not help either as the other side, non-operated, might need to be treated by dopaminergic medication which, most of the time, clearly induces dyskinesias on the STN-stimulated side. The reported data tending to support unilateral stimulation have been obtained in fact in patients in whom the degree of advancement of the Parkinson's disease is still moderate, when the symptoms may be highly asymmetrical and in whom a mild improvement on one side might be sufficient. But one may wonder whether or not it is legitimate to propose deep brain stimulation in patients in whom the degree of disability is still low and does not warrant the risks and complications, which are independent of the clinical status, for obtaining benefits which will necessarily be small. This last consideration applies also for bilaterally implanted patients, such as in recent comparative studies, where on average the UPDRS was low and the percentage of improvement was also low.

COSTS

The problem of the cost of the method is multidimensional.

It depends first on the price that is set by the industrial companies, mostly based on marketing studies rather than on the cost of development and production. In all social security systems, either based on a governmental support or being taken care of by insurances, the price of the device is a strong limiting factor which introduces money as a strong selection criterion. I fear that the development of the indications and the multiplication of the companies would not lead rapidly to a reasonable decrease of prices, the market being comparatively smaller than the market of cardiac pacemakers. Nevertheless, when patients are sufficiently advanced in their diseases, the cost of the medication is extremely high, particularly when dopamine agonists are used. DBS at high frequency in STN (this is not true in the internal pallidum), when it significantly improves the patients (which means that they had been correctly selected and correctly operated), the improvement of the symptoms allows medication to be reduced by about two-thirds of the doses, which similarly decreases the levodopa-induced dyskinesias, side effects of the medical treatment. This induces an equivalent decrease on drug expenses. It has been proven already 10 years ago by a comparative study in France that this reduction of expensive drugs, such as dopamine agonists, by itself is sufficient to reimburse the cost of the stimulator (which is usually functioning for 5–7 years) in two years. This is of course true if surgery is done in advanced patients and if the chosen target is STN.

EVIDENCE-BASED MEDICINE (EBM) AND CLINICAL TRIALS

EBM is not science, it is a methodology, it is a timely evaluation of how efficient a method is in a given circumstance (country, practice, social security system, public or private), providing a judgment about the benefit to the public of the practice of this method by a non-selected medical community.

The rules of evidence-based medicine were established some years ago to try to regulate the publication of results, namely of therapeutic outcome following pharmacological treatments. The goal was, and has been achieved, to produce rules which would be the framework of publications to report results in a more objective manner than the previous statements as 'mild, moderate, good, excellent'. But the goal was also to introduce the systematic comparison with non-treated groups in order to provide a standard basis. Evidence-based medicine requires also a double-blind approach, aimed at avoiding subjective evaluations from the investigators as well as the placebo effect based on the expectations and interpretations from the patients.

All this makes EBM an incomparable tool for evaluation of practices, comparison of methods, demonstration of the significance of observed benefits or of drawbacks or side effects.

But the dogmatic application of the EBM methods might sometimes be deleterious.

Innovation might be ignored or rejected if not presented according to EBM standards. EBM discourages reporting a serendipitous discovery in a fortuitous situation, or data based on preliminary results, short series. EBM provides answers only to the questions asked in the clinical trial, which cannot lead to innovative discoveries. EBM, as is, is not suited for surgery. An EB surgery (or EB neurosurgery) should be designed to take into account the specificities of surgical clinical trials. Over several years, it is clear that this evidence-based medicine method has difficulties when applied to surgical methods. This is particularly due to the difficulty to have a comparison group which might be usable in a double-blind study. Surgery is generally not reversible and does not permit cross-overs. Operated and control patients differ by the scar, the pain, some evident consequences of the surgical act (such as quandrantanopia in temporal lobectomy, etc). This has led to sham surgery which sometimes cannot be really a control group and which may also raise ethical problems.

In this respect, deep brain stimulation has opened new ways for evidence-based medicine: the possibility to reverse the effects as well as the side effects by turning on and off the stimulators provides this unique and unprecedented situation where the treated group and the control group may be the same and serving as control to each other by this activation/inactivation of the stimulator. This also allows cross-overs as this can be done with pharmacological treatments.

To this point, the principles of evidence-based medicine can be applied to functional neurosurgery.

The problem is not yet solved correctly, when it comes to large multi-center trials, which imply the participation of several surgical teams. The application of the protocol is strongly biased by the variability of the skills of the participating surgeons. Having a surgical procedure performed by ten neurosurgeons cannot be compared to the distribution of drugs by ten physicians in a pharmacological trial. In medical pharmacological treatments, the administration of drugs according to a well-established protocol (dose, number of doses, associations with other treatments, etc.) is easily applicable with little influence of the physicians or teams who are in charge of the clinical trial in different institutions. The delivery of treatment by surgeons of different teams is, in theory, impossible to conceive as strictly identical, given the fact that the surgical procedure still has, even in stereotactic conditions, a high variability which is operator dependent. Not to speak about the difficulties which are encountered when setting a clinical trial where all teams and surgeons would use strictly the same method (microelectrode recording or not, simple or multiple tracks, ventriculography or not, extent of the preoperative imaging methods, general or local anesthesia) which are known to be difficult, if not impossible, to obtain. This often ends up in an agreement that 'each team will perform the procedure with the tools they are the most familiar with'. The main consequence of these specific aspects of surgical clinical trials is that the standard deviation of the quality of application of the method is highly significantly increased. As a comparison, this would be like if the physician might have their own degree of variability of the doses and the regimen of administration of drugs in a pharmacological trial.

This creates necessarily an average outcome which encompasses the best and the worst teams and provides results which cannot be considered as the typical feature of the method. This does not mean that clinical trials should not be used; one should simply clearly state what are the purposes and the objectives. Large multicenter clinical trials clearly provide a global picture, a landscape of the therapeutic community, and an evaluation of how a given method can be efficiently used on a large scale as a common treatment method. This may eventually lead to the conclusion that the method does not provide better results than others (such as pharmacological treatments for instance) in a given global circumstance depending om the country social security system, the repartition of the practitioners, their training, etc. and may ultimately lead to the proposal to stop performing this method in these given circumstances.

This is totally different when one wants to compare the ultimate efficacy of a method (or a target) to another, to judge the specific merits of these methods, regardless of the circumstances of their application. This requires that all conditions are the same: strict identical surgical procedures (one knows how difficult this is to obtain, under the pretext that 'every team should use the method they feel most comfortable with'), same surgical team, same protocol, and same follow up, same postoperative management, (particularly for post-op programming). Such a comparison requires that the outcomes are the best, which means performed by the best teams with the best technological support and follow ups. Comparing suboptimal results does not make sense, even when they are published in the best impact factor ranking reviews. Two large clinical trials corresponding to what is described above have been published recently. One is concerning resective surgery for epilepsy and the other one, most recent, to compare deep brain stimulation for Parkinson's disease in the subthalamic nucleus and in the pallidum GPi. In both cases, the global outcomes of the methods were much lower than what has been published, particularly by several teams considered as experts in the field. Outcomes of 35% of improvement of UPDRS in DBS therapy of Parkinson's disease are not satisfactory results. They cannot be considered as the average standard value that a patient must expect. They cast doubt about the expertise of the teams participating in the study, or more probably express the variability of expertise between these teams. Therefore, as a consequence for this particular study, comparing two largely suboptimal results does not allow the provision of any valid conclusion, for instance that the two methods are equal and should be considered similarly as similar options.

ALTERNATIVE METHODS AND CONCLUSION

The fate of a method is to be replaced by better ones, and this will be for the benefit of the patients. At the present time, although limitations exist (cost, surgical skills, embargo, etc) DBS-HF is becoming the first choice method to treat movement disorders, some mental disorders, and an increasing number of indications; alternatives are being developed which, in the future, might become the preferred methods. Local delivery of drugs, growth factors, agonists or inhibitors, although attempts with GDNF have failed, will be improved and may find a specific field of application. Gene therapy is the ultimate solution. Several clinical trials are in progress and, even though there will be difficulties, the future might be there. Cell transplants are benefiting from the research of hundreds of laboratories and success is still awaited. Stem cells are currently the new hope in this field. One should not forget the possibility to design a dopaminergic agonist without dyskinesias. In all cases, the challenger of DBS-HF will have to be successful, reversible, adaptable, and even safer.

Commentary on Neuromodulation Perspectives

Erwin B. Montgomery Jr., MD

University of Alabama at Birmingham, Department of Neurology, Birmingham, AL, USA

The recent history of deep brain stimulation (DBS) since its popularization following the report by Benabid and colleagues in 1987 [1] has been both gratifying and simultaneously, disconcerting. The remarkable expansion in the number of patients treated with DBS for an increasing range of neurological and psychiatric disorders has been a wonder. Since the resurgence of interest in neurosurgical methods to treat movement disorders, many tens of thousands of patients have had the quality of their lives greatly improved. But it has not been the revolution that it should have been, at least not yet. Perhaps the bittersweet success is because the truly revolutionary nature of DBS has not been appreciated. Early adopters of the technology may have viewed DBS through outdated perspectives or at least perspectives whose relevance to DBS could be questioned.

The longer history of DBS is complicated. There are significant undercurrents that have affected the course of DBS development, undercurrents not appreciated by most practitioners and scientists of DBS. Yet, if DBS is to reach its full and, as yet, untapped potential, the deeper dynamics need to be understood.

HISTORY AS THE FUTURE

Historical perspectives are important. George Santayana, wrote: 'Those who cannot remember the past are condemned to repeat it' [2]. The recent history of DBS, in terms of innovation, is mixed. While patients have been rescued from failed pharmacological or no therapies, have the technologies employed and the scientific understanding radically changed that much? Where will the next breakthroughs come if not from radical change as current perspectives and approaches are approaching exhaustion? The value of clear historical analyses is that by knowing where one came from, one might get a better idea where one is headed and perhaps a change in direction would be in order.

As reviewed by Hariz et al [3], the use of electrical stimulation within the brain goes back to 1947. Indeed, the first description of the term 'Medtronic' (the manufacturer of the implanted systems) conjoined with 'DBS' goes back to Dieckmann [4] in 1979 for psychiatric disorders and to Cooper et al [5] in 1980 for movement disorders, over 7 years before Benabid et al. The critical question becomes why were Benabid and colleagues successful in popularizing DBS in 1987 when Dieckmann and Cooper and his colleagues were not.

Some of the standard answers include the limitations of pharmacological therapies for Parkinson's disease did not become recognized until the late 1980s and that is what drove DBS and other surgeries for Parkinson's disease. While it is true that the introduction of levodopa was a blessing, the complications and difficulties maintaining adequate clinical control were well known in the 1970s leading neurologistst then to avoid the use of levodopa in favor of bromocriptine that was introduced in the mid-1970s. Pallidotomies for movement disorders since the 1950s and pallidotomies and thalamotomies continued thorough the 1980s. In the case of Parkinson's disease, the continued use of surgical therapies implicitly acknowledged the limitations of pharmacological therapies and the need for surgical therapies.

The question is why did it take decades since the introduction of levodopa into clinical practice in the late 1960s for the disillusionment to set in, at least to the point where more neurologists would consider referral to surgery. Perhaps it was not the degree of disillusionment but the lack of consideration of surgical alternatives. Perhaps it was not a lack of consideration of surgical alternatives but rather a concern over the potential for irreversible complications. However, this would not explain the resurgence of interest in ablative surgical techniques in the 1980s as the surgical risks were little different from those previously.

Could there be an inherent bias against surgical therapies and overestimations of the efficacy of pharmacological therapies? While pharmacological treatments often are safer than surgical treatments, the inference from the risk ratio is often biased towards pharmacological treatments even when they fail to provide adequate relief. Evidence of this is seen in the not uncommon statement by neurologists to patients that the patient's disease is just not bad enough for surgery. What does this mean and how can a neurologist know other than by projecting his or her beliefs onto the patient?

Could this bias be reflected as a bias towards biological (such as stem cells or fet al dopamine cell transplants) and gene therapies, particularly when considered as alternatives to DBS? Certainly, most surgical techniques for stem cell, other cells, and gene therapies involve more risk than DBS because of the greater number of penetrations of the brain usually required and the possible long-term biological complications which are rare in DBS. It is the opinion of this author that the bias towards these pharmaceutical and biological therapies aimed at dopaminergic mechanisms is the result of mistakenly conflating pathoetiology with the pathophysiology. By pathoetiology, it is meant those mechanisms or agents that are the proximate cause of the initial departure from normalcy. With respect to the motoric symptoms of Parkinson's disease, these reflect degeneration of the dopaminergic neuron of the substantia nigra pars compacta; whether due to genetic causes, exogenous or endogenous toxins, or some combination of genetics and toxins.

By pathophysiology, it is meant those changes in the neuronal physiology that result in dysfunction manifesting itself in disability. Thus, pathophysiology need not be the most proximate change such as degeneration of dopaminergic neurons in the substantia nigra pars compacta. Rather, it can be downstream effects that more directly relate to the abnormalities of motor unit recruitment, which is comprised of the single lower motor neuron in the spinal cord and the individual muscle fibers it innervates. Ultimately, any pathophysiological theory or intervention must involve the motor unit, as it is the fundamental unit of behavior. Note, the downstream effects relate to a temporal or dynamical sense and not necessarily an anatomical sense as is inherent in the Globus Pallidus Interna Rate theory of pathophysiology that focuses on activity in the globus pallidus interna.

Parkinsonian motor features need not be associated with degeneration of the dopaminergic neurons in the substantia nigra pars compacta at all. Damage to the globus pallidus external, the supplementary motor area, and the putamen (see [6]) can produce parkinsonism in which their motor symptoms and signs are indistinguishable from those of idiopathic Parkinson's disease associated with degeneration of the dopaminergic neurons of the substantia nigra pars compacta. Anticholinergic medications dating back to Charcot at the Salpêtrière in the late 1800s demonstrate that dopaminergic mechanisms are not a necessary condition for improvement of the pathophysiology. The remarkable efficacy of DBS for Parkinson's disease, greater than the best pharmacological (predominantly and effectively dopamine replacement) treatments [7], and the demonstration that the DBS therapeutic effect is not mediated by dopamine [8] clearly demonstrates that there are other non-dopaminergic mechanisms that can be targeted even if these are only secondarily consequent to degeneration of the dopaminergic neurons of the substantia nigra pars compacta.

While it may be intuitive, risking simplistic, to think that reversing the demonstrated lack of dopamine should make things better, the question is what is being replaced. Intact neurons release dopamine in a precise manner in space and over time periods of approximately 100 ms and in response to very precise electrical signals [9,10]. It is not likely that therapies that operate over wide regions of the brain or over long time domains, such as pharmacological and biological treatments, are likely to restore completely the normal dynamics despite the remarkable promise of this current era of molecular neurobiology.

Pathoetiology and pathophysiology are not synonymous and recognition of their differences will entail a radical revision of our understanding of pathophysiology from past inferences based predominantly, if not exclusively, on the anatomy and chemistry compared to a basis on the actual physiology [6,11] There is the danger that DBS could be viewed primarily as an alternative pharmacological technique and such a conceptualization would hinder future developments. The brain is basically an electrical device that processes, stores, and transmits information electronically. Neurotransmitters are the messenger not the message. Certainly, neurotransmitters are important just as electrons are important in an electronic computer. But one cannot say that the properties of electrons, per se, are sufficient to explain the operations of a computer. Viewing neurological and psychiatric disorders as electronic misinformation in the brain amenable to electrophysiological techniques greatly expands the opportunities.

Others have argued that advances in neurosurgical techniques were responsible for the resurgence of interest in surgical therapies for Parkinson's disease and other movement disorders. However, this is not true. Indeed, the early methods of surgical navigation used by Dr Benabid and colleagues, specifically ventriculography and microelectrode recordings, were available well before the 1980s and, in fact, were used by Cooper et al in their description of DBS in 1980 [5]. Thus, the question is what are the other, and perhaps more important, reasons for the resurgence of surgical procedures for movement disorders.

An editorial by Goetz et al in 1993 [12] retrospectively assessing the field is telling. Central to the editorial by Goetz et al in 1993 [12] were the discussions of a theory of neuronal pathophysiology that provided a ready rationale for the surgical therapies. This theory, known as the Globus Pallidus Interna Rate theory, posits neuronal overactivity of the globus pallidus interna as causal to parkinsonism, from which the therapeutic mechanisms of pallidotomy seem intuitively or logically to fall. Unfortunately, the Globus Pallidus Interna Rate theory is wrong [11] with continuing adverse scientific consequences, but nonetheless it provided 'cover' for the more aggressive pursuit of surgical interventions such as pallidotomy and DBS was the beneficiary. It is not at all clear whether Benabid and his colleagues would have been as successful had the previous theory of Parkinson pathophysiology, that being the Cholinergic/Dopaminergic Imbalance theory, prevailed.

History is not without irony. Pallidotomy and DBS have been right for the wrong reasons. Perhaps given the nature of the process of science, the benefits of first pallidotomy and now DBS may forgive the wrong reasons but only if this apology is clear and widely known. Otherwise, the wrong reasons will be seen as right and the search for the real right reasons will be stymied. But the implication is clear and important. Theories of neuronal pathophysiology are critical to the development of new DBS-like therapies and, indeed, it is unlikely that there will be significant progress in the latter if there is not progress in the former.

THE DBS FREQUENCY STORY

The notion of differential effects of different DBS frequencies is interesting, with some frequencies improving while other frequencies worsening various symptoms. Most past discussions clearly delineated high versus low DBS frequencies. Despite the historical false dichotomization into high and low frequencies, the nature of the DBS frequency dependence of symptom relief may yet emerge as a key to future innovations. To be sure, it is now clear that such dichotomization of DBS frequencies is counterproductive [13]. However, the dichotomization early of DBS effects into those associated with high frequency and those associated with low frequency resonated with the dichotomization of clinical phenomenology, such as hypokinetic and hyperkinetic disorders. The dichotomization of DBS frequency effects resonates with the dichotomization of behavioral control by the globus pallidus interna as opening or closing gates to movements through the ventrolateral thalamus by underactivity or overactivity of the neurons of the globus pallidus interna and its associated subthalamic nuclues, respectively. These notions are inherent in the Globus Pallidus Interna Rate theory and its derivative Action Selection/Focused Attention theory [11].

Characterization of Parkinson's disease as a hypokinetic disorder, while having heuristic value for educating the unsophisticated, much in the manner of a white lie, is inconsistent with the tremor, rigidity and dystonia affecting many untreated patients with Parkinson's disease. Further, the subthalamic nucleus neuronal activity is no greater in patients with Parkinson's disease than in patients with epilepsy [14] or in the globus pallidus interna of non-human primates carefully made parkinsonian with n-methyl-4-phenyl-1, 2, 3, 6-tetrahydropyridine (MPTP) compared to normal non-human primates [15,16]. Further, patients with Huntington's disease are bradykinetic in purposeful movements similarly to patients with Parkinson's disease [17] and neuronal activity in their globus pallidus interna is no different from that in patients with Parkinson's disease [18].

The dichotomizations described above are part and parcel with one-dimensional push–pull dynamics based on reciprocal interactions inherent in the concepts of basal ganglia pathophysiology and physiology arising in the 1980s contemporaneously with the popularization of DBS. Globus pallidus interna overactivity closes the gate to movement producing hypokinesia, while the opposite, underactivity, opens the gate producing involuntary movements. These reciprocal dichotomizations are instantiated in and revolve around the presumptive inhibitory effects of globus pallidus interna neurons on the activity of neurons in the ventrolateral thalamus and subsequent decreased drive of the motor cortex. However, neuronal recordings before and after MPTP-induced parkinsonism demonstrate no changes in baseline ventrolateral thalamic neuronal activity [15], supplementary motor area [19] or motor cortical neuronal activity [20,21]. Further, recordings of ventrolateral thalamic neuronal activity in humans associated with globus pallidus interna DBS demonstrate that most ventrolateral thalamic neurons show post-inhibitory rebound increase activity, that for many neurons results in a net increase in activity [22]. Thus, globus pallidus interna activity, rather than being inhibitory, may be accurately characterized as delayed excitation. It is important to note that a great many neurons within the basal ganglia display significant post-inhibitory rebound increased excitability. Indeed, some neural networks in invertebrates are made up entirely of 'inhibitory' neurons yet the dynamics do not collapse into nothing because of post-inhibitory rebound excitation [23]. There are significant implications here for theories of basal ganglia physiology and pathophysiology where the putative inhibitory interactions are so narrowly conceived, such as the Globus Pallidus Interna Rate and the Action Selection/Focused Attention theories.

Again, either history is not without irony or humans have remarkable powers of selective consciousness. The convergence of the dichotomizations of clinical phenomenology, globus pallidus interna neuronal pathophysiology, and notion of positive and negative symptoms inculcated by Hughlings Jackson, and the fulcrum of those dichotomizations in the inhibitory influence of the globus pallidus interna onto the ventrolateral thalamus provided a sense of consilience. This sense of consilience presupposes that such a conjunction by chance was just impossible and, consequently, the conjunction had to be right. There is incredible power in a story that just seems to be right and hangs together despite any amount of contravening data [24].

The seductive power of such a consilience leads to category errors of logic [25]. The inference was made that, because pallidotomy and pallidal high-frequency DBS produced symptomatic improvement as did thalamotomy and thalamic high-frequency DBS, then the mechanism of actions were the same. This is a category error as will be demonstrated. A stroke and curare both produce paralysis but it would be a mistake to claim they produce paralysis in the same manner. But it was vociferously argued that high-frequency DBS inhibits the stimulated structure despite the fact that most experience in electrophysiology would not support such a claim [26]. Indeed, various investigators demonstrating contrary data encountered considerable resistance (personal observation). To be sure, there are those who maintain that DBS produces local inhibition by activating the presynaptic terminals, which are predominantly inhibitory, and releasing large amounts of inhibitory neurotransmitters [27]. However, this is not the sense in which high frequency DBS initially was thought to be inhibitory.

DBS is clearly the stepchild of ablative surgical techniques, which was expedient in the short term but not without potentially compromising future innovations. Just as the Globus Pallidus Interna Rate theory survived what should have been devastating contrary evidence that pallidotomy improved hyperkinesias (the antithesis of the predictions of the Globus Pallidus Rate theory), so too did the notion that high-frequency globus pallidus interna DBS inhibits the globus pallidus interna which should have worsened rather than improve hyperkinetic disorders. That the Globus Pallidus Interna Rate theory co-survived with the notion that high-frequency globus pallidus interna DBS as inhibitory is interesting, if not troublesome.

There is a wealth of observations now that DBS effects cannot be so neatly dichotomized. While high-frequency DBS in patients with Parkinson's disease may improve upper extremity function, it may worsen gait [28] and speech [29]. Low-frequency stimulation may worsen upper extremity function yet improve gait [28] and speech [30]. Intermediate-frequency DBS (90–100 pps) for essential tremor is just as effective as high-frequency DBS (160–170 pps) [31].

The critical question is whether the conceptual presuppositions that caused scientists to think in terms of false dichotomizations, such as high- or low-frequency DBS, and one-dimensional push–pull dynamics, such as inhibition or excitation, will give way to better notions that will allow conceptual breakthroughs that will then lead to technological and therapeutic breakthroughs. Some theories are in the offing [6,11]. It may evolve that specific dysfunctions will be targeted by specific frequencies (or bandwidths of noisy stimulation for stochastic resonance) and that multiplexed frequencies may improve a wider range of disabilities than the current trade-offs, such as control of upper extremity function at the expense of gait or speech that, currently, many patients with Parkinson's disease and essential tremor must accept.

THE SURGICAL PROCEDURE, INCLUDING ANESTHESIA, FROM THE NEUROPHYSIOLOGIST/NEUROLOGIST PERSPECTIVE

Reason, even in the presence of direct data in the form of randomized controlled trials [32], must prevail and even more so when direct data are lacking. There is a set of assumptions that may be reasonable guides. The quickest, easiest, safest and most sure way of getting to the target is better; the least time and expense of the surgery, the better. But, after that, who really knows what is the best approach; be it ventriculography, direct targeting based on MRI, targeting based on the anterior and posterior commissures (AC–PC), or with or without microelectrode recordings. However, this is not an academic question for neurosurgeons and neurologists who must take some responsibility for the surgical methods to be used when referring to a particular neurosurgeon. In the case of DBS for neurological (psychiatric) disorders, the neurologist (psychiatrist) is often the one that has to deal with the sequela.

Using the actual length of the AC–PC line and the height of the thalamus may reduce the inter-subject variance, but it cannot be excluded that using these landmarks may actually introduce more variance. There may be some other set of internal landmarks visible by a range of neuroimaging techniques that may have a more consistent spatial orientation to the best targets (whatever those may be) and thus, using the AC–PC coordinates may actually decrease accuracy. If using the microelectrode recordings to define best target, the variance related to the AC–PC is large relative to the effective radius of the DBS stimulation [33].

The neurologist, when listening to or reading the discussions among neurosurgeons, is struck by how the debate seems to center on accuracy. To this neurologist, the issue of targeting is analogous to a diagnostic test; that is can we diagnosis the optimal target? In this case, accuracy is conceptualized as specificity and sensitivity. In some ways, specificity and sensitivity are analogous to accuracy, in the sense of how close it comes to the target, and to precision, how often the same point is hit irrespective of the accuracy. However, in diagnostic tests, specificity and sensitivity alone are insufficient to assure the diagnostic value. Rather, percent positive and negative predictive values that relate specificity and sensitivity to prior probabilities are critical. For example, a test that is 95% sensitive and 95% specific still will have as many false positives as true positives if the prevalence (prior probability) of the condition being diagnosed is only 5% of those who would be screened. In the case of targeting DBS, the prior probability would be the percent of patients whose brains conform to those used to construct the various available surgical atlases which currently is unknown.

The more troublesome question is how is this going to be resolved beyond experts stating their preference. One could say do a randomized control trial but this has been difficult for many physicians and surgeons to buy into. As Joseph Fins points out, the surgeon often does not have the advantage of equipoise, which is a truly neutral stance relative to the surgical options [34]. The neurologist can say with all equanimity that he or she does not know if the study drug will work or not and, consequently, referral to active study drug or placebo is less problematic. Generally, surgeons have very strong opinions as to the superiority of the methods they specifically have chosen and therefore, often it is personally ethically uncomfortable to refer patients to alternative methods.

Further, challenges to any type of evidence-based medicine (multiple randomized control trials) resolution as to surgical methods are complicated by the outcomes measures that reflect the best practice. In the absence of any suitable surrogate marker for actual clinical consequences, clinically-based outcomes measures would be required. This immediately raises statistical concerns; especially if the one surgeon's claim is that his or her technique is just as good as another´s. Thus, the statistical effect size would be zero and regardless of the variance of the measure, the sample size would be impossible. Even if a difference in outcomes between surgical techniques would be 10%, based on previous studies of Parkinson's disease using the unified Parkinson disease rating scales (UPDRS), the sample size necessary to have an 80% probability of avoiding a type II statistical error (not finding a difference when one truly exists) at the $P < 0.05$ level, would require a sample size of over 300 subjects. Consequently, surgeons commenting on their personal experience and claiming comparable outcomes with different techniques are not credible unless their validated experience is over 300 cases, assuming the surgeon use the UPDRS as the outcomes measure prospectively.

THE ART, AND HOPEFULLY, THE SCIENCE OF CHOOSING THE BEST PRACTICE

Recent technology has outstripped the philosophical, political, social, ethical, and moral conceptual scaffolding necessary for the rational care of patients. In the past, our failures to help patients were because we did not know how.

Now, and regrettably in the future, failures may be not that we cannot but we are prevented from helping. This is the tragedy of our technological success. Unfortunately, increasing sophistication of care also carries with it limiting access for those who need the care. The schizophrenic attitudes toward medical care, where one day *laissez faire* capitalism is championed to cure the ills and yet, innovation and orphan indications are abandoned because they may not immediately increase shareholder value. On the other day, there is ham-handed governmental interference. The Balkanization of health care often means that short-term costs borne by one insurer are not offset by the greater long-term benefits for other insurers or society. The result is care dictated by short-term and often counterproductive concerns.

The resistance to governmental regulation in some countries means that others, presumably physicians, surgeons and the clinic/hospital systems for whom they work, would exercise the necessary discretion. At the meeting of the external panel advising the United States Food and Drug Administration (FDA) regarding approval of subthalamic nucleus and globus pallidus interna DBS for Parkinson's disease, the panel questioned what methods the manufacturer would take to train and, by implication, regulate DBS surgery. Several experts testifying urged the FDA to allow clinics and hospitals to regulate DBS surgery through their credentialing process. One wonders whether in retrospect, this was a mistake. This neurologist cannot understand how a surgeon performing fewer than 10 DBS surgeries per year can maintain the necessary expertise to provide his or her patients with the best chance for the best outcome; yet there is little to stop the surgeon, including the surgeon's conscience.

To say that, currently, there is an art to choosing the best practice is necessarily to presuppose some rationality. Art, as the ability to discern between things based on some sense of value, does not describe well the current situation with the delivery of DBS care. That there is no real science to choosing the best practice is already evident in the discussions above and in Dr Benabid's chapter. Indeed, in the field of DBS, habit has been confused as knowledge. This problem is compounded by bureaucracies that instantiate habits as standards of care by using epidemiological approaches to analyzing health-care delivery, which are inherently descriptive but are taken as normative.

There is more that can be done than just to curse the darkness. Careful quantitative and operationalized approaches to characterize the delivery of health care and the benefits derived are being developed and applied. Indeed, much needed health-care reform in the USA has taken tentative steps by calling for comparative effectiveness studies. These steps will be necessary as no society can afford unrestrained and irrational allocation of health-care resources even if resource diversions, such as war and corporate subsidization, were to fade away. Some may see this incursion into the prerogatives of physicians and surgeons, whether by governments or private health-care insurers, as a pit; but it is one that physicians and surgeons either can leap into or be pushed.

When Solomon-like wisdom and courage are lacking, the convenient escape is to play a zero-sum game and default to pitting the participants against each other. Relative value units (RVUs) or its equivalents used by some insurers to reimburse, and thus, control health-care delivery, determine merely the percent or slice of the pie and not the size of the pie and therefore, the actual size of the reimbursement piece. Consequently, the health-care providers are pitted against each other. This is not an antidote for the irrationality of health-care delivery but it allows insurers to claim victory over raising health-care costs. This can occur in the context of capitated care to health-care provider organizations or in assigning RVUs based on recommendations of professional organizations in which politics often prevails. In some situations, it is the spine surgeon who is the greatest threat to functional and stereotatic neurosurgeons and thus the delivery of DBS.

A terrible result is or will be that physicians and surgeons will be stuck with having to determine cost-effectiveness and then attempt some allocation of resources. There is no problem with health-care professionals determining effectiveness; that is fundamental to the professions. However, health-care professionals have no business determining cost, as it is highly unlikely that the health-care professionals uniquely pay the costs, other than in some the zero-sum game of capitated care and RVUs. Placing the responsibility of costs uniquely and directly on health-care professionals immediately places the health-care professional at a dangerous conflict of interest pitting financial responsibilities to the health-care providing institution against the best interests of the patient. It would seem that physicians and surgeons would be the first to object.

REFLECTING PROSPECTIVELY

There is good reason to celebrate the successes of DBS. Many thousands of patients have received a new life as a consequence. But it is a bittersweet victory as there should have been much more. If one believes that the brain has more in common with a computer motherboard than a stew of chemicals; that is the brain processes and conveys information electronically and that neurotransmitters and neuromodulators are the messengers and not the message, the potential for electrophysiologically-based treatments would be nearly limitless. Yet, any historical analysis that is not snowed by focusing on personalities as other historical accounts have done, perhaps for popular entertainment value, one cannot help but be disappointed.

The primary conceptual driving force has been equating high-frequency DBS as equivalent to surgical ablation, although scientific research demonstrates that this is not the case [35]. This narrow view of DBS has led to 'chasing lesions' as the modus operandi of DBS translational and clinical research. Surgeons predominantly place the DBS lead where previously they would destroy. Soon, if not already, DBS surgeons will have exhausted the number of ablative targets.

Even the development of new targets not previously subjected to ablation, nevertheless, follows from the same notion as 'chasing lesions'. Consider targeting area 25 of the subgenu cingulum for depression [36]. Area 25 was targeted because positron emission tomography (PET) scans in medically refractive depression demonstrated increased metabolism in area 25. Note that other areas demonstrated decreased metabolism. Base on the now arguably outdated notion that high-frequency DBS inhibits, area 25 was chosen. It is unknown and indeed would have been interesting to know if DBS of the areas with decreased metabolism also would have helped depression. These comments in no way are meant to detract from the remarkable and important work of Mayberg and colleagues. The courage it took to offer this help in spite of the prevalent hostility borne of previous attempts at 'psychosurgery' is greatly appreciated by all and especially those suffering from severe medically refractory depression.

The notable exception to 'chasing lesions' or inventing lesions for DBS, is the work of Nicholas Schiff and his colleagues leading up to the successful use of DBS in patients with minimally conscious states [37]. After careful studies of patients with minimally conscious states identifying the possible role of the interlaminar nuclei of the thalamus and relevant experiments in non-human primates, they proceeded to apply DBS to the interlaminar nuclei of the thalamus in patients with minimally conscious states. The presupposition was that DBS would enhance the physiology rather than acting as a lesion, the latter probably would have worsened the condition.

In many ways at the conceptual level, the prevalent notions of the underlying physiology and pathophysiology of movement disorders, particularly Parkinson's disease, and the past and, in some cases, current notions the mechanisms of action of DBS are less sophisticated than Morse code. At least in Morse code the sequences of electronic 'dots' and 'dashes' carry far more complexity, and hence potential for information, than does the putative roles of the basal ganglia according to the Globus Pallidus Interna Rate and the Action Selection/Focused Attention theories and more dynamics than low-frequency DBS as exciting (analogous to just holding the telegraph key used in Morse code in the down or closed position) and high-frequency DBS as inhibiting (holding the telegraph key open).

The lessons of history are clear. The scientific context or weltanschauung (world view) is critical to progress in any discipline. At the beginning of this chapter, it was demonstrated, hopefully, that it was the change in the conceptual understanding of basal ganglia pathophysiology in the early 1980s that allowed DBS to become popular, whereas just a few years before, the same DBS languished. There are considerable challenges to developing an appropriately sophisticated conceptual understanding of basal ganglia physiology and pathophysiology. To this day, Parkinson's disease is thought to be a neurotransmitter deficiency despite the fact that all manners of neurotransmitter replacements, be it pill or cell transplant, have not demonstrated benefit that approaches that which can be achieved by DBS. The understanding of basal ganglia physiology and pathophysiolgy needs to catch up to the promise of DBS.

REFERENCES

1. Benabid A-L, Pollak P, Louveau A, et al. Combined (thalamotomy and stimulation) stereotactic surgery of the VIM thalamic nucleus for bilateral Parkinson disease. *Appl Neurophysiol.* 1987;50:344–346.

2. Santayana, G. (1905). The life of reason, Vol. 1. Reason in common sense. http://www.gutenberg.org/files/15000/15000-h/vol1.html.

3. Hariz MI, Blomstedt P, Zrinzo L. Deep brain stimulation between 1947 and 1987: the untold story. *Neurosurg Focus.* 2010;29:E1.

4. Dieckmann G. Chronic mediothalamic stimulation for control of phobias. In: Hitchcock ER, Ballantine HT, Myerson BA, eds. *Modern concepts in psychiatric surgery.* Amsterdam: Elsevier; 1979:85–93.

5. Cooper IS, Upton AR, Amin I. Reversibility of chronic neurologic deficits. Some effects of electrical stimulation of the thalamus and internal capsule in man. *Appl Neurophysiol.* 1980;43:244–258.

6. Montgomery EB Jr. Dynamically coupled, high-frequency reentrant, non-linear oscillators embedded in scale-free basal ganglia-thalamic-cortical networks mediating function and deep brain stimulation effects. *Nonlinear Stud.* 2004;11:385–421.

7. Weaver FM, Follett K, Stern MB, et al. Bilateral deep brain stimulation vs best medical therapy for patients with advanced Parkinson disease: a randomized controlled trial. *J Am Med Assoc.* 2009;301:63–73.

8. Hilker R, Voges J, Ghaemi M, et al. Deep brain stimulation of the subthalamic nucleus does not increase the striatal dopamine concentration in parkinsonian humans. *Mov Disord.* 2003;18:41–48.

9. Montgomery EB Jr. Pharmacokinetics and pharmacodynamics of levodopa. *Neurology Suppl.* 1992;1:17–21.

10. Arbuthnott GW, Wickens J. Space, time and dopamine. *Trend Neurosci.* 2007;30:62–69.

11. Montgomery EB Jr. Basal ganglia physiology and pathophysiology: a reappraisal. *Parkinsonism Relat Disord.* 2007;13:455–465.

12. Goetz CG, DeLong MR, Penn RD, et al. Neurosurgical horizons in Parkinson´s disease. *Neurology.* 1993;43:1–7.

13. Montgomery EB Jr. *Deep brain stimulation programming: principles and practice.* Oxford: Oxford University Press; 2010.

14. Montgomery EB Jr. Subthalamic nucleus neuronal activity in Parkinson's disease and epilepsy patients. *Parkinsonism Relat Disord.* 2008;14:120–125.

15. Montgomery EB Jr, Buchholz SR, Delitto A, et al. *Alterations in basal ganglia physiology following MPTP in monkeys.* New York: Academic Press; 1986.

16. Wang Z, Jensen A, Baker KB, et al. Neurophysiological changes in the basal ganglia in mild parkinsonism: a study in the non-human primate model of Parkinson's disease. Program No. 828.9. Neuroscience Meeting Planner. Chicago, IL. Society for Neuroscience, 2009. Online.

17. Sánchez-Pernaute R, Künig G, del Barrio A, et al. Bradykinesia in early Huntington´s disease. *Neurology.* 2000;54:119–125.

18. Tang J, Moro E, Lozano A, et al. Firing rates of pallidal neurons are similar in Huntington's and Parkinson´s disease patients. *Exp Brain Res.* 2005;166:230–236.

19. Watts RL, Mandir AS, Montgomery Jr EB. Abnormalities of the supplementary motor area (SMA) neuronal activity in MPTP parkinsonism. *Soc Neurosci.* 1989; Abst 15787.

20. Mandir AS, Watts RL, Buchholz SR, et al. Changes in motor cortex neuronal activity associated with increased reaction time in MPTP parkinsonism. *Soc Neurosci.* 1989;15 Abst 787.

21. Doudet DJ, Gross C, Arluison M, et al. Modifications of precentral cortex discharge and EMG activity in monkeys with MPTP-induced lesions of DA nigral neurons. *Exp Brain Res.* 1990;80:177–188.

22. Montgomery EB Jr. Effects of GPi stimulation on human thalamic neuronal activity. *Clin Neurophysiol.* 2006;117:2691–2702.

23. Marder E, Calabresi RL. Principles of rhythmic motor pattern generation. *Physioll Rev.* 1996;76:687.

24. Johnson-Laird PN. *How we reason.* New York: Oxford University Press; 2006.

25. Ryle G. *The concept of mind.* London: Penguin Classic; 2000.

26. Montgomery EB Jr, Baker KB. Mechanisms of deep brain stimulation and future technical developments. *Neurol Res.* 2000;22:259–266.

27. Dostrovsky JO, Levy R, Wu JP, et al. Microstimulaiton-induced inhibition of neuronal firing in human globus pallidus. *J Neurophysiol.* 2000;84:570–574.

28. Moreau C, Defebvre L, Destée A, et al. STN-DBS frequency effects on freezing of gait in advanced Parkinson disease. *Neurology.* 2008;71:80–84.

29. Tornqvist AL, Schalen L, Rehncrona S. Effects of different electrical parameter settings on the intelligibility of speech in patients with Parkinson´s disease treated with subthalamic deep brain stimulation. *Mov Disord.* 2005;20:416–423.

30. Wojtecki L, Timmermann L, Jorgens S, et al. Frequency-dependent reciprocal modulation of verbal fluency and motor functions in subthalamic deep brain stimulation. *Arch Neurol.* 2006;63:1273–1276.

31. Kuncel AM, Cooper SE, Wolgamuth BR, et al. Clinical response to varying the stimulus parameters in deep brain stimulation for essential tremor. *Mov Disord.* 2008;21:1920–1928.

32. Montgomery EB Jr, Turkstra LS. Evidenced based medicine: let´s be reasonable. *J Med Speech Language Pathol.* 2003;11:ix–xii.

33. Montgomery EB Jr, Baker KB. Ninety-nine percent confidence volume and area for optimal DBS location in the subthalamic nucleus (STN). 7th International Congress of Parkinson´s Disease and Movement Disorders: Abst. no. 656; 2002.

34. Fins JJ. Surgical innovation and ethical dilemmas: precautions and proximity. *Cleveland Clin J Med.* 2008;75(Suppl. 6):S7–S12.

35. Montgomery EB Jr, Gale JT. Mechanisms of action of deep brain stimulation (DBS). *Neurosci Biobehav Rev.* 2008;32:388–407.

36. Mayberg HS, Lozano AM, Voon V, et al. Deep brain stimulation for treatment-resistant depression. *Neuron.* 2005;45:651–660.

37. Schiff ND, Giacino JT, Kalmar K, et al. Behavioural improvements with thalamic stimulation after severe traumatic brain injury. *Nature.* 2007;448:600–603.

Neuromodulation involves the implantation of biomedical devices to modulate nervous system function. The mainstay of this field has been the electrode and implantable pulse generators, though it also includes implantable medication pumps and catheters. Although electrodes and generators are also being used in devices treating epilepsy (vagus nerve stimulation by Cyberonics and the NeuroPace devices) we have compiled a complete listing of all devices used for DBS, SCS, MCS, and PNS in the following pages of this Appendix as these indications currently include the vast majority of procedures performed in the field. Devices are listed by manufacturer (Medtronic, St. Jude Medical, and Boston Scientific) and reveal standard dimensional data and basic parameter ranges for comparison.

RESTOREULTRA®

Specifications for RestoreUltra, Model 37712

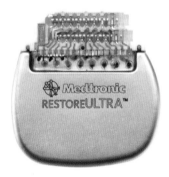

Height	2.1 in (54 mm)
Length	2.1 in (54 mm)
Thickness	0.4 in (10 mm)
Weight	1.6 oz (45 g)
Volume	22 cc
Battery Life	9 years
Maximum electrodes	16
Rate	2–1200 Hz
Pulse Width	60–1000 μsec
Amplitude	0–10.5V
Rechargeable	Yes
Days between recharge (sessions at medium setting)	17 Days
Recharge Coil Position	Inside can
Groups	8
Programs	16
Lead Array options	1–4
Extensions	Optional
Implant Depth	≤ 1 cm

RESTOREULTRA®

Specifications for RestoreUltra, Model 37712

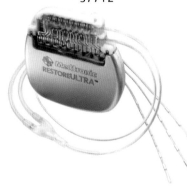

Height	2.1 in (54 mm)
Length	2.1 in (54 mm)
Thickness	0.4 in (10 mm)
Weight	1.6 oz (45 g)
Volume	22 cc
Battery Life	9 years
Maximum electrodes	16
Rate	2–1200 Hz
Pulse Width	60–1000 μsec
Amplitude	0–10.5V
Rechargeable	Yes
Days between recharge (sessions at medium setting)	17 Days
Recharge Coil Position	Inside can
Groups	8
Programs	16
Lead Array options	1–4
Extensions	Optional
Implant Depth	≤ 1 cm

RESTOREADVANCED®

Specifications for
RestoreAdvanced,
Model 37713

Height	2.6 in (65 mm)
Length	1.9 in (49 mm)
Thickness	0.6 in (15 mm)
Weight	2.5 oz (72 g)
Volume	39 cc
Battery Life	9 years
Maximum electrodes	16
Rate	2–130 Hz
Pulse Width	60–450 µsec
Amplitude	0–10.5V
Rechargeable	Yes
Days between recharge (sessions at medium setting)	31.1 Days
Recharge Coil Position	Outside can
Groups	26
Programs	32
Lead Array options	1–4
Extensions	Optional
Implant Depth	≤ 1 cm

RESTOREADVANCED®

Specifications for
RestoreAdvanced,
Model 37713

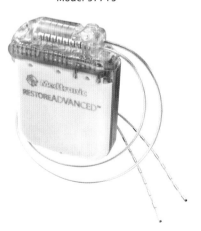

Height	2.6 in (65 mm)
Length	1.9 in (49 mm)
Thickness	0.6 in (15 mm)
Weight	2.5 oz (72 g)
Volume	39 cc
Battery Life	9 years
Maximum electrodes	16
Rate	2–130 Hz
Pulse Width	60–450 µsec
Amplitude	0–10.5V
Rechargeable	Yes
Days between recharge (sessions at medium setting)	31.1 Days
Recharge Coil Position	Outside can
Groups	26
Programs	32
Lead Array options	1–4
Extensions	Optional
Implant Depth	≤ 1 cm

PRIMEADVANCED®

Specifications for PrimeAdvanced,
Model 37702

Height	2.6 in (65 mm)
Height	1.9 in (49 mm)
Thickness	0.6 in (15 mm)
Weight	2.4 oz (67 g)
Volume	39 cc
Battery Life	Depends on settings and use.
Maximum electrodes	16
Rate	3-130 Hz
Pulse Width	60-450 μsec
Amplitude	0-10.5V
Rechargeable	No
Groups	26
Programs	32
Lead Array options	1-4
Extensions	Optional
Implant depth	≤ 4 cm

Itrel® 3 implantable neurostimulator (INS)

© 2010 Medtronic, Inc.

Programmable parameter	Operating
Amplitude	
Normal resolution Upper limit Lower limit	0.1 V steps 10.5 V maximum 0.0 V minimum
Fine resolution Upper limit Lower limit	0.05 V steps 6.35 V maximum 0.0 V minimum
Rate	49 values
Upper limit Lower limit	(from 2.1 to 130 Hz) 130 Hz maximum 2.1 Hz minimum
Pulse width Upper limit Lower limit	increments of 30 μs 450 μs maximum 60 μs minimum
Operation mode	Continuous or cycling
Cycle ON/OFF Time W/O SoftStart/Stop With SoftStart/Stop	0.1 s to 24 h 1 s to 24 h
SoftStart/Stop	1 s, 2 s, 4 s, 8 s, or off
Ramp	15, 20, 25, 30 s, or off at start of stimulation
Dose time	15, 30, 45, 60, 75 m, or off stimulation
	periods
Dose lockout time	1.0, 1.5, 2.0, 2.5, 3.0, 3.5, or 4.0 h
Output On/Offc	On or off
Magnet control	On or off

Appendix

Synergy® Model 7427 and Synergy Versitrel ® Model 7427V Neurostimulator Operating Range

© 2010 Medtronic, Inc.

Programmable Parameters	Values
Amplitude (Peak Voltage) and Limits for Programs 1 and 2	0 to 10.5 volts, Programmable, Normal Resolution (100 mV steps) 0 to 6.35 volts Programmable, Fine Resolution (50 mV steps)
Pulse Width and Limits for Programs 1 and 2	60, 90, 120, 150, 180, 210, 240, 270 300, 330, 360, 390, 420, 450 μsec
Mode	Continuous or Cycling
Rate and Limits	36 values from 3 to 130 Hz
Cycle ON/OFF Time	0.1 second to 24 hours
Cycle ON/OFF/Time w/ SoftStart/Stop	2 seconds to 24 hours
SoftStart/Stop	Allows selection of 1, 2, 4, or 8 seconds ramp increasing gradually from 0 (zero) to the selected amplitude and vice versa for Stop
Dual-Program or Single-Program Operation	Either Dual-Program or Single-Program Operation Selected
Electrode Polarity for Programs 1 and 2	Electrodes 0, 1, 2, 3, 4, 5, 6, and 7: Off, Negative, or Positive
Day Cycling ON/OFF Time	30 minutes to 24 hours
Power ON Reset (POR)	1.9 to 2.1 volts

Activa® PC

Specifications for Activa PC, Model 37601

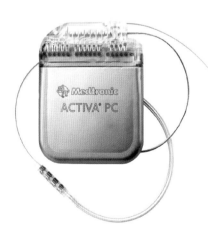

Height (mm)	65
Length (mm)	49
Thickness (mm)	15
Weight (g)	67
Volume (cm³)	39
Battery Life	9
Electrodes	1 to 4 / lead
Rate	2 - 250 Hz
Pulse Width	60 - 450 μsec
Amplitude	0 - 10.5 V 0 -25.5 mA
Rechargeable	No
Days between recharge (sessions at medium setting)	N/A
Recharge Coil Position	N/A
Groups	4
Programs	4
Lead Array options	
Extensions	Optional
Implant Depth	< 1 cm

Activia® RC

Specifications for Activa PC,
Model 37612

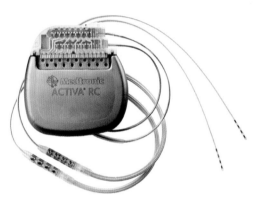

Height (mm)	54
Length (mm)	54
Thickness (mm)	11
Weight (g)	40
Volume (cm³)	22
Battery Life	
Electrodes	1 to 4 / lead
Rate	2 – 250 Hz
Pulse Width	60 – 450 µsec
Amplitude	0 – 10.5 V 0 –25.5 mA
Rechargeable	No
Days between recharge (sessions at medium setting)	N/A
Recharge Coil Position	N/A
Groups	4
Programs	4
Lead Array options	
Extensions	Optional
Implant Depth	< 4 cm

Kinetra®

Specifications for Kinetra,
Model 7428

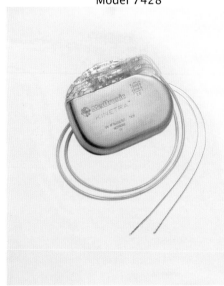

Height (mm)	61
Length (mm)	76
Thickness (mm)	15
Weight (g)	83
Volume (cm³)	51
Battery Life	5 years
Electrodes	2 leads 4electrodes Each
Rate	3 to 250 Hz
Pulse Width	60 to 450 uSec
Amplitude	0 to 10.5 V
Rechargeable	No
Days between recharge (sessions at medium setting)	N/A
Recharge Coil Position	N/A
Groups	1
Programs	1
Lead Array options	2
Extensions	Yes
Implant Depth	< 4 cm

Soletra®

Specifications for Soletra,
Model 7426

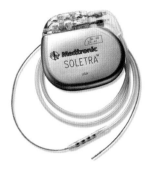

Height (mm)	55
Length (mm)	60
Thickness (mm)	10
Weight (g)	42
Volume (cm³)	22
Battery Life	5 years
Electrodes	4
Rate	2 to 185 Hz
Pulse Width	60 to 450 uSec
Amplitude	0 to 10.5 V
Rechargeable	No
Days between recharge (sessions at medium setting)	N/A
Recharge Coil Position	N/A
Groups	1
Programs	1
Lead Array options	2
Extensions	Yes
Implant Depth	< 4 cm

Percutaneous Leads

	3487A - Pisces® Standard	3887 - Pisces Compact	3888 - Pisces Plus	3777 - 1×8 Standard	3778 - 1×8 Compact	3776 - 1×8 Sub-compact (SC)
ELECTRODE						
Number	4	4	4	8	8	8
Length (mm)	3	3	6	3	3	3
Individual Surface Area (mm)	12	12	24	12	12	12
Inter-Electrode Spacing: Edge to Edge (mm)	6	4	12	6	4	1.5
Inter-Electrode Spacing: Center to Center (mm)	9	7	18	9	7	4.5
Array Length (mm)	30	24	60	66	52	35
LEAD						
Length (cm)	28 33 45 56	28 33 45 56	28 33 45 56	45 60 75	45 60 75	45 60 75
Diameter (mm)	1.27	1.27	1.27	1.32	1.32	1.32

39286
Specify™
2×8 Surgical Leads

PADDLE	
Length (mm)	56.4
Width (mm)	7.6
Thickness (mm)	1.9
ELECTRODE	
Number	16
Shape	Rectangle
Length (mm)	4
Width (mm)	1.5
Longitudinal Spacing: Edge to Edge (mm)	1.0
Longitudinal Spacing: Center to Center (mm)	5.5
Lateral Spacing: Edge to Edge (mm)	1.0
Array Length (mm)	43.0
Array Width (mm)	4.5
LEAD	
Length (cm)	30.65
Diameter (mm)	1.3

39565 Specify™
5-6-5 Surgical Leads

PADDLE	
Length (mm)	64.2
Width (mm)	10
Thickness (mm)	2.0
ELECTRODE	
Number	16
Shape	Rectangle
Length (mm)	4
Width (mm)	1.5
Longitudinal Spacing: Edge to Edge (mm)	4.5
Longitudinal Spacing: Center to Center (mm)	9.0
Lateral Spacing: Edge to Edge (mm)	1.0
Array Length (mm)	49.0
Array Width (mm)	7.5
LEAD	
Length (cm)	30
Diameter (mm)	1.3

3987A On-Point® Surgical Lead

© 2010 Medtronic, Inc.

Paddle	
Length (mm)	44.0 (excl. mesh)
Width (mm)	6.6 (excl. mesh)
Thickness (mm)	1.4
Electrode	
Number	4
Shape	Circular
Length (mm)	4.0
Width (mm)	4.0
Longitudinal Spacing (mm) (edge to edge)	6.2
Longitudinal Spacing (mm) (center to center)	10.2
Lateral Spacing (mm) (edge to edge)	NA
Array Length (mm)	34.6
Array Width (mm)	4.0
Lead	
Length (cm)	25, 60
Diameter (mm)	1.3

3587A Resume II®
Surgical Lead

Paddle	
Length (mm)	44.0
Width (mm)	8.0
Thickness (mm)	1.8
Electrode	
Number	4
Shape	Circular
Length (mm)	4.0
Width (mm)	4.0
Longitudinal Spacing (mm) (edge to edge)	6.2
Longitudinal Spacing (mm) (center to center)	10.2
Lateral Spacing (mm) (edge to edge)	NA
Array Length (mm)	34.6
Array Width (mm)	4.0
Lead	
Length (cm)	25
Diameter (mm)	1.3

3986A Resume® TL
Surgical Lead

Paddle	3986A Resume TL
Length (mm)	44.0
Width (mm)	6.6
Thickness (mm)	1.4
Electrode	
Number	4
Shape	Circular
Length (mm)	4.0
Width (mm)	4.0
Longitudinal Spacing (mm) (edge to edge)	6.2
Longitudinal Spacing (mm) (center to center)	10.2
Lateral Spacing (mm) (edge to edge)	NA
Array Length (mm)	34.6
Array Width (mm)	4.0
Lead	
Length (cm)	25, 45, 60, 70
Diameter (mm)	1.3

2X4 Hinged Surgical Lead

© 2010 Medtronic, Inc.

Paddle	
Length (mm)	41.0
Width (mm)	9.9
Thickness (mm)	1.8
Electrode	
Number	8
Shape	Rectangle
Length (mm)	3.0
Width (mm)	2.0
Longitudinal Spacing (mm) (edge to edge)	3.3
Longitudinal Spacing (mm) (center to center)	6.3
Lateral Spacing (mm) (edge to edge)	3.5
Array Length (mm)	28.2
Array Width (mm)	7.5
Lead	
Length (cm)	30, 45, 60
Diameter (mm)	1.3

3387 Deep Brain Stimulation Lead

Cylindrical	
Lead Length (mm)	250
Lead Diameter (mm)	1.27
Number	4
Shape	Cylindrical
Electrode Length (mm)	1.5
Electrode Diameter (mm)	1.27
Longitudinal Spacing (mm) (edge to edge)	1.5
Array Length (mm)	10.5

3389 Deep Brain Stimulation Lead

Cylindrical	
Lead Length (mm)	250
Lead Diameter (mm)	1.27
Number	4
Shape	Cylindrical
Electrode Length (mm)	1.5
Electrode Diameter (mm)	1.27
Longitudinal Spacing (mm) (edge to edge)	0.5
Array Length (mm)	7.5

Boston Scientific
Precision Plus

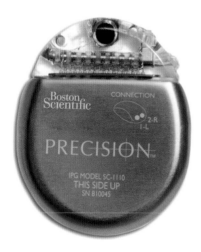

Height	55 mm
Length	45 mm
Thickness	11 mm
Weight	33 g
Volume	22 cc
Battery Life	At least 5 years
Maximum electrodes	16
Rate	2 – 1200 Hz
Pulse Width	0 – 1000 uS
Amplitude	16 ind. CC sources up to 20 mA
Rechargeable	Yes
Days between recharge	3 to 30 days
Recharge Coil Position	Inside can
Groups	4
Programs	4
Lead Array options	1 to 2
Extensions	Optional
Implant Depth	< 2.0 cm

Artisan Surgical Leads

Lead	
Length (cm)	50, 70
Diameter (mm)	1.3
Electrode Configuration	2x8 Paddle
Paddle Width (mm)	8
Electrode	
Number	16
Shape	Circular
Length (mm)	3.0
Contact Width (mm)	2
Longitudinal Spacing (mm) (edge to edge)	1.0
Lateral Spacing (mm) (edge to edge)	
Array Length (mm)	34.6
Array Width (mm)	4.0

Linear Series Leads

Lead	
Length (cm)	30, 50, 70
Diameter (mm)	1.3
Electrode	
Number	8
Shape	Circular
Length (mm)	3.0
Longitudinal Spacing (mm) (edge to edge)	1.0, 4.0, 6.0
Lateral Spacing (mm) (edge to edge)	NA
Array Length (mm)	34.6
Array Width (mm)	4.0

St. Jude

Eon

Height	59
Length	58
Thickness	16
Weight	
Volume	42 cc
Battery Life	10 y
Maximum electrodes	16
Rate	2 – 1200 Hz
Pulse Width	50 – 500 uS
Amplitude	0 – 25.5 mA
Rechargeable	Yes
Days between recharge (medium usage)	56
Recharge Coil Position	
Groups	8
Programs	24
Lead Array options	19
Extensions	
Implant Depth	< 2.5 cm

Eon^C

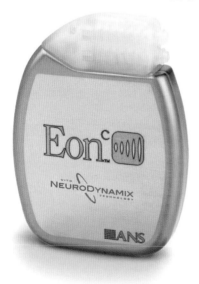

Height (mm)	
Length (mm)	
Thickness (mm)	
Weight	
Volume	
Battery Life	
Maximum electrodes	16
Rate	
Pulse Width	
Amplitude	
Rechargeable	No
Days between recharge	N/A
Recharge Coil Position	N/A
Groups	
Programs	
Lead Array options	
Extensions	
Implant Depth	< 4

Eon *mini*

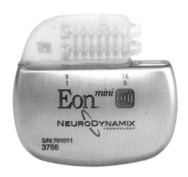

Height	
Length	
Thickness	
Weight	29 g
Volume	18 cc
Battery Life	10 yrs
Maximum electrodes	16
Rate	2 – 1200 Hz
Pulse Width	50 – 500 uS
Amplitude	0 – 25.5 mA
Rechargeable	Yes
Days between recharge (medium usage)	22
Recharge Coil Position	In Can
Groups	8
Programs	24
Lead Array options	
Extensions	
Implant Depth	< 2.5 cm

Genesis

Height (mm)	50
Length (mm)	54
Thickness (mm)	14
Weight	
Volume	
Battery Life	
Maximum electrodes	
Rate	
Pulse Width	
Amplitude	
Rechargeable	
Days between recharge (medium usage)	
Recharge Coil Position	
Groups	
Programs	
Lead Array options	
Extensions	
Implant Depth	

Penta

PADDLE	
Length (mm)	46
Width (mm)	11
Thickness (mm)	2
ELECTRODE	
Number	16
Shape	Rectangle
Length (mm)	4
Width (mm)	1
Longitudinal Spacing: Edge to Edge (mm)	3
Lateral Spacing: Edge to Edge (mm)	1
Array Length (mm)	25
Array Width (mm)	9.0
LEAD	
Length (cm)	60
Diameter (mm)	1.4

Lamitrode Tripole 16c

PADDLE	
Length (mm)	57
Width (mm)	10 or 13
Thickness (mm)	2.0
ELECTRODE	
Number	16
Shape	Rectangle
Length (mm)	4 (center) 6 (outer)
Width (mm)	2.5 (center) 1.8 (outer)
Longitudinal Spacing: Edge to Edge (mm)	3 (center) 1 (outer)
Lateral Spacing: Edge to Edge (mm)	1
Array Length (mm)	40
Array Width (mm)	8
LEAD	
Length (cm)	60
Diameter (mm)	1.4

Percutaneous Leads

Product Name w/Trademark	Lead Diameter (mm)	Contact Length (mm)	Inter-Contact Distance (mm)	# of Contacts	Model Numbers
Quatrode™ Lead (3/4)	1.4	3	4	4	3143, 3146, 3149, 3141
Octode™ Lead (3/4)	1.4	3	4	8	3183, 3186, 3189, 3181
Quattrode™ Lead(3/6)	1.4	3	6	4	3153, 3156, 3159, 3151
Wide-Spaced Quatrode™	1.4	3	11, 18, 11	4	3163, 3166, 3169, 3161
Axxess™ Quad Lead (3/4)	0.9	3	4	4	4131, 4146
Axxess™ Quad Lead (3/6)	0.9	3	6	4	4153, 4156

Paddle Leads

Lamitrode Series

Style	S-4	4	5-8	8	44	44C	8	8C	Exclaim	88	88C
Paddle											
Length (mm)	39	51	67	71	51	51	64	57	33	79	79
Width (mm)	4	8	4	7.5	10	13	10	13	9.5	10	13
Thickness (mm)	1.8	1.7	1.8	1.7	1.7	2	1.5	2	2	1.7	2
ELECTRODE											
Number	4	4	8	8	8	8	8	8	8	16	16
Length (mm) Cent./outer	4	4	4	4	4	4	4 / 6	4 / 6	5.8 rect 2.2	4	4
Width (mm) Cent./outer	2.5	4	2.5	3	2.5	2.5	2.5 / 1.8	2.5 / 1.8	button 1.8 rect	2.5	2.5
Longitudinal Spacing: Edge to Edge (mm)	3	6	3	3.4	3	3	3 / 1	3 / 1	1.6	3	3
Lateral Spacing: Edge to Edge (mm)	N/A	N/A	N/A		1	2	1	1	1	2	2
Array Length (mm)	25	35	53	56	28	28	39	39	21	56	56
Array Width (mm)	2.5	4	2.5	3	6	7	8	8	8	7	7
LEAD											
Length (cm)	30 60 90	30 60	30 60 90	60	60 90	60 90	60	60	60 90	60	60
Diameter (mm)	1.4	1.4	1.4	1.4	1.4	1.4	1.4	1.4	1.4	1.4	1.4

Index